BUILDING SKILLS IN

ORGANIC CHEMISTRY

π-BONDS AND PROBLEM SOLVING

MARK C. ELLIOTT

Cardiff University

© 2023 Mark Christopher Elliott

The moral rights of the author have been asserted.

First published in 2023.

All rights reserved. No part of this book may be reproduced or used in any manner without the prior permission in writing of the author.

Limit of Liability and Disclaimer of Warranty: While the author has used their best efforts in preparing this book, they make no representations or warranties with respect to the accuracy or completeness of the contents of this book and specifically disclaim any implied warranties of fitness for a particular purpose. The advice and strategies contained herein may not be suitable for you situation. You should consult with a professional where appropriate. The author shall not be liable for any loss of profit or any other commercial damages, including but not limited to special, incidental, consequential, or other damages.

ISBN 979-8-39-426937-0 (paperback)

ISBN 979-8-39-718137-2 (hardback)

Image on cover design © 2022 Elizaveta Ivanova. Used with permission.

Foreword i
Acknowledgements iii
The Organization of this Book v

Section 1
BASIC THEORY AND REACTIVITY 1
RECAP 1 Bonding in Organic Compounds 3
RECAP 2 Stability of Carbocations 10
FUNDAMENTAL REACTION TYPE 1 Addition Reactions of Alkenes 19
RECAP 3 Relative Reactivity of Substituted Alkenes 22
REACTION DETAIL 1 Hydrohalogenation and Halogenation of Alkenes 28
RECAP 4 Chemoselectivity and Regioselectivity 34
FUNDAMENTAL REACTION TYPE 2 Nucleophilic Addition to the Carbonyl Group 36
REACTION DETAIL 2 Formation of Cyanohydrins 41
APPLICATIONS 1 Relative Reactivity of Carbonyl Compounds 43
RECAP 5 Stability of Simple Carbanions 51
BASICS 1 The Difference Between Carbonyl Compounds and Alkenes 52
BASICS 2 Acid Catalysis in Carbonyl Addition Reactions 56
EXERCISE 1 Bond Dissociation Energies and Predicting Outcome of Reactions 59
REACTION DETAIL 3 Formation of Hydrates 61
RECAP 6 Conjugation and Aromaticity 64
REACTION DETAIL 4 Hydrohalogenation and Halogenation of Dienes 68
FUNDAMENTAL REACTION TYPE 3 Aromatic Electrophilic Substitution 74
RECAP 7 Stability of More Complex Carbanions 80
FUNDAMENTAL REACTION TYPE 4 Enolates and Enols 87
RECAP 8 And that's it! 92

Section 2
THE CHEMISTRY OF ALKENES, BENZENE RINGS AND ALKYNES 99
REACTION DETAIL 5 Further Aspects of Hydration of Alkenes 101
BASICS 3 Oxidation States in Organic Compounds 106
REACTION DETAIL 6 Oxidation of Alkenes 110
STEREOCHEMISTRY 1 Conformers of Cyclohexanes and Cyclohexenes 119
STEREOCHEMISTRY 2 Stereoisomers and Energy 123
STEREOCHEMISTRY 3 Stereochemistry of Alkene Bromination 127
STEREOCHEMISTRY 4 Diastereoselective Epoxidation of Cyclic Alkenes 133
REACTION DETAIL 7 Hydrogenation of Alkenes, Alkynes and Aromatic Compounds
.... 138
BASICS 4 Regioselectivity in Aromatic Chemistry 142
REACTION DETAIL 8 Aromatic Electrophilic Substitution 146
REACTION DETAIL 9 Friedel-Crafts Alkylation 161
REACTION DETAIL 10 Cross-Coupling 167
REACTION DETAIL 11 Structure and Addition Reactions of Alkynes 170
STEREOCHEMISTRY 5 Conformers—Rotation of Bonds in Acyclic Systems 175
STEREOCHEMISTRY 6 Diastereoselective Epoxidation of Acyclic Alkenes 183

Contents

STEREOCHEMISTRY 7 Catalytic Asymmetric Alkene Epoxidation 189
STEREOCHEMISTRY 8 Asymmetric Dihydroxylation of Alkenes 196
STEREOCHEMISTRY 9 Asymmetric Hydrogenation 198

Section 3
NUCLEOPHILIC ADDITION TO THE CARBONYL GROUP 201

BASICS 5 Linking Carbanions and Organometallic Reagents 203
REACTION DETAIL 12 Addition of Carbon Nucleophiles to Carbonyls 205
REACTION DETAIL 13 Acetylide Chemistry 209
REACTION DETAIL 14 Addition of Hydrogen Nucleophiles to Carbonyls 212
STEREOCHEMISTRY 10 Diastereoselective Addition of Nucleophiles to Carbonyl Compounds 214
STEREOCHEMISTRY 11 Some Important Stereochemical Definitions 225
STEREOCHEMISTRY 12 The Origin of Chirality in Life 229
EXERCISE 2 Bond Dissociation Energies and Predicting Outcome of Reactions 233
REACTION DETAIL 15 Formation of Acetals and Ketals 237
APPLICATIONS 2 Formation of Imines and Enamines 244
REACTION DETAIL 16 Synthesis and Hydrolysis of Esters and Amides 250
APPLICATIONS 3 Lithium Aluminium Hydride Reduction of Esters and Amides 260
REACTION DETAIL 17 Formation of Acid Chlorides 262
REACTION DETAIL 18 Friedel-Crafts Acylation 264
APPLICATIONS 4 The Wolff-Kishner Reduction 270
REACTION DETAIL 19 Hydride is Never a Leaving Group—Except When It is! 274
REACTION DETAIL 20 A Special Elimination—Oxidation of Alcohols 281

Section 4
CARBONYL COMPOUNDS AS NUCLEOPHILES 287

REACTION DETAIL 21 Alkylation of Enolates 289
STEREOCHEMISTRY 13 Diastereoselective Alkylation of Enolates 293
EXERCISE 3 C- or O-Alkylation of Enolates? 300
APPLICATIONS 5 Kinetic and Thermodynamic Enolates 303
REACTION DETAIL 22 Halogenation of Enols and Enolates 310
APPLICATIONS 6 Predicting the Acidity of Enols 316
REACTION DETAIL 23 Aldol Reactions Under Basic Conditions 319
BASICS 6 Ring Strain and Ring-Forming Reactions 327
APPLICATIONS 7 Aldol Reactions Under Acidic Conditions 333
STEREOCHEMISTRY 14 Diastereoselective Aldol Reactions 335
EXERCISE 4 Combining Aldol Transition States and Felkin-Anh Stereochemistry 339
APPLICATIONS 8 Aldol Reactions with Evans Oxazolidinones 342
REACTION DETAIL 24 Dianion Chemistry and Decarboxylation 345
REACTION DETAIL 25 More Complicated Aldol Reactions 349
REACTION DETAIL 26 Alkene formation—the Wittig Reaction 359
REACTION DETAIL 27 Alkene Formation—Metathesis 367
REACTION DETAIL 28 Conjugate Addition Reactions 371
REACTION DETAIL 29 Aromatic Nucleophilic Substitution 377
BASICS 7 Heterocyclic Compounds are Important 382

APPLICATIONS 9 Imines, Enamines and Heterocycles 384
BASICS 8 The Curtin-Hammett Principle 392
STEREOCHEMISTRY 15 Improving Stereoselective Enolate Reactions 399
REACTION DETAIL 30 The Favorskii Rearrangement 403

Section 5
REARRANGEMENT REACTIONS **407**
FUNDAMENTAL REACTION TYPE 5 Rearrangement Reactions of Carbocations 409
REACTION DETAIL 31 Carbocation Rearrangements 414
REACTION DETAIL 32 Rearrangement to Electron-Deficient Nitrogen and Oxygen
 418
REACTION DETAIL 33 Borate Rearrangements 430

Section 6
PERICYCLIC REACTIONS **435**
FUNDAMENTAL REACTION TYPE 6 Pericyclic Processes 437
REACTION DETAIL 34 Allowed and Forbidden Cycloadditions 442
STEREOCHEMISTRY 16 Diels-Alder Reactions 450
REACTION DETAIL 35 Electrocyclic Processes 456
REACTION DETAIL 36 Sigmatropic Hydride Shifts 462
REACTION DETAIL 37 [3,3]-Sigmatropic (Claisen and Cope) Rearrangements 466

Section 7
TOTAL SYNTHESIS **471**
TOTAL SYNTHESIS 1 Synthesis of Strychnine 473
TOTAL SYNTHESIS 2 Synthesis of Prostaglandin F2α 484
TOTAL SYNTHESIS 3 Synthesis of Laurenene 494
TOTAL SYNTHESIS 4 Synthesis of Ionomycin 502
TOTAL SYNTHESIS 5 Synthesis of (−)-7-Deacetoxyalcyonin Acetate 508
Afterword 515
Further Reading 516
Previous Reading 517
Index 519

Preamble

FOREWORD

When I started writing How to Succeed in Organic Chemistry (Oxford University Press, 2020, ISBN 0198851294), my ambition was to write a 'Sykes' for the next generation. 'Sykes' refers to the classic 'A Guidebook to Mechanism in Organic Chemistry', first published in 1961. The sixth edition (1986) manages, in only 400 pages, to cover pretty much all the reaction mechanisms you will ever encounter in organic chemistry.

I very quickly came to realize that if I wanted to present these same mechanisms in a modern and pedagogically-sound way, it could not be done in a book of that size—or at least *I* couldn't do it the way I wanted to do it. For all its elegance, 'Sykes' has a lot of dense text and does not give the level of repetition of principles that will allow a student to develop their skills.

'How to Succeed' therefore only includes a couple of reactions, but it covers them in great depth and with much emphasis on correct drawing of structure, shape and stereochemistry. All these ideas can (and must!) be applied rigorously to every other reaction you encounter.

Of course, by the time I realized I couldn't cover 'everything' I had a large amount of material that didn't make it into the first book. So, here it is! I put too much work into it to not tidy it up. What I have tried to do here is to sort the reactions into a logical order so that the subject builds naturally.

I hope you are going to be able to work through this book, and to build your knowledge in layers. When you see a new reaction, you need to be able to see how it relates to the reactions you already know—otherwise it becomes a memory exercise. I want to keep emphasizing common principles—molecular orbitals, stability, pK_a, curly arrows—to show you how to develop your skills so that you can see a new reaction and understand how to draw a plausible reaction mechanism.

Do I Need the First Book?

You definitely don't need to own a copy of How to Succeed in Organic Chemistry to get the benefit from this book. Where I need to present concepts that were covered in detail in the first book, I will provide Recap chapters which cover the same material, generally more concisely.

> *I will be assuming some prior knowledge of chemical bonding, orbitals, thermodynamics and kinetics, but you will have encountered this during the course of your studies.*

You will have encountered substitution and elimination reactions before you started your university course, and you will have seen enough of these reactions to follow the discussion here. If you need a 'top up', there are many good resources out there.

Throughout this book, I will give cross-references to chapters in the previous book as **HTSIOC** followed by the chapter name in that book.

Preamble

At the end of the book, under the heading of 'Previous Reading' I have also provided a list of chapters in which the same information can be found in the 2nd edition of 'Organic Chemistry' by Clayden, Greeves and Warren (Oxford University Press, 2012).

ACKNOWLEDGEMENTS

In many ways, this book has been a 'COVID project', and therefore quite insular. Nevertheless, there are some people I do need to thank.

Dr Chris Ashling read the entire manuscript, and provided many helpful comments. Dr Niek Buurma provided assistance in a number of areas. The book is much better for their contributions.

I am grateful to Elizaveta Ivanova for allowing me to use one of her designs as part of the cover. I don't quite know why, but it seems to fit the book perfectly!

Throughout the development of this book, my students have had access to various draft copies. Their kind feedback has made this project worthwhile.

And finally, thank you to my family who have tolerated me spending far too much of my time in the evenings and weekends working on this book.

Preamble

THE ORGANIZATION OF THIS BOOK

Sections

In contrast to How to Succeed in Organic Chemistry (from here on referred to as HTSIOC) I have taken a more traditional approach with sections covering examples of the most important types of reactivity of π-bonded functional groups. Section 1 is an exception, since it covers the most basic theory, along with relatively simple examples of each type of reactivity.

Section 2 covers most of the reactions of alkenes, alkynes and benzene rings. I have included the stereochemical outcome of reactions almost immediately after introduction of the reaction itself. I feel very strongly that organic chemistry is all about the shape of molecules. To leave the description of stereoselective reactions until 'later' would not encourage you to fully integrate this aspect of the chemistry into your toolbox.

Section 3 covers the addition of nucleophiles to carbonyls. There are a lot of nucleophiles and a lot of carbonyls, but everything works according to the principles introduced in Section 1. I've drawn all the mechanisms, but you need to copy them out as well! Friedel-Crafts acylation is in this section, despite it being aromatic electrophilic substitution. The problem with organic chemistry is that everything is very connected.

> Of course, this is also the beauty of organic chemistry. If you focus on the guiding principles, it becomes a lot easier to learn.

In Section 4, we find that carbonyl compounds can also act as nucleophiles. This is chemistry that really confused me when I was learning, so I have tried to explain why I was confused, and what you might do differently to avoid confusion. There are a few chapters in this section that don't include this type of reactivity, but they fit better in that section in terms of the 'flow' of knowledge and skills.

There are a number of rearrangement reactions that don't all feature reactions of π-bonded functional groups. Nevertheless, they are important in their own right, so they need to be here. These are covered in Section 5.

We start Section 1 with a discussion of orbitals, and in Section 6 we see reactions they are controlled by orbital symmetry.

Section 7 then shows you a small selection of total syntheses with examples of how (and why!) the reactions in the earlier sections have been used.

Chapters

I have tried to classify chapters according to the type of material they contain. We start with a few 'Recap' chapters which cover some of the earlier material. The absolute essentials of a reaction is covered in a 'Fundamental Reaction Type' chapter.

> *It's useful for you to be able to refer back quickly.*

We have 'Reaction Detail' chapters for adding more depth. These are the chapters where the bulk of the content is delivered.

Once we have covered a reaction type, there are a few instances where a similar reaction is described. I have classified these as 'Applications'. I think it is useful that you can see that there is no new chemistry being introduced.

Stereochemistry is fundamental to organic chemistry. I have tried to keep stereochemical aspects of reactions close to the coverage of the reaction itself. However, it makes sense to have dedicated chapters for the stereochemistry.

There are some aspects that transcend individual reactions, and I would like you to view them as the basic principles that underpin all of organic chemistry. These chapters are imaginatively titled 'Basics'.

Finally, in Section 7, all the chapters are 'Total Synthesis'.

Layout

It is routine within the organic chemistry literature to number compounds and to then refer to the structures by number within the text. Compound numbers are only continuous within the one chapter.

> *I've used a bit of additional formatting to break up the text. These boxes are used to pull out important points, and to nag/encourage you.*

> This type of box is used to ask you questions or get you to do the work. Make sure you take the time to do this. These exercises are there to help you consolidate your learning. I don't give you all the answers!

How to Study?

I hate to break it to you, but there is a simple truth! It doesn't really matter which book you use—you will need to draw lots of structures and reaction mechanisms. Curly arrows are great, but at first you will have curly arrows going in the wrong direction, or simply not representing the electron flow for the reaction you are drawing. After a while, but ***only*** with lots of practice, you will draw correct curly arrows every single time.

Preamble

At first you will struggle to remember all the different reactions.

> *I did!*

And then you will realize that there actually aren't as many reactions as you thought there were.

It doesn't matter how many times I tell you. The first time through, you will see a list of reactions. When you find two reactions that have strong similarities and apparent differences, draw them side-by-side. Compare the mechanisms, and ensure that you focus on **why** the reactions do different things.

> *The more you understand, the less you have to remember.*

Perhaps you don't see the point in drawing out something over and over again, especially if you are finding it hard to understand. Your brain is able to work simultaneously on multiple levels. By getting the drawing process into your 'implicit memory' you free up your 'declarative memory' to learn facts. And then you place the facts into a broader context and they move into your implicit memory, and so the process goes on.

> *Don't worry about this, just draw the mechanisms again!*

Honestly, I know you won't find this very encouraging, but at least it gives you a plan. You don't want to be sitting in an exam and not knowing how to solve a problem. What I am trying to do is help you define individual steps that will lead you to the answer.

> *And when all else fails, ask your tutor for help!*

I have also created a Facebook group to go along with this book. You can find it at **https://www.facebook.com/groups/159669516998134**

Feel free to join the group, and to ask questions there. Perhaps I will signpost you to useful resources. Perhaps I will make some videos to answer questions. Even better, perhaps you will answer questions from each other, with occasional comments from me!

Preamble

SECTION 1

BASIC THEORY AND REACTIVITY

SECTION 1 BASIC THEORY AND REACTIVITY

INTRODUCTION

In this first section, we will be laying down the basics. We will recap some material covered in my previous book, **How to Succeed in Organic Chemistry**. We will then see how it can be applied to the fundamental reactivity of alkenes, benzene rings and carbonyl groups.

> *Let's start with some good news.*

Once you get to the end of this section, you'll have seen all the principles of alkene and aromatic chemistry, and quite a bit of carbonyl chemistry.

> *That isn't to say that you will be able to work out what happens to a given compound under given reaction conditions. Organic chemistry is a very 'predictive' discipline, but you do need to see enough 'examples' to build the patterns.*

Building these patterns is what we will be doing in **Section 2**, **Section 3**, and **Section 4**. I really want to cover this material so that it doesn't come across as a list of reactions that you need to memorize.

Take your time with these sections. Make sure you understand the reactions. Make sure that after reading them, you can draw out the entire mechanism for yourself. Make sure you can do the same, a couple of months later!

Organic chemistry gets a lot of bad press! There is a perception of that you have to remember long lists of reactions.

> *Sadly, there is some truth in this.*

At first, you will learn a few reactions. As you keep adding more and more reactions, you can rapidly become overloaded. The way to not become overloaded is to understand the reactions, so that they are either easy to remember, or easy to predict. I have taught many thousands of students over the last 25 years. Every single one of them that became a good organic chemist initially believed that they would 'never get organic chemistry'.

> *They did it, and so can you.*

SECTION 1 BASIC THEORY AND REACTIVITY

RECAP 1

Bonding in Organic Compounds

WHY?

I am going to start quite a few chapters with a 'why' section. It would be really easy to give you an explanation of a particular theory/reaction without telling you why you need it.

Hybridization is a bonding model. Although we will draw lots of curly arrow mechanisms, they all have a meaning in terms of molecular orbitals. We need to know what the orbitals look like. It turns out that an understanding of molecular orbitals will tell us why some reactions work, and some do not. It will tell us why we get a particular outcome in a reaction.

> *This chapter is stuff you absolutely need to understand!*

HYBRIDIZATION AND ORBITALS

All we are going to do here is give a brief overview of hybridization. Refer back to **HTSIOC Basics 6** for a more detailed description.

> *I had a lot of trouble accepting hybridization. I found it to be quite random, and I didn't see why the orbitals would undergo this process.*

The short answer is that they do not! That is, hybridization is not a physical reality.

> *This doesn't mean it isn't useful. It is useful. Indeed, it is absolutely essential!*

The longer answer is that molecular orbitals are solutions to the Schrödinger equation.

> *This is the most important equation in quantum mechanics, but you will never need to solve the equation to do organic chemistry. I am mentioning it to give context, nothing more.*

There are solutions to the Schrödinger equation in which each orbital spans more than two atoms, and contributes to several individual bonds. But here's the nice thing. If we take any linear combination of the solutions for the Schrödinger equation, this combination will also be a solution to the Schrödinger equation. Mathematically, the hybridized orbital are valid solutions to the Schrödinger equation.

> *In short, hybridization is 'mathematically acceptable', and that's all that really matters.*

And because each hybridized orbital contributes to one (and only one!) bond, we have an entity that we can think of as 'the bonding orbital'. This is incredibly convenient.

There are other solutions to the Schrödinger equation, and the orbitals 'look different'. These orbitals do not have the 'arbitrary' nature of the hybridized orbitals.

If you are interested, you can have a look at HTSIOC Perspective 3 to see what these orbitals look like. But it is not absolutely necessary.

ALKANES—sp³ HYBRIDIZATION

Recall that the electronic configuration of a carbon atom, in group 4 of the periodic table, is

$$1s^2 2s^2 2p^2$$

Since the 1s shell does not participate in bonding, so we only need to consider the 4 'outer' electrons. The '2' shell can hold 8 electrons (an octet). It therefore requires 4 more electrons to become filled. One way it can do this is to gain one electron from each of four other atoms (most commonly hydrogen, another carbon atom, or nitrogen or oxygen). As a result of this, carbon can form four 'single' bonds.

In the hybridization bonding model, we would like each of these bonds to be a discrete entity. The three p orbitals are at 90° to one another. The s orbital is spherically symmetrical. Methane is tetrahedral. We can make four 'new' orbitals by 'mixing up' the atomic orbitals. The way to look at this is to consider promoting an electron from the 2s orbital to a 2p orbital. We will ensure that we put one electron in each of the p orbitals. This gives us a carbon atom with the following electronic configuration:

$$1s^2 2s^1 2p_x^1 2p_y^1 2p_z^1$$

Now we get to the mathematical bit. Each orbital (s,p) is a 'wavefunction'. We take these four wavefunctions, add them up and divide by four.

It's a bit more complicated than this, but fortunately we don't need to do the maths.

This gives us 4 new orbitals. They are described as sp³ hybrid orbitals, as each of them has a contribution from one s and three p orbitals. Each of them contains 25% of an s orbital and 75% of a p orbital, from the perspective of the carbon atom. Our effective electronic configuration is now

$$1s^2 2(sp^3)^4$$

We can form a bond from each of these orbitals to a 1s orbital on hydrogen. One of the sp³ hybrid orbitals is shown on the structure of methane below. Because the orbital is partly derived from a p orbital, we have a **node** at the carbon atom.

SECTION 1 BASIC THEORY AND REACTIVITY

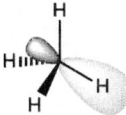

*A node is a region in which there is zero probability of finding an electron. These are accompanied by a change of **symmetry**. Orbital symmetry is quite an abstract concept. We will see how we apply it as organic chemists in **Section 6**.*

The p orbital symmetry is apparent in the shape of the hybrid orbital. However, there is more electron density between the carbon and hydrogen atoms. The orbital lobe behind the carbon atom is small.

The hybrid orbitals are arranged tetrahedrally (bond angle 109.5°) in contrast to the original atomic orbitals that were at 90° to one another (p orbitals) or spherically symmetrical (s orbital).

It took me a long time to fully appreciate that this is a result of the maths, rather than it physically being 'something that the orbitals actually do'.

It is absolutely fine to use the hybridized molecular orbitals in bonding rather than the 'original' molecular orbitals. They simply represent a different, but mathematically equivalent, solution to the quantum mechanical equations.

This is a **bonding orbital**. It is "the bond". However, when the C and H orbitals overlap, there is another way to do this that results in a second, antibonding, orbital.

BONDING AND ANTIBONDING ORBITALS—σ BONDS

Let's have a deeper look at the overlap of the sp^3 hybridized orbital on carbon with the s orbital of hydrogen. These orbitals are shown on the left in the diagram below.

There are two 'ways' for these orbitals to overlap, and they differ in symmetry. The bonding (σ) orbital, as we have just seen, has overlap between the C and the H. The antibonding orbital (σ*) has a node between the C and the H, accompanied by a change in symmetry. It also has a significant lobe (orbital coefficient) behind the atom.

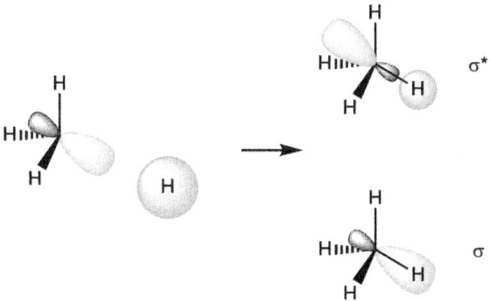

SECTION 1 BASIC THEORY AND REACTIVITY

> *Recall (HTSIOC Reaction Detail 2) that the shape of this orbital explains why an S_N2 substitution reaction proceeds with inversion of stereochemical configuration.*

Now let's continue our look at hybridization.

ALKENES—sp² HYBRIDIZATION

An alkene double bond is a combination of a σ bond and a π bond. We will see the π bond, which only has a contribution from one p orbitals on each carbon atom, in a moment.

Recall that when we considered sp^3 hybridization, we generated the following electronic configuration for carbon after we promoted an electron:

$$1s^2 2s^1 2p_x^1 2p_y^1 2p_z^1$$

We need a p orbital on each carbon atom to form the π bond in ethene. We will assume this is the $2p_z$ orbital (it doesn't really matter which we 'reserve'). This leaves us with the 2s, $2p_x$ and $2p_y$ orbitals for σ bonding. We will take the mathematical average of these to give three new orbitals that are described as sp^2 hybrid orbitals. Each of them is 33% s and 67% p.

$$1s^2 2(sp^2)^3 2p_z^1$$

There's one important point we can make. These three orbitals don't have any contribution from the $2p_z$ orbital, so they have no component in the z direction. They are trigonal planar, which is why alkenes are flat. This time, we will use two of the orbitals to bond to hydrogen atoms, and we will overlap the remaining one with an identical orbital on another carbon atom.[1] Qualitatively, the sp^2 hybridized orbitals look the same as sp^3 hybrid orbitals, apart from being trigonal planar.

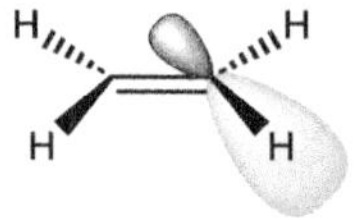

Of course, this doesn't just apply to alkenes. It applies to benzene rings and to carbonyl groups as well. We can apply hybridization to oxygen in the same way.

π BONDS

Now let's come back to the $2p_z$ orbital that we reserved on each carbon atom. We can overlap these two orbitals to form two new molecular orbitals. The lower energy orbital will be bonding, and there will be an antibonding orbital that is higher in energy. We can represent this as shown below.

---

[1] We need two carbon atoms (or a carbon atom and an oxygen atom) for a double bond.

SECTION 1 BASIC THEORY AND REACTIVITY

The bonding (π) orbital has overlap of the p orbitals with the same symmetry on the two carbon atoms. The antibonding (π*) orbital has a change of symmetry (a node) between the carbon atoms. Of course, there is also a node in the plane of the double bond because it is derived from p orbitals. In a stable molecule, only the bonding orbital is occupied.

> Antibonding orbitals are really important when we start to consider reaction mechanisms.

We could also draw the bonding orbital like this, showing the original p orbitals on carbon. Most organic chemists would use these representations interchangeably, although we see in Section 6 that there are situations where one representation works better than another.

HYBRIDIZATION IN ALKYNES—sp HYBRIDIZATION

An alkyne has a triple bond. This is made up of two π bonds and a σ bond. For the π bond, we have to reserve two of the p orbitals. This leaves us with one p orbital and one s orbital. When we hybridize these, we get two new sp hybrid orbitals that are pointing in opposite directions (*i.e.* at a 180° angle). Since we only have one p orbital and one s orbital, there is no other possibility.

As above, we then form two π bonds by overlapping the remaining p orbitals. We will look at the chemistry of the alkyne bond in Reaction Detail 11. I pretty much *had* to draw the individual p orbitals here, or it would have looked very messy.

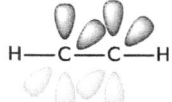

7

BOND STRENGTHS, LENGTHS AND ANGLES

Let's use this model to explain some apparent peculiarities of bond strengths in ethane (**1**), ethene (**2**) and acetylene (**3**). The bond lengths are in Angstrom units (Å) and the bond dissociation energies are in kJ mol^{-1}.

I think it is really important that you know how long/strong a chemical bond is. Chemical bonds in organic compounds are between 1 and 2 Å long. A typical single (σ) bond in an organic compound has a bond dissociation energy between 350 kJ mol^{-1} and 400 kJ mol^{-1}.

> These numbers will serve as your reference point for other distances/strengths of interactions.

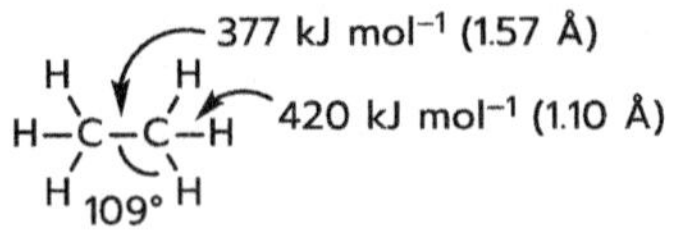

Ethane (1): 377 kJ mol^{-1} (1.57 Å), 420 kJ mol^{-1} (1.10 Å), 109°
Ethene (2): 728 kJ mol^{-1} (1.35 Å), 458 kJ mol^{-1} (1.07 Å), 120°
Acetylene (3): 954 kJ mol^{-1} (1.21 Å), 549 kJ mol^{-1} (1.06 Å), 180°

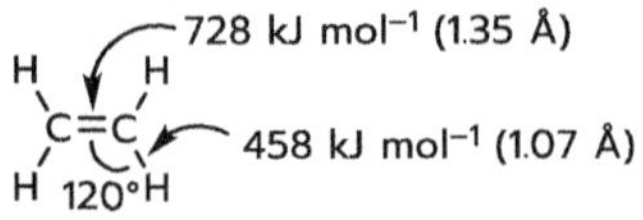

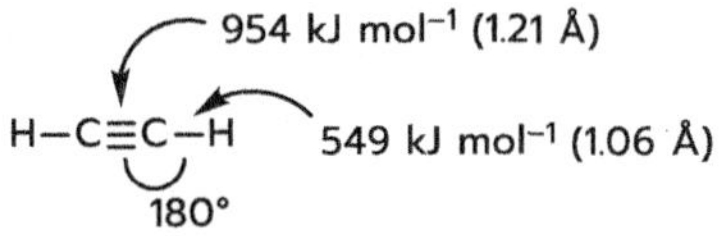

Comparing the carbon-carbon bonds, we can see that a C–C double bond is about 1.9 times as strong as a C–C single bond. It is also a bit shorter. A C–C triple bond is even shorter, and even stronger—about 2.6 times as strong as a C–C single bond. The second π bond doesn't 'count' as much as the first.

When we look at how the C–H bonds change with hybridization, we see an interesting trend. Remember that s orbitals are lower in energy (more tightly held by the nucleus) than p orbitals. As a result of this, a bond with more s character will be stronger. The strength of a C–H bond therefore follows the trend:

alkane (25% s, 75% p) > alkene (33% s, 67% p) > alkyne (50% s, 50% p)

We have already mentioned the bond angles, but they are also summarized on the diagram above. Ethane is tetrahedral, with bond angles of 109.5°. Ethene is trigonal planar, with approximate angles of 120°. Acetylene (ethyne) is linear.

THE STRENGTH OF π BONDS

How strong is the π bond in ethene? It is tempting to say that it is 351 kJ mol^{-1} (the difference between the strength of the C=C bond in ethene and the strength of the C–C bond in ethane). This isn't quite right, because (following the reasoning above for the C–H bond strengths) the C–C σ bond in ethene (which would not be easy to measure directly) is stronger than the C–C σ bond in ethane. In fact, the π bond in ethene has a bond dissociation energy quite a bit lower than 351 kJ mol^{-1}.

> *This is important, in a book that (mostly) focuses on the reactions of π bonds.*

DEFINITION—HOMO AND LUMO

The HOMO is the **highest occupied molecular orbital**. For an alkene, the HOMO will be the π-orbital. The σ bond orbitals are lower in energy. When a compound reacts as a nucleophile,[2] it will use 'the most available' electrons.

> *This is important. If we know the shape of the HOMO, we can make predictions about the reactivity.*

The LUMO is the **lowest unoccupied molecular orbital**. When a compound reacts *with* a nucleophile, the nucleophile will 'react with' the LUMO.

If we are dealing with an alkyl chloride (sp^3 hybridized), the LUMO is the σ* orbital of the C–Cl bond. We have seen that the shape of this orbital dictates the direction of nucleophilic attack, and hence the stereochemical outcome in S_N2 substitution reactions.

We will soon see (**Fundamental Reaction Type 2**) that carbonyl compounds are attacked by nucleophiles as well. The LUMO of a carbonyl compound will be the π* orbital. If we know the shape of this orbital, we can predict the direction of attack. This will have important stereochemical consequences which we will start to explore in **Stereochemistry 10**.

> *It's nice when this happens. We have a single guiding principle.*

If you want more information, have a look at **HTSIOC Basics 9**.

The shape and symmetry of molecular orbitals is really important! We will develop these ideas throughout the book, leading to **Section 6** where we find that orbitals are the ***only*** way to understand some aspects of reactivity.

DEFINITION—FRONTIER MOLECULAR ORBITALS

The HOMO and LUMO are referred to as the Frontier Orbitals. We can generally simplify our molecular orbital considerations to the HOMO of one component and the LUMO of the other component.

Even for the reactions in **Section 6**, where a full treatment would require consideration of all molecular orbitals (we won't do this!), we will still see a useful representation involving only the Frontier Orbitals.

---

[2] Another way to put this is 'when it reacts with an electrophile'. We will see lots of reactions of alkenes with electrophiles.

SECTION 1 BASIC THEORY AND REACTIVITY

RECAP 2

Stability of Carbocations

WHY?

Many of the reactions we are going to talk about will feature carbocation intermediates. In most cases, we could get more than one possible carbocation.

We generally get the one that is more stable.

To predict the outcome of a given reaction, we need to be able to look at two (or more) different carbocations and to be able to determine which is the most stable.

Notice I said 'predict' and 'determine', not 'learn' or 'remember'!

INTRODUCTION

In HTSIOC Perspective 2, we quantified the stability of carbocations using computational data on a hypothetical reaction with a hydride anion.

$$R^{\oplus} + H^{\ominus} \rightleftharpoons R-H$$

We call this the **hydride ion affinity**. I'm going to use a reference point where the *t*-butyl carbocation has a 'standard' energy of zero. A carbocation with negative energy is more stable than the *t*-butyl carbocation. A carbocation with positive energy is less stable. The size of the number gives us the relative stability.

Personally, I find this much more useful than simply stating whether a carbocation is stable or not.

ALKYL SUBSTITUTED CARBOCATIONS

A carbocation carbon atom is sp² hybridized. The positive charge can be considered to be an empty p orbital. Carbocations are therefore **planar**, similar to alkenes.

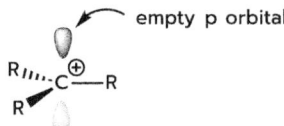

The nice thing about calculations is that you can calculate the energy of something that doesn't exist. In this case, I have calculated the hypothetical energy of 'tetrahedral' methyl and *t*-butyl carbocations, and I have compared them with the planar carbocations.

10

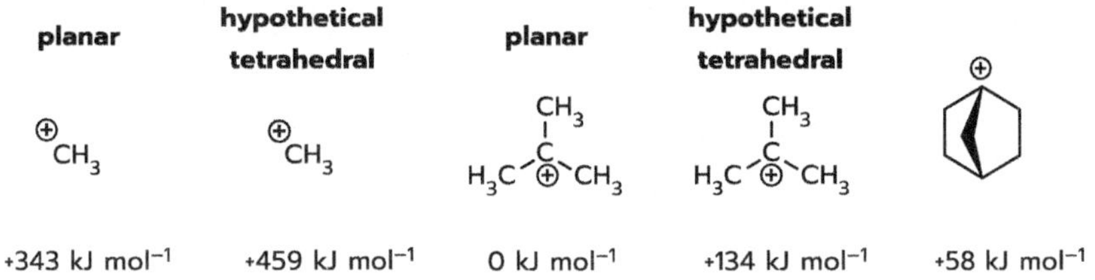

SECTION 1 BASIC THEORY AND REACTIVITY

planar	hypothetical tetrahedral	planar	hypothetical tetrahedral	
⊕CH₃	⊕CH₃	(CH₃)₃C⊕	(CH₃)₃C⊕ tetrahedral	cyclohexenyl cation ⊕
+343 kJ mol⁻¹	+459 kJ mol⁻¹	0 kJ mol⁻¹	+134 kJ mol⁻¹	+58 kJ mol⁻¹

In both cases, the hypothetical tetrahedral (= sp³ hybridized) carbocations are roughly 120 – 130 kJ mol⁻¹ less stable than the preferred planar structures.

> We saw in **Recap 1** that a C–H bond to sp² carbon is about 40 kJ mol⁻¹ stronger than a C–H bond to sp³ carbon. Here we have three of them (3 × 40 kJ mol⁻¹). It's nice when the numbers 'match up'!

On the right, we have a real carbocation that cannot be planar. It turns out that it can derive some stabilization, which is why it is only 58 kJ mol⁻¹ less stable than the *t*-butyl carbocation. The reasons for this are beyond the scope of this book.

What we can also see is that a methyl carbocation is **much** less stable than a *t*-butyl carbocation. We can complete this trend as follows.

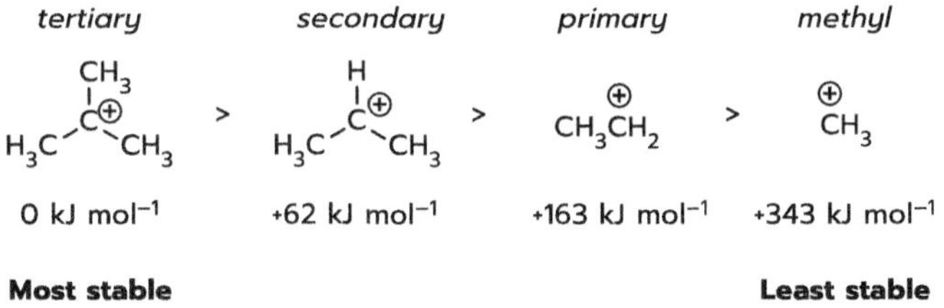

tertiary	secondary	primary	methyl
(CH₃)₃C⊕	(CH₃)₂CH⊕	CH₃CH₂⊕	⊕CH₃
0 kJ mol⁻¹	+62 kJ mol⁻¹	+163 kJ mol⁻¹	+343 kJ mol⁻¹
Most stable			**Least stable**

Each additional methyl substituent provides stabilization to the carbocation. But the second methyl group doesn't quite provide as much stabilization as the first.

The mechanism for stabilization is known as **hyperconjugation**, and it can be represented as shown below.

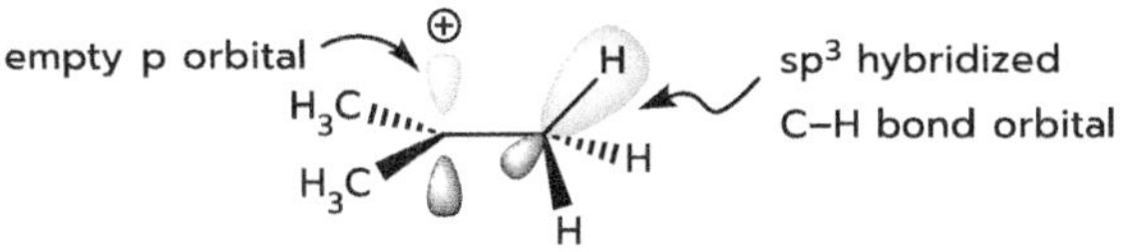

empty p orbital ⊕   sp³ hybridized C–H bond orbital

We have donation of electron-density from the C–H bond into the empty p-orbital. This can be shown using an orbital energy diagram as follows.

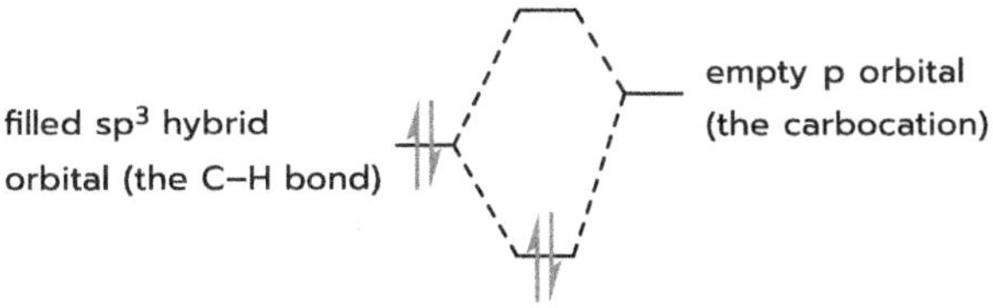

filled sp³ hybrid orbital (the C–H bond)   empty p orbital (the carbocation)

We can also show this by drawing resonance forms as below. Of course, it looks as though we are breaking a C–H bond.

This is not the case. Remember what resonance forms mean. Each individual structure has no independent existence. The **real** structure is somewhere between the two (or more) individual structures. It is described as a resonance hybrid of the canonical forms.

> *It is really important that you really understand resonance forms. I will be repeating the explanation at every opportunity!*

Remember, the curly arrow, whether it is used to show resonance forms or reactions, represents the movement/sharing of electrons. We start this curly arrow at the C–H bond **because this is where the electrons are**. The curly arrow finishes at the C–C bond **because this is where the electrons are going**!

> *Whenever you see curly arrows, make sure you understand what they are showing you. Take your time to do this at the start, and you will quickly find that you start to draw them correctly without needing to think about it too much. If you don't take the time to do this now, you will be trying to remember curly arrow mechanisms without understanding 'the rules'.*

CARBOCATIONS ARE STABILIZED BY π-BONDS

Carbocations are stabilized by things that are electron-donating. π-Bonds have electrons! Here is the allyl cation, shown with resonance forms below. Think of this as a methyl carbocation which has an added alkene substituent. One alkene provides almost as much stabilization as two methyl groups.

allyl cation, +91 kJ mol^{-1}

Let's remind ourselves again what the resonance forms in the square brackets mean (I did warn you!). In this case, we have two identical canonical forms. The 'real structure' has half a positive charge on each end; both C–C bonds are '1.5 bonds'.

Benzene rings have electrons as well. Here are the resonance forms. One benzene ring provides almost as much stabilization as three methyl groups.

benzyl cation, +11 kJ mol^{-1}

SECTION 1 BASIC THEORY AND REACTIVITY

Here is an alternative, molecular orbital, representation for the stabilization.

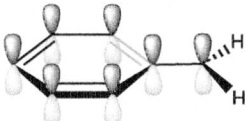

> The key point here is that the positive charge is an empty p orbital, and the π orbitals of the benzene ring have the correct symmetry to overlap.

As with methyl groups, each additional benzene ring contributes less and less to the stabilization. The first benzene ring provides 332 kJ mol^{-1} of stabilization. The second provides 104 kJ mol^{-1}, and the third only provides 65 kJ mol^{-1}.

Ph$_3$C$^+$	Ph$_2$CH$^+$	PhCH$_2^+$	CH$_3^+$
−158 kJ mol^{-1}	−93 kJ mol^{-1}	+11 kJ mol^{-1}	+343 kJ mol^{-1}

We can see why this is the case when we consider the three-dimensional shape of the triphenylmethyl carbocation. There is absolutely no way that the three phenyl rings can be in one plane. Therefore, each benzene ring is twisted (like a propellor) relative to the p-orbital of the carbocation. Quite a lot of orbital overlap is lost.

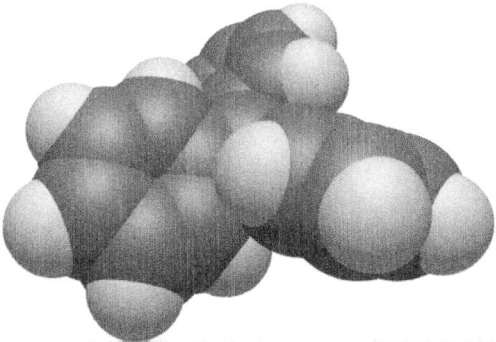

The diphenylmethyl carbocation (below) is essentially planar. We manage to keep the orbital overlap, but it comes at a cost. The two benzene rings are pushed apart slightly, so that the bond angles are not all 120°. The key C–C–C bond angle is 136°.

> Work out for yourself which angle I am talking about!

While we clearly lose some energy from this distortion, we gain more from orbital overlap.

SECTION 1 BASIC THEORY AND REACTIVITY

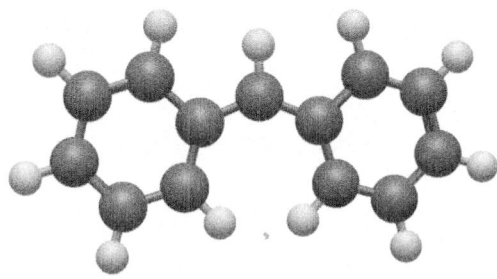

There are a couple of messages we can take from this situation. The first is that the rotation of bonds in the triphenylmethyl carbocation and the distortion in the diphenylmethyl carbocation are both steric effects. However, they can have implications for orbital overlap, which is an electronic effect.

> *The molecules don't care. They will just do whatever is needed to get to the lowest energy point.*

The second message is that we could have made much the same argument for the methyl-substituted carbocations, although it wasn't as easy to see. Go back and have another look at those structures. Make a molecular model. See how many C–H bonds 'can overlap' with the p orbital at any given time.

STABILIZATION BY LONE PAIRS

The following three compounds allow us to make a direct comparison of **inductive** and **mesomeric** electronic effects.

$\overset{\oplus}{CH_3}$ $H_3CO\text{–}\overset{\oplus}{CH_2}$ $(H_3C)_2N\text{–}\overset{\oplus}{CH_2}$

+343 kJ mol⁻¹   +32 kJ mol⁻¹   –130 kJ mol⁻¹

We have two compounds with an electronegative element (O, N) attached to a carbocation carbon. The **inductive** effect (**HTSIOC Basics 5**) will destabilize the carbocation, since it is electron-withdrawing—it will make the carbocation carbon even more positive.

> *An inductive effect is a polarization of the σ-bond framework due to the differing electronegativity of elements. A more electronegative element will pull electron-density towards itself, leading to polarization as follows for the corresponding hydrocarbon.*

$H_3CO\overset{\delta+}{\text{–}}CH_3$ $(H_3C)_2N\overset{\delta+}{\text{–}}CH_3$
   δ–   δ–

> *The amount of δ– will be larger for oxygen than for nitrogen, because oxygen is more electronegative. This effect, which is easier to show as bond polarization in the hydrocarbon, leads to an increase in the amount of positive charge on the carbocation carbon atom. Anything that increases this charge will destabilize it.*

SECTION 1 BASIC THEORY AND REACTIVITY

There is a corresponding electron-donating (stabilizing) effect called a **mesomeric** effect. This is a sharing of lone pair electrons with the positively charged carbon, and can be shown using resonance forms as follows.

$$\left[H_3C\ddot{O}-\overset{\oplus}{C}H_2 \longleftrightarrow H_3C\overset{\oplus}{O}=CH_2 \right] \quad \left[(H_3C)_2\ddot{N}-\overset{\oplus}{C}H_2 \longleftrightarrow (H_3C)_2\overset{\oplus}{N}=CH_2 \right]$$

Compare this with **hyperconjugation**. It's the same thing, but using electrons that are not being used in bonding.

> *It's a bigger effect!*

In both cases the net effect is that the carbocation is stabilized.

But, oxygen is considerably more electronegative than nitrogen. It has a greater inductive electron-withdrawing effect, and a lesser mesomeric electron-donating effect. The net effect of this is that both atoms provide significant stabilization to the positive charge, but nitrogen, with its greater mesomeric electron-donating effect and lesser inductive electron-withdrawing effect, provides more stabilization than oxygen.

> *It is impossible to consider these two effects in isolation. You need to get used to considering 'everything'. We have already seen steric effects that lead to loss of orbital overlap. Now we see two electronic effects that 'disagree', and we need to determine 'which one wins'.*

You might find this overwhelming at first. Try not to over-think it. If you draw the structures out for yourself, and take the time to think it all through, you'll get it. You probably won't get it the first time, and maybe not even the second or third time, but you will get it. And then there will be less to remember, and the new material will make sense.

DOUBLE BONDS CAN DESTABILIZE CARBOCATIONS

Now we have seen the factors that stabilize carbocations, we can look at one instance in which a carbocation is **destabilized**. I haven't calculated the energy of the carbocation in this section. In Recap 8 we will see that a carbonyl group stabilizes an adjacent **negative** charge. It most definitely does not stabilize an adjacent positive charge.

$$H_3C-\underset{O}{\overset{O}{C}}-\overset{\oplus}{C}H_2$$

Here are the resonance curly arrows that we drew for the allyl cation, now applied to the above structure.

15

SECTION 1 BASIC THEORY AND REACTIVITY

$$\left[\underset{H_3C}{\overset{O}{\vphantom{X}}}\!\!-\!\!\underset{\oplus}{CH_2} \longleftrightarrow \underset{H_3C}{\overset{\overset{\oplus}{O}}{\vphantom{X}}}\!\!=\!\!CH_2 \right]$$

> *This is BAD!*

At a first glance it looks okay. You are used to seeing a positive charge on oxygen. But this one is different. Normally, when you see a positive charge on oxygen, it is because you have protonated it.

> *It still has a share in eight outer electrons.*

The one above does not. It has a positive charge because we have removed a share in two electrons. It is left with only six outer electrons. And oxygen is electronegative. It **really** wants those electrons.

> *This carbocation is horrifically unstable!*

There is another way to look at it, but it doesn't get any better. Here are the resonance forms we draw to show the electron distribution in a carbonyl group.

In the right-hand resonance form, we have a positive charge on two adjacent carbon atoms. This resonance form would not make any meaningful contribution.

Remember what resonance forms actually mean! For the π bond to stabilize this carbocation, there would have to be donation of electron density from the π orbital into the empty p orbital. With an oxygen atom on the bond, the carbonyl group does not want to donate (share) those electrons.

> *At the heart of this is an important question. How would you know to 'do' this? The best answer I can give is that I will show you the same few methods, applied to a range of problems. In time, you will learn that there aren't actually that many things you can 'do' to help you understand a problem. In taking the time to 'do' these things over and over again, you will find that you start to do them subconsciously.*

CARBOCATIONS AREN'T REALLY STABLE!

In general, **any** carbocation will be less stable than the neutral molecule that we form it from. Here is a typical reaction profile for carbocation formation from a generic alkyl halide, R–X.

SECTION 1 BASIC THEORY AND REACTIVITY

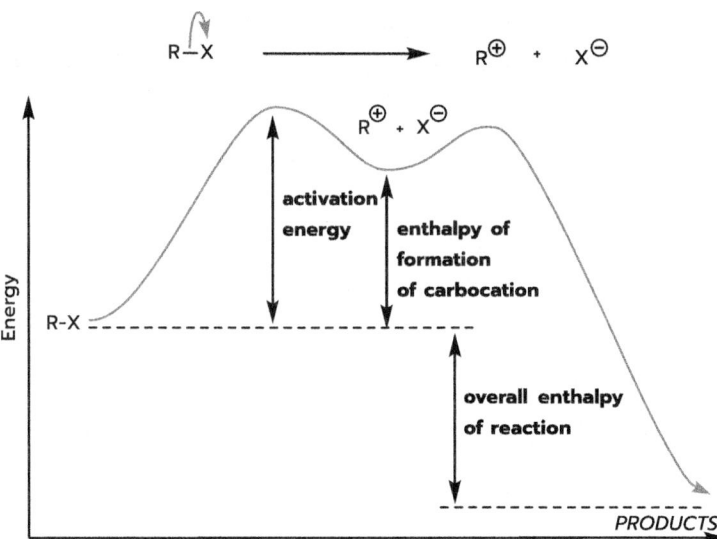

Let's make some important connections. The rate of reaction is determined by the activation energy. The rate of the reverse reaction, formation of carbocation from the product, which then re-combines with $X^\ominus$ to give starting material, will depend on the activation energy for the reverse reaction.

> Find the activation energy for the reverse reaction on the diagram.

The more stable (less unstable) a carbocation is, the faster it will be formed.

> How do we know this?

There is wonderful 'idea' called the **Hammond Postulate**. If you've read **HTSIOC**, you'll know how much I love the Hammond Postulate. Here is the usual wording.

> *"If two states, as, for example, a transition state and an unstable intermediate, occur consecutively during a reaction process and have nearly the same energy content, their interconversion will involve only a small reorganization of the molecular structures."*

The implications of this are subtle but important. There is only a small reorganization of the molecular structure in going from the transition state to the carbocation intermediate. This means that something that stabilizes the carbocation intermediate will also stabilize the transition state that precedes it on the reaction coordinate—the activation energy will be lowered!

We will look at lots of reactions with intermediates. The Hammond Postulate tells us that the more stable the intermediate, the faster it will be formed. It applies to all intermediates, not just carbocations.

> I will be reinforcing this point a lot, so that you get used to it quicker than I did!

Rather than talking about some carbocations being more stable than others, perhaps we should talk about them being less unstable! According to the reaction profile

17

above, the carbocation will be formed slowly, and will react quickly to form the products. This overall reaction is *exothermic* (ΔH is negative). But formation of the carbocation intermediate is *endothermic*.

DOES IT NEED TO BE A FULL POSITIVE CHARGE?

Many of the reactions we look at will involve a carbocation intermediate—a full positive charge. Some of the reactions we will encounter will have a transition state where we build up partial positive (or indeed negative) charge. Now we know which substituents stabilize positive charges, we will be able to identify which substituents will stabilize a developing positive charge in a transition state.

> *This isn't quite the same as applying the Hammond Postulate, but it does use some of the same reasoning.*

We will come back to this idea in due course!

FUNDAMENTAL REACTION TYPE 1

Addition Reactions of Alkenes

BONDING IN ALKENES

We looked at the hybridization bonding model in Recap 1. It is important that you know the approximate bond strengths in organic compounds, but it is even more important that you understand the trends. It's always easier to remember trends that you understand.

	C–C	C–O
Bond length	1.5 Å	1.4 Å
Bond dissociation energy	350 kJ mol^{-1}	350 kJ mol^{-1}

	C=C	C=O
Bond length	1.3 Å	1.2 Å
Bond dissociation energy	611 kJ mol^{-1}	732 kJ mol^{-1}

The double bond in an alkene is about 260 kJ mol^{-1} stronger than a C–C single bond, while the double bond in a carbonyl is about 380 kJ mol^{-1} stronger than a C–O single bond.

We also saw in Recap 1 that the σ-part of the C–C double bond is slightly stronger than an alkane σ-bond. This is because there is more s-character in the sp^2 hybridization in an alkene compared to the sp^3 hybridization in an alkane. We can estimate that the π-bond in an alkene is considerably weaker (approx. 220 kJ mol^{-1}) than the σ-bond (approx. 390 kJ mol^{-1}).

Recall that alkenes can exist as geometrical isomers. For a 1,2-disubstituted alkene, these can simply be referred to as *trans* (*E*, entgegen) and *cis* (*Z*, zussamen). *E* and *Z* can be applied to any alkene in which the substituents at the ends are different.

In general, more substituted alkenes are more stable. This doesn't actually mean they are less reactive. The reasons for this are important and quite complex at first. We will look at this in Recap 3.

SECTION 1 BASIC THEORY AND REACTIVITY

GENERAL REACTIVITY OF ALKENES

Alkenes generally undergo addition reactions. The defining characteristic of an alkene is the π-electrons. An alkene is electron-rich, and therefore reacts with electron-deficient (electrophilic) species.

The addition of a generic electrophile E⊕ to an alkene will give a species with a positive charge formally on carbon. This is almost invariably followed by addition of a generic nucleophile, Nu⊖, as shown below in the general case.

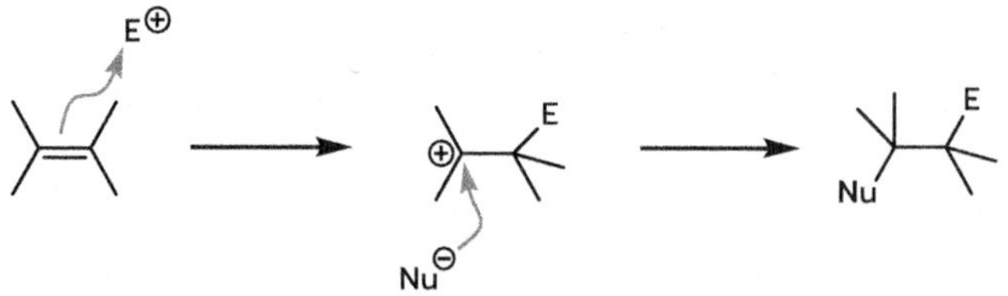

> In this way, the weak π-bond is lost, and two σ-bonds (to E and to Nu) are formed, so that energetically this tends to be favoured.

HYDRATION OF ALKENES

For now, we are going to look at the simplest reaction of alkenes that will allow us to discuss one key principle, regioselectivity, which we will define more formally in **Recap 4**.

Hydration is the addition of water. In the case of an alkene, it is the addition of a hydrogen atom to one of the alkene carbon atoms and addition of a OH group to the other alkene carbon atom. If the alkene is unsymmetrically substituted, two different products can (in principle) be formed. In all structures in this section, "R" is a simple alkyl group.

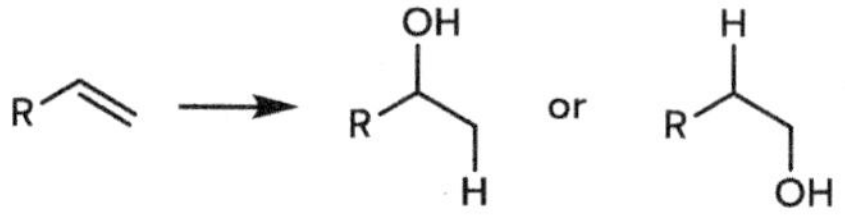

It is important that we can understand the regiochemical outcome of a reaction. If we have a given alkene, and specified reaction conditions, we need to be able to predict which product will be formed, or which will predominate.

> Once we understand the inherent selectivity in a reaction, we can then start to think of creative ways to enhance it, or even to over-ride it.

The best way to understand how this can be achieved is to look at the mechanism. For any given reaction type, we need to understand the common features, and the differences. We need to understand the fundamental reactivity type of the organic functional group, as this will never change.

SECTION 1 BASIC THEORY AND REACTIVITY

UNDERSTANDING THE SIMPLE HYDRATION OF ALKENES

In the simplest possible mechanism, we can have addition of $H^{\oplus}$ and $HO^{\ominus}$. We have already established that alkenes react with electrophiles, and in this case, the electrophile is a proton. Protonation will take place at the less substituted end of the alkene, so that the more stable *secondary* carbocation is formed.

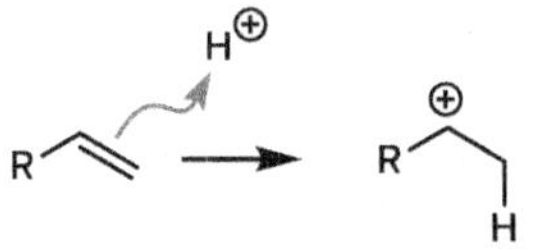

Here is the alternative outcome. We protonate on the left, giving a *primary* carbocation, which is much less stable. We quantified this in **Recap 2**.

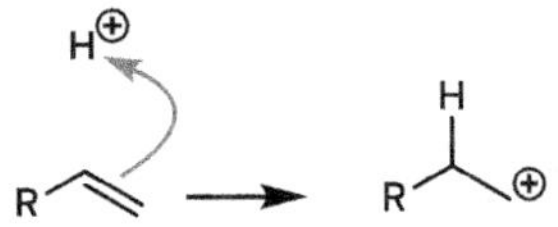

> I'd like to encourage you to make another refinement at this point. Try not to think "*secondary* is more stable than *primary*". Instead, try to look at the structure and identify which groups are providing stabilization to the carbocation.

Now, water (not hydroxide—we are working under acidic conditions) will act as a nucleophile. Here is the rest of the mechanism.

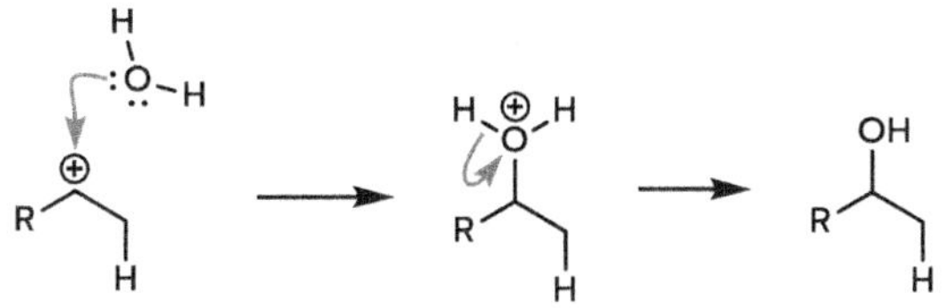

From this mechanism, it is relatively easy to see why the product is formed as above, in which the OH group bonds to the more substituted end of the alkene.

> Every addition to an alkene, featuring an electrophile and then a nucleophile, will follow the same principles as hydration. If you can understand this, you can understand all the rest. More importantly, you can predict the outcome of a reaction you haven't seen, without first having to memorize it.

SECTION 1 BASIC THEORY AND REACTIVITY

RECAP 3

Relative Reactivity of Substituted Alkenes

INTRODUCTION

We are going to be spending quite a lot of time discussing the chemistry of alkenes. If we have two different alkene bonds in a compound, which one will react fastest?

> *This is a really important question, as it will make a difference in any synthesis you plan and undertake. This is an example of chemoselectivity, which we will define in Recap 4.*

In **Fundamental Reaction Type 1**, we saw that when we add an electrophile to an alkene, we get a carbocation intermediate. In **Recap 2**, we looked at the relative energies of a range of carbocations, and we also saw the reaction profile for formation of a carbocation from a stable neutral precursor (such as an alkene!).

Here are two reactions to consider. Protonation of alkene **1** gives a *tertiary* carbocation **2**. Protonation of alkene **3** gives a *secondary* carbocation **4**. We can expect (**Recap 2**) carbocation **4** to be about 60 kJ mol^{-1} less stable than carbocation **2**.

Now let's start with a simple fact. Alkenes **1** and **3** are constitutional isomers. We would not expect them to have the same stability. One will be more stable than the other. We can draw a reaction profile. I am assuming, for now, that alkene **1** is more stable than alkene **3**.

> *Here is the profile. Let's work through the problem.*

I have labelled three key energy differences as "a", "b" and "c".

> *For now, I have deliberately drawn all the energy differences the same.*

We know that "c" is about 60 kJ mol^{-1}. The energy difference between the transition states, "b" will be close to 60 kJ mol^{-1}.

> *This is the Hammond Postulate (Recap 2).*

SECTION 1 BASIC THEORY AND REACTIVITY

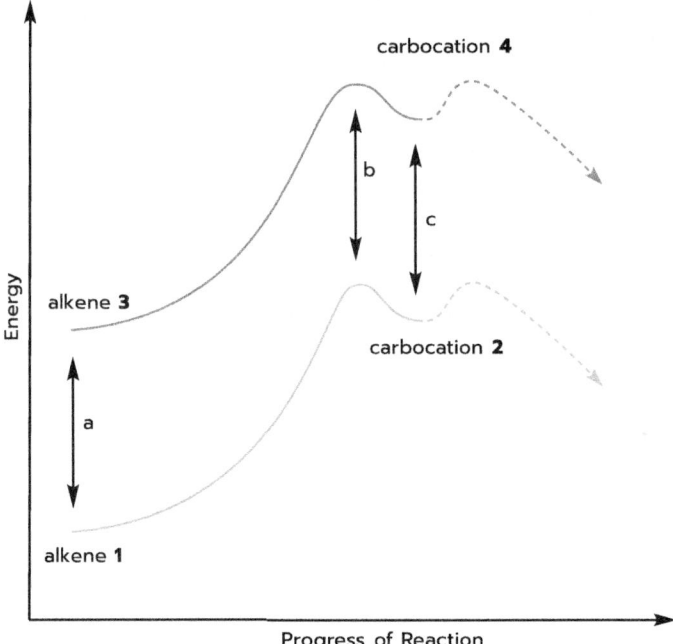

Two species that occur sequentially along the reaction coordinate and are similar in energy will be similar in structure. If something stabilizes carbocation **2**, it will stabilize the transition state that precedes it.

We are now in a position to make a statement. If "a" is less than 60 kJ mol^{-1}, the activation energy for reaction of alkene **1** will be smaller than that for alkene **2**.

> I haven't drawn the activation energy. Look back at **Recap 2** and draw on the activation energy for each reaction.

Conversely, if "a" is more than 60 kJ mol^{-1}, the activation energy for reaction of alkene **3** will be smaller than that for alkene **1**.

So, in order to determine which alkene reacts the fastest, we need to compare the relative energies of the two alkenes, and the relative energies of the carbocation intermediates produced from them.

TAKING STOCK

What we have done can be applied to any reaction. It doesn't even need to produce an intermediate. If the reaction proceeds *via* a transition state (*e.g.* S$_N$2 substitution) we can follow the same process, but we need to be able to directly compare transition state energies.

ALKYL GROUPS STABILIZE ALKENES

We have reached the point where we need one key piece of information. Here it is! When we consider alkenes that are substituted only by alkyl groups, a more substituted alkene is more stable.

SECTION 1 BASIC THEORY AND REACTIVITY

> *Now, we need to know the mechanism for this stabilization, and how much stabilization is provided.*

The stabilization is an electronic effect. In **Recap 1** we saw antibonding orbitals of alkenes. The π bond of the alkene has a filled orbital, but it also has a corresponding π* orbital which is empty. The sp³ hybridized C–H bond orbital of a methyl group has the correct symmetry to overlap with the π* orbital of the alkene. Since this is a filled orbital overlapping with an empty orbital, it leads to a net lowering in energy of the system. I have shown the symmetry on the left in the diagram below, and the energy stabilization on the right.

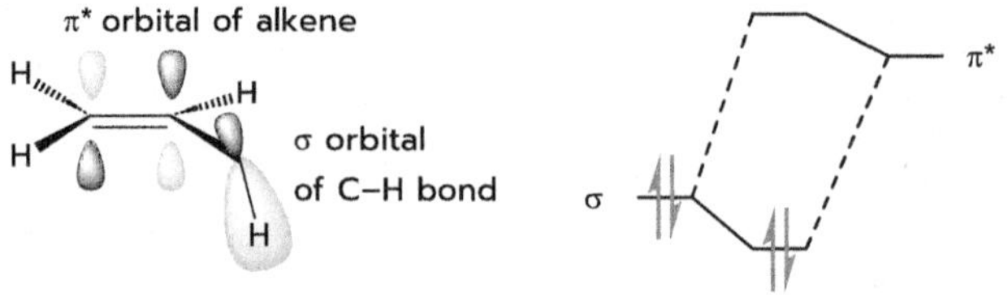

Recall that we did exactly the same thing in **Recap 2** for hyperconjugation for the stabilization of a carbocation. We have a filled orbital overlapping with an empty orbital **with the correct symmetry**.

> *There would be no point drawing an orbital energy diagram for an interaction between two orbitals that did not have the correct symmetry to overlap.*

In this case they do. As with hyperconjugation, we are dealing with a filled orbital with 'p orbital character' (sp³ hybridized) overlapping with an orbital (in this case π*) also with symmetry derived from a p orbital.

> *This is good! We are seeing two types of stabilization, but in reality they are two applications of the same principles.*

Now we need to quantify this effect. This is the point where you **do** need to learn some numbers. There are seven possible alkenes that only have methyl substituents. Alkene **1** is included!

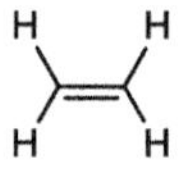

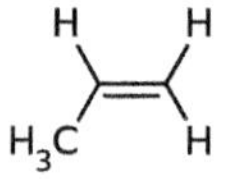

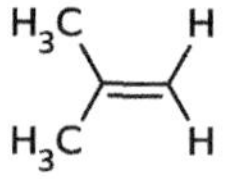

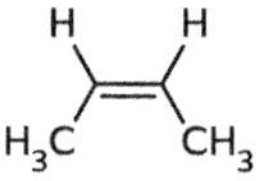

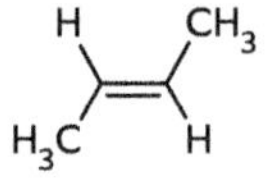

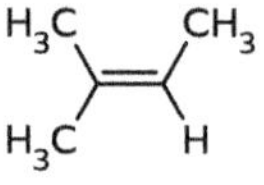

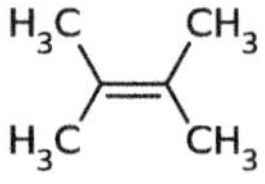

SECTION 1 BASIC THEORY AND REACTIVITY

We discussed the origin of the stability data in **HTSIOC Basics 34**. For now, we will simply take the data as read. Structure **6** is about 10 kJ mol^{-1} more stable than structure **5**.

> *This is the stabilization provided by the first methyl group.*

There are two points to make here. First, the amount of stabilization that a methyl group provides to an alkene (10 kJ mol^{-1}) is considerably less than the amount of stabilization that a methyl group provides to a carbocation (60 kJ mol^{-1} in the example above).

> Go back to **Recap 2** and find this energy difference. Check that you understand which pair of carbocations to compare.[3]

Structure **10** is about 2 kJ mol^{-1} more stable than structure **1**.

By the time we get to the fourth methyl group, we are getting much less stabilization. We saw the same with carbocations—each additional methyl group contributes less.

> *There are two reasons for this. One is electronic. There is only so much stabilization a group can provide. The second is steric. The alkene is getting pretty crowded.*

It is important to always consider electronic and steric effects. It is very rare to be able to make a change that only has one effect. Let's consider one such case. Alkenes **8** and **9** both have two substituents, but in one case they are on the same side (**8**, Z) while in the other case they are on opposite sides (**9**, E).

> *Assignment of E and Z alkene geometry can be found in* **HTSIOC Habit 6**.

Alkene **9** is about 4 kJ mol^{-1} more stable than alkene **8**. The 1,1-disubstituted alkene **7** has about the same stability as **8**.

> *So, when we look at the energy profile above, we will find that "a" will always be smaller than "b" or "c".*

Take a bit of time with this. We do have some facts, but we have a framework into which to place the facts.

A COMMON ERROR

It is very common at this point to make the connection that "more stable" equals "less reactive". If we had two alkenes that were able to give the same intermediate, this would definitely be the case. However, as we have seen above, we don't usually

---

[3] I know this is tedious, but the constant reinforcement is how you will learn. You may think it is 'getting in the way', but it is actually more efficient in the long run.

SECTION 1 BASIC THEORY AND REACTIVITY

get the same carbocation, so there is no reason for the more stable alkene to be less reactive.

FINISHING THE JOB

Here are the two reactions we were looking at earlier.

Alkene **1** is more stable than alkene **3**. It has two additional alkyl substituents directly attached to the alkene bond, so we might estimate an additional 15 kJ mol^{-1} of stability. However, carbocation **2** is *tertiary*, which is considerably more stable than the *secondary* carbocation **4**. The energy difference is about 60 kJ mol^{-1}.

Here is a reaction energy profile for this process, showing the relative energies more realistically. As before, it isn't complete, as it doesn't show the fate of the carbocation.

> It also doesn't show the activation energy. Once again, add this yourself. The activation energy is the step 'up' from the energy of the starting material, not from the horizontal axis.

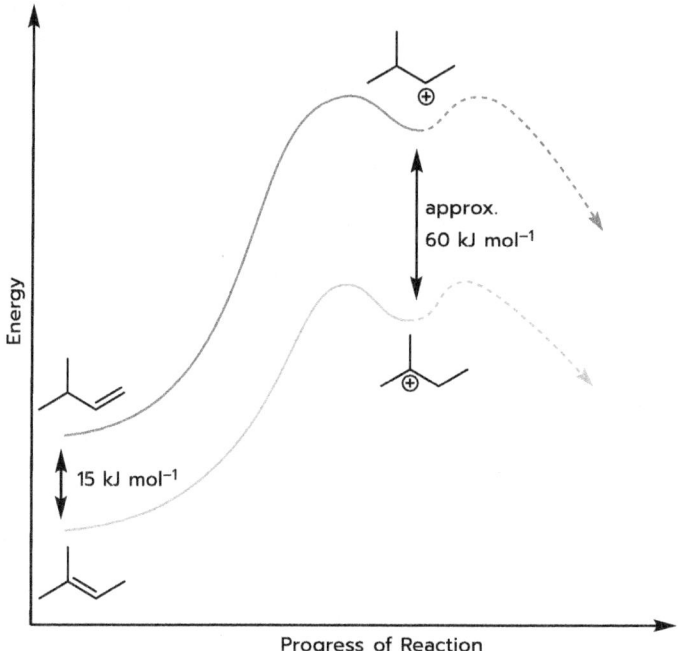

26

> *As we have established, the more stable alkene is definitely more reactive, because the intermediates they form are considerably different in energy.*

Don't get too hung up on this. I don't want to confuse you at this early stage. But I really want to prepare you to make the adjustments you need to make. It is all too easy to equate "more stable" with "less reactive".

MAKING MORE CONNECTIONS

We are going to consider the relative reactivity of different 'types' of carbonyl group in **Applications 1**. Hopefully you won't be surprised that we follow the same process.

However, we will find that the intermediates we form there are of similar energy, so in that case, 'more stable' ***does*** equate to 'less reactive'.

SECTION 1 BASIC THEORY AND REACTIVITY

REACTION DETAIL 1

Hydrohalogenation and Halogenation of Alkenes

WHY?

You can make alkenes relatively easily. Addition reactions of alkenes provide a whole range of useful organic compounds. They can be used to make pharmaceuticals, materials, and a whole host of other products. In some cases, the alkene addition reaction is just one step on the way.

> *For now, we are covering this material because it's important. For each of the key functional groups, you need to know what they react with, how they react, and what selectivity is involved.*

INTRODUCTION

We encountered addition reactions of alkenes in **Fundamental Reaction Type 1**. It is now time to add a little detail to the basic ideas.

In **Fundamental Reaction Type 3** we will see that addition reactions of alkenes tend to be *exothermic*. This is the case with most (but not all) electrophile/nucleophile combinations. We are losing a relatively weak π-bond and forming two new σ-bonds.

The defining characteristic of an alkene is that it is electron-rich. Attack of the electrophile will happen first. Assuming the electrophile has a positive charge, then this will give a cationic intermediate.

> *It will give the most stable cationic intermediate, where there is more than one possible outcome.*

Most of this chapter is concerned with carbocation stability. We will take all the ideas that we saw in **Recap 2**, and we will also add a little refinement.

REGIOSELECTIVITY IN THE ADDITION OF HBr TO ALKENES

If we consider a specific reaction, the addition of HBr to an unsymmetrical alkene, there is only really one thing we need to consider. The electrophilic H$^\oplus$ could be added to either end of the alkene, and then the nucleophilic Br$^\ominus$ will then add to the other end. The two possibilities are shown below.

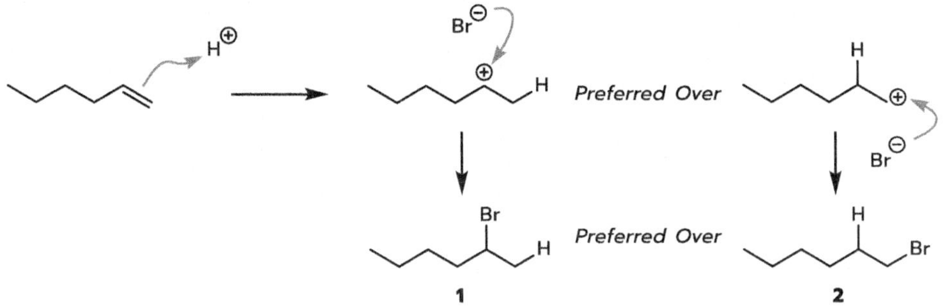

SECTION 1 BASIC THEORY AND REACTIVITY

In the left-hand case, the positive charge is a *secondary* carbocation, and this is more stable than the right-hand, *primary*, carbocation. Therefore, this is the pathway that is followed, and compound **1** is the preferred product.

> This is a simple example. A proton is the simplest possible electrophile, so we don't have any added complications.

Let's add a little more complication, and we will see how it changes things.

ADDITION OF Br$_2$ TO ALKENES—DETAILED STRUCTURE OF INTERMEDIATES

If we add bromine, Br$_2$, to an alkene, then we add one bromine atom to each end of the alkene. Here is the overall process for 1-hexene.

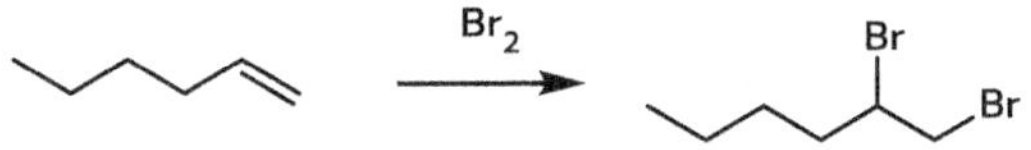

There are a number of questions we need to address. First of all, which Br adds first, or do they both add at the same time?

> You might think this doesn't matter. It does! If we don't know how this reaction works, how will we be able to apply our understanding to other reactions?

I'm going to take a short detour into nucleophilic substitution, a reaction we covered in **HTSIOC Reaction Detail 2**. That one is actually a chapter on the stereochemistry of nucleophilic substitution.

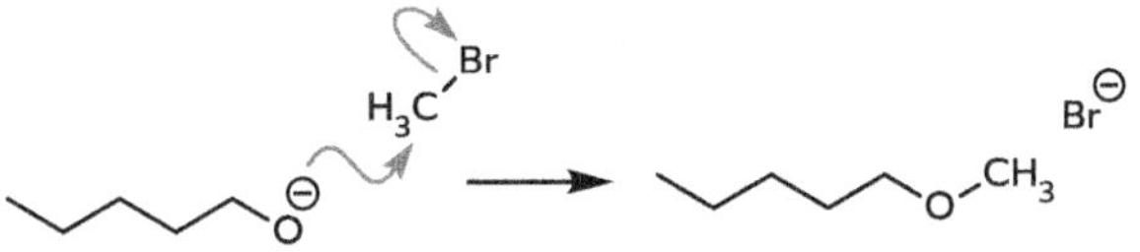

We have an electron-rich alkoxide (anionic oxygen) nucleophile attacking the carbon atom of bromomethane. I'm not going to recap everything here, but this reaction takes place because the HOMO of the alkoxide (the electrons corresponding to the negative charge) donates electron-density into the LUMO of the C–Br bond, breaking that bond.

SECTION 1 BASIC THEORY AND REACTIVITY

> We looked at the LUMO of an sp³ hybridized C–H bond in **Recap 1**. Go back and have another look at it.

Now let's see the same for bromination of an alkene. Here are the same curly arrows, but this time, we are donating from the π-bond of the alkene into the σ* orbital of the Br–Br bond.

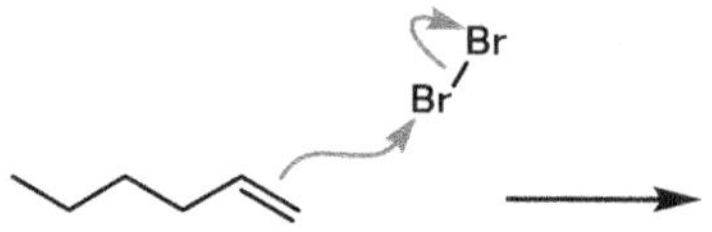

Here is the big question—what do we draw on the right of the arrow? When you have a problem like this, start with something 'obvious'. We are kicking out Br with a pair of electrons. We will have a bromide anion, exactly as we had above.

> *What this means is the 'other thing' will have a positive charge!*

This leads us to the next question—where is the positive charge?

> *I really like this approach of breaking problems down. It allows us to apply knowledge in manageable chunks.*

Let's refine the question we just asked. We will make an assumption that the positive charge is on a carbon atom. Now we can ask 'which one'? Here are the two possibilities.

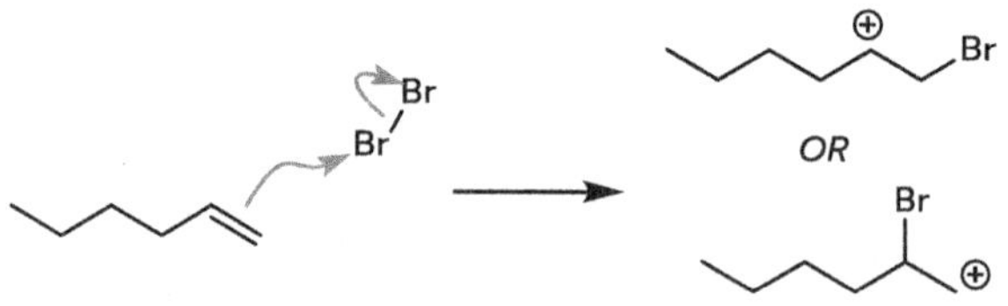

> *I think this gets us somewhere. We know about stabilization of positive charges—the top one 'wins'.*

Now let's ask another question—what is stabilizing this carbocation?

> *I suspect you are getting a bit irritated by this slow approach. Of course the top one is more stable—it's a secondary carbocation. The bottom one is a primary carbocation.*

In fact, there is more to it. Positive charges can be stabilized by **anything with electrons**. The Br atom has lone pairs. We can show this stabilization from the top carbocation (**A**) and the bottom carbocation (**C**) along with a further species **B** which is not a carbocation.

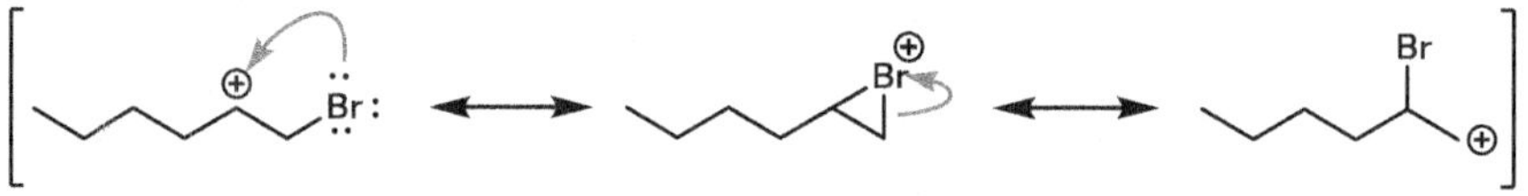

A   **B**   **C**

SECTION 1 BASIC THEORY AND REACTIVITY

These are **resonance forms**. We should remind ourselves what this actually means.

> We are **NOT** talking about a situation where the structure is equilibrating between these extreme forms. The structure **IS** the structure. That might sound like a meaningless statement, but it is the simple truth. We could be really smart and try to isolate the intermediate and determine its structure. It is far easier to calculate the structure, which removes the experimental difficulties.

The structure isn't **A**, **B**, or **C**. In fact, structure **B** is the closest to the reality. Here is the calculated structure of the intermediate. The C1–Br bond is 2.02 Å. The C2–Br bond is 2.20 Å—about 10% longer. This reflects a contribution of structure **A**. This is what you would expect (or what you need to train yourself to expect!). Since the positive charge is more stabilized on the *secondary* carbon, the C–Br bond to this carbon is slightly longer, with a partial positive charge on this carbon atom. There is no meaningful contribution from the *primary* carbocation **C**.

> We refer to the intermediate as a bromonium ion, as this is a better description of its structure. There is actually a partial positive charge on C1 as well, since Br is electronegative, but it is not as much of a positive charge so I haven't drawn it on. It is better to focus on the things that matter the most.

We can take this a step further. Let's add another substituent to make the structure equivalent to **A** a *tertiary* carbocation. I've used a simpler structure, shown below.

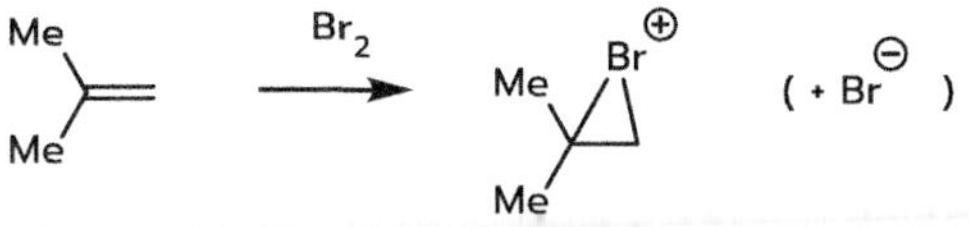

The calculated geometry of this intermediate is shown below. The longer of the C–Br bond lengths is now 2.40 Å, while the shorter one is unchanged at 2.0 Å. Hopefully you expected this change.

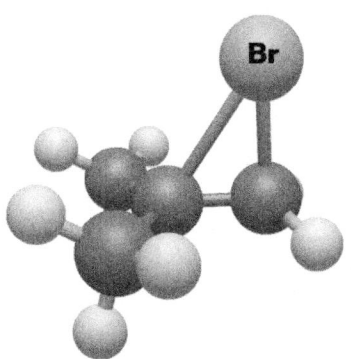

THE IMPLICATIONS OF BROMONIUM IONS— FINISHING THE JOB

So far, we have seen 'addition' of Br$^{\oplus}$ to the alkene, but the bromonium ion that we form in this way is only an intermediate along the reaction profile. It isn't the final product. We need to consider the next (final) step, and the only thing that makes sense is to draw the bromonium ion and consider its likely reactivity. Here is the outcome.

> *We now need to ask the question 'why does the bromide anion attack at C2?'*

We already established that the C2–Br bond is longer than the C1–Br bond. This means that it is weaker. Also, the contribution from the resonance form **A** above means that there is more positive charge on C2 in the 'real' structure.

> *If bromide attacks at C1, we would be breaking the C1-Br bond, and at the same time shortening the C2-Br bond. Attack at C2 simply 'finishes what was already started'— breaking a weak bond.*

What we have here is an S$_N$2 substitution at C2. We will see the stereochemical consequences of this in **Stereochemistry 3** in the next section. We should also note that this is an S$_N$2 substitution with some S$_N$1 character. We saw this in **HTSIOC Perspective 4**. The idea of a pure S$_N$2 or pure S$_N$1 mechanism is not adequate.

PROVING WHERE THE NUCLEOPHILE ATTACKS

If we simply add Br$_2$ to an alkene, we can't easily prove which end the 'electrophilic Br' added to and which end the 'nucleophilic Br' added to.

However, if we use bromine and sodium hydroxide, the situation is much clearer. We will add Br$^{\oplus}$ and HO$^{\ominus}$. We would not add the hydroxide anion as an electrophile!

This regiochemical outcome is totally consistent with the above.

> Draw the mechanism!

> *Start with the bromination mechanism, but when you need a nucleophile, use hydroxide instead. Don't panic when I ask you to draw a mechanism. What I am trying to do is get you to consolidate what you know, but also to extend it a little.*

Oh, and on the scheme above, I have added NaBr in brackets as the other product. Organic chemists would very often not drawn the by-product in a reaction such as this. It is worth looking for the by-products in reactions, but don't expect them to be drawn every time.

STEREOCHEMISTRY OF BROMINATION

There's one more interesting implication. This is a natural consequence of the structure of the bromonium ion and the S_N2 nature of the attack of the nucleophile.

> *Don't think of this as something else to learn!*

When we add a nucleophile to the bromonium ion, it attacks from the opposite side to the Br. In the reaction shown above, we can't tell that this has happened—there are no observable consequences. However, with some alkenes, there are consequences. We will come back to this in **Stereochemistry 3**, but I want to mention it now so that it is already in the back of your mind.

SECTION 1 BASIC THEORY AND REACTIVITY

RECAP 4

Chemoselectivity and Regioselectivity

REGIOSELECTIVITY

We have used the word 'regioselective' several times already, and you have almost certainly understood it from the context. However, it is important enough to state the definition clearly.

We looked at regioselectivity in the context of alkene hydration in **Fundamental Reaction Type 1**. Here is another one, which we will start to look at in **Fundamental Reaction Type 3**. If we react methylbenzene with a mixture of nitric and sulfuric acids, there are three possible nitration products (assuming we only nitrate once).

> If we nitrate toluene three times, we make trinitrotoluene—TNT! As we will see in Reaction Detail 8, the second and third nitration reactions will be slower, so we have to heat it quite a bit if we want to make TNT. Just think about that for a moment!

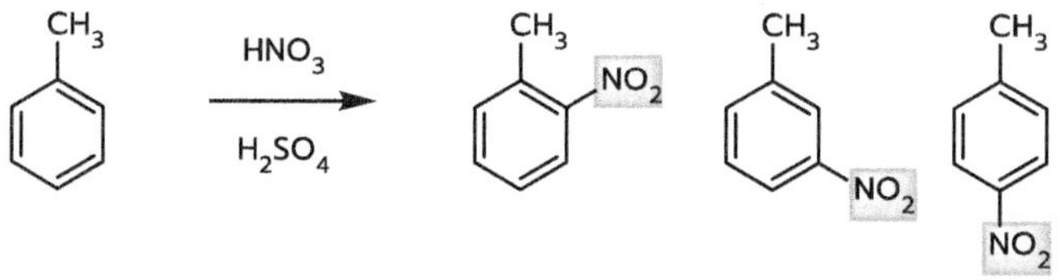

The three products are **constitutional isomers**. That is, they are isomers that have the same molecular formula but different connectivity of atoms. In this case, they have the same functional groups, but connected differently.

> If a given reaction can produce two or more products as a result of reaction at two or more different sites, this is regioselectivity.

It is quite common to describe the products formed as **regioisomers**. It is okay to do this, but bear in mind that the term only applies to the outcome of a reaction.

> It would not be correct to describe the products shown above as regioisomers without giving the reaction in which they might be formed.

When you are first learning organic chemistry, it is quite common to see a question on one type of selectivity, and to answer it with a different type of selectivity. The answer will be wrong!

> You would probably answer the question correctly if you were confident of your definitions.

Make sure you understand when you are looking at regioselectivity.

SECTION 1 BASIC THEORY AND REACTIVITY

CHEMOSELECTIVITY

We actually need to consider two different definitions of chemoselectivity.

> *A **chemoselective** reaction is one in which one functional group reacts preferentially to another in a particular reaction.*
>
> *A **chemoselective** reaction is one in which a functional group reacts preferentially to give one of two (or more) possible outcomes.*

Here is an example that shows the first definition. The starting material has two alkene bonds. One of them is much more reactive than the other. We looked at alkene bromination in **Reaction Detail 1**.

I've drawn the product with the correct stereochemistry, even though we haven't covered this aspect yet. We will cover this in **Stereochemistry 3**.

> *This is actually quite a complex example. To rationalize the chemoselectivity, you need to have a look at **Recap 3**, but also to read ahead to **Recap 6**. For now, just focus on the fact that we have two alkene bonds, and only one of them reacts.*

Here is an example of the second definition. Cyclohexanone can, in principle, give either of the two products shown. We will look at these processes in **Applications 5**.

In practice, it's not quite as simple as this, but it's a nice example to show you the second definition of chemoselectivity.

> *Don't worry about this reaction just yet. I simply want to signpost you to the relevant material. If you're on the second reading of this book, you've seen it and you understand.*

If you are trying to make a particular organic compound, you need to understand the selectivity (regioselectivity, chemoselectivity, stereoselectivity) exhibited by every reaction you propose to use. If the reactions won't give the desired selectivity, your synthesis is doomed to fail.

> *We will look at a selection of total syntheses in **Section 7**.*

SECTION 1 BASIC THEORY AND REACTIVITY

FUNDAMENTAL REACTION TYPE 2

Nucleophilic Addition to the Carbonyl Group

WHY?

Where do I even start? There are lots of nucleophiles. Depending on which nucleophile we add to a carbonyl group, we can form C–C, C–O, C–N, and indeed many other bonds. We tend to be left with an OH group in the product which we can do other things with. In many cases, we end up with a chiral product, and we can even find ways to make these as single enantiomers/diastereoisomers.

> *In short, the carbonyl group is without doubt the most versatile functional group in organic chemistry, and yet it only does two things.*

Addition of a nucleophile is one of them. You need to fully understand this reaction, so that when you see 'yet another example' of a carbonyl addition reaction, you don't over-think it.

STARTING WITH THE BASICS

Drawing mechanisms is our 'key skill'. Carbonyl groups only do two things. Let's start with one of them—addition of a nucleophile. Here are the curly arrows, without worrying about which type of nucleophile we are using.

In this case, we are starting with a negatively charged nucleophile. We add this nucleophile to the carbonyl carbon, pushing the π-bond electrons out onto oxygen.

> *If the $O^{\ominus}$ is more stable than the $Nu^{\ominus}$, this will be an energetically-favorable process. We know how to quantify the stability of anions—we use pK_a.* **(HTSIOC Basics 18,** *summarized in* **Recap 7)**.

Oxygen is electronegative, so the $O^{\ominus}$ is relatively stable. Now this isn't quite finished. The 'product' has a negative charge. We need to get rid of that, and the simplest way to do this is to protonate it. Here it is again, with the second step now shown.

SECTION 1 BASIC THEORY AND REACTIVITY

Note that I said the **simplest** way to deal with the negative charge on oxygen is to protonate it. It's not the only way. There are several other things we could do, and which one occurs will depend on the reaction conditions. We will deal with all of this detail in due course.

Now let's take a moment to look at the stability of our intermediate $O^\ominus$.

I am going to make a gross approximation—every $O^\ominus$ has the same stability. The pK_a for the compound on the left is approximately 16.

We will look at pK_a in **Recap 7**. An 'acid' with a lower pK_a value is more acidic. Water has a pK_a of 15.7, and generates hydroxide, $^\ominus OH$, when it dissociates.

What we are saying here is that every O minus will have the same stability as hydroxide. This is where '16' comes from. Of course, this approximation is not really correct, but unless there is 'something interesting' stabilizing the negative charge, it isn't so far off.

Most of the 'detail' of carbonyl chemistry focuses on the following questions:

- What is the nucleophile?
- How do we generate the nucleophile?
- How reactive is the nucleophile?

We will look at very reactive nucleophiles, as well as less reactive nucleophiles. We will look at neutral nucleophiles as well as negatively charged nucleophiles.[4] Before we do that, I'm going to add a bit more detail in the form of theory.

BONDING IN CARBONYL COMPOUNDS

We looked at the hybridization bonding model in **Recap 1**. Let's use it to understand why carbonyl groups react as they do. Remember that double bonds are stronger than single bonds of the same type.

It is important that you know the approximate bond strengths in organic compounds, but it is even more important that you understand the trends. It's always easier to remember trends that you understand.

The double bond in an alkene is about 260 kJ mol^{-1} stronger than a C–C single bond,

---

[4] Where are the electrons in the nucleophile if there isn't a negative charge?

37

SECTION 1 BASIC THEORY AND REACTIVITY

while the double bond in a carbonyl is about 380 kJ mol⁻¹ stronger than a C–O single bond.

	C–C	C–O
Bond length	1.5 Å	1.4 Å
Bond dissociation energy	350 kJ mol⁻¹	350 kJ mol⁻¹

	C=C	C=O
Bond length	1.3 Å	1.2 Å
Bond dissociation energy	611 kJ mol⁻¹	732 kJ mol⁻¹

We saw in **Recap 1** that π-bonds are weaker than σ-bonds, so what is happening? It turns out that this is still true for carbonyl groups, but we need to remember that the σ-bond part of a carbonyl (sp² hybridized) bond is stronger than the C–O σ-bond of an alcohol (sp³ hybridized). Have another look at **Recap 1** if this confuses you.

CARBONYLS, RESONANCE AND CURLY ARROWS

The early part of this book is focused on helping you develop the skills. Resonance forms are really important.

> *You need to think about resonance forms all the time, and to identify when a structure has a number of different resonance forms. You need to understand what they mean, and when to draw them. I often find that students write half a page of text to try to explain something that would have been better explained by drawing two resonance forms.*

Here are two extreme structures for the carbonyl group.

On the left, we have the 'normal' carbonyl group. The curly arrow shows movement of electrons from the π-bond onto the oxygen atom.

> *This makes sense because the oxygen atom is electronegative. We would never draw a resonance form in which we push the electrons onto carbon.*

Neither structure fully represents the carbonyl group. What we are trying to show

SECTION 1 BASIC THEORY AND REACTIVITY

with these resonance forms is that:

> 1. The carbonyl group is a double bonded functional group.
>
> 2. The carbonyl carbon has a partial positive charge and the oxygen a partial negative charge.

We can represent the carbonyl group as shown below.

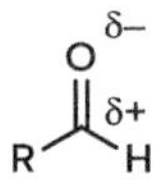

Most of the time, we don't put the δ+ and δ− on the structure, but you need to know they are there. With practice, you will find you just get used to this.

Since we have a partial positive charge on the carbonyl carbon atom, there is every reason for a nucleophile to be 'attracted' to this position. However, there must be a 'reason' in order for this 'attraction' to lead to a reaction.

BREAKING BONDS BY ADDING ELECTRONS

Let's recap one more idea that we encountered in **HTSIOC Basics 9**. When a nucleophile attacks, we need to think about where the electrons are going.

Recall that whenever two atomic orbitals overlap, two new molecular orbitals are formed—if we have two electrons, one of these is bonding and one is antibonding. This is shown below for a carbonyl group.

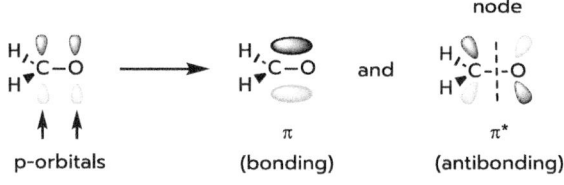

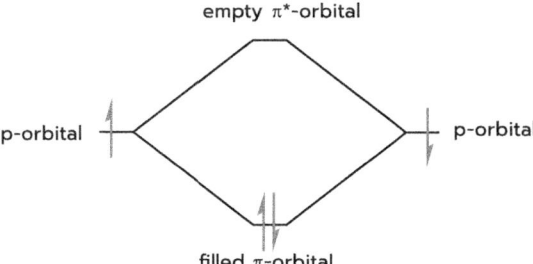

If we were to put two electrons into the antibonding orbital, then there would be no net bonding, so we would have effectively broken the π part of the bond.

> The nucleophile donates electron density into the π* orbital of the carbonyl bond. This results in the π-bond being broken.

SECTION 1 BASIC THEORY AND REACTIVITY

This has implications that we will come to in due course (**Stereochemistry 10**). For now, I just want to introduce this as a basic idea.

TAKING STOCK OF THE SITUATION

Do you really need to think about the orbitals every time you draw a carbonyl reaction?

> *Absolutely not!*

What you need is to be able to return to the arguments about electronegativity, resonance forms and orbital interactions as and when you need to. That way, you can extrapolate from this reaction to other reactions you have not yet encountered.

Most of all, you need to be able to confidently draw reaction mechanisms for carbonyl reactions. You need to ensure that the curly arrows always go the right way, and that you practice enough for this to become a habit.

The ***big*** problem with carbonyl chemistry is that there are quite a few different types of carbonyl group. We will look at a few of these, and we will predict/rationalize their relative reactivity in **Applications 1**. There are also a huge number of different nucleophiles that we could use. This can lead to students thinking that there are a lot of reactions to learn.[5]

---

[5] I know this can be true, because this was my own experience. With the benefit of lots and lots of practice, I can now see that all the 'different' reactions are actually aspects of the same reactivity.

SECTION 1 BASIC THEORY AND REACTIVITY

REACTION DETAIL 2

Formation of Cyanohydrins

We looked at addition of nucleophiles to carbonyls in **Fundamental Reaction Type 2**. Organic chemistry is concerned with the chemistry of compounds with carbon-carbon bonds. Let's now see how to use nucleophilic addition to form a carbon-carbon bond.

FORMATION OF CYANOHYDRINS

A cyanohydrin is a compound in which a cyano group and a hydroxyl group are bonded to the same carbon atom. They are formed by reaction of a carbonyl compound with HCN.

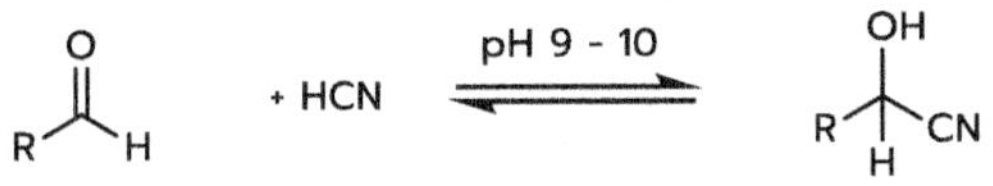

Cyanohydrin formation is carried out under weakly basic conditions. HCN has a pK_a of approximately 9.2, which makes it weakly acidic.

> Recall that water has a pK_a of 15.7, so that anything less than this is acidic. However, acetic acid has a pK_a of 4.8, and acetic acid still isn't a strong acid.

Under weakly basic conditions, there will be cyanide anion present as the base will 'mop up' the H⊕ formed in the following equilibrium.

Now we can add the cyanide nucleophile to the carbonyl group, as shown below. In a second step, we then protonate the negatively charged oxygen atom in the intermediate. I have chosen to use a second molecule of HCN to do this, which will regenerate more cyanide anion.

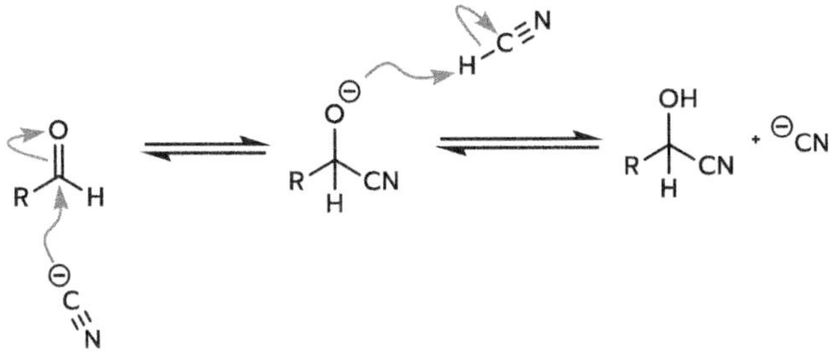

Now, the reality is that most organic chemists would not worry too much about what was actually delivering the proton. They would take the mental shortcut that at pH 9 – 10, when protonating a basic (pK_a 16) alkoxide anion, there will be a suitable

41

source of protons. Therefore, they would generally draw the slightly abbreviated mechanism as follows.

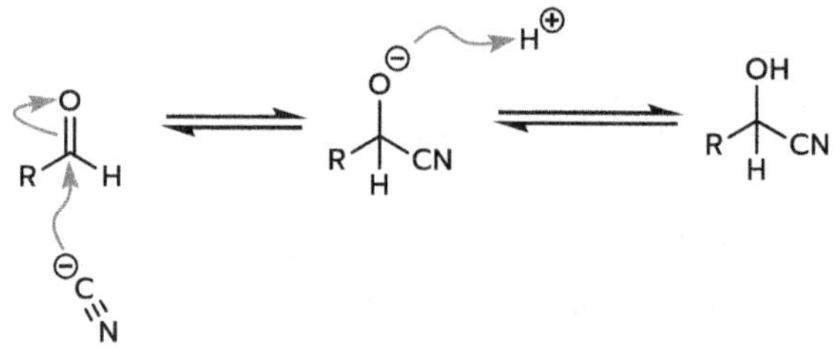

MAKING CONNECTIONS

For our first example, I chose cyanide as a nucleophile because you have seen the cyanide anion before, mainly in inorganic chemistry. You are familiar with the cyanide anion because it is stable, but we should ask a more fundamental question.

> *Why is the cyanide anion stable?*

Hydrogen cyanide has an sp hybridized carbon atom. The C–H bond is 50% s as far as the carbon atom is concerned. An s orbital is held more tightly by the nucleus.

> *We saw in Recap 1 that this makes the C–H bond stronger.*

In what may seem to be a contradiction, it doesn't actually make it harder to break. In **HTSIOC Perspective 1** we saw that a bond-dissociation energy relates to formation of free-radicals. Since the electrons in an sp hybridized system are more tightly held by the nucleus, this stabilizes a negative charge on the carbon atom. We will come back to this in **Recap 7**.

> *We are going to see the implications of the same effect for acetylide anions in Reaction Detail 13.*

WHERE IS THIS GOING?

There are several different types of carbonyl compound, and they all react in fundamentally the same way. However, they do react at different rates. We will explore this next, in **Applications 1**.

SECTION 1 BASIC THEORY AND REACTIVITY

APPLICATIONS 1

Relative Reactivity of Carbonyl Compounds

We looked at the relative reactivity/stability of substituted alkenes in **Recap 3**. We are now going to do something similar for carbonyl compounds. This one is an 'Applications' chapter, as there isn't much new information. Instead, we will focus on applying fundamental principles to predict and rationalize the relative reactivity of two slightly different functional groups towards the same reagent. In this case, we are going to look at reaction of an aldehyde and a ketone with cyanide to form a cyanohydrin (**Reaction Detail 2**).

> *This chapter isn't really about cyanohydrin formation. It's about the principles that you can apply to any reaction!*

STARTING WITH ALDEHYDES AND KETONES

We just encountered cyanohydrin formation in **Reaction Detail 2**. Here's a quick recap of the curly arrow mechanism, as before, with an aldehyde.

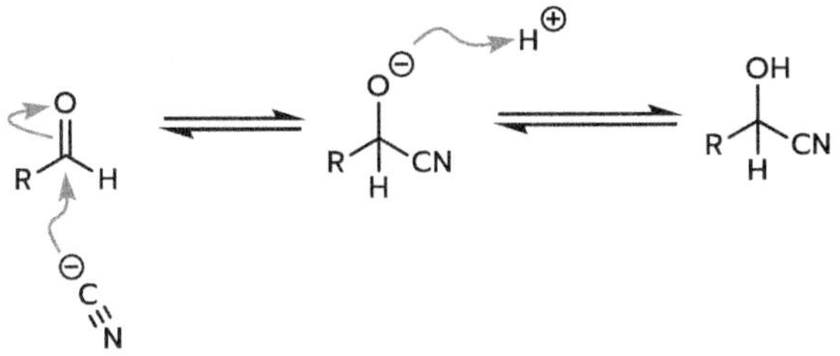

It turns out that aldehydes are much more reactive towards nucleophiles than are ketones. We need to try to understand why this is. We will compare two reactions, cyanohydrin formation from propanal and from acetone. Propanal and acetone are isomers, so this is a valid comparison.

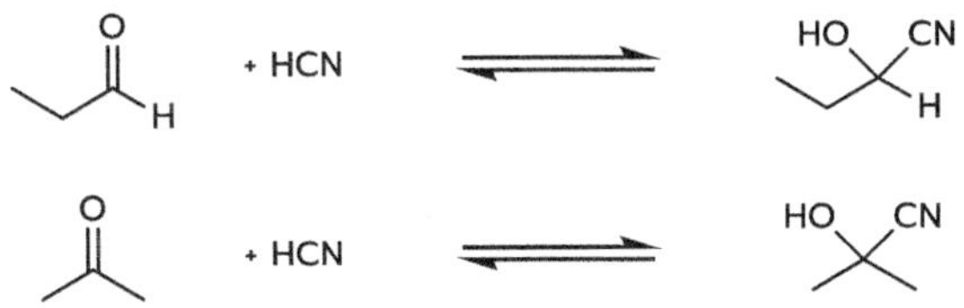

CONSIDERATION OF THERMODYNAMICS

In stating that we are looking at relative reactivity, we are implying that we are looking at rates of reaction (kinetics). However, we should also consider the position of the equilibria shown above.

If we were to calculate the enthalpy of reaction using the method in **HTSIOC Basics 13**, we would find they are identical.

SECTION 1 BASIC THEORY AND REACTIVITY

Here are all the data you need.

Average Bond Dissociation Energy / kJ mol^{-1}			
C–C	350	C=O	732
C–H	410		
H–O	460		
C–O	350		

> Do this calculation. Remember, we treat it as if you have completely broken the C=O bond and formed a new C–O bond.

The two reactions give the same number because we are breaking and forming the same bonds in each case, and we are using 'average' bond dissociation energies in our calculation. The point we are now making is that the bonds are *not* identical. One way around this problem is to measure, experimentally, the heat of formation of starting materials and products. As an alternative (and this is what I have done), it is possible to calculate these energies using quantum chemical calculations.[6]

Here are the calculated enthalpy changes of reaction.

CH₃CHO + HCN ⇌ CH₃CH(OH)CN $\Delta H = -44.0$ kJ mol^{-1}

(CH₃)₂CO + HCN ⇌ (CH₃)₂C(OH)CN $\Delta H = -27.2$ kJ mol^{-1}

Both reactions are calculated to be *exothermic*, but the cyanohydrin formation from the aldehyde is *more* exothermic. We would expect the equilibrium mixture to contain more cyanohydrin in the aldehyde case than the ketone case.

> More of the aldehyde is converted into the cyanohydrin if the reaction reaches equilibrium. Does this mean the aldehyde is 'more reactive'? At one level it does, but we really need to consider rates of reaction.

First of all, let's quickly delve into *why* these reaction enthalpies are different.

(CH₃)₂CO   is more stable than   CH₃CHO   by 29.1 kJ mol^{-1}

(CH₃)₂C(OH)CN   is more stable than   CH₃CH(OH)CN   by 12.3 kJ mol^{-1}

---

[6] As with other calculations in this book, I used Density Functional Theory (DFT) at the B3LYP/6-31+G* level. This is a pretty standard level of theory for organic chemistry calculations.

44

SECTION 1 BASIC THEORY AND REACTIVITY

> The difference in energy between the aldehyde and the ketone is more significant than the difference in energy between the two cyanohydrins.

This is all well and good, but it isn't quite what we set out to investigate. To make sensible predictions about the rates of reaction, we need to be looking at activation energies, so we need to start thinking about transition states.

SIMPLIFYING THE PROBLEM

We are starting with hydrogen cyanide. We first need to deprotonate the hydrogen cyanide to make the cyanide anion. Let's use hydroxide for this. We also need to protonate the intermediate, and we could use water for this. Here is a more complete reaction scheme, for a generic aldehyde.

The barrier to deprotonation of the cyanide will be the same in each case. Therefore, the rate of this step will be the same. We might also assume that protonation of the alkoxide intermediate will not be rate-determining.[7] Therefore, the only information we are missing is the energies of the transition states for attack of the cyanide anion.

Calculating the transition states for the addition of cyanide to the two carbonyl compounds is not all that difficult. The transition state for cyanohydrin formation from acetone is calculated to be 16.0 kJ mol^{-1} more stable than that from propanal.

Here are reaction profiles for the two processes, both significantly simplified.

We could not have predicted the exact value, but for an *exothermic* reaction we would expect the transition state to resemble the starting material than the product. This is the Hammond Postulate (**Recap 2**) again.

> It really does get everywhere!

---

[7] We can always reassess these assumptions if the data do not explain the outcome.

45

SECTION 1 BASIC THEORY AND REACTIVITY

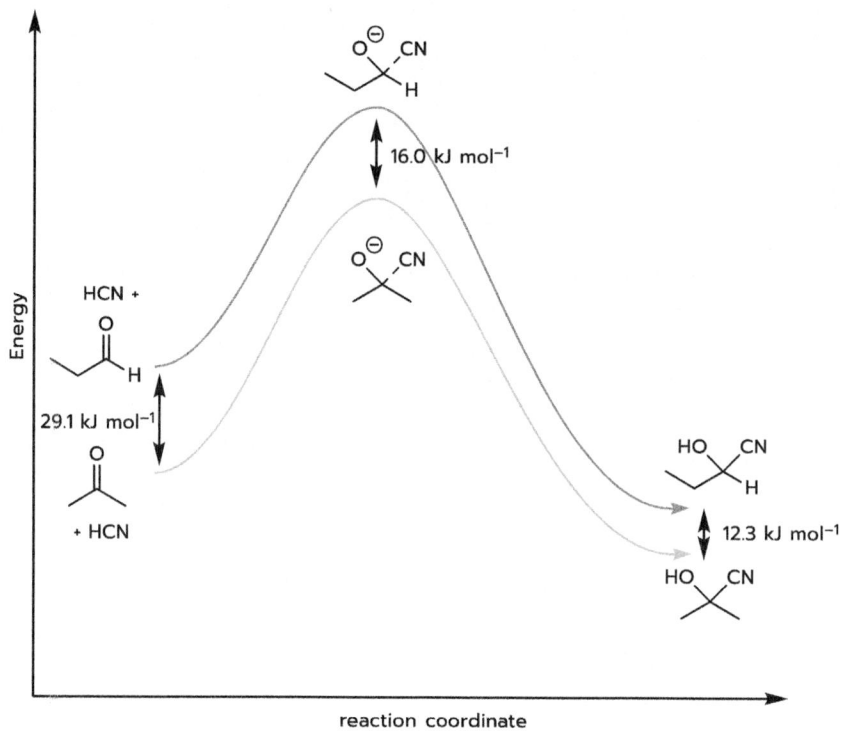

reaction coordinate

You would expect the difference to be less than 29 kJ mol⁻¹, but perhaps not much less.

> For each of the processes, identify the activation energy on the reaction profile. Then have a look at the footnote.[8]

We said we were simplifying the problem. The simplifications are as follows. First of all, we haven't worried about making the cyanide anion from HCN.

> Assuming we use hydroxide, this will be exothermic.

We have the transition state for addition of cyanide, but we do not have the intermediate with the full negative charge.

> This will be a step down from the transition state, but not too much. Have a look at the reaction profile at the end of **Recap 2**. The alkoxide intermediate here will be very similar in energy to the carbocation in that case.

We also don't have the subsequent transition state for protonation of the alkoxide intermediate.

---

[8] Did you draw it as the energy difference between each starting material and the transition state? Or did you draw it as the energy difference from the axis or from the same point? The energy from each starting material to each transition state is correct.

46

SECTION 1 BASIC THEORY AND REACTIVITY

> Draw a reaction profile that includes these.

As a result of these simplifications, we cannot determine an absolute activation energy for the reaction. However, we **can** see that the barrier to cyanohydrin formation from the ketone will be significantly higher (approximately 13 kJ mol^{-1}) than the same barrier from the aldehyde. This is because the ketone is considerably more stable than the aldehyde.

> *Drawing the full reaction profile will help you consolidate your understanding.*

WHY ARE KETONES MORE STABLE THAN ALDEHYDES?

We should consider whether this is a steric effect or an electronic effect. It is unlikely to be a steric effect in the absence of particularly bulky substituents. Therefore, we need to ask the question 'what does the extra methyl group do?'

We know (**Recap 2**) that methyl groups stabilize carbocations. Here are the resonance forms for propanal and acetone.

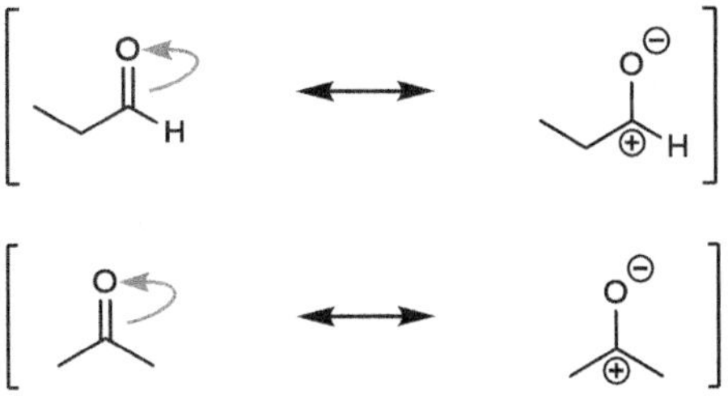

In the case of acetone, the right-hand resonance form is stabilized by the additional methyl group. This can be considered to cause stabilization of the compound.

When we consider a molecular orbital explanation for this stabilization, we should be thinking about where the electrons from the methyl group (hyperconjugation) are overlapping. There is only one sensible candidate—the π* orbital of the carbonyl bond.

> *This is exactly the explanation we used for why a more substituted alkene is more stable in **Recap 3**.*

This interaction can be seen on the structure and in the orbital energy diagram below. Recall that when a filled and an empty orbital with the correct symmetry overlap, in effect we get two new orbitals, one of which is lower and the other higher in energy.

47

SECTION 1 BASIC THEORY AND REACTIVITY

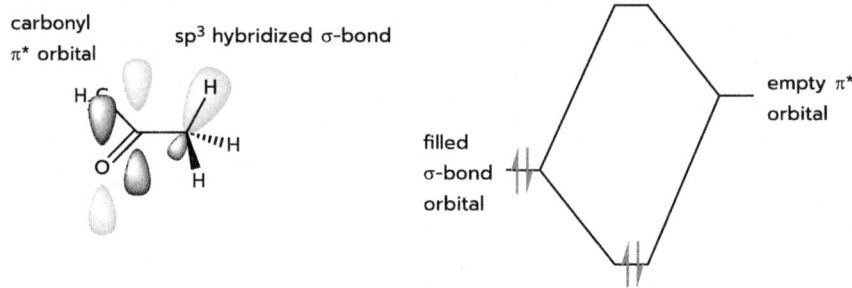

WHY ISN'T THERE MUCH DIFFERENCE BEWEEN PRODUCTS?

In this case, we can either consider the final, protonated, product, or we can consider the initial product with a negative charge on oxygen.

> It doesn't matter!

In the latter case we have an $O^{\ominus}$. It isn't particularly stabilized. More importantly, one is not stabilized relative to the other. The extra methyl group doesn't do anything to stabilize or destabilize the product from the ketone reaction.

> Did I just hear you ask what would happen if we had a ketone (or an aldehyde) with **really big groups**?

Well, when we go from the carbonyl starting material to the alkoxide product, the bond angles change from 120° to 109.5°. It is becoming more crowded. Once we get to substituents that are big enough, we will have a steric effect that destabilizes the product.

> This is good. We made a prediction.

Aldehydes are less stable than ketones, and hence they are more reactive. But when the aldehyde becomes really crowded (see example below), it becomes less reactive. We might simply say that attack of the nucleophile is hindered by the large group, or we could discuss it (as we have just done) in terms of the steric crowding in the product.

> They are the same argument.

WHAT ABOUT OTHER CARBONYLS?

We started with a rigorous consideration of energies of starting materials, products, and transition states. We found, rather quickly, that we only really need to consider the energy of the different types of carbonyl compounds.

SECTION 1 BASIC THEORY AND REACTIVITY

> At least for aldehydes and ketones, we can do this using the resonance forms we are getting used to drawing.

Let's try to do the same for other types of carbonyl compound. What about esters, amides and acid chlorides? As always, draw the structures and the resonance forms, and have a look!

> You might think I am trivialising this. I'm not! What I am trying to do is give you 'something to do' when you face a problem. The fact is, most of the time, it's the same thing—this is good!

Here are the resonance forms.

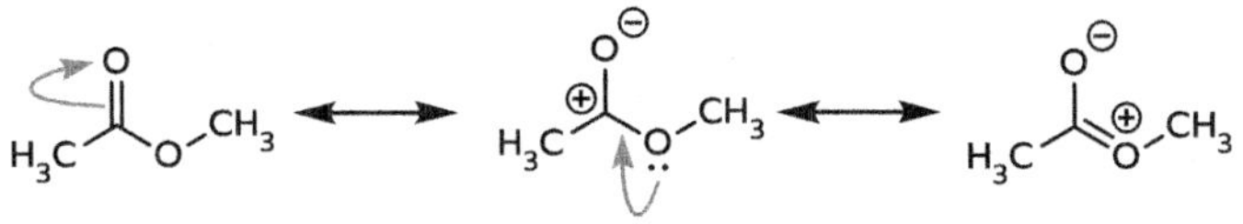

[Resonance structures for ester (methyl acetate): neutral form ↔ A (O⁻ on carbonyl O, O⁺ on OCH₃) ↔ B]

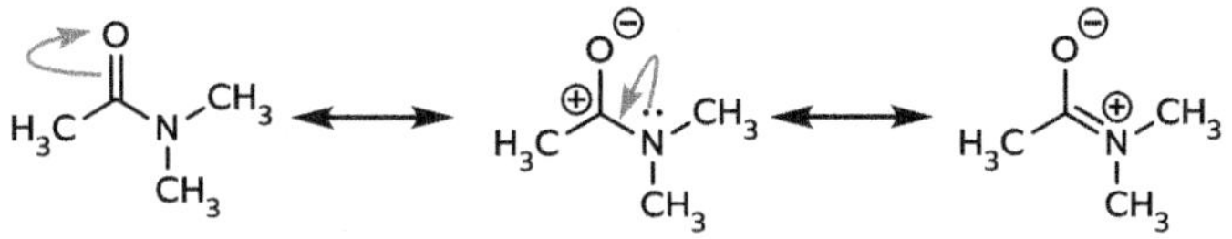

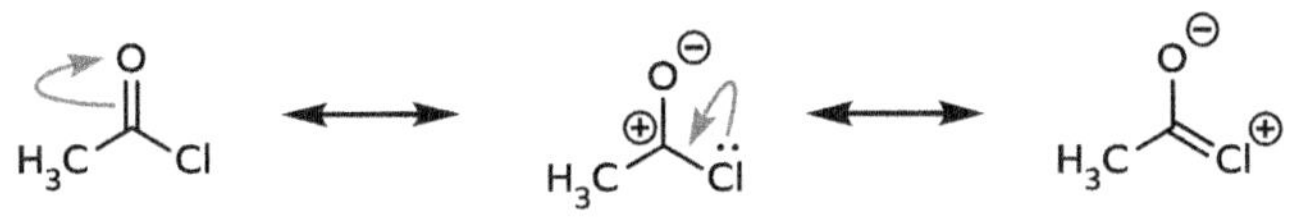

[Resonance structures for amide (N,N-dimethyl acetamide): neutral form ↔ A ↔ B]

[Resonance structures for acid chloride (acetyl chloride): neutral form ↔ A ↔ B (with Cl⁺)]

It almost doesn't matter what order you do these in, but of course I have a preference!

For the acid chloride, resonance form **B** doesn't do anything for us! The curly arrow on resonance form **A** shows overlap of the chlorine lone pair with the π* orbital of the carbonyl group. Chlorine is not in the same row of the periodic table as carbon and oxygen.

> This orbital overlap does not give us much, if any, stabilization.

In addition, we have an inductive electron-withdrawing effect that destabilizes resonance form **A**.

> Acid chlorides are not stabilized—they are reactive. In fact, acid chlorides are among the most reactive carbonyl groups you will encounter. We will see the implications of this in **Reaction Detail 17** and **Reaction Detail 18**.

Oxygen has the same electronegativity as chlorine. But it is in the same row of the periodic table as carbon.

49

SECTION 1 BASIC THEORY AND REACTIVITY

> *The orbital overlap in resonance form B is more effective for the ester. Esters are more stable (less reactive) than acid chlorides.*

Nitrogen is less electronegative than oxygen. The inductive effect destabilizing resonance form **A** is less for the amide.

> *Because nitrogen is less electronegative, the orbital overlap (resonance form B) is more effective. This stabilizes the amide. Amides are (much) less reactive than esters.*

It is no coincidence that Nature has 'chosen' the amide bond for enzymes/proteins. It is a nice stable functional group.

ONE MORE POINT

We can readily compare the relative reactivity of aldehydes and ketones. We can also readily compare the relative reactivity of esters, amides and acid chlorides.

> *What we cannot do so easily is work out where aldehyde and ketone reactivity 'fits in' with the ester, amide and acid chloride reactivity.*

It is important to recognize the limitations of our ability to predict. There **are** some things you need to learn.

> *In most cases, aldehydes and ketones are more reactive than esters, but considerably less reactive than acid chlorides.*

… SECTION 1 BASIC THEORY AND REACTIVITY

RECAP 5

Stability of Simple Carbanions

WHY?

We are going to spend a lot of time talking about carbanions. 'Simple' alkyl carbanions are incredibly unstable. I cannot think of any reactions where they are directly involved. However, these carbanions allow us to understand organometallic reagents, as we will see in Basics 5.

More fundamentally, we need to start somewhere. We won't understand the more complex, and more useful, carbanions if we don't understand the 'simple' ones.

SIMPLE CARBANIONS—STRUCTURE AND STABILITY

A carbanion has a negative charge on a carbon atom. When we only consider carbanions with simple alkyl substituents, we see the following stability trend, which is the opposite to what we saw for carbocations in Recap 2.

$$\underset{\text{Least stable}}{\underset{\text{tertiary}}{(CH_3)_3C^{\ominus}}} < \underset{\text{secondary}}{(CH_3)_2CH^{\ominus}} < \underset{\text{primary}}{CH_3CH_2^{\ominus}} < \underset{\text{Most stable}}{\underset{\text{methyl}}{{}^{\ominus}CH_3}}$$

This is what we would expect to see. Since a methyl group stabilizes a positive charge, it will destabilize a negative charge. Also, carbanions are tetrahedral, so that the more (bulky) substituents are present, the more crowded the carbanion.

> *There are two points to make here. First of all, none of these carbanions is stable enough to be considered an intermediate in a reaction. Secondly, we should always consider steric and electronic factors that affect stability.*

WHAT NEXT?

These are not the only carbanions you need to know about. In a sense, they are not even the most important. We will look at a range of carbanions in Recap 7, in which the negative charge is stabilized by various different groups.

> *We will also see that some substituents are able to stabilize both positive and negative charges.*

SECTION 1 BASIC THEORY AND REACTIVITY

BASICS 1

The Difference Between Carbonyl Compounds and Alkenes

We have now seen the typical reactivity of alkenes and of carbonyl compounds. They are different! Perhaps this is surprising—after all, alkenes and carbonyl groups are both characterized by the presence of a π-bond. They are inherently electron-rich.

And yet while we have seen that alkenes are attacked by electrophiles, we found that carbonyl groups tend to be attacked by nucleophiles.

> We need to fully and instinctively understand why this is the case.

The main purpose of this short chapter is to help you identify strategies that will improve your problem-solving ability.

WHY ARE ALKENES NOT ATTACKED BY NUCLEOPHILES?

This may seem like a simple enough question. Alkenes are electron-rich. Nucleophiles are electron-rich. Why would one electron-rich species attack another?

> And yet, on the face of it, this is exactly what happens with carbonyl compounds. Clearly this answer isn't good enough.

We need to identify a better strategy. Try drawing a possible reaction. With a simple monosubstituted alkene, there are two possible outcomes.

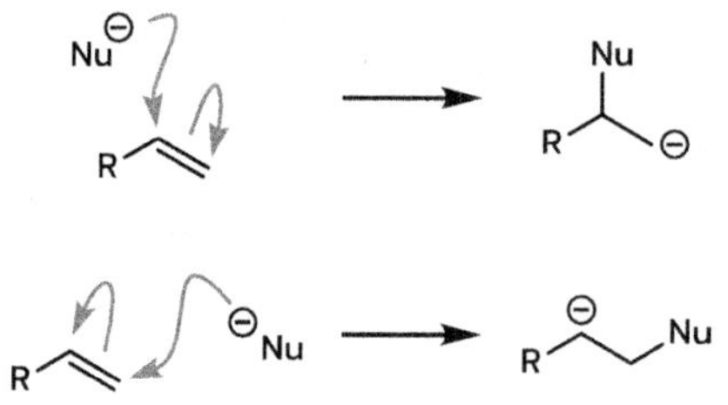

Assuming R is a simple alkyl group, neither of the two products, carbanions, are particularly stable.

> Yes, the top one is primary, so it is the more stable of the two. But they are both bad. It's hard to imagine a nucleophile that is reactive enough to allow this to be favourable.

This is a productive analysis of the problem, and it then allows us to speculate. We have said that the reason this doesn't happen is because the carbanions are not stable. This reasoning has a consequence. *If* the carbanions *were* stable, we would actually expect this process to be feasible. We will come back to this in a moment.

AN IMPORTANT POINT

In **Fundamental Reaction Type 1** we have alkenes reacting with electrophiles to give carbocation intermediates. Now, I am telling you that alkenes do not react with nucleophiles to give carbanion intermediates.

> *How do you know that one charge is better than the other?*

The short answer is that you should not simply expect to know this. But you know now, because I am telling you! As long as we are only considering carbocations/carbanions with simple alkyl groups as substituents, carbocations are much more stable.

> *There are only a few reactions that involve alkyl-substituted carbanions. But there are many reactions that involve alkyl-substituted carbocations.*

SOMETHING TO THINK ABOUT

Let's look at the lower reaction again.

A simple alkyl group will definitely *not* stabilize the carbanion. But we can imagine plenty of things that will. If 'R' is a group that strongly stabilizes the carbanion, then this reaction should work.

> *Guess what—it does! We will come back to this in **Reaction Detail 28**. We will need to see some other substituents that stabilize carbanions in **Recap 7**. When you get there, have a think about what 'R' might have to be to provide a lot of stabilization.*

Of course, we have already seen that nucleophiles **do** attack carbonyl compounds, and this is because we get an $O^{\ominus}$ instead of a $C^{\ominus}$. This is ***massively*** more stable! We can quantify this by considering the following equilibrium.

$$R\text{—}CH_2^{\ominus} \;+\; R\text{—}OH \;\rightleftharpoons\; R\text{—}CH_3 \;+\; R\text{—}O^{\ominus}$$

$$pK_a \; 16 \qquad\qquad pK_a \; 51$$

With the pK_a values given, this equilibrium has $K_a = 10^{35}$. To put this another way, you would have approximately 1.66×10^{11} ***moles*** of the alkoxide anion for every ***molecule*** of the alkyl anion.

> *If the alcohol was ethanol, we would have approximately 7.5 million tonnes of ethoxide for every **molecule** of the alkyl anion!*

SECTION 1 BASIC THEORY AND REACTIVITY

WHY ARE CARBONYL GROUPS NOT ATTACKED BY ELECTROPHILES?

We can undertake the same type of analysis that we used above. Whenever I want to draw the attack of an electrophile, I always use a proton—you can't get a simpler electrophile than this.

Here is the reaction. Notice that I have drawn this in a way that puts the positive charge on the carbon atom.

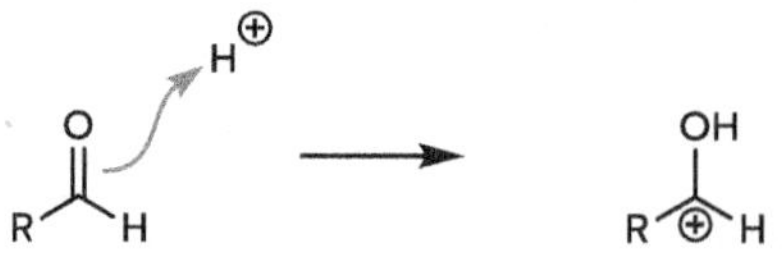

So, why wouldn't this happen? What's wrong with the structure on the right?

> The short answer is that the structure on the right is actually pretty stable, and this can happen. But we need to add a little refinement.

How might we represent the stability of this carbocation? Resonance forms, of course!

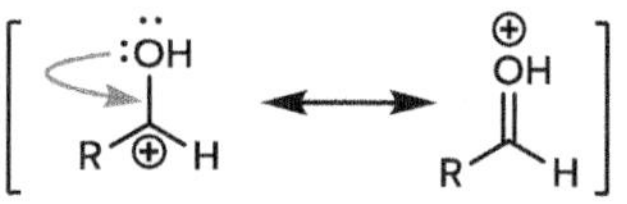

In order to draw this, we needed to add the lone pairs onto oxygen. Of course, they were there all the time. The right-hand structure is a better representation. We have protonated the carbonyl oxygen atom. A better way to draw this process is as follows.

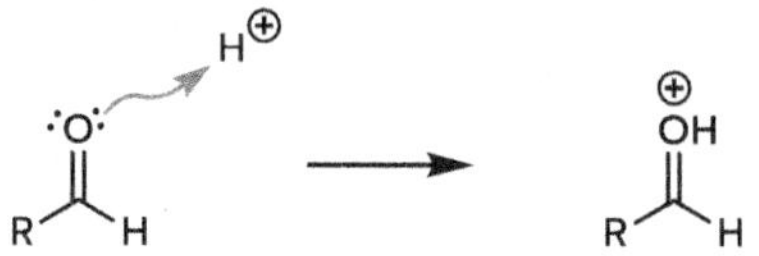

And what would happen next? A nucleophile can attack this species. I'm going to use water as the nucleophile.

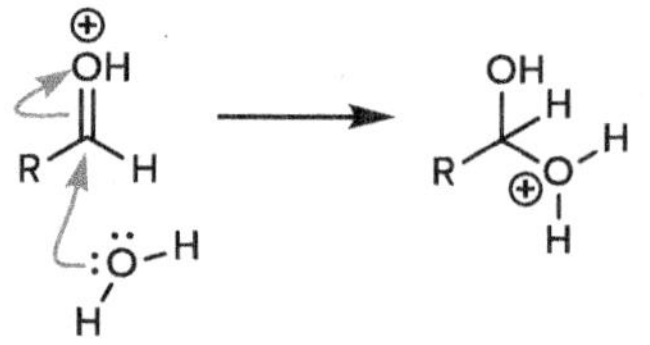

We will see this again in **Reaction Detail 3**. Before we get to that point, we will look at some more of the principles in the next chapter.

> First, let me leave you with a question. Which of the following two species will be most readily attacked by water?

We will answer this question in the next chapter.

IN SUMMARY

We asked why a carbonyl group is attacked by a nucleophile rather than by an electrophile. It turns out that a carbonyl group **can** be attacked by an electrophile, and this can then be followed by attack by a nucleophile.

Initially, in **Fundamental Reaction Type 2** we looked at addition of a nucleophile followed by addition of an electrophile (especially H⊕).

> *The order of the steps can easily be reversed.*

We also asked why an alkene is attacked by an electrophile rather than by a nucleophile. We found that this is because we would form a carbanion intermediate which is extremely unstable.

> *And then we identified a scenario where this might not actually be the case.*

We will only be able to do this properly if we understand which carbanions are more stable, and which are less stable.

> *It's really important that you understand (and can quantify) the stability of carbocations and carbanions. You won't get very far in organic chemistry without this.*

SECTION 1 BASIC THEORY AND REACTIVITY

BASICS 2

Acid Catalysis in Carbonyl Addition Reactions

WHY?

At the moment, it might look like we are adding facts with no context. Some carbonyl reactions can be catalysed by addition of acid. Some cannot. It's important to understand why this is the case.

There are two aspects to this. If you are actually *doing* the reaction, you've almost certainly looked up a suitable recipe.

> We never make up our own recipes from scratch, if we can find suitable conditions in the chemical literature.

But if you are drawing a reaction that *needs* an acid catalyst, and you draw the mechanism without one, you will end up drawing a very unstable structure at some point—something that would never be formed. And that's not good, because we are trying to build the patterns based on what is right, and what is reasonable.

INTRODUCTION

In **Fundamental Reaction Type 2**, we saw addition of nucleophiles onto carbonyl groups. We considered a generic negatively charged nucleophile, as in the following equilibrium.

We considered the stability of the negatively charged nucleophile compared to the product alkoxide. The position of the above equilibrium will depend on this stability. It will also depend on the reactivity of the carbonyl group itself (**Applications 1**).

> In some cases, the negatively charged nucleophile won't be reactive enough. In other cases, it might be impossible or inconvenient to use a negatively charged nucleophile.

In **Basics 1**, we found that we could protonate a carbonyl oxygen atom, and we asked what this would do to the reactivity. Let's finish the job!

We will compare two equilibria. For consistency, we will draw the nucleophile as a neutral species with a lone pair. We will look at the reasons for this in a moment, and we will consider specific examples in **Reaction Detail 15** and in **Applications 2**.

SECTION 1 BASIC THEORY AND REACTIVITY

Here is the first equilibrium. This reaction produces a **zwitterionic** species—one with a positive and negative charge within the same molecule. This charge separation is not good.

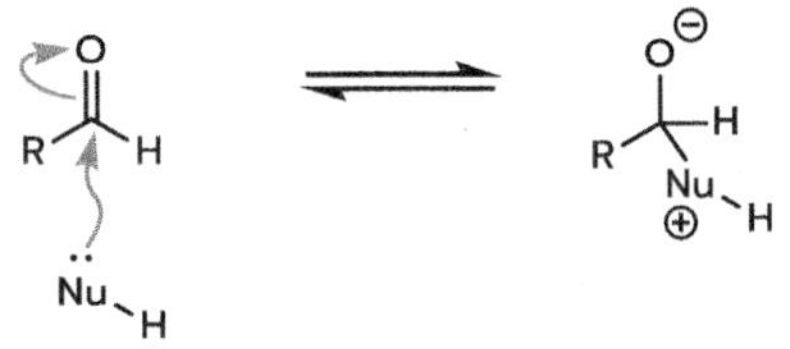

What about the second equilibrium, below, in which we have first protonated the carbonyl oxygen atom? By avoiding production of a negative charge and charge separation, this equilibrium is shifted to the right compared to the one above.

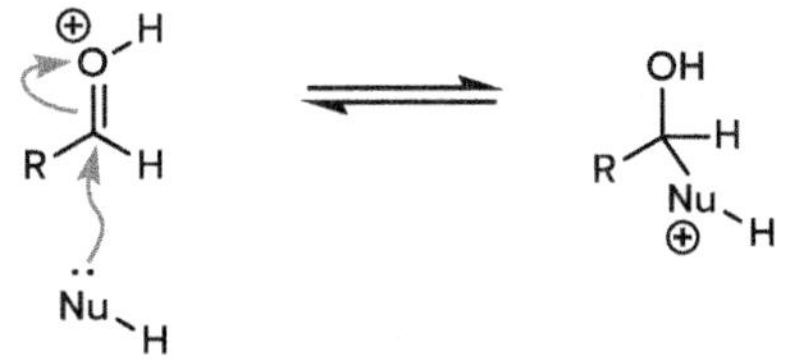

> Could we avoid charge separation by using a negatively charged nucleophile?

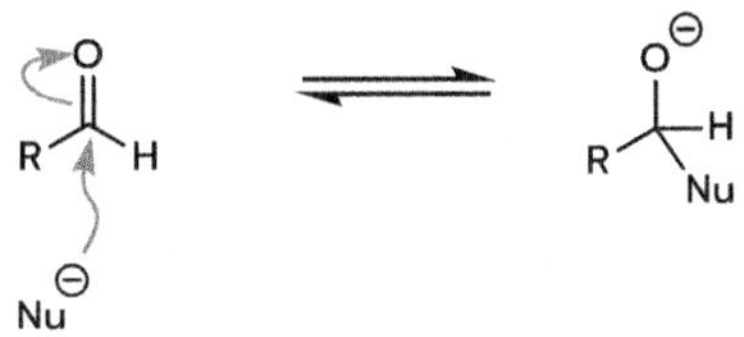

Absolutely! If you had a nucleophile with a negative charge, it would be more reactive (compared to the equivalent neutral nucleophile—e.g. hydroxide versus water). We might not actually need the catalyst at all. However, in this case, we wouldn't be able to use an acid catalyst.

> Why not? Well, acids protonate things. Let's say you have a nice stable neutral carbonyl compound, and a rather unstable, negatively charged basic compound. You add an acid.

> Which one of the two do you think will be protonated first?

CONSIDERATION OF RESONANCE FORMS

We can look at this another way, which is equivalent, but may be easier to see. We have seen the resonance forms for the carbonyl group itself.

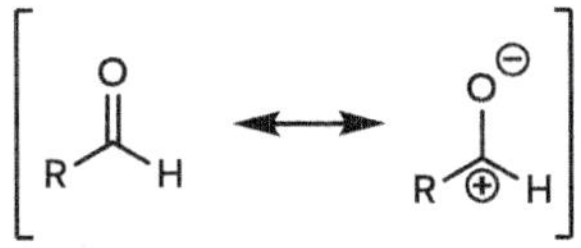

What happens when we protonate the oxygen atom?

57

SECTION 1 BASIC THEORY AND REACTIVITY

The species in square brackets, represented by two extreme resonance forms, is positively charged. The right-hand resonance form is 'more stable' in the lower scheme compared to the upper, due to it not having separation of two unstable charges. Therefore, we increase the contribution of this resonance form. By doing this, we increase the positive charge on the carbonyl carbon atom, making it easier to add a nucleophile at this position.

> We are not talking about shifting an equilibrium to favour the right-hand structure. We are talking about an increase in the contribution of the right-hand resonance structure to the 'real structure'.

The real structure might look something like the following, with the dotted bond representing more than a single bond and less than a full double bond. And we need to put a δ+ on both atoms, without being able to specify **how much** partial charge is on each.

> Overall, it is much better to get used to drawing resonance forms, and fully understanding what they mean.

WHEN CAN YOU USE ACID CATALYSIS?

It is probably better to ask when the use of acid catalysis is not applicable. If the nucleophile will undergo irreversible reaction with the acid (*i.e.* it will be destroyed by the acid) then you cannot use acid catalysis.

For example, a Grignard reagent (we will see these in Reaction Detail 12) will react with acid. In the following example, we are specifically using HBr as the acid.

$$R-Mg-Br \quad + \quad HBr \longrightarrow R-H \quad + \quad MgBr_2$$

So, we could never expect to add a Brønsted acid (H⊕) to a Grignard reagent. Fortunately, we would never need to, as these nucleophiles are reactive enough to not need catalysis.

Once again, we are considering the position of the following equilibrium.

SECTION 1 BASIC THEORY AND REACTIVITY

EXERCISE 1

Bond Dissociation Energies and Predicting Outcome of Reactions

In the next chapter, we will look at hydrate formation from aldehydes and ketones. Later, in **Reaction Detail 15**, we will look at acetal and ketal formation.

What we will do here, as a worked exercise, is to calculate the enthalpy change for both processes using average bond-dissociation energies. We covered the method in **HTSIOC Basics 13**, and we saw an extension of the method in **HTSIOC Basics 35**. It doesn't matter that you haven't seen the reaction yet. You can still count the bonds broken and formed.

Reaction 1 – Hydrate Formation

Reaction 2 – Ketal Formation

> Calculate the enthalpy of reaction for these two reactions using the data in the following Table.

I've deliberately kept the Table simple, and just given you what you need. Work them out before looking at the answers below.

Bond Dissociation Energy / kJ mol⁻¹			
C–O	350	C=O	732
O–H	460		

REACTION 1—HYDRATE FORMATION

Here, we break a C=O bond, and we form two C–O bonds. We cannot consider simply breaking the π-bond and forming a new σ-bond using this method, so we have to 'completely break' the C=O and form two new bonds.

SECTION 1 BASIC THEORY AND REACTIVITY

We have two O–H bonds on the left (H$_2$O) and on the right (the hydrate), so there is no net change.

$$\Delta H = 732 - (2 \times 350) = +32 \text{ kJ mol}^{-1}$$

This reaction is calculated to be **endothermic**.

REACTION 2—KETAL FORMATION

Here, we break two O–H bonds (methanol), but we also form two O–H bonds (H$_2$O). We can assume that these cancel out. In addition, we break a C=O bond, and form two C–O bonds, just as before.

$$\Delta H = 732 - (2 \times 350) = +32 \text{ kJ mol}^{-1}$$

This reaction is also calculated to be **endothermic**, by exactly the same amount.

FURTHER PERSPECTIVE

There is always more than one way to look at a problem. Here is another representation of the first reaction, but I've added another water on each side to make it look more like the second reaction. Perhaps it is easier to see here why ΔH is calculated to be the same as in the second reaction.

Why is this *endothermic*? Remember that alkenes undergo addition reactions. We break a weak π bond and we form two σ bonds. We will look more closely at the energetics of this in **Fundamental Reaction Type 3**.

The carbonyl bond is stronger than the alkene bond (**Fundamental Reaction Type 1**). Here, we are breaking the stronger π bond of the carbonyl group. We are also replacing the strong σ component of the double bond (sp^2) with a weaker sp^3 C–O bond (**Recap 1**).

We are going to look at hydrate formation next (**Reaction Detail 3**). This is an equilibrium that generally favours the carbonyl compound.

We will look at ketal formation in **Reaction Detail 15**. The ketal is stable, so although it might be less stable, we can devise reaction conditions that favour its formation (removal of water).

SECTION 1 BASIC THEORY AND REACTIVITY

REACTION DETAIL 3
Formation of Hydrates

INTRODUCTION

In **Fundamental Reaction Type 2** we considered attack of nucleophiles onto carbonyl groups in the general sense. In **Reaction Detail 2** we looked at a simple anionic nucleophile—cyanide. In **Basics 2** we established that it is possible to use acid-catalysis to promote addition of nucleophiles, but only if the nucleophile is compatible with acidic reaction conditions. Now we will consider addition of oxygen nucleophiles.

> *The focus for now is to develop our understanding of the processes involved, and the type of reactivity.*

FORMATION OF HYDRATES

Let's start with the simplest possible oxygen nucleophile—water! The product is called a hydrate.

We saw in **Exercise 1** that hydrate formation is *endothermic*. In addition, ΔS for this reaction is definitely negative, so the reaction will also be *endergonic* ($\Delta G > 0$).

Of course, we saw in **Applications 1** that the average bond-dissociation energies are limited. A reaction that is *endothermic* by 32 kJ mol^{-1} **on average** could very well be *exothermic* for a more reactive (less stable) aldehyde.[9]

> *After all, these numbers are only an indication!*

Whenever a carbonyl compound is exposed to water, we will have some hydrate present—it's an equilibrium. The position of the equilibrium will depend on the carbonyl compound. A more reactive (less stable) carbonyl compound may actually favour hydrate formation.

$$\underset{R\;\;\;H}{\overset{O}{\|}} + H_2O \;\rightleftharpoons\; \underset{R\;\;\;H}{\overset{HO\;\;OH}{\vee}}$$

First, we are going to consider the mechanism of hydrate formation. This reaction is acid catalysed. We are going to approach this from a very fundamental perspective, so that the lessons you get from this mechanism can be applied to other mechanisms.

> *It's all about the questions you ask yourself.*

---

[9] It's going to be an aldehyde! They are more reactive than ketones in general.

SECTION 1 BASIC THEORY AND REACTIVITY

In this case, if we add an acid to a mixture of an aldehyde and water, which will the acid protonate preferentially?

In the first case, we form a bond from sp² hybridized oxygen to the proton. In the second case, we form a bond from sp³ hybridized oxygen to the proton. We know from **Recap 1** that a bond to an sp² hybridized atom is stronger than the corresponding bond to an sp³ hybridized atom. This is still true when the atom is oxygen.

> *In reality, we are dealing with equilibria, so that both species will be protonated to some extent. However, the fact remains that the carbonyl oxygen atom is more basic (more easily protonated).*

Now, we can draw a mechanism for hydrate formation under acidic conditions.

The carbonyl group is protonated by the acid. This makes the carbonyl carbon more electrophilic, as we saw in **Basics 2**. Water is a nucleophile, so it can attack as shown. Finally, we lose a proton to give the hydrate.

> *Pay attention to the direction of each curly arrow. This is where mistakes tend to creep in.*

Hydrate formation is rarely a synthetically useful process. However, it is important that you can draw a mechanism for this reaction and consider the position of the hydration equilibrium.

> *Draw the mechanism out for yourself. You might think copying a mechanism won't help you learn. You would be wrong!*

POSITION OF EQUILIBRIUM

Now let's look at the position of the equilibrium for three hydrates. Here are the first two—ethanal and propanone. These are equilibrium proportions of the hydrate with water as the solvent.

SECTION 1 BASIC THEORY AND REACTIVITY

> *There is a **lot** of water in there!*

$$H_3C-CHO + H_2O \rightleftharpoons H_3C-CH(OH)_2$$
ethanal → 58% hydrated

$$H_3C-CO-CH_3 + H_2O \rightleftharpoons H_3C-C(OH)_2-CH_3$$
propanone (acetone) → 0.1% hydrated

To put this another way, if we only had one equivalent of water,[10] we would only have a small amount of the hydrate with ethanal, and essentially no hydrate with propanone.

> *Have another look at Applications 1. Make sure you understand why the aldehyde is more reactive (less stable) than the ketone, and why it leads to this experimental observation.*

Now let's look at an extreme case, 2,2,2-trichloroacetaldehyde, commonly known as chloral.

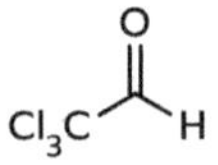

$$Cl_3C-CHO$$
chloral

> *Draw the two resonance forms for chloral. Consider whether the Cl₃C group will stabilize or destabilize the resonance form with a positive charge on the carbonyl carbon.*

The knee-jerk reaction is that this is an alkyl group and alkyl groups are electron-releasing. However, you have three electronegative atoms. There is a strong inductive electron withdrawing effect which will destabilize the positive charge.

> *Chloral is less stable (more reactive) than ethanal.*

Chloral in water is 100% hydrated.

---

[10] *i.e.* the same number of moles of water as we have of the aldehyde or ketone.

63

SECTION 1 BASIC THEORY AND REACTIVITY

RECAP 6

Conjugation and Aromaticity

WHY?

This book being all about π-bonds, and conjugation is something that can happen when we have multiple π-bonds. It's a really important something! We saw conjugation and aromaticity in **HTSIOC Basics 10**.

MOLECULAR ORBITALS FOR BUTADIENE

Let's start with the molecular orbitals. We looked at the π molecular orbitals of ethene in **Recap 1**. The following compound is buta-1,3-diene. The two double bonds are in conjugation.

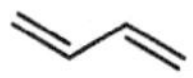

Let's start with the ethene HOMO (the π orbital—bonding). When we have two double bonds in conjugation, we can overlap these two double bonds in two ways, differing in symmetry. The lowest energy orbital has overlap across all four carbon atoms. The next orbital has a node between the second and the third carbon atom. These two orbitals are filled. We can label them ψ_1 and ψ_2.

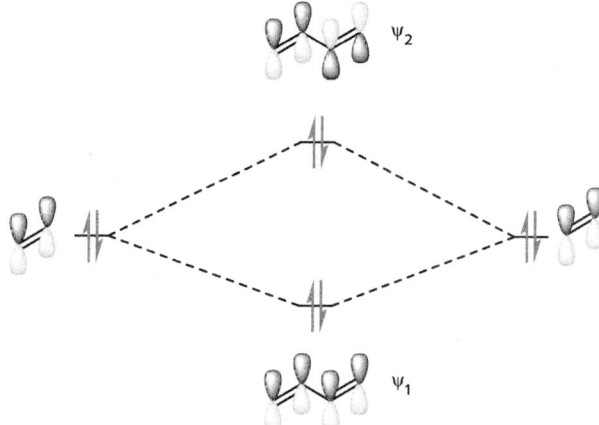

I haven't put the orbital energies on the diagram. If you look at the diagram above, you might think that ψ_1 is going down in energy exactly the same as ψ_2 is going up, and as such, there would be no overall stabilization. In fact, this is incorrect.

> The conjugation (orbital overlap) in buta-1,3-diene provides about 20 kJ mol^{-1} of stabilization.

We can do the same with the ethene LUMO (the π* orbital—antibonding). We get two more orbitals (ψ_3 and ψ_4) as shown below, again differing in symmetry. These orbitals have two and three nodes respectively. We will start to look at some of the implications of orbital symmetry in Section 6. Don't worry about the fact that these orbitals don't have any electrons in them.

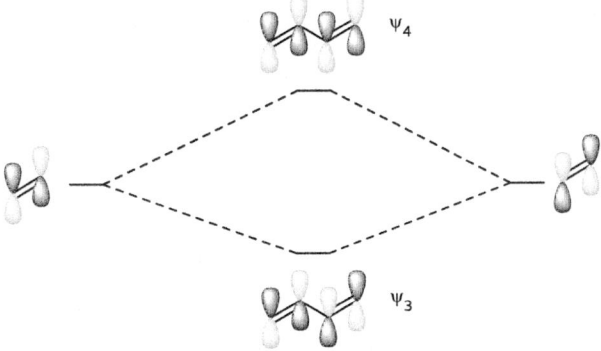

The complete orbital energy diagram is shown on the next page.

THE IMPLICATIONS OF CONJUGATION

As we have just seen, a conjugated compound is more stable than an otherwise equivalent non-conjugated compound. We can see this with the following two example compounds.

NON-CONJUGATED   **CONJUGATED**

In the compound on the left, we cannot get overlap of orbitals across both alkene bonds—there is a CH$_2$ group in the way. As a result of orbital overlap, the conjugated compound is approximately 20 kJ mol^{-1} more stable than the non-conjugated compound.

Of course, following the logic used in Recap 3, it isn't necessarily true that a (more stable) conjugated compound is less reactive than a (less stable) non-conjugated compound. In many cases, conjugation provides new modes of reactivity. We will look at this next, in Reaction Detail 4.

We also need to look at the bond lengths. I'm going to start with the C1–C2 bond length in hexa-2,4-diene (1.52 Å). This is shorter than the C–C bond length in ethane (1.57 Å). We saw in Recap 1 that a σ bond to an sp^2 carbon is shorter than one to an sp^3 carbon. We also saw in Recap 3 that a more substituted alkene is more stable. The overlap of the C–H bond of the methyl group with the π* orbital of the double bond makes the C1–C2 bond shorter.

1.52 Å 1.47 Å

1.35 Å

SECTION 1 BASIC THEORY AND REACTIVITY

Of course, we also see that the C3–C4 bond is even shorter, due to the overlap between the two π bonds.

Here is the complete orbital energy diagram.

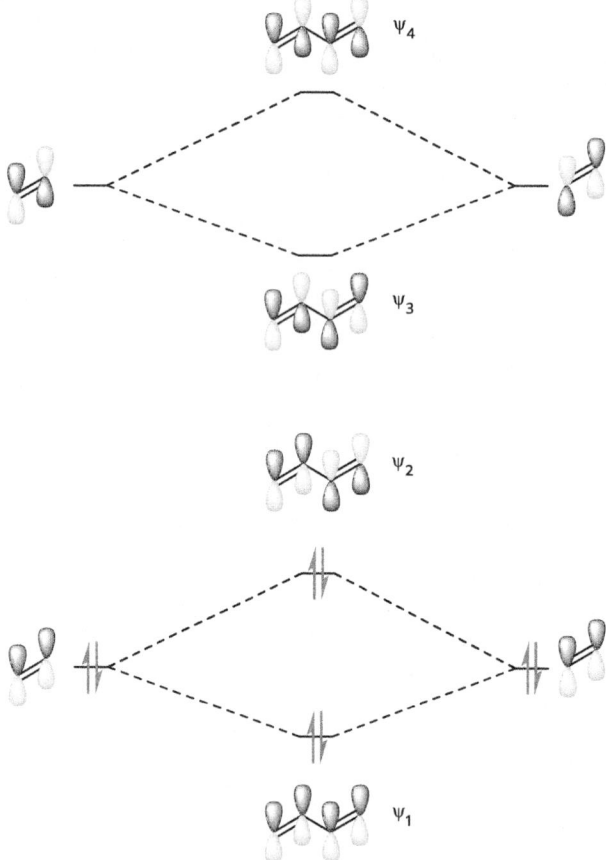

AROMATIC COMPOUNDS ARE VERY STABLE

Now we get to the 'extreme' form of conjugation. Remember that benzene is roughly 150 kJ mol^{-1} more stable than it would be if it didn't have the property known as 'aromaticity'. This is a lot of energy! It isn't as much as a σ-bond or a π-bond, but it's getting there!

This stabilization dominates the chemistry of benzene and related compounds. We will start to explore this in **Fundamental Reaction Type 3**.

For now, let's focus on the terminology. We tend to use terms such as 'conjugated' and 'delocalized' more or less interchangeably. When we look at the molecular orbital diagram for buta-1,3-diene, we can see that the electrons are not associated

SECTION 1 BASIC THEORY AND REACTIVITY

(localized) with one or the other double bond. They are **de**localized. We can represent the orbital overlap in benzene using resonance forms as follows.

I'm going to take yet another opportunity to remind you what resonance forms mean. The arrow is a resonance arrow, not an equilibrium arrow. These two structures do not have any individual existence. Each C–C bond in benzene is a '1.5 bond' and the 'real structure' is somewhere between the two resonance forms.

> To put is another way, benzene is definitely **not** resonating between the two structures. Do not use the word 'resonating'!!!

It used to be quite common to draw benzene rings with a circle to represent the π bonded electrons. We need to be able to draw curly arrow mechanisms for reactions, and we cannot properly do that with this representation. Don't do it!

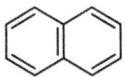

> What you absolutely need to do is see and draw enough benzene rings and resonance forms so that when you encounter a benzene ring, you **know** that all the bonds are the same length, and you know how to draw good curly arrows to show the possible reactions.

Of course, benzene isn't the only aromatic compound. We will encounter heterocyclic aromatic compounds in **Basics 7**. Some of them will have six-membered rings, like benzene, but many of them will have five-membered rings. They still have six π or n (lone pair) electrons fully conjugated in a ring.

Other aromatic compounds have more electrons and rings. Naphthalene has five double bonds (10 π electrons) in two rings.

naphthalene

It turns out that there is a more general definition of aromaticity—the Hückel rule. A fully conjugated (alternating single and double bonds) planar cyclic molecule with **4n + 2** π electrons (where n is an integer) will be aromatic. In the case of benzene, we have n = 1. For naphthalene, with 10 π electrons (5 double bonds), we have n = 2, so it is also aromatic.

You may have noticed that I didn't include the molecular orbitals of benzene here. They were in **HTSIOC**. I decided not to include the diagram here, as we won't be doing anything with the orbitals.

SECTION 1 BASIC THEORY AND REACTIVITY

REACTION DETAIL 4

Hydrohalogenation and Halogenation of Dienes

WHY?

This chapter is about reinforcing some of the basic ideas (**Recap 3, Reaction Detail 1**) and adding a little more detail. This detail should not be 'more to learn'. It should reduce the burden by reassuring you that even though we have added something to the example, the underlying principles have not changed.

ADDITION OF HBr TO A DIENE

We saw in **Recap 3** that although a more substituted alkene is more stable, it can actually be more reactive. We will now see the same for conjugated dienes. Let's start with the classic example, diene **1**.

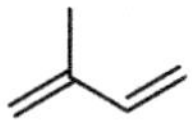

1

Let's think about adding HBr to this diene. We know by now that we need to consider this in two steps. First of all, we will protonate diene **1**.

> We have four carbon atoms that we could protonate. There are four different carbocations we could form.

Let's get two of these out of the way quickly.

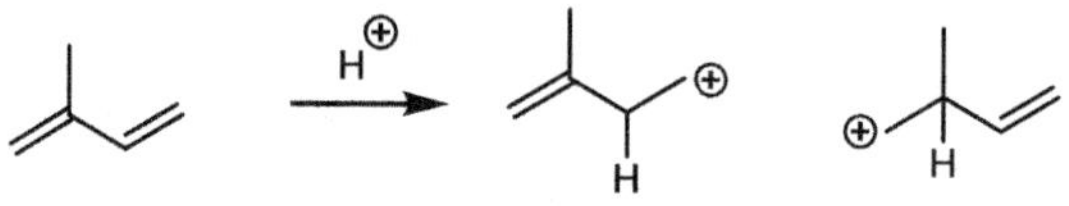

1   2   3

> Draw a curly arrow mechanism for the formation of each of these carbocations. Note that I have drawn the hydrogen atom I added. There are other hydrogen atoms in all the structures.

Carbocations **2** and **3** are *primary*. The carbocation CH_2 group has only a single alkyl substituent providing stabilization. The double bond doesn't provide stabilization.

> Try drawing curly arrows to use the double bond to stabilize the carbocation. They won't be correct!

Now let's look at the other two possibilities.

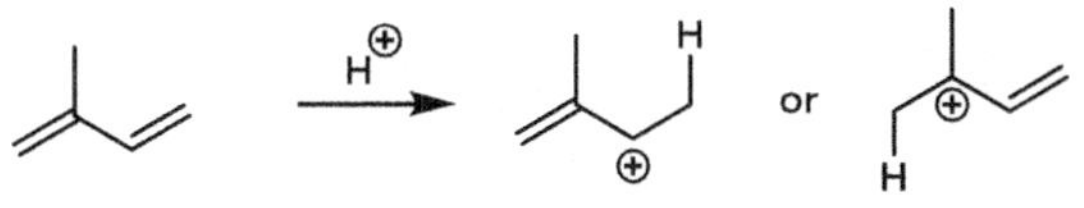

SECTION 1 BASIC THEORY AND REACTIVITY

Both of these carbocations are allylic.

> The positive charge is adjacent to a C=C double bond. You get orbital overlap from the π-bond which stabilizes the positive charge. You can draw resonance forms to show the stabilization.

Basically, I said the same thing three times in different ways. It's important to understand that these are equivalent statements.

> The first one is recognizing the pattern (where the positive charge is with respect to the double bond. The second one is the underlying reason. The third one is how we commonly represent it.

Now, we can consider which of the carbocations, **4** or **5**, is more stable. In carbocation **4**, the methyl group is attached to the double bond. This doesn't affect the carbocation to any significant extent.[11] In structure **5**, the methyl group is directly attached to the carbocation carbon atom, so it provides stabilization by hyperconjugation.

> The simpler way to say this is that carbocation 5 is *tertiary* allylic and carbocation 4 is *secondary* allylic. We can only see this in the resonance forms shown above for carbocations 4 and 5. Had I drawn 'the other' resonance form, it would have been less obvious.

> Draw both resonance forms in each case.

So, we have finally got there. When we protonate diene **1**, we will get carbocation **5**, because it is the most favourable of all possible outcomes. Now to the next question. If we are adding HBr, where does the Br add?

To answer this, we need to fully understand what resonance forms mean. Here is carbocation **5** again.

---

[11] I can't decide if I love it or hate it when this happens! You see, this methyl group does affect the double bond. It affects the stability, which is another way of saying it affects the π bond orbital energy (**Recap 3**). So, the methyl group will have an effect, but it will be much smaller than the effect it has on the stability of carbocation **5**, so that we can ignore it.

69

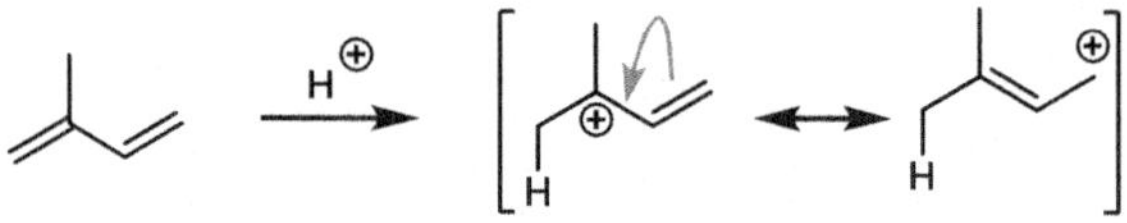

SECTION 1 BASIC THEORY AND REACTIVITY

We have two resonance forms. They differ in the position of the double bond and the location of the positive charge. Let's look at some flawed logic. Remember, the next statement is wrong!!!

> *The resonance form on the left has a tertiary carbocation, so it is more stable. Therefore, the Br will attack this resonance form.*

Let's analyse why this is wrong. The two resonance structures for carbocation **5** have no independent existence. Carbocation **5** is a single entity with a single energy.

> *The two resonance forms are not interconverting. The carbocation is certainly not resonating between the two structures.*[12]

The word 'resonance' is unfortunate, in that it has a conventional meaning that can then be translated to the structures we draw. You just have to treat it as an entirely new word, and not bring any preconceptions with it. Let's see a better analysis of this problem.

We could have Br attack on the left to give product **6**, or on the right to give product **7**. Which is more favoured?

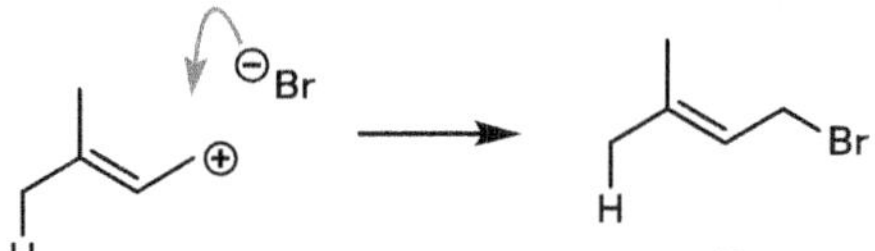

> *In structure 6 we have a monosubstituted alkene. In structure 7 we have a trisubstituted alkene. More substituted alkenes are more stable (Recap 3).*

So, we have one advantage for product **7**.

> *What happens when the Br approaches each of the carbon atoms? Well, for structure 6, it is approaching a tertiary carbon, and as we form the product, we are increasing the level of steric crowding.*

---

[12] I might have mentioned this before, once or twice!

Structure **7** looks better on steric grounds as well. Another way to analyse this would be to look at the reaction in reverse. If we had compound **6** and compound **7**, we would expect carbocation formation to occur more rapidly from compound **6**.

So, let's summarize the outcome.

The H and Br are four carbon atoms apart. We refer to this as a 1,4-addition. You couldn't get this process with a 'simple' alkene. But, as we can see, the outcome can readily be predicted if we apply basic principles.

> Right, over to you. You know which of compounds **6** and **7** is more stable. You know that structure **5** has only a single energy, despite being able to draw to resonance forms. Draw a full reaction energy profile from compound **1** to intermediate **5** and then to structures **6** and **7**. Make sure you can see where the activation energy is for each step.

Doing this will take a little time, but it will be time well spent as it will reinforce the learning.

ADDITION OF Br$_2$ TO A DIENE

When we looked at alkene addition reactions, we found that when we added bromine, rather than HBr, there was an alternative structure for the intermediate. We can expect the same for dienes. Let's look at bromination of compound **1**.

We will use the same approach we used before, but now we will go straight to the bromonium ion intermediates.

I have deliberately drawn these with one C–Br bond longer than the other. We saw the reasons for this in **Reaction Detail 1**.

> Which of these bromonium ions is most stable?

We could draw out the resonance forms and look at the individual contribution to each resonance form. The methyl group in bromonium ion **8** stabilizes one resonance form. Then again, bromonium ion **8** is more crowded. Working out which is more stable is far from trivial.

> It turns out that bromonium ion **8** is actually more stable.

SECTION 1 BASIC THEORY AND REACTIVITY

We then need to consider the reactivity of this bromonium ion. We could, at least in principle, have the bromide attack at carbon 2 or carbon 4. We don't get any attack at carbon 2, presumably because it is quite crowded. Also, as we have just seen, it would give a less-stable product.

Here are the curly arrows for attack at carbon 4. At first, they may look a little strange. This does look like we are attacking an electron-rich double bond with a nucleophile.

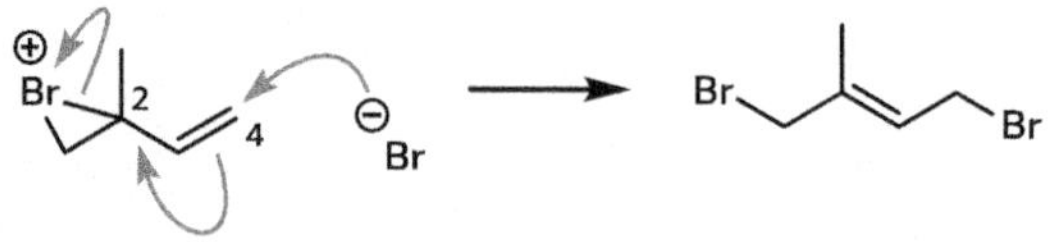

To help you understand this, let's look at resonance forms of cation **8**. We have a contribution from **8a**—this is why I drew the C2–Br bond in structure **8** as being quite long. Resonance form **8a** is *tertiary*, but it is also allylic. The curly arrow and resonance form **8b** show this stabilization.

> So, carbon 4 in structure **8** is electrophilic. Attack by a nucleophile is reasonable. And we cannot actually tell which double bond the bromine reacted with first.

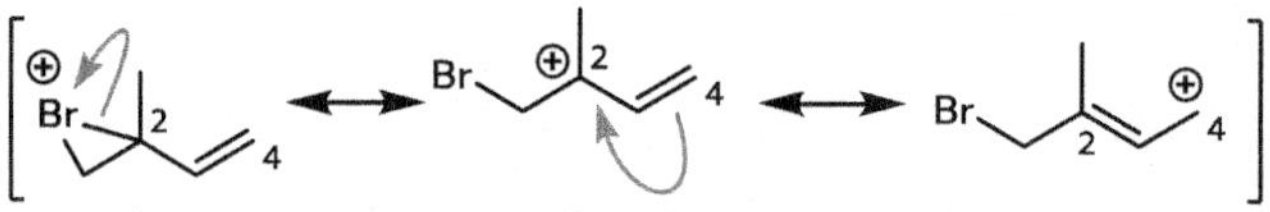

The methyl group in compound **1** adds a complication we don't need. If we didn't have it, carbon 2 would be less crowded, and we wouldn't have to worry about which bromonium ion we form—the double bonds would be identical. Here is the first step.

We could now attack carbon 4, to give compound **12**.

Alternatively, we could attack carbon 2 directly, to give compound **13**.

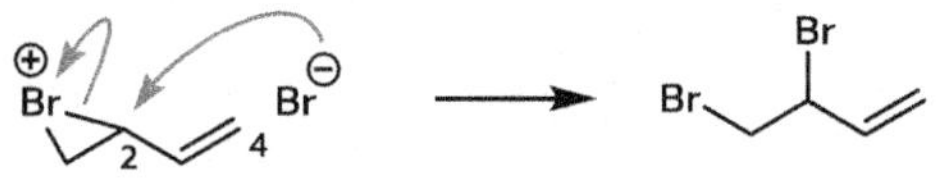

SECTION 1 BASIC THEORY AND REACTIVITY

11 → **13**

> Carbon 2 is less crowded than in compound 8.

There is no doubt that compound **12** is more stable. We have seen the reasoning for this earlier in the chapter. It turns out that compound **13** is formed fastest.

> The reaction to form product **13** must have a lower activation energy!

But if you were to subject product **13** to the reaction conditions, and warm it up, it is converted into product **12**.

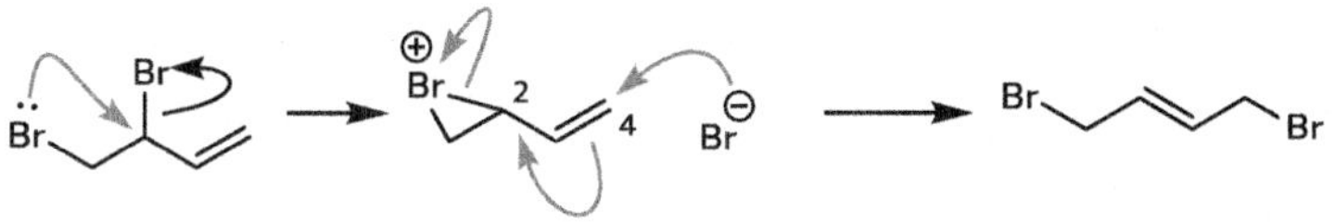

13 → **11** → **12**

I'll be honest—I'm not convinced this is something you could have predicted. Perhaps this is a rare occasion when learning a fact, rather than practising a skill, is essential.

> Draw a reaction energy profile for the reaction of compound **11** to give bromonium ion **11**, and then to products **12** and **13**. Does it look like the reaction profile you drew for reaction of compound **1**? It should. Did you draw bromonium ion **8** higher or lower in energy than carbocation **5**?

In general, I am not going to give you the answers to these problems. You will probably find this frustrating at first, but I have given you enough information (and hints) to fully solve the problem. I tend to find that when you draw a right answer, you will know that it is the right answer. If you are 'winging it' you will know this as well. Just don't carry on winging it and hoping for the best. Reach out to your tutors at the earliest opportunity and your confidence will increase dramatically.

SECTION 1 BASIC THEORY AND REACTIVITY

FUNDAMENTAL REACTION TYPE 3

Aromatic Electrophilic Substitution

WHY?

There are very few pharmaceuticals that do not contain an aromatic ring. We looked at regioselectivity in Recap 4. If we put two (or more) substituents on a benzene ring, we know *exactly* where they are. By that, I mean they cannot move out of the way by rotating bonds. Putting substituents in specific places around a benzene ring is really important. So it is really important that we understand this reaction, and the factors at play. Fortunately, these factors are very fundamental.

> This doesn't mean they are easy to learn. But once you understand them, you'll be able to predict the outcome of these reactions with confidence.

INTRODUCTION

A benzene ring has three double bonds. We have established how alkene double bonds react. While the molecular orbitals of benzene differ slightly from those of alkenes, there are more similarities than differences.

The chemistry of aromatic compounds tends to be dominated by their electron-rich nature. We will encounter exceptions to this, but they are for good reasons and we are going to leave them until you have learned and understood the basics.

We would expect aromatic compounds to react with **electrophiles**, just as alkenes do. Here are the reactions side-by-side. I am using 'E⊕' as a generic electrophile for now.

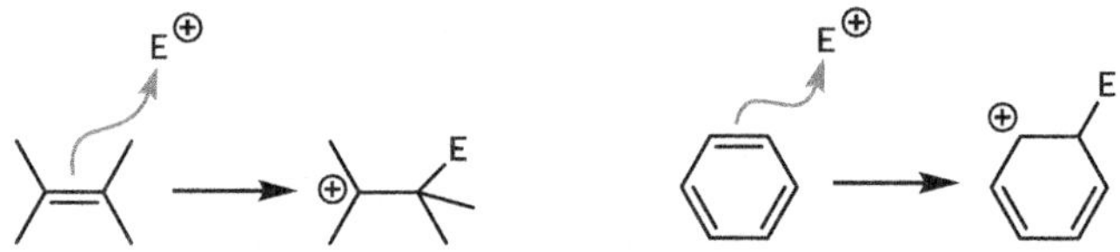

They do look rather similar! You would expect this.

> The reaction on the left gives a carbocation—in this case a *tertiary* carbocation. The reaction on the right also gives a carbocation—in this case an allylic carbocation which is stabilized by two double bonds.

SECTION 1 BASIC THEORY AND REACTIVITY

In a moment, we are going to look at these processes in terms of calculation of enthalpy of reaction.[13] We just need to consider one more thing before we do this.

> *Aromatic compounds are aromatic! This means that they are more stable than they would be if they were not aromatic. Yes, I do know how ridiculous that all sounds, but I'm sure you get the point. Remember that we quantified the effect of aromaticity before. A benzene ring is about 150 kJ mol^{-1} more stable than it would be if it was not aromatic.*

This is one clear case where we can confidently say that a more stable compound will be less reactive.

Consider the following two reactions—bromination of a double bond, which we have just looked at in **Reaction Detail 1**.

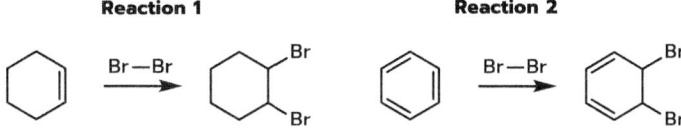

Here are the bond-dissociation energies that you will need. I have also included a rather nebulous value for 'aromaticity'. We looked at this in **Recap 6**.

> Use this data to calculate the enthalpy change in each of the above reactions.

I know it would be quicker to simply read ahead for the answers, but this won't internalize the principles as much for you.

Bond	Energy / kJ mol^{-1}
Br–Br	193
H–Br	366
C–Br	270
C–C	350
C–H	410
C=C	611
"aromaticity"	150

---

[13] What! You mean I am taking an opportunity to look at a reaction and to apply basic principles to it to give you some practise and help you internalize the ideas? Shocking!

SECTION 1 BASIC THEORY AND REACTIVITY

WORKED SOLUTION—ALKENE BROMINATION

We have formed two C–Br bonds, and a C–C single bond. We have lost a C=C double bond and a Br–Br single bond.

$$\Delta H = (193 + 611) - (270 + 270 + 350) = -86 \text{ kJ mol}^{-1}$$

bonds broken   bonds formed

This reaction is *exothermic*.

WORKED SOLUTION—BENZENE BROMINATION (ADDITION)

We can take a short-cut here. All the bonds formed/broken are the same. We have just lost aromaticity.

$$\Delta H = -86 + 150 = +64 \text{ kJ mol}^{-1}$$

This reaction is predicted to be *endothermic*.

WHAT'S THE ALTERNATIVE?

Let's take the intermediate we drew above, in the general sense. We have recognized that there is a fundamental problem. If we simply add two groups to a double bond in benzene, the reaction is unlikely to be favourable. We want to regain aromaticity. We can do this by losing a proton!

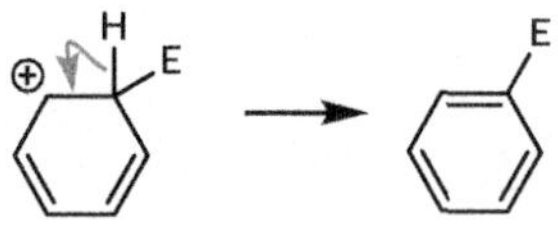

Here is the overall reaction for bromination as an overall substitution reaction.

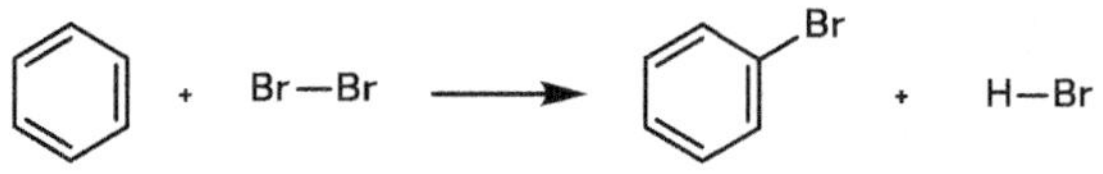

As above, we can calculate the enthalpy change for this reaction. We have broken a C–H bond and a Br–Br bond, and formed a C–Br bond and a H–Br bond.

$$\Delta H = (410 + 193) - (270 + 366) = -33 \text{ kJ mol}^{-1}$$

bonds broken   bonds formed

This is now *exothermic*.

Overall, the chemistry of aromatic compounds is similar to that of alkenes, but with one important difference. The added stability associated with aromatic compounds means that there is a driving force for them to regain aromaticity. This means that they undergo substitution reactions rather than addition reactions.

AROMATIC ELECTROPHILIC SUBSTITUTION REACTIONS

Here is the overall process.

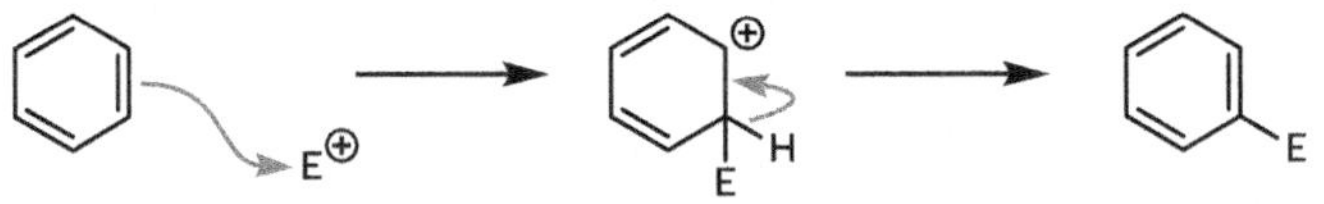

After the first step, we end up forming a bond to the lower carbon atom, and so we put a positive change on the upper carbon atom. After all, this is the carbon atom that has lost a share in the double bonded electrons. Count the bonds if you aren't convinced. On the left, the lower carbon atom has four bonds—three to carbon atoms and one to a hydrogen atom that isn't shown. Don't forget about this hydrogen atom. It is drawn in the second structure because now we need to use it (lose it!).

> Make sure you understand the fundamental reaction type, and the reasoning. Remember why we get **regioselectivity** in addition reactions of alkenes—we form the most stable carbocation preferentially.

What would happen when we add an electrophile to a substituted benzene ring. For example, let's look at methoxybenzene (trivial name, anisole).

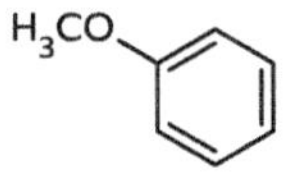

When we add an electrophile, we can get three different carbocation (**Recap 2**) structures depending on where the electrophile adds. These structures have different energies, and you get the more stable ones. Two of them are more stable than the third.

> Draw resonance forms and see if you can work out which are the more stable.

If you can do this, you can predict the regioselectivity (which of the different possible sites the electrophile attacks, **Recap 4**) of aromatic electrophilic substitution reactions.

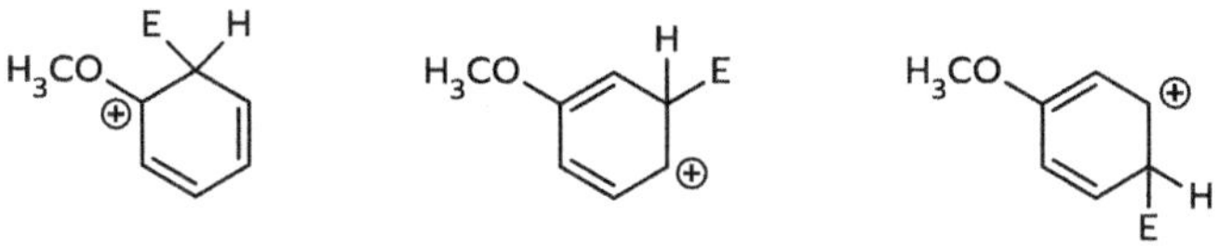

> Remember, 'E' is a generic electrophile. It isn't a specific element or functional group.

Selectivity in aromatic substitution is governed by the same principles as selectivity in alkene addition. It seems more complicated because there are more resonance forms to draw.

SECTION 1 BASIC THEORY AND REACTIVITY

> If you are good at applying the principles (carbocation stability and resonance), the fact that you have a bigger structure with more resonance forms won't worry you. You'll just start drawing resonance forms, and see what happens. If you have practised enough, this will become natural. We will do lots of practice!

We will come back to this process in Reaction Detail 8 and Reaction Detail 9. The answer to the above problem will be found there.

ALKENES VERSUS BENZENES—RELATIVE REACTIVITY

Here is a reaction profile we have seen before. It works for any reaction that involves formation of an intermediate.

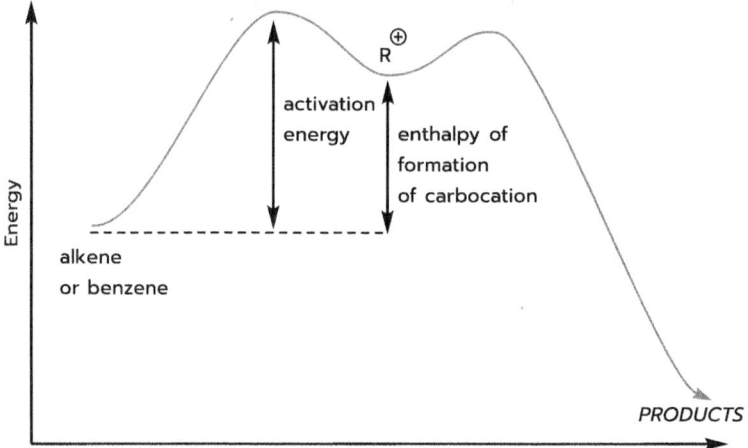

When we consider relative reactivity, we are comparing the activation energy for reaction of an alkene with activation energy for reaction of an aromatic compound.

In order to make this comparison, we first of all need to compare the energy (stability) of the alkene with the aromatic compound. Aromaticity provides about 150 kJ mol^{-1} of additional stability.

> That is, benzene is about 150 kJ mol^{-1} more stable than it would be if it was not aromatic. This isn't quite the same as saying that *any* aromatic compound is 150 kJ mol^{-1} more stable than *any* non-aromatic compound, but we can use this number as an approximation.

What about the carbocations? For the sake of simplicity, let's assume we are talking about reaction of an alkene which will give a *tertiary* carbocation. These are more stable than *primary* or *secondary* carbocations.

Here are some relative energies of carbocations, taken from Recap 2.

SECTION 1 BASIC THEORY AND REACTIVITY

$^{\oplus}CH_3$ $CH_2=\overset{\oplus}{C}H_2$ (allyl) $Ph-\overset{\oplus}{C}H_2$ $Ph-\overset{H}{\underset{\oplus}{C}}-Ph$ $H_3C-\overset{CH_3}{\underset{CH_3}{\overset{|}{\underset{|}{C}}}}-CH_3$ (with ⊕)

+343 kJ mol⁻¹   +91 kJ mol⁻¹   +11 kJ mol⁻¹   −93 kJ mol⁻¹   +0 kJ mol⁻¹

On the right, we have our *tertiary* alkyl carbocation. We set this as our reference point for comparisons. We need to compare this to a carbocation that is stabilized by two alkene double bonds.

> We don't have that in **Recap 2**.

Of course, I could calculate the energy of that system directly. But let's see how we might estimate it.

We saw that one benzene ring contributes 332 kJ mol⁻¹ of stability (the difference between +343 kJ mol⁻¹ and +11 kJ mol⁻¹), while a second benzene ring contributes a further 104 kJ mol⁻¹.

> Here, we are starting with the methyl carbocation and adding the substituent. Make sure you can see where the numbers are coming from.

One alkene contributes 252 kJ mol⁻¹ of stability (the difference between +343 kJ mol⁻¹ and +91 kJ mol⁻¹). Since alkenes contribute less than benzene rings, we can assume that the second alkene won't contribute more than 100 kJ mol⁻¹.

Since the allyl cation 'has an energy' (on this scale) of +91 kJ mol⁻¹, a second double bond in conjugation will bring this down to somewhere in the region of 0 kJ mol⁻¹. We would therefore estimate that the carbocation below, which would be formed by attack of a generic electrophile onto benzene itself, will be of **broadly** similar stability to a *tertiary* alkyl carbocation.

[cyclohexadienyl cation with H and E substituents]

Almost there now! The benzene starting material is more stable than the alkene starting material. The carbocation intermediates are about the same energy. Therefore, the activation energy will be higher for reaction of the aromatic compound.

> Of course, this uses the Hammond Postulate (**Recap 2**) to connect the stabilization of the transition state with the stabilization of the carbocation.

So, we conclude that an aromatic compound will be less reactive than an alkene.

SECTION 1 BASIC THEORY AND REACTIVITY

RECAP 7

Stability of More Complex Carbanions

WHY?

We are going to spend a lot of time in this book talking about nucleophiles. In some cases the nucleophiles are carbanions. We need to know how stable they are.

> A more stable carbanion will be less reactive. A less stable carbanion will be more reactive.

INTRODUCTION

We looked at 'simple' carbanions in **Recap 5**. These are carbanions that are destabilized by increasing numbers of simple alkyl groups. We will see how these 'simple' anions link to organometallic reagents when we get to **Basics 5**, but there are far more interesting carbanions out there.

> To put that another way, there are many other things in organic molecules that can stabilize a negative charge.

This is a recap of material in **HTSIOC Basics 18**, but with some different examples.

OUR TABLE OF pK_a VALUES

I'm going to give you a Table of pK_a values. I've got some bad news for you.

> You need to know these values. I'm going to go through them all, which I hope will make them easier to learn, but there's no escaping the fact that you need to know them.

There's some good news though. If you know a few of them, you will be able to predict where the other ones are (higher or lower, a bit, or a lot). And as you use them, you will find that you remember them much more easily. Let's start with water (Entry 6). Water is neutral. Of course, this doesn't mean it doesn't dissociate. Here's the equilibrium, with the equilibrium constant and the pK_a value.

$$H_2O \quad \underset{}{\overset{K_a = 10^{-15.7}}{\rightleftharpoons}} \quad H^{\oplus} \quad + \quad {}^{\ominus}OH \qquad pK_a = 15.7$$

Recall that

$$pK_a = -\log_{10} K_a$$

So, anything with pK_a higher than 15.7 is less acidic than water. Anything with pK_a lower than 15.7 is more acidic.

> *Learn this number first. It's your reference point.*

SECTION 1 BASIC THEORY AND REACTIVITY

Entry	Compound	pK$_a$	Anion
1	CH$_4$	48	$^\ominus$CH$_3$
2	Ph$_3$P$^\oplus$–CH$_3$ Br$^\ominus$	30	Ph$_3$P$^\oplus$–CH$_2^\ominus$
3	H$_3$C–≡–H	25	H$_3$C–≡$^\ominus$
4	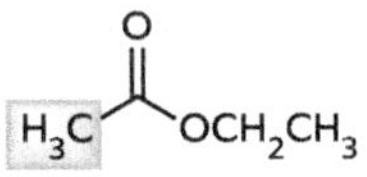 H$_3$C–C(=O)–OCH$_2$CH$_3$	25	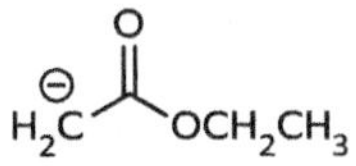 H$_2$C$^\ominus$–C(=O)–OCH$_2$CH$_3$
5	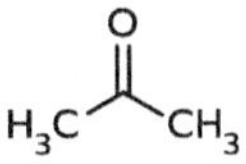 H$_3$C–C(=O)–CH$_3$	19	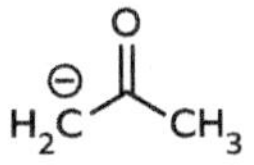 H$_2$C$^\ominus$–C(=O)–CH$_3$
6	H$_2$O	15.7	$^\ominus$OH
7	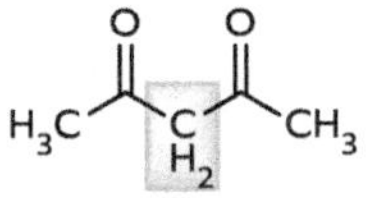 H$_3$C–C(=O)–CH$_2$–C(=O)–CH$_3$	13	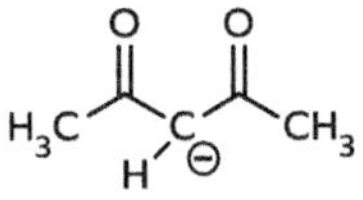 H$_3$C–C(=O)–CH$^\ominus$–C(=O)–CH$_3$
8	H$_3$C–NO$_2$	10	H$_2$C$^\ominus$–NO$_2$
9	H–C≡N	9.2	$^\ominus$C≡N

Let's look at this in a slightly different way. We know that hydroxide is reasonably basic. Any of the anions above hydroxide in the Table are more basic than hydroxide. Any anions below hydroxide are less basic.

Basicity and nucleophilicity are basically[14] the same thing.

> A base is a nucleophile that attacks hydrogen. A nucleophile is a base that attacks something else (usually carbon).

Right, let's work through the Table.

CYANIDE AND ACETYLIDE

We have seen the cyanide anion (Entry 9) in cyanohydrin formation in **Reaction Detail 2**. We didn't need very strongly basic conditions to form the cyanide anion. Cyanide is much more stable (less reactive) than hydroxide.

We saw why the cyanide anion is relatively stable. The carbon atom in cyanide is sp hybridized. The electrons are held tightly by the nucleus. This stabilizes the anion.

---

[14] See what I did there!

SECTION 1 BASIC THEORY AND REACTIVITY

Entry 3 is an acetylide anion. We have directly deprotonated an alkyne carbon atom to give the following species.

$$H_3C-{\equiv}{:}^{\ominus}$$

The pK_a of the alkyne is about 25.

> We often refer to the corresponding anion as having a pK_a of 25. This isn't quite accurate, but it's clear enough once you get used to it.

The acetylide anion is much less stable than the cyanide anion. In the cyanide anion, we also have an inductive electron-withdrawing effect pulling the excess of negative charge towards nitrogen.

On the other hand, we can compare the acetylide anion (entry 3) with the methyl anion (Entry 1). The latter has a pK_a of 48. The acetylide anion is 23 pK_a units, or a factor of 10^{23}, more stable.

> This is massive!

I want to make one more connection. In **Recap 1** we saw that the C–H bond in an alkyne is considerably stronger than the C–H bond in an alkane (go back and look at the numbers again!).

> We discussed bond dissociation energies at length in HTSIOC Perspective 1. These are *homolytic* bond dissociation energies. Here, we are forming an anion, not a free-radical.

What this means is that bond dissociation energies are not always a good indicator of how easy or difficult it is to break a bond. We have the same factors stabilizing the carbanion and making the C–H bond stronger. In this case, it is easier to break the C–H bond, as long as we do it 'the right way'.

> The right way is to form a carbanion.

ANIONS α- TO CARBONYL GROUPS

In carbonyl chemistry, we refer to the carbon atom adjacent to the carbonyl group as the α- (alpha) carbon. Here is the anion resulting from dissociation of acetone (propanone, Entry 5). We generally show this stabilization by drawing resonance forms. We refer to this anion as an enolate.

propanone
pK_a 19

an enolate

With a pK_a of 19, propanone (Entry 5) isn't very acidic. On the other hand, it's a million times more acidic than an alkyne, which we considered above.

SECTION 1 BASIC THEORY AND REACTIVITY

On balance, we could consider an enolate to be a stable anion. We don't mean that it is **really** stable. But it's stable enough that we can make it relatively easily by deprotonating propanone with a strong base.

> There aren't any bases strong enough to deprotonate methane.

It's also unstable (reactive!) enough to react with things. It's a nucleophile. We will start to look at enolate chemistry in **Fundamental Reaction Type 4**.

We should take this opportunity to remind ourselves what resonance forms mean.[15] The enolate doesn't have two **different** structures with the charges on different atoms. In the **real structure** there is some negative charge on carbon and some negative charge on oxygen.

The curly arrows show the donation of the negative charge on carbon towards the carbonyl group, forming a double bond, and then breaking the π bond of the carbonyl group to put the negative charge onto oxygen.

Let's see how we represent the molecular orbitals to show this stabilization. The pair of electrons corresponding to the negative charge overlaps with the π* orbital of the carbonyl bond as follows.

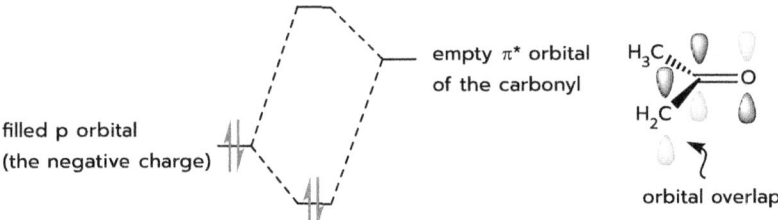

Recall (**Fundamental Reaction Type 2**) that donation of electron density into an antibonding orbital breaks the bond.

> We didn't mention it in Recap 5, but a simple anion is generally sp³ hybridized. We saw in Recap 2 that simple carbocations are sp² hybridized.

The key difference is that for carbocations, we gain more stabilization by having the stronger sp² hybridized C–H bonds **and** having the alkyl groups far apart. For a simple carbanion, having the pair of electrons (that represents the negative charge) in a higher energy p orbital is unfavourable. The most stable situation is sp³ hybridization.

> Remember that hybridization is just a **model** for the bonding, and we would reach the same conclusions, perhaps not as easily, using other bonding models.

Now, with the enolate, we stabilize the p orbital in the sp² hybridized carbon atom so that the planar geometry is more favoured.

---

[15] Again? Really?

SECTION 1 BASIC THEORY AND REACTIVITY

So, the orbitals and the resonance forms tell us the same thing. That's reassuring.

With entry 7, we can see the effect of a second carbonyl group. It provides about a million times (6 pK_a units) more stabilization.

> Draw the resonance forms for Entry 7. You will understand them when they are drawn for you, but you need to practise.

Conversely, ethyl ethanoate (Entry 4) is a million times *less* acidic than propanone. We definitely need to try to understand this difference. First of all, we can draw the same resonance forms for stabilization of the enolate anion.

$$\left[\underset{H_2C}{\overset{O}{\underset{\ominus}{\bigvee}}} OCH_2CH_3 \quad \longleftrightarrow \quad \underset{H_2C}{\overset{O^{\ominus}}{\bigvee}} OCH_2CH_3 \right]$$

It isn't at all clear whether this stabilization will be more effective or less effective than we saw with propanone.

> Remember, we don't just draw the resonance forms. We need to start building an instinctive understanding of whether we get more stabilization in one compound than in another.

But ethyl ethanoate is also stabilized by resonance. The oxygen lone-pairs provide stabilization as shown below.

$$\left[\underset{H_3C}{\overset{O}{\bigvee}} \ddot{O}CH_2CH_3 \quad \longleftrightarrow \quad \underset{H_3C}{\overset{O^{\ominus}}{\bigvee}} \overset{\oplus}{O}CH_2CH_3 \right]$$

What we appear to be saying is that there isn't much difference in the stability of the anions derived from propanone (Entry 5) and ethyl ethanoate (Entry 4).

> This is exactly what we are saying.

But there is a difference in stability between the ketone and the ester.

> This is correct! We saw this in **Applications 1**.

And the acidity of the compound (pK_a) reflects the ease of forming the anion (the conjugate base) which is the difference in stability between the compound and its conjugate base.

> Draw a reaction energy profile showing this energy difference. If you are feeling really brave, you could try to calculate the energy difference. You know that a difference in pK_a of 6 units corresponds to a difference in K_a (the equilibrium constant) of 10^6. The physical chemistry equation you need is $\Delta G = -RT \ln K$.

PHOSPHORUS STABILIZING ANIONS

The anion in Entry 2 is a bit different to the other examples. To be fair, a pK_a of 30 isn't **very** stable, but it's stable enough to be formed and reactive enough to be useful. We will see reactions of these anions in **Reaction Detail 26**.

As always, we need to ask what is stabilizing the negative charge.

> *Obviously, it's phosphorus!*

The more important question is **how** it is stabilizing the negative charge. It can't simply be an inductive effect, or nitrogen (above phosphorus in the periodic table) would stabilize a negative charge even more. Here is a resonance form to represent the stabilization.

$$[Ph_3\overset{\oplus}{P}-\overset{\ominus}{C}H_2 \longleftrightarrow Ph_3P=CH_2]$$

The phosphorus atom has five bonds. That's okay. Phosphorus is in the third row of the periodic table. It has d orbitals.

> *The electrons have to go somewhere!*

Nitrogen cannot do the same thing. But I bet you can now come up with a list of other elements that can!

ONE MORE EXAMPLE

When we compared the stability of cyanide with acetylide, we found that the nitrogen atom had an inductive effect that stabilized the anion. The oxygen atom in an enolate stabilizes the anion by a combination of inductive and mesomeric (resonance) effects.

When we look at nitromethane (Entry 8), we see that we can draw resonance forms for the anion as shown below. It is definitely stabilized.

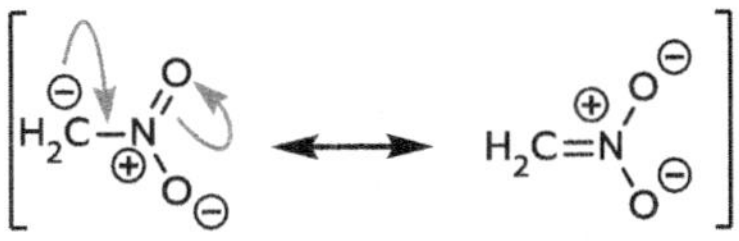

$$\left[\overset{\ominus}{H_2C}-\overset{\oplus}{N}\overset{O}{\underset{O^\ominus}{\diagup}} \longleftrightarrow H_2C=\overset{\oplus}{N}\overset{O^\ominus}{\underset{O^\ominus}{\diagup}}\right]$$

To draw these resonance forms, we needed to draw the nitro group out 'fully'.

> *How do you know when you need to do this?*

That question doesn't have a simple answer. The best I can give is that if you cannot see a convincing explanation, try drawing something a different way. So much of your success in organic chemistry will depend on you 'just seeing stuff'.

SECTION 1 BASIC THEORY AND REACTIVITY

> *Practise enough and your brain will take care of the rest.*[16]

But if the mesomeric (resonance) effect was the only factor, this anion would not be so much more stable (a billion times more stable) than the anion next to a ketone carbonyl group in Entry 5. In addition, we have an electronegative element directly attached to the carbanion carbon, and this is going to stabilize the negative charge through an electron-withdrawing inductive effect.

$$H_2\overset{\ominus}{C}-\overset{\oplus}{N}\begin{smallmatrix}O\\O^{\ominus}\end{smallmatrix}$$

The mesomeric (resonance) effect is more significant.

> *It usually is!*

But the inductive effect is important too.

pK_a DEPENDS ON SOLVENT

When an acid, HA, dissociates, we are looking at the following equilibrium.

$$HA \overset{K_a}{\rightleftharpoons} H^{\oplus} + A^{\ominus}$$

When something dissolves in a solvent, the individual molecules are separated, and are stabilized by interaction with the solvent. We refer to this stabilization as solvation. In the case of ionic species, the individual ions can be separately solvated.

If we have a solvent that solvates (stabilizes) $H^{\oplus}$ and $A^{\ominus}$ more than it solvates HA, then the equilibrium will be further to the right. This will increase K_a. If it increases K_a, it will decrease pK_a.

Solvents dissolve things! They do this by solvating them. Different things are solvated to different extents.

> *You already know that 'like dissolves like'. Any acid will be 'more acidic' in water than in hexane. With experience, you will start to think about where other solvents fit into this.*

We aren't going to be looking at reactions in so much detail that pK_a in different solvents becomes a factor. But I want to make sure you are prepared for when you need this information.

---

[16] Don't think I am trivializing this. When you draw structures, you are reinforcing pathways in your brain. For a while, you'll feel frustrated that you aren't 'getting it', and once you do, you'll feel frustrated that it took so long. That's normal. Just accept it, do the work and don't worry!

SECTION 1 BASIC THEORY AND REACTIVITY

FUNDAMENTAL REACTION TYPE 4

Enolates and Enols

WHY?

Organic molecules have C–C bonds, and they have very useful properties (medicines, materials, *etc.*). We need good ways to make these materials, which means we need good ways to form C–C bonds. Enolate chemistry allows us to do this. Section 4 of this book is just about the chemistry of enols and enolates. This *must* be really important chemistry.

> It is!

That's why you need to understand the fundamental reactivity of these useful intermediates.

INTRODUCTION

In Recap 7 we saw a number of carbanions. Formally, these are species with a negative charge on carbon.

> As we saw, sometimes we draw resonance forms where the negative charge is no longer on carbon.

There is usually an associated cation, which complicates things a little bit. However, a reagent with a negative charge is looking a lot like a potential nucleophile!

ENOLATES AS NUCLEOPHILES

We saw that a ketone has a pK_a of about 19. The α-hydrogen atoms can be removed, but not too easily.

> We don't want the hydrogen atoms to be too acidic. If they were, that would mean that the corresponding anion is too stable/too unreactive.

Here we have an enolate being alkylated using an alkyl halide.

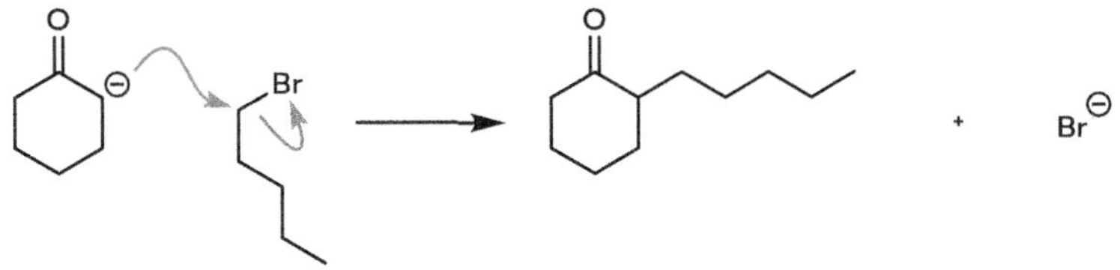

SECTION 1 BASIC THEORY AND REACTIVITY

We could, of course, look at this reaction from the perspective of the alkyl halide. It is being attacked by a nucleophile. This is a one-step substitution reaction, so this would be an S_N2 mechanism. We looked at S_N1/S_N2 mechanisms in detail in **HTSIOC**.

> It would never be S_N1. The reaction conditions you would need to form a carbocation are not compatible with the conditions required to form an enolate, and the carbocation you would need in this case would be *primary*, and very unstable **(Recap 2)**.

HOW DO WE FORM ENOLATES?

We looked at the α-acidity of carbonyl compounds in **Recap 7**. If you have an acidic hydrogen, you need a base to remove it. We established that hydrogen atoms α- to carbonyls are acidic compared to other hydrogen atoms attached to carbon, but they still aren't very acidic. We need a pretty strong base.

> Let's be more specific. Hydrogen atoms α- to ketones have a pK_a in the region of 19. To fully remove such a hydrogen atom, we need a base with a pK_a significantly greater than 19. Hydroxide won't work!

Introducing LDA—Lithium diisopropylamide! The 'amide' bit of the name might be a bit misleading. There isn't an amide functional group present. In this case, this refers to a negative charge on nitrogen.

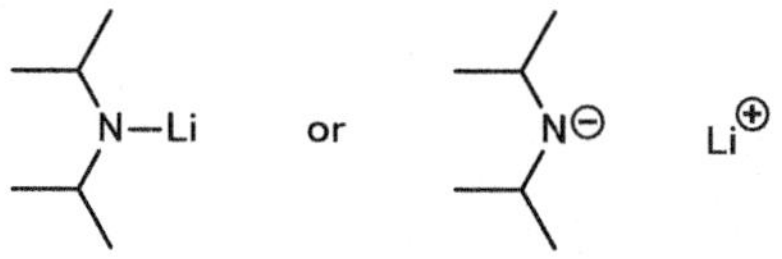

The great thing about LDA is that the two isopropyl groups on nitrogen make it bulky, so it can attack something small, like a hydrogen, but nothing else. That is, it can act as a base, but not as a nucleophile. This is important, as you wouldn't want your base to react directly with your electrophile.

> It's important to consider these points. When you need to carry out a reaction that has never been carried out before, you will need to consider which reagents are likely to give the best outcomes. We are only just scratching the surface here. There are many 'additives' that we include in reactions to modify the reactivity of reagents. We aren't going to get bogged down with this in detail. The fundamentals never change. With practice and experience, you get used to finding the most suitable reaction conditions.

Putting this together, we have the following reaction to form an enolate.

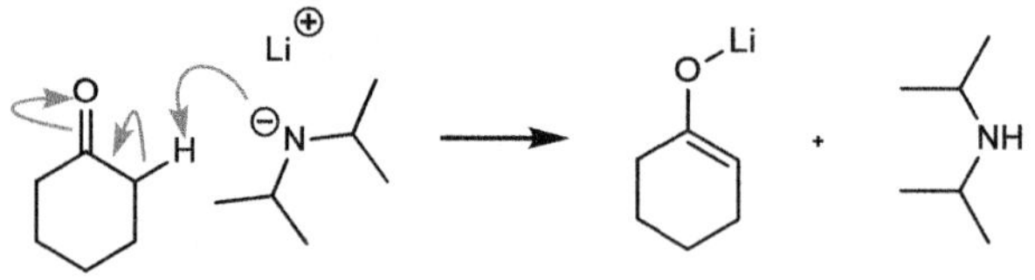

SECTION 1 BASIC THEORY AND REACTIVITY

I've slipped in one more detail. The product is the lithium enolate, in which the enolate anion is closely associated with the lithium from the LDA. When you draw the alkylation reaction, you can draw it in three different ways.

1. Using the resonance form with a negative charge on carbon.

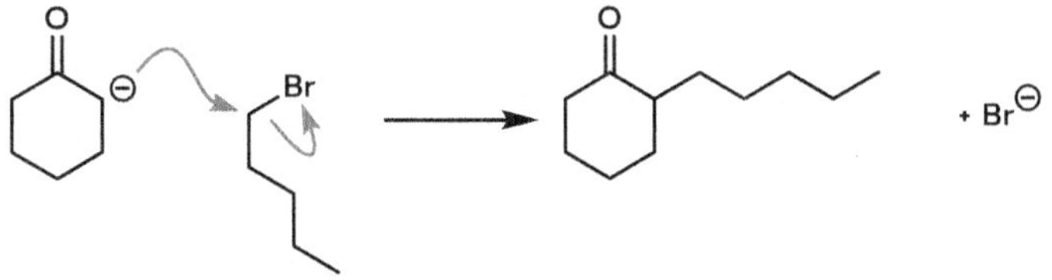

2. Using the resonance form with a negative charge on oxygen.

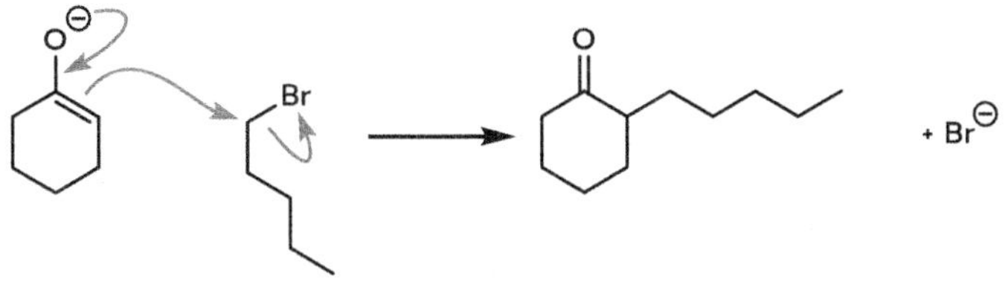

3. Using the lithium enolate.

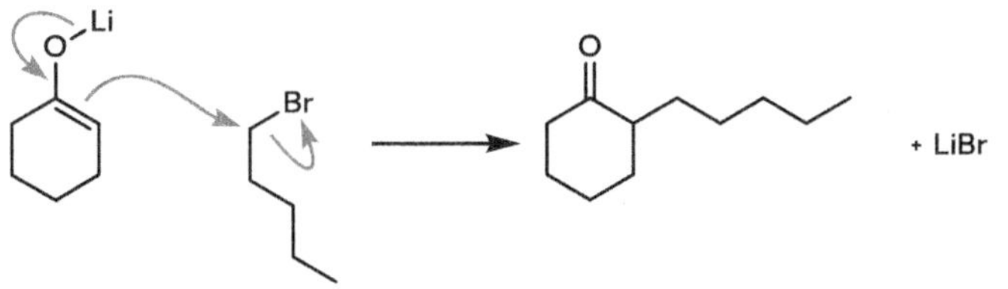

These are all equivalent, and you will get used to seeing all of them.

As an added note, I don't think the above reaction has ever been done using LDA as base. But it has been done, and we are more interested in general mechanisms at this stage.

THERE IS A COMPLICATION!

When we look at the resonance forms for an enolate, we can see that we have two resonance forms, one with a negative charge on oxygen and one with a negative charge on carbon. We recognize that the enolate is stabilized because oxygen is electronegative, and we can delocalize the negative charge onto the oxygen atom.

With this in mind, couldn't the enolate act as an oxygen nucleophile instead?

The short answer is yes, but it depends on the electrophile. With some electrophiles, you will get reaction on carbon, and with other electrophiles you will get reaction on oxygen. **Exercise 3** gives you an opportunity to practise calculating enthalpy

89

KETO-ENOL TAUTOMERIZATION

Here is a really important process. It has a great many implications, but for now we are just going to focus on the process.

> First of all, we need to define a new type of isomer—tautomers. Tautomers are constitutional isomers that can be readily interconverted, normally by moving a proton from one atom to another.

Here is the mechanism.

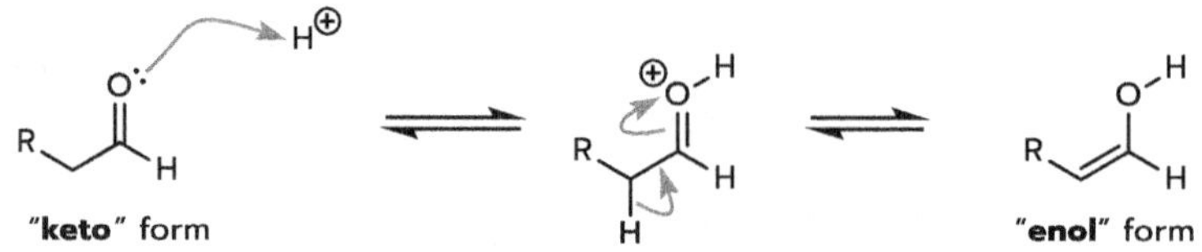

"keto" form ⇌ ⇌ "enol" form

We do this under acidic conditions. The carbonyl oxygen atom is protonated, and we can then lose a proton from the α-carbon. As we have already seen, a hydrogen α- to a carbonyl is acidic. It is even more acidic once the oxygen is protonated, because we are now losing a proton from a positively charged species to form a neutral species.

> These are the same reasons we saw in **Basics 2**.

There are a couple of points.

> We call the carbonyl compound the 'keto' form, even when it isn't a ketone. Live with it!
>
> The position of the equilibrium will depend on the relative stability of the two tautomers.[17]

ENOLS ARE NUCLEOPHILES

The enol looks rather a lot like the lithium enolate we saw on the previous page. It reacts in the same way too. Here it is with a generic electrophile.

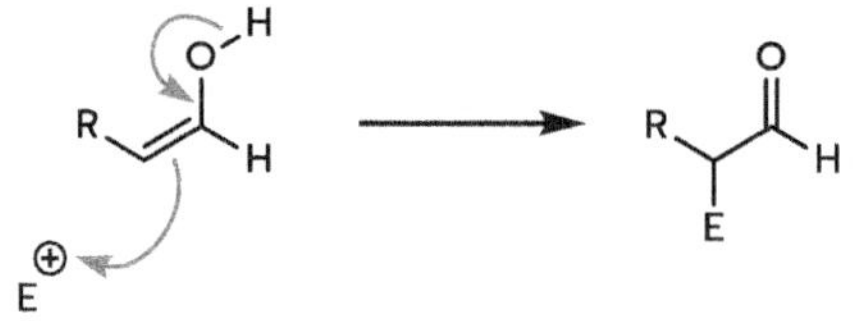

---

[17] Not exactly a surprise. Why should this process be any different!

Of course, the enol isn't quite as reactive as the enolate. All other things being equal, a negatively charged nucleophile will always be more reactive than a neutral nucleophile.

WHERE IS THIS LEADING?

There are many reactions in which enolates, enols and related species act as nucleophiles. In a lot of these cases, the electrophile is a carbonyl compound. We will come back to these examples starting in Reaction Detail 21. For now, we will simply consider the application of enolates, **and things like enolates**,[18] in nucleophilic substitution reactions.

NOW A PHILOSOPHICAL QUESTION

Is this really a fundamental reaction type? Isn't it just an S_N2 reaction with a particular nucleophile?

> *There really aren't that many different reactions. There are just a huge number of ways to put them together, and a huge number of variations on the basic themes.*

---

[18] Once again, we will build connections. If it looks a *bit* like an enolate, it probably reacts a *lot* like an enolate.

SECTION 1 BASIC THEORY AND REACTIVITY

RECAP 8

And that's it!

INTRODUCTION

This chapter is a recap of material in this book, not of material covered in the previous book. It's quite early for a recap, but I think it's important that we do this. The key point is that once we've got past this chapter, (almost) everything else you see will be applications of things you already know.

The majority of the chemistry of alkenes and aromatic compounds concerns their reaction with electrophiles. There are other types of reaction, and by definition, they are **different**. What we need to do is get you to the point where you are **so** familiar with the most common reactivity types that you recognize the 'other' reactivity when you encounter it.

> And that you do this by understanding principles rather than by learning lists of reactions.

We are quickly going to recap reactions of alkenes and aromatic compounds with electrophiles, just reiterate the point that they are guided by the same principles.

> The emphasis here is on what you need to 'do' with this information.

Here is the reaction of an alkene with an electrophile. I don't care what the electrophile is, so let's call it E⊕.

Here is a reaction energy profile for this process. We had two choices, **A** and **B**. The reaction will follow the lowest energy pathway, which will be the one in which the intermediate has the lowest energy.

You know about carbocation stability. As long as you **really** understand carbocation stability,[19] you can predict the outcome with **any** electrophile.

> As before, 'R' is a simple alkyl group which will stabilize an adjacent carbocation.

---

[19] By this I mean that you have replaced 'more substituted = more stable' with a more refined 'what does each substituent do?'

SECTION 1 BASIC THEORY AND REACTIVITY

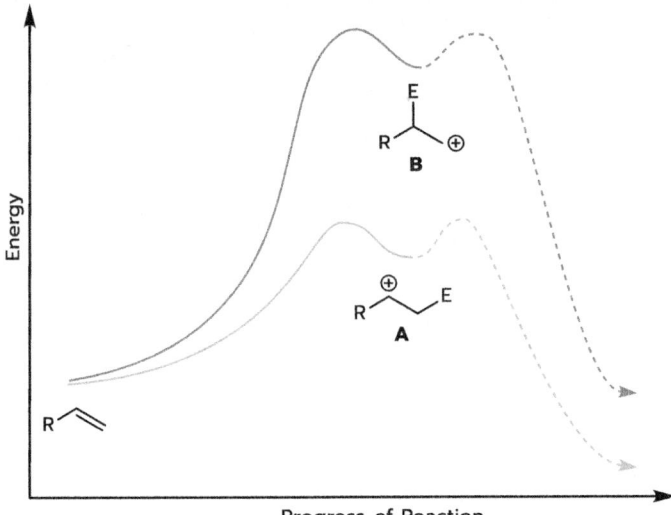

Now we reiterate the point about bromination. The intermediate is a carbocation—but it isn't. It is stabilized by donation of electrons from the bromine lone pair.

Of course, this doesn't really change all that much. Instead of talking about where the positive charge is, we need to talk about where 'most of the positive charge is'. In **HTSIOC Basics 16**, we recognized that anything that can stabilize a positive charge can stabilize a partial positive charge.

We had to replace the idea of 'which end the electrophile adds to' with 'where does the nucleophile attack in the intermediate', but we are using the same ideas.

ONE MORE STEP

Here is a reaction we will see again in **Reaction Detail 5**. We add a boron to one end of an alkene and a hydrogen to the other end. The problem is, they happen at the same time. Here are two possibilities.

SECTION 1 BASIC THEORY AND REACTIVITY

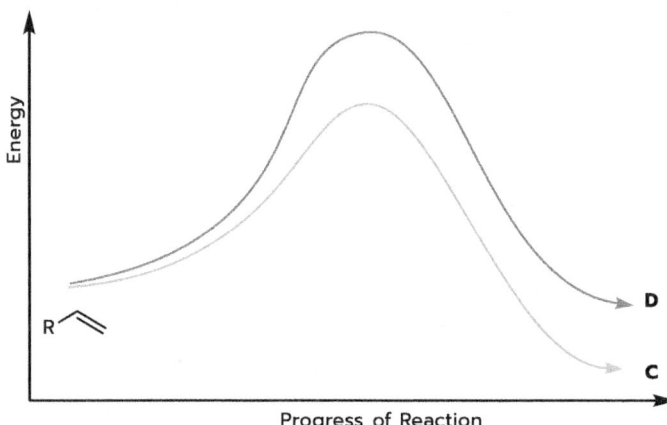

It turns out that we start to form the C–B bond a bit ahead of the C–H bond. So, we will build up some positive charge on the alkene carbon atom that is not attacked by the electrophilic boron. Which carbon will this be?

The one that can best stabilize the positive charge, of course!

This is subtle. But have a look at the bromination reaction above. We are adding the Br to **BOTH** alkene carbon atoms, but we are doing so in a way that is not symmetric. Now, we are adding a B–H bond across an alkene in a similarly asymmetric way. Here is a reaction profile for the two possibilities, **C** and **D**.

Note that here, we have transition states rather than intermediates, but it's basically the same thing. The reaction with the lowest energy transition state has the lowest activation energy, by definition. If we know where charge develops during the reaction, we can predict whether substituents in the molecule will stabilize or destabilize this developing charge. This is almost like the Hammond Postulate (**Recap 2**) but not quite.

Don't worry. We will come back to this. It's not something you should expect to 'get' the first time.

WHY NOW?

If this is so subtle and 'complicated', why am I introducing these ideas so early? I have a good reason for doing this. I need you to develop the way you think about these reactions. In some cases, it is 'just a carbocation', and you need to make sure you fully understand carbocation stability. In other cases, the structure of the intermediate is not quite a carbocation, but its stability is determined by the ability of various groups to stabilize a build-up of positive charge. You *still* need to understand carbocation stability. In yet other cases, it is a transition state with a build-up of partial positive charge on one atom or another, and you *still* need to understand carbocation stability.

> At any point, in any reaction, you need to understand the factors that stabilize a full or a partial positive or negative charge.

WHAT ABOUT AROMATIC ELECTROPHILIC SUBSTITUTION?

Well, the intermediate is a carbocation. If you understand carbocation stabilization, you don't have anything to learn. We will leave this for now. I tend to find that the problems with this chemistry often stem from poor drawing of curly arrows. If the curly arrows are wrong, you might actually be looking at the wrong carbocation.

> We will deal with this in **Reaction Detail 8**.

RECAP—CURLY ARROWS FOR CARBONYL ADDITION REACTIONS

We are going to encounter a lot of carbonyl reactions. My own experience was that this was when organic chemistry became that mythical 'list of reactions to learn'. It's important to be able to see through the morass of reactions and focus on the patterns.

> Carbonyl groups do two things!

In **Fundamental Reaction Type 2**, we saw the following.

This is relatively straightforward, because the nucleophilic centre has a negative charge that represents a pair of electrons.

> It doesn't have to be a negative charge—it could be a lone pair of electrons.

We saw this in **Reaction Detail 3**, although in that case we protonated the carbonyl oxygen atom first (**Basics 2**).

Let's take a step back. In the above reaction, we generate an $O^{\ominus}$. Most of the time, we protonate the negatively charged oxygen to form a stable neutral molecule, but

95

SECTION 1 BASIC THEORY AND REACTIVITY

sometimes the O$^\ominus$ can do other things. We will see this in **Reaction Detail 16**, and in quite a few other places.

Now let's ask a really important question.

> Apart from a negative charge, or a lone pair of electrons, where else might we find a pair of electrons?

In a bond, of course! In **Reaction Detail 12** and **Reaction Detail 14**, we will see the following curly arrows, which represent the transfer of a σ-bonded nucleophile to a carbonyl group.

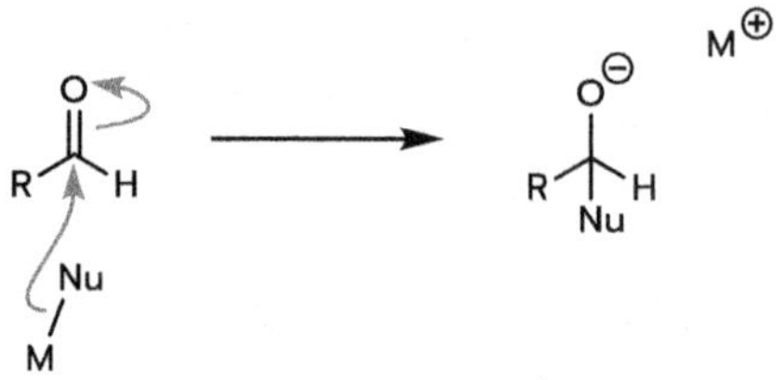

In this case, we will see that the atom is nucleophilic as a result of it being attached to something electropositive (M = a metal).

> It is important, when you encounter new carbonyl addition reactions, to be able to understand the curly arrows.

RECAP—ENOLS AND ENOLATES

The second type of reactivity for carbonyl compounds is the removal of a proton from the carbon atom α- to the carbonyl group. We have only just seen this, in **Fundamental Reaction Type 4**. We can do this under basic conditions as follows.

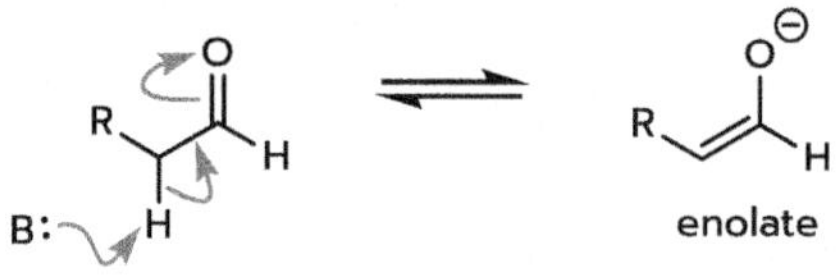

Here, I have used 'B' with a lone pair of electrons to indicate a generic base.

Alternatively, we can do this under acidic conditions, so that we form the enol rather than the enolate.

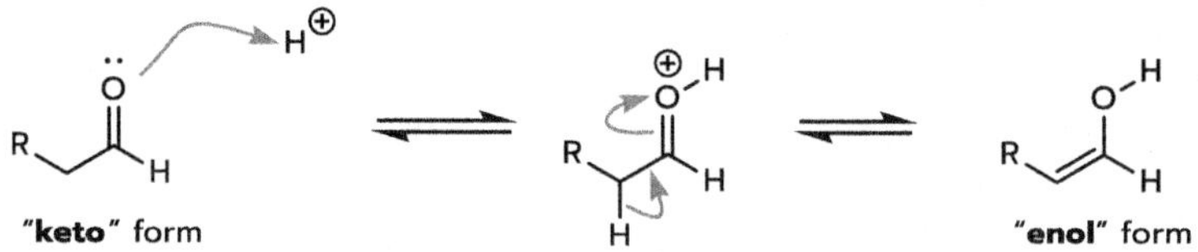

"keto" form   "enol" form

The point is **not** that we can form an enolate or an enol. The point is what they can do. In both cases, they can act as nucleophiles. Here it is with an enolate.

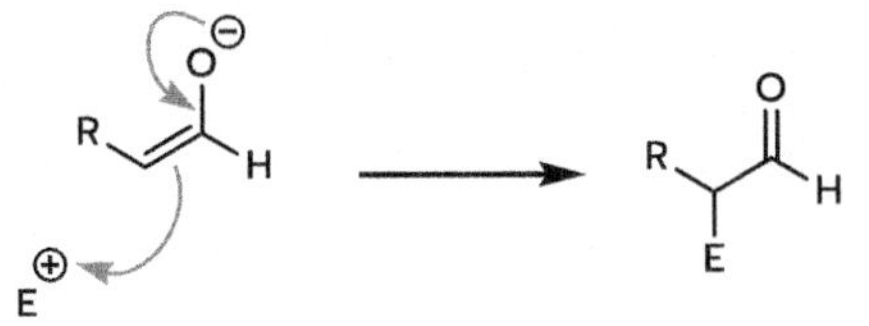

SECTION 1 BASIC THEORY AND REACTIVITY

And here is the same thing with an enol.

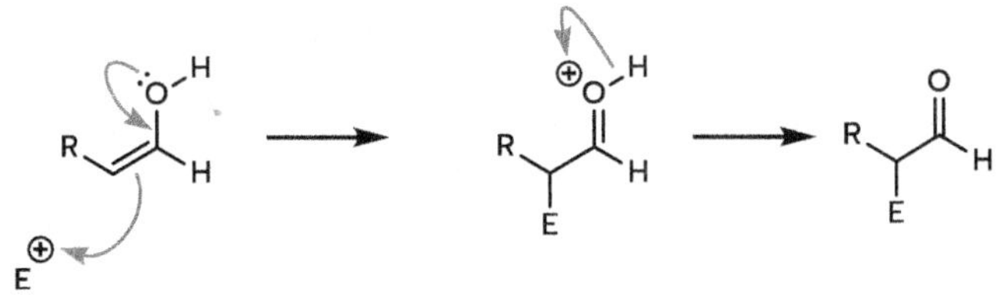

In many cases, the electrophile (which I have simply shown as 'E⊕') will be a carbon atom. We saw this in **Fundamental Reaction Type 4** with an alkyl halide.

> We formed a carbon-carbon bond—this is what organic chemists like to do!

There are plenty of other "carbon electrophiles". We have seen that aldehydes are good carbon electrophiles. They can be attacked by nucleophiles.

> Draw the reaction of an enolate (as nucleophile) with an aldehyde (as electrophile). Make sure you use the curly arrows in this chapter.

Congratulations! You've just 'invented' the aldol reaction (**Reaction Detail 23**). Depending on the enolate and aldehyde structures you used, you might have formed one or two stereogenic (chiral) centres in the product.

> Organic chemists also like to form stereogenic centres.

IN CONCLUSION

We have established the fundamental reactivity of alkenes, benzene rings and carbonyl compounds.

In the coming chapters, you will see a lot of reactions. You will think they are all different.

> That's when you need to come back to this chapter!

And then you will (eventually) realize that they are all the same. And then it will become easier to learn.

There is a lot of nagging in this book. It might be tedious, but it is well-intentioned. My own experience with this chemistry is that I kept drawing out mechanisms until they 'stuck'. It took a while, and then I wondered **why** it had taken a while, because it all seemed so obvious.

SECTION 1 BASIC THEORY AND REACTIVITY

There are several reasons why learning happens like this. The most fundamental reason is that in studying, you are building new connections between neurons, that allow you to process information in more efficient ways. You don't 'get' something right up until the point where you make a new connection in your brain.

> *You are 'writing the program'.*

And then you get it. And it seems obvious. I want to point this out to you, because I have had too many students tell me they will never be able to 'get' organic chemistry. They feel they are 'not clever enough'. I want to convince you that all you need to do is focus on the patterns, and if that doesn't work, draw the mechanisms again.[20]

> *You will get there!*

---

[20] In Japanese martial arts, when a student is struggling to do something correctly, they might be told 'mō ichido'. This literally means 'one more time'. It's always the answer!

SECTION 2

THE CHEMISTRY OF ALKENES, BENZENE RINGS AND ALKYNES

SECTION 2 THE CHEMISTRY OF ALKENES, BENZENE RINGS AND ALKYNES

INTRODUCTION

In many respects, we have 'done' alkene chemistry and aromatic chemistry in **Section 1**. We have seen that you add an electrophile to an alkene to generate a carbocation (sometimes it is a bit more complicated) and you then add a nucleophile. We have seen that when an alkene is unsymmetrically substituted, there is a regiochemical preference, and this is dictated by the formation of the most stable intermediate.

Of course, you can use lots of electrophiles and nucleophiles. This gives a large number of reactions, and if you approach them in the wrong way, you end up with a lot to remember. We are going to add some more examples in this section, but they will always follow the principles we have introduced.

We also looked at aromatic electrophilic substitution in **Fundamental Reaction Type 3**. We introduced the idea of regioselectivity there, but we did not draw any resonance forms. Because these reactions have resonance-stabilized carbocations as intermediates, the ***only*** way to explain these reactions is by drawing resonance forms. We will look at a few examples in this section.

> *If you are not confident drawing resonance forms, you will always be trying to remember what 'type' each substituent is. If you are confident drawing resonance forms, you can simply predict the outcome of a given reaction based on your understanding.*

Have I convinced you to do what you need to become good at drawing resonance forms? Practice, practice and more practice!

Finally, we will look at some reactions of alkynes. They are the same reactions we looked at with alkenes, but they give a different outcome. We will see why this is the case.

SECTION 2 THE CHEMISTRY OF ALKENES, BENZENE RINGS AND ALKYNES

REACTION DETAIL 5

Further Aspects of Hydration of Alkenes

WHY?

In **Fundamental Reaction Type 1** we looked at hydration reactions of alkenes. We saw that you add acid and water. The acid is the electrophile, and it adds with regioselectivity such that the most stable carbocation is formed.

There are two problems with this. First of all, simply adding water and acid to an alkene doesn't actually work all that well.

> *Very few alkenes dissolve well enough in water for this to be an effective process.*

Secondly, what if the alcohol product regioisomer you want is not the one that is formed *via* the most stable carbocation?

> *How can we force the alternative outcome? We cannot change the principles, so we must change the strategy.*

There is a risk that you will look at this chapter as 'learning two reactions'. If you approach the subject in this way, there will always be lots to learn. I have chosen these two reactions because they give two different regiochemical outcomes, but for good reasons.

> *Your first goal is to understand the reasons.*

Understanding why you get this outcome in these reactions will allow you to make predictions about reactions you have not yet encountered.

> *If you can predict the outcome of a new reaction, there is less to learn.*

Finally, we will make connections with reactions you have seen, so that these 'new reactions' will already look familiar.

OXYMERCURATION OF ALKENES

Let's get this out of the way up front. This is not a 'nice' reaction. It uses a mercury salt, which is toxic. Not only that, it uses a stoichiometric amount of the mercury salt. This process doesn't get used all that often!

This is a reaction you can use if you want the same regioisomer as would be produced by 'simple' hydration (acid and water), but you want a higher yield. Oxymercuration tends to be carried out in an organic solvent such as diethyl ether, and with water to act as a nucleophile.

SECTION 2 THE CHEMISTRY OF ALKENES, BENZENE RINGS AND ALKYNES

Let's work through this mechanism in a couple of stages. The first step is one of the rare reactions where the intermediate we draw isn't quite the one implied by the curly arrows.

This three-membered 'mercurinium' ion (also referred to as a mercuronium ion) looks very similar to the bromonium ion that we saw in **Reaction Detail 1**.

> We could draw two additional resonance forms for this structure in the same way. Do it!

Although we didn't do it in **Reaction Detail 1**, we could have modified the mechanism to use a lone pair on bromine so that the curly arrows *do* give the product. I don't have a problem with that, but when we do it for the mercurinium ion, we have to draw the following. It doesn't look right, as we are not used to drawing lone pairs on transition metals.

> *The problem is very fundamental. Curly arrows are a way of drawing mechanisms. The molecules react together by an interaction of a filled orbital on one species and an empty orbital on the other species.[21] The orbital interaction gives the product. No-one told the molecules about curly arrows!*

Either way, since the 'mercurinium' ion looks very similar to the bromonium ion that we saw in **Reaction Detail 1**, It also reacts in a similar way.

> Draw the mercurinium ion with more realistic relative bond lengths, based on your understanding of the corresponding bromonium ion. Make sure you understand why the water attacks at the more substituted carbon atom.

From this point, we need to replace the mercury atom with a hydrogen atom. We do this reductively using sodium borohydride. We will see this reagent again, in

---

[21] Even this is a simplification, and does not cover all reaction types.

Reaction Detail 14. For now, I have just drawn the borohydride anion, which has a negative charge on boron.

> This negative charge is not a pair of electrons. The boron has a negative charge as a result of having four bonds. All the electrons are in the bonds.

So that's where the curly arrow starts!

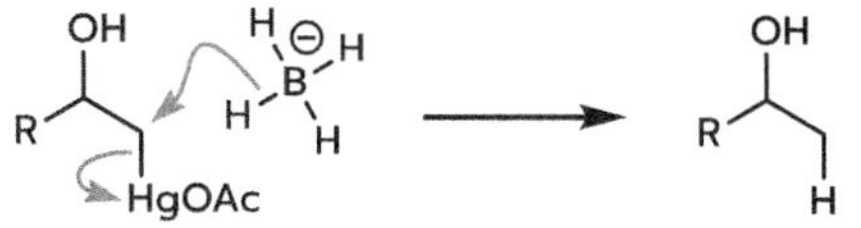

This gives us the same regioisomer as 'simple' hydration, and for the same reason—stabilization of positive charge. I haven't spelled everything out fully in this chapter. You need to have another look at Reaction Detail 1. The more you can see parallels between 'different' reactions, the easier it becomes to learn new reactions.

> I want to make one more point about the product of the above reaction. I have drawn one hydrogen atom on the right-hand carbon. Of course, there are two more hydrogen atoms that I have not drawn. You still need to know they are there.

HYDROBORATION OF ALKENES

What happens if we want to put the OH group onto the less substituted carbon atom of the alkene? We would need to do something like this.

It turns out that doing this directly would be quite difficult. I want to focus on the main reason—there isn't a good reagent that can act as an oxygen electrophile.

> In fact, the situation (as always!) is a little more complicated. We would need to think about the structure of the carbocation on the right, and it wouldn't be a simple secondary carbocation.[22] Then we would need to add a hydride nucleophile to the carbocation intermediate. It wouldn't be easy to find one that is compatible with this hypothetical first step.

Let's think our way around the problem. If we cannot use an oxygen electrophile, we need to use an electrophile that can subsequently be replaced by an oxygen atom.

I like problem-solving, but coming up with a solution to this problem is not straightforward, so I will just present the reaction. We will then analyse the mechanism in a bit more detail.

---

[22] Draw out some resonance structures and see what you can come up with!

SECTION 2 THE CHEMISTRY OF ALKENES, BENZENE RINGS AND ALKYNES

It turns out that we use borane, BH_3.[23]

> Borane tends to exist as the dimeric structure, B_2H_6. This is in equilibrium with the monomer, BH_3.

In the monomeric structure, the boron atom is electron-deficient. It can act as an electrophile. We have just seen that a B–H bond is able to deliver a hydride anion. The only piece of the jigsaw we do not yet have is how to replace boron with oxygen at the end.

Let's look at two representations of the mechanism.

We add a boron electrophile in order to give the more stable of the two possible carbocation intermediates. This gives a borohydride anion, and we already know that these can deliver a hydride anion. This then gives us a *primary* alkylborane.[24]

In fact, this isn't quite right. Instead of forming the new C–B and C–H bonds in two steps, they are formed in one step. However, they are not formed at the same rates. The C–B bond is formed 'a little more' than the C–H bond in the transition state. The formation of the two different bonds is **concerted** (in the same step) but **asynchronous** (not to the same extent at all points along the reaction coordinate).

> What this means is that we develop partial positive charge on carbon, and this is better stabilized on the more substituted carbon atom. We did say back in Recap 2 that all the discussion about stabilization of full charges also applied to partial charges. This is why!

> If you are having trouble getting your head around these ideas, just think of it as the second mechanism, but with a bit of the first mechanism thrown in to explain the regioselectivity. There's no point worrying about it too much!

In this case, since we used borane, BH_3, the initial product still has B–H bonds, so it can hydroborate more alkenes, as follows. I won't draw the full mechanism.

> You should though!

---

[23] Or a substituted borane reagent. We will encounter these in **Reaction Detail 11**.
[24] This reaction involves hydrogen, carbon and boron. Rather ironically, the symbols for these elements are the initials of the pioneer of this chemistry, Herbert C. Brown, who shared the Nobel Prize in Chemistry in 1979 for this (and other related) work.

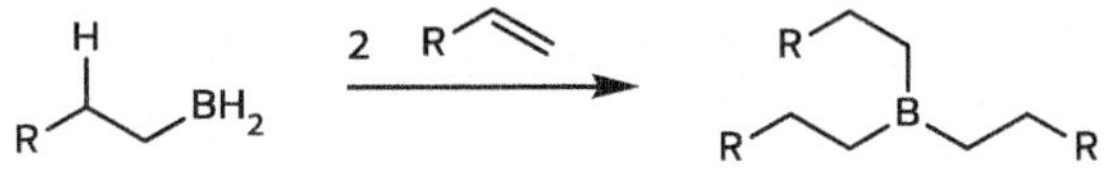

SECTION 2 THE CHEMISTRY OF ALKENES, BENZENE RINGS AND ALKYNES

Once all the alkene has reacted, we add a mixture of sodium hydroxide and hydrogen peroxide to oxidize the B–C bonds, to give the alcohol product.

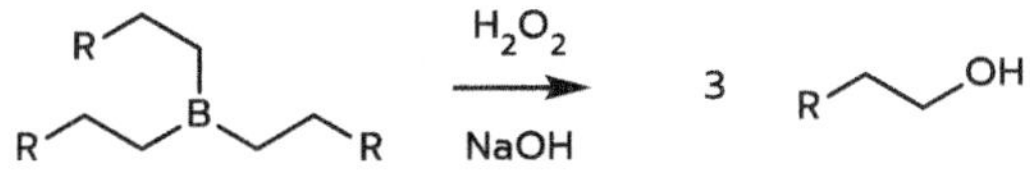

We are not going to explain this part of the reaction yet. We need to see a couple more reactions first. We will come back to this in **Reaction Detail 33**.

SUMMARY AND PERSPECTIVE

There's quite a lot going on in this chapter, but it's all very fundamental. The key point is understanding the regiochemical outcome of reactions—which end of the alkene is functionalized.

It is very unlikely that your examination will feature a question of the type "Draw the mechanism of the X reaction". It is much more likely that you will be given a reaction and asked to draw the mechanism or to explain the outcome.

> One generally leads to the other.

Or you might be given starting materials and asked to draw the product.

> If you can draw the mechanism, this is easy!

So, your priority is to understand the fundamental reactivity of the organic functional groups and of the reagents they react with.

SECTION 2 THE CHEMISTRY OF ALKENES, BENZENE RINGS AND ALKYNES

BASICS 3

Oxidation States in Organic Compounds

It is very useful to consider oxidation states in organic compounds.

Let's not forget why we are learning organic chemistry. We want to be able to make molecules. Molecules are really important in so many areas of life and society. Ultimately, when we plan a synthesis, we need to decide what reactions will be used and what reagents we will use. All these ideas will be developed in due course, but there is one aspect that is so fundamental that it is worth getting it in at the start.

> *If a reaction is an oxidation, you will need an oxidizing agent. If a reaction is a reduction, you will need a reducing agent!*

When you are proposing reagents for a synthesis, you need to be able to identify whether an oxidation or a reduction is needed. You need to be able to compare carbon atoms in compounds, and determine whether they are at the same oxidation state, or whether they are at different oxidation states.

Carbon is more electronegative than hydrogen. Not by much, but enough to be considered different. Therefore, we would consider the carbon atom in methane to have an oxidation state of −4. We can think of it as a "4$\ominus$" carbon attached to four times H$\oplus$.

Chlorine is more electronegative than carbon, so we would consider the carbon atom in tetrachloromethane to have an oxidation state of +4. Effectively, it is a "4$\oplus$" carbon attached to four Cl$\ominus$. These are the two extremes.

methane
oxidation state −4

tetrachloromethane
oxidation state +4

Let's now fill in the gaps with a few examples. Where there could be more than one carbon atom, the relevant carbon atom is indicated. You will see that different carbon atoms in a molecule can have different oxidation states.

SECTION 2 THE CHEMISTRY OF ALKENES, BENZENE RINGS AND ALKYNES

Oxidation State	Compounds		
−4	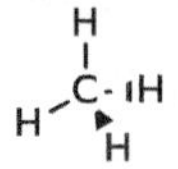 methane	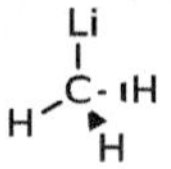 methyllithium	
−3	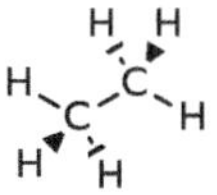 ethane	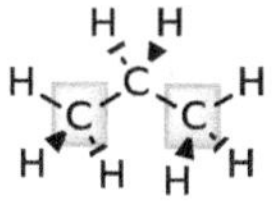 propane	
−2	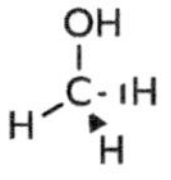 methanol	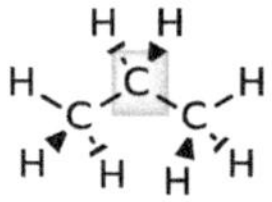 propane	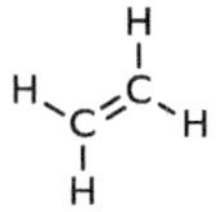 ethene
−1	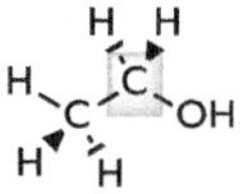 ethanol	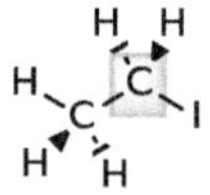 iodoethane	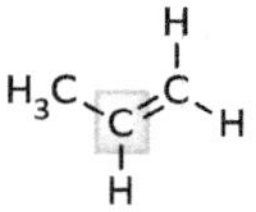 propene
0	but-2-yne	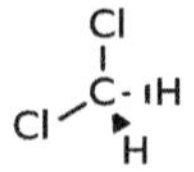 dichloromethane	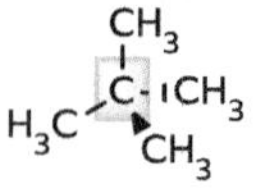 2,2-dimethylpropane
+1	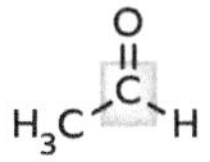 ethanal	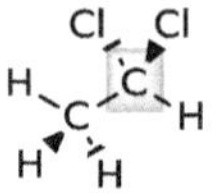 1,1-dichloroethane	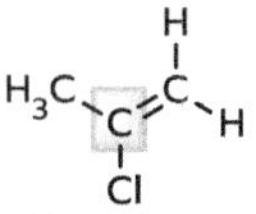 2-chloropropene
+2	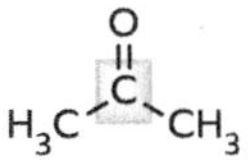 propanone	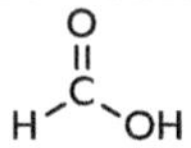 methanoic acid	
+3	ethanoic acid	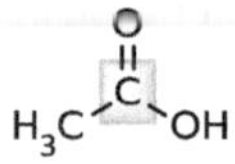	
+4	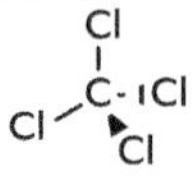 tetrachloromethane	carbon dioxide	

Don't get too bogged down by this. A bond to carbon "doesn't count". A bond to an element that is less electronegative than carbon reduces the oxidation state of the carbon by one. A bond to an element that is more electronegative than carbon increases the oxidation state by one. A double bond counts twice as much as a single bond.

SECTION 2 THE CHEMISTRY OF ALKENES, BENZENE RINGS AND ALKYNES

Let's see how we would use this information. Propane has two "−3" carbons and one "−2" carbon, so the total is −8. Propene has a "−3" carbon.

> I didn't highlight this one—find it!

It also has a "−2" carbon and a "−1" carbon, so the total is −6. If we go from propene to propane, the oxidation state is reduced. It's a reduction reaction!

In the previous chapter, encountered sodium borohydride. In **Reaction Detail 14** we will see it reacting with aldehydes and ketones. This is also a reduction reaction. Let's prove it! Here is the reaction.

$$R-CHO \xrightarrow{NaBH_4 \text{ then } H_2O} R-CH(OH)-H$$

Assume that the 'R' group is carbon. Actually, it doesn't matter. What matters is that it is the same on both sides of the reaction arrow. The aldehyde carbon atom has an oxidation state of +1. The alcohol carbon has an oxidation state of −1. Therefore, this is a reduction.

Let's keep a sense of perspective about this though. Yes, conversion of methane to ethane is **formally** an oxidation. However, it requires the formation of a new C−C bond, and you would not expect to treat methane with an oxidizing agent and get ethane.

Similarly, I would consider methanoic acid and ethanoic acid to be closely related—they have the same functional group, but a different oxidation state because methanoic acid has a hydrogen atom attached to the carbonyl carbon atom. Conversely, methanoic acid and propanone are quite different, although they both contain a carbonyl group at the same oxidation state.

> *Comparison of oxidation state is most useful when following changes in functional group through a reaction sequence.*

If we look at oxidation state 0, we see that there are three **very** different compounds. There are saturated alkanes in the table with oxidation states −4, −3, −2 and 0, and we could have included one with oxidation state −1 if we had wanted to.

> Work it out!

A REALLY IMPORTANT POINT!

Let's look at ethene and ethanol again. They are in different rows of the table for different oxidation states, but that isn't the whole story. Ethene has two identical carbon atoms, both with oxidation state −2. The total is −4. Ethanol has a "−3" carbon (not in the table—find it!) and a "−1" carbon, so the total is also −4.

If we wanted to convert ethene into ethanol, we would need to add water. We saw this reaction in **Fundamental Reaction Type 1**, and in **Reaction Detail 5**. The above calculation shows us that this is not an oxidation, although we are adding oxygen! Don't get blinkered by the oxygen atom.

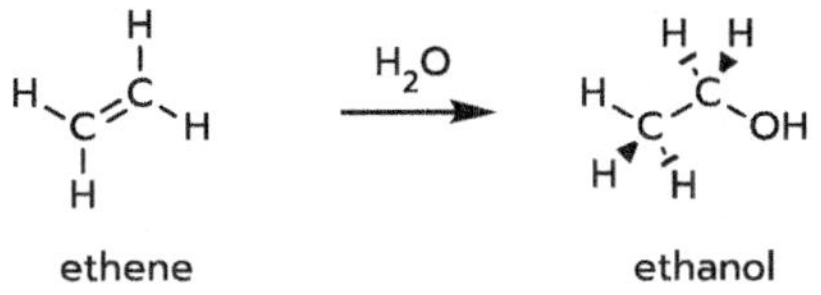

$$\text{ethene} \xrightarrow{H_2O} \text{ethanol}$$

The next chapter focuses on reactions of alkenes that *are* oxidations. It is important that you appreciate the distinction.

OXIDATION STATES AND OXIDATION LEVELS

Although we won't do it here, it is also possible to consider oxidation **levels**. In effect, these simply consider the number of bonds to electronegative elements, and they make no distinction between H and C. An aldehyde and a ketone have the same oxidation level, but not the same oxidation state.

SECTION 2 THE CHEMISTRY OF ALKENES, BENZENE RINGS AND ALKYNES

REACTION DETAIL 6
Oxidation of Alkenes

WHY?
Alkenes are relatively easy to make. Oxidation reactions of alkenes are really important, as they add functionality that can be further manipulated. In many cases, stereochemistry can also be added, and this can be controlled. We will encounter this, initially in **Stereochemistry 4**.

For now, we will simply focus on the basic transformations and their mechanisms.

> We have already encountered hydration (addition of water) reactions of alkenes in **Fundamental Reaction Type 1** and we saw more in **Reaction Detail 5**. We have just seen, in **Basics 3**, that addition of water is not an oxidation, despite the fact that we are forming a C–O bond.

As I write more of these 'why' sections, I find that the reason is very often 'because you can make useful things with this chemistry'. That's a pretty good reason. If there is a particular target you have in mind, and there's a neat reaction that will allow you to make it, it's a problem if you don't know about that reaction.

> But if you know a good selection of reactions and the underlying principles, you can very often 'predict' the existence of such a reaction.

CAUTION
The reaction mechanisms in this chapter are horrible. You should not expect to understand everything the first time through. And if you don't understand it, you'll find it harder to remember. I want to manage your expectations.

> Just keep drawing the mechanisms out and check that your curly arrows make sense. As you become familiar with some of the curly arrows, the rest will fit into place.

ALKENE EPOXIDATION
An epoxide is a compound with a three-membered ring containing two carbon atoms and one oxygen atom. They are really useful and quite easy to make by reaction of an alkene with a peracid.[25]

This is like a carboxylic acid but with an extra oxygen as shown below. The peracid we tend to use is *meta*-chloroperbenzoic acid.

---

[25] Hydrogen peroxide is H_2O_2. It's like water but with one more oxygen. This is why a peracid is like a carboxylic acid but with one more oxygen. The O–O bond is weak, which makes these compounds quite reactive.

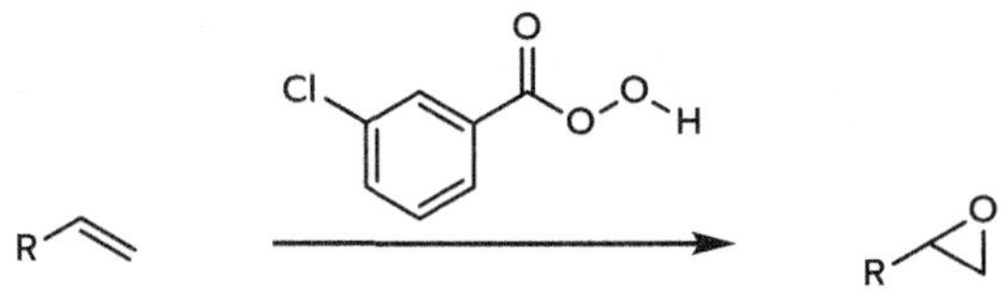

SECTION 2 THE CHEMISTRY OF ALKENES, BENZENE RINGS AND ALKYNES

There are many other reagents and conditions used for making epoxides, but we need to start somewhere. The focus for now is on understanding the transformation and being able to draw a mechanism.

The mechanism of this reaction is a bit of a nightmare to draw, so we will break it down. Effectively you have the oxygen behaving as a "⊕" and as a "⊖" at the same time. If we draw a simplified version, this is easier to see. Follow curly arrow (1) going from the alkene to the oxygen atom, and then we have an arrow (2) going back to the alkene to complete the reaction.

Now we need to look at this for the full reaction. We still start with an arrow (1) going from the alkene π-bond to the oxygen.

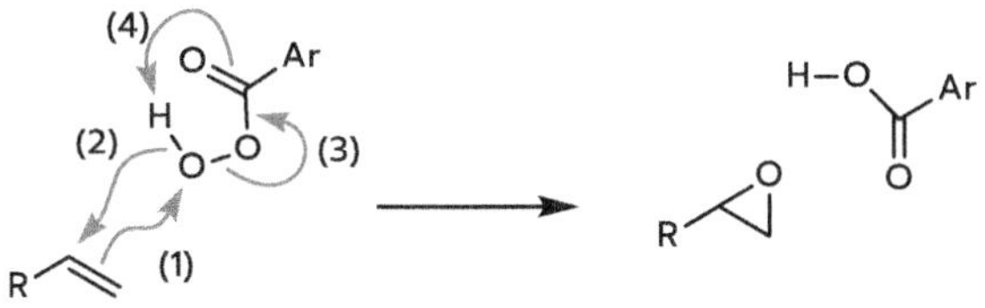

Now we get to the driving force for this reaction—and why the peracid is an oxidizing agent. We break the weak O–O bond. Curly arrow (3) shows us taking away the two electrons that we had given to this oxygen atom with the previous curly arrow.

Remember that this O has a share in 8 outer electrons, so if we try to give it a share in two more, we have to take some away.

Curly arrow (4) shows the transfer of a proton to the carbonyl group. Of course, there would already be a hydrogen bond there.

The final curly arrow, (2), forms the second C–O bond. In this case, it comes from the O–H bond. This is because we need to break this bond.

This is not an easy mechanism to understand. Take your time. Follow the curly arrows and make sure you can see why the curly arrows above result in the bond-forming and bond-breaking in this reaction.[26]

---

[26] Remember, Ar is a commonly used abbreviation for any aromatic (aryl) group.

SECTION 2 THE CHEMISTRY OF ALKENES, BENZENE RINGS AND ALKYNES

> *Remember that curly arrows are a way of representing the bond-breaking and bond-forming processes. The 'real' interactions are between molecular orbitals in the two reacting species. Both new C–O bonds are formed in the same step. This is a concerted process.*

ALKENE OZONOLYSIS

Here is another important reaction, and one with a rather complicated mechanism. First of all, we will look at the overall reaction, for a simple 1,2-disubstituted alkene.

Ozone is O_3. We break the alkene double bond completely, to give two molecules of an aldehyde (in this case, because the alkene also has two hydrogen atoms). The third oxygen atom of ozone oxidizes dimethyl sulfide to give dimethylsulfoxide.

The mechanism has a lot of steps, and it is hard to learn. It also involves some reactions that you have not seen yet. We will take this slowly.

> *Let's start with absolutes. We have a reagent, ozone, made up only of three oxygen atoms. We should look at the structure of this reagent.*

Ozone, O_3, is an important small molecule. We are constantly hearing about the ozone layer, and in particular the damage that various organic molecules do to it. The smell of ozone is the 'burnt air' that you often notice when you stand close to a photocopier. The three oxygen atoms are not all in a line, so that the molecule is bent. For any possible structure we draw, there will be an oxygen atom with a positive charge and an oxygen atom with a negative charge. Here are the resonance forms. For most purposes, we only need to consider the first two.

> *In the third resonance form, we have an oxygen atom on the left with only six outer electrons. This would be less stable, and so contributes less. The net effect of this is that the central oxygen atom has a partial positive charge, and each terminal oxygen atom has a partial negative charge.*

Right, now let's move to our next absolute. We form a bond from each carbon atom of the alkene to one oxygen atom of ozone in each case.

> *Which oxygen atoms?*

It turns out that they are the oxygen atoms on the ends, not the one in the middle. In the key resonance forms, the central oxygen atom has a positive charge, but this does not actually make it electrophilic.

112

SECTION 2 THE CHEMISTRY OF ALKENES, BENZENE RINGS AND ALKYNES

It still has a share in eight outer electrons—a complete octet.

> On to the next question. Which alkene carbon atom reacts first?

For the alkene we are using, it is symmetrical. Assuming one C–O bond formed first, forming a carbocation intermediate, it simply wouldn't matter. Let's see how this might happen!

We would now have a ⊕ and a ⊖ in the same structure, and they would react with one another.

Okay, so we have both alkene carbon atoms now bonded to oxygen. It's looking pretty good.

> There's a problem. This isn't quite how it happens! The carbocation is high in energy. The O⊖ is also high in energy. To have them both in the same molecule would be incredibly unstable. The reaction manages to find another way that is lower in energy.

Instead of forming the positive and negative charges, and then combining them, we form the second C–O bond at the same time. The intermediate is known as a **molozonide** (molecular ozonide).

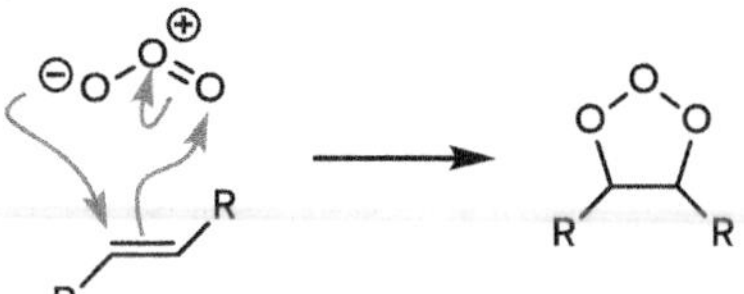

molozonide

Let's just take a step back for a moment and recap some of what we have seen.

When we add HBr to an alkene, we get a carbocation intermediate that is then attacked by a bromide nucleophile. When we add Br₂ to an alkene, we form a cyclic bromonium ion which is then attacked by a bromide nucleophile. This happens because the cyclic bromonium ion is more stable than the non-cyclic 'simple carbocation'.

> Things happen for very good, and simple, reasons!

When we added BH₃ to an alkene, we didn't form a carbocation intermediate. We developed some partial positive charge, but the nucleophilic hydride 'didn't wait' until the carbocation had formed.

113

SECTION 2 THE CHEMISTRY OF ALKENES, BENZENE RINGS AND ALKYNES

> We have the fundamental electrophilic nature of borane balanced against the energy of an intermediate in a hypothetical stepwise reaction.

Now, we have an entirely concerted reaction. We form both C–O bonds in the same step. This reaction is described as a **cycloaddition**. This is a new reaction type which we will encounter in Fundamental Reaction Type 6. The molecular orbitals for a cycloaddition look a little different to those we have seen before, but we will get to that in Section 6.

Unfortunately for this mechanism, we are not quite finished. The molozonide is quite unstable. It has two O–O bonds, and these are weak. The molozonide can fragment as follows to give an aldehyde (in the example we are using) and a carbonyl oxide. This is actually the reverse of a cycloaddition, but we aren't going to worry about that for now.

a carbonyl oxide

> We aren't adding any hydrogen atoms in this step—they were there all the time!

> Draw the curly arrows for the reverse reaction of the carbonyl oxide plus aldehyde to form the molozonide. Doing this will help you see why this is the same type of process as the reaction of ozone with the alkene.

These can then react together (same curly arrows, different orientation!) to give an intermediate we call an **ozonide**.

ozonide

Let's have a quick look at why this is favoured. The molozonide has two O–O bonds. The ozonide only has one O–O bond. It looks more stable. You might also think about the curly arrows above. We have a nucleophile adding to the carbonyl carbon atom. This is 'typical' carbonyl reactivity.[27]

---

[27] Actually, this last part isn't the strongest argument, but we are trying to build patterns, so it's worth emphasizing this point.

SECTION 2 THE CHEMISTRY OF ALKENES, BENZENE RINGS AND ALKYNES

> Formation of the ozonide isn't usually the end point of the reaction. It can be, if you really want to make an ozonide, but most of them are unstable, and they can be explosive.

Yes, this is a really long reaction mechanism. In many ways, it highlights the nature of the challenge you face. You cannot simply memorize the steps. You do really need to understand each step, so it fits into a framework.

What we now need is to convert the ozonide into the two equivalents of aldehyde. We need to get rid of an oxygen atom, and something needs to take it. This is where the dimethyl sulfide comes in. Here is the overall process.

In this respect, the dimethyl sulfide is acting as a reducing agent. It is being oxidized. We looked at oxidation states of carbon atoms in **Basics 3**. You can do exactly the same thing for sulfur.

> When we think about the mechanism for this process, the key question is 'how does dimethyl sulfide act as a reducing agent?' Dimethyl sulfide is a nucleophile. In this case, it attacks an oxygen atom.

These curly arrows are not easy to follow. Let's have a closer look at the oxygen atom indicated. Curly arrow (1) from sulfur would give this oxygen atom a negative charge, but we then have curly arrow (2) taking electrons away, breaking the O–O bond. This means that at this stage this oxygen atom is neutral again. We have now formed our first C–O double bond. Curly arrow (3) breaks a C–O single bond, forming the other C–O double bond. Finally, curly arrow (4) pushes electrons back to the first oxygen atom. It therefore ends up with one bond and a negative charge.

> Compare these curly arrows with those for alkene epoxidation earlier in this chapter. They are rather similar.

We can draw dimethylsulfoxide as a double bonded resonance form. It wouldn't be very easy to draw the above mechanism to form this directly.

If you want, you can also use sodium borohydride in the reduction step. This gives the products as alcohols rather than aldehydes (or ketones depending on the number of substituents on the alkene).

We will see, in Reaction Detail 14, that aldehydes and ketones are reduced by sodium borohydride. Then, this aspect of ozonolysis won't be too surprising.

A SHORT BREAK

Wow, that was a long and difficult mechanism. I don't think it is reasonable for you to expect to be able to remember this mechanism after drawing it a couple of times. What is hopefully more realistic is that you will be able to understand the curly arrows, and see why this mechanism is reasonable.

> After a while, because you understand it, you'll find that you do remember it.

Ozonolysis is an historically important reaction. Before NMR spectroscopy was widely used, it would be common to use a reagent such as bromine to confirm the presence of an alkene in a compound (decolourization of bromine water) and then ozonolysis to cleave the alkene into smaller fragments that might be identified. Ozone can also cause damage to polymers that contain double bonds.

DIHYDROXYLATION OF ALKENES

There is another important oxidation reaction of alkenes, in which we convert an alkene into a 1,2-diol.

> We will see an enantioselective version of this reaction in Stereochemistry 8. We will see some useful chemistry of 1,2-diols in Reaction Detail 31. Of course, it could simply be that you want the 1,2-diol itself. Either way, these are useful compounds, so that reactions that allow you to make them are important.

We use osmium tetroxide to oxidize an alkene. We can draw curly arrows for this process as follows.

These curly arrows look very similar to the curly arrows for ozonolysis. There is something really important to note here. The osmium starts out with eight bonds, and it ends up with six bonds. We have gone from a +8 oxidation state to a +6 oxidation state. The osmium tetroxide has been reduced as the alkene has been oxidized.

SECTION 2 THE CHEMISTRY OF ALKENES, BENZENE RINGS AND ALKYNES

> We don't often draw curly arrow mechanisms for reactions involving transition metals. This is actually one that works well. But if you drew the same curly arrows with carbon instead of osmium, it wouldn't work—you would have the wrong number of bonds.

You need to have water present. We are going to keep the number of bonds to osmium the same. This represents a simplification, but that's okay.

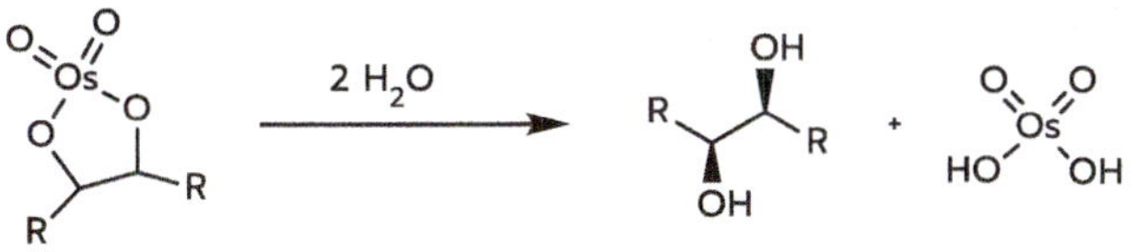

We get the 1,2-diol product, and the two oxygen atoms have been added to the same side of the alkene. In the structure above, I have only drawn one stereoisomer, where both oxygen atoms have reacted on the top face of the alkene. In the structure below, both oxygen atoms have reacted on the bottom face of the alkene.

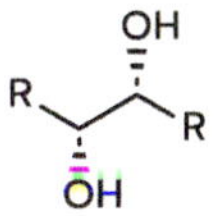

In energetic terms, 'top face' and 'bottom face' are equivalent. We get a 1:1 mixture of these two products. They are enantiomers, and we call this a **racemic mixture** (or racemate, **HTSIOC Basics 29**).

> The key point is that we need to draw some stereochemistry so that we know that the product arises from attack of both oxygen atoms onto the same face of the alkene.

We then need to oxidize the osmium back to the +8 oxidation state. There are a number of possible reagents that will do this, but one common one is N-morpholine N-oxide. Here is the overall equation. We won't worry about the mechanism.

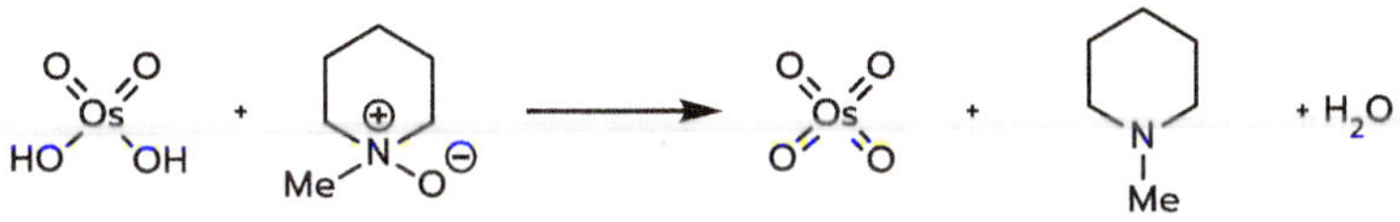

> Now the osmium tetroxide can be used to oxidize another molecule of the alkene. That's right, we only need a catalytic amount of the osmium tetroxide. This is important, as osmium tetroxide is expensive and very toxic.

WHERE IS THIS GOING?

Many epoxides and diols are chiral, and many methods are available for carrying out epoxidation and dihydroxylation reactions to form single enantiomers of the epoxides and diols. We will encounter some of these reactions in **Stereochemistry 7** and **Stereochemistry 8**. Epoxidation is a really useful way of taking a flat inexpensive molecule and converting it into a useful chiral product. So is dihydroxylation.

SECTION 2 THE CHEMISTRY OF ALKENES, BENZENE RINGS AND ALKYNES

Because an epoxide possesses a strained three-membered ring (Basics 6), it is relatively easy to get a nucleophile to attack an epoxide. It's an S_N2 substitution, in which you have used the strain present in the ring to make the oxygen a better leaving group.

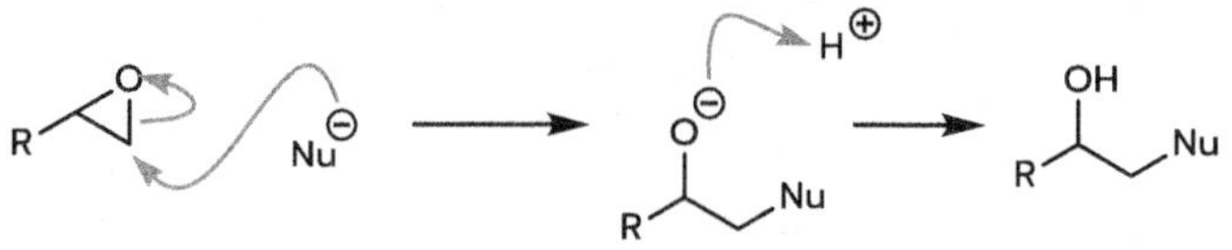

We looked at epoxide opening reactions in HTSIOC Worked Problem 5. They are S_N2 substitution reactions, with all that this entails.

Ozonolysis is a neat reaction for cleaving alkenes to give carbonyl compounds. It is carried out under relatively mild conditions (no acid, no base, no heating) so it can be used for a broad range of compounds. We have already seen how useful the carbonyl group is, and we have barely scratched the surface. Any reaction that produces a carbonyl group is useful!

SECTION 2 THE CHEMISTRY OF ALKENES, BENZENE RINGS AND ALKYNES

STEREOCHEMISTRY 1

Conformers of Cyclohexanes and Cyclohexenes

INTRODUCTION

We are going to be doing quite a lot with cyclohexanes in this book, so you need to get good at looking at any of the various representations of cyclohexanes we use, and seeing the shape of the molecule in your head.

> Use your molecular models to help build the connections between structures on a page and three-dimensional molecules.

Remember, cyclohexane isn't flat. The lowest energy conformation of cyclohexane is a chair.

> Recall that conformers are structures that can be interconverted by rotating bonds.

Here is the same structure drawn as a stick model in the chair conformer. Recall that we have two different spatial orientations for the hydrogen atoms—axial (bold) and equatorial (normal).

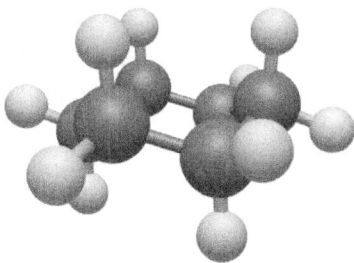

The same applies when we replace the hydrogen atoms with something bigger.

INTERCONVERTING THE CYCLOHEXANE CHAIR CONFORMERS

Now let's remind ourselves that we have two chair conformers that we can interconvert as follows.

> Do this with your molecular model. It's the only way to 'see' it.

SECTION 2 THE CHEMISTRY OF ALKENES, BENZENE RINGS AND ALKYNES

Pull the carbon atom on extreme left upwards. Then pull the carbon atom on the right downwards. This is indicated on the structure below. You get another chair structure, but all the hydrogen atoms that were equatorial are now axial, and all the hydrogen atoms that were axial are now equatorial.

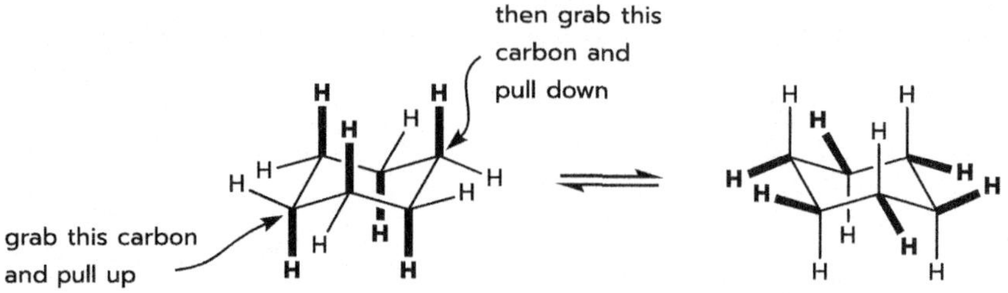

The two conformers here have only hydrogen atoms, so they each have six equatorial hydrogen atoms and six axial hydrogen atoms.

> They have the same energy.

This is not the case when we start replacing the hydrogen atoms with 'more interesting' substituents. Let's have a look at this.

EQUATORIAL IS FAVOURED

Here is methylcyclohexane. We have done exactly the same as we did above, but now we have one conformer in which the methyl group is equatorial and one conformer in which the methyl group is axial.

The conformer on the left is more stable. In the conformer on the right, the methyl group is close to two axial hydrogen atoms. We describe this as a **1,3-diaxial interaction**.

> Look at the structures and work out why this is!

With a methyl substituent, both conformers are present at equilibrium, to some extent. When you get to a *t*-Bu group ($(CH_3)_3C$) it is simply too big—you only ever have the conformation in which the *t*-Bu group is equatorial.

t-butyl equatorial
stable

horrifically unstable!

t-Butyl cyclohexanes have played an important role in the development of our understanding of stereochemistry. Because we know the *t*-Bu group will be equatorial, any other substituents on the cyclohexane ring will be in defined orientations—sometimes equatorial, sometimes axial—and this matters for the outcome of some reactions as we will see, starting in **Stereochemistry 10**.

CYCLOHEXANE BOATS

Go back to your molecular model and have another go at interconverting the chair forms. Unless you are very good, you probably cannot pull one carbon atom up and the other down at exactly the same time. Let's look more closely at this.

Take the cyclohexane chair below, and just pull the carbon atom on the left upwards. This is what happens. We get a new conformer, which we call a boat.

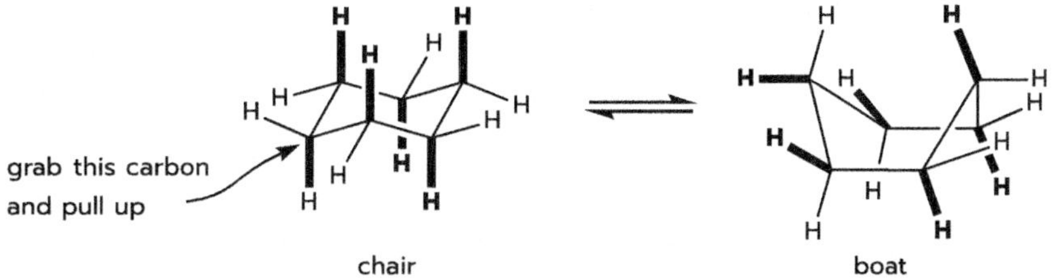

All other things being equal, boat conformers are less stable than chair conformers. In the boat conformer, we have a number of bonds that are eclipsed. We will look at the energies for these interactions in **Stereochemistry 5**. There is one further interaction that we need to consider. This is the **flagpole** interaction which is shown below. Because the hydrogen atoms are pointing towards one another, this is worse than a 1,3-diaxial interaction.

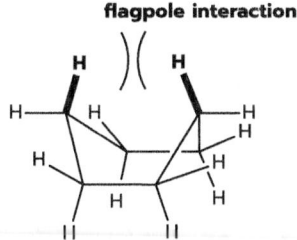

CYCLOHEXENE

The focus of this book is the chemistry of π-bonds. We need to know how the presence of a double bond affects the shape of the molecule. Here is cyclohexene.

> Make it with your molecular model.

Now look at it from the direction shown in the following diagram.

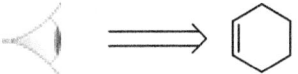

121

SECTION 2 THE CHEMISTRY OF ALKENES, BENZENE RINGS AND ALKYNES

Since this is a double bond, we have four carbon atoms in the same plane—the alkene carbon atoms and the two attached to them. The remaining two carbon atoms of the cyclohexene ring are out of the plane.

It should look like this.

In these structures, the hydrogen atoms are not quite axial and equatorial. In this case, we use the terms "*pseudo*-axial" and "*pseudo*-equatorial". I bet you already guessed that it is energetically more favourable for a substituent to be *pseudo*-equatorial rather than *pseudo*-axial!

> At first, I didn't find this easy to see. If you make a molecular model of cyclohexene, it will be easier to see why these diagrams are 'right'.

WHERE ARE WE GOING WITH THIS?

In **Stereochemistry 4** we will start looking at the stereoselective reactions of cyclohexenes. You do need to know which substituents are axial and which are equatorial, but it's much more important to develop an instinctive understanding of why a reagent will approach one face of an alkene because something is blocking the other face.

> *It's not quite as simple as this, but there's no point 'remembering' axial and equatorial without understanding what is making an axial substituent less stable.*

STEREOCHEMISTRY 2

Stereoisomers and Energy

WHY?

Different stereoisomers of drug molecules can possess vastly different biological activity. In extreme cases, one stereoisomer can cure a disease, but another stereoisomer could kill the patient.

> This is **not** a hypothetical scenario!

You are given two molecular structures. You have two choices. Either they are the same stereoisomer, or they are not the same stereoisomer.

If they are not the same stereoisomer, then they are either mirror images (enantiomers) or they are not mirror images (diastereoisomers).

> And that's it. There are no other possibilities.

You need to be able to tell the difference between stereoisomers before you can start devising strategies to make one stereoisomer in preference to another.

STEREOISOMERS

We covered stereochemistry and stereoisomers at length in HTSIOC. Here, we will apply what we learned. The important thing now is to develop a deep understanding of the basic definitions, so that you can glance at two structures at tell how they are related.

The key word here is 'glance', rather than needing to look carefully at two structures for five minutes and comparing every feature.

> How do you get to the point where you can do this? You look carefully at two structures for five minutes and compare every feature, until you no-longer need to do this.

Let's start with the basic definition. Stereoisomers are isomers (same formula) with the same connectivity but a different arrangement of those atoms in space. When we compare two structures, we have three possibilities.

THE SAME

We are going to do this in the context of reactions. Here is cyclohexene reacting with *m*-CPBA. This is an abbreviation for *meta*-chloroperbenzoic acid, which we saw in Reaction Detail 6.

SECTION 2 THE CHEMISTRY OF ALKENES, BENZENE RINGS AND ALKYNES

Taking cyclohexene, we can draw products in which epoxidation has taken place on the top face (wedged bonds) or on the bottom face (dashed bonds). The important thing to recognize is that the two products are identical. If we take the structure on the right and rotate it by 180° around a horizontal line through the centre of the molecule, we get the structure on the left.

> *This is the simplest case, but it is important to recognize. Otherwise, you will be drawing two 'different' products and talking about which one is preferred, when in fact they are the same.*

ENANTIOMERS

Here is the dihydroxylation reaction, which we also saw in **Reaction Detail 6**.

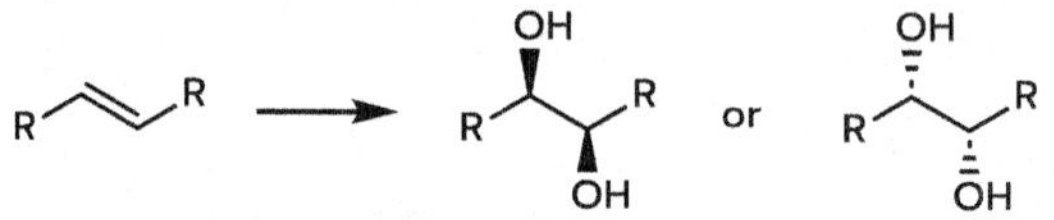

In this case, the products of reaction on the top face and on the bottom face are not the same.

> *How do you prove this? Good question!*

You can try to rotate them and overlay them. You could assign the stereochemistry of each stereogenic centre (**HTSIOC Habit 6**) to see if they are the same. Ultimately, though, your end goal is to be able to glance at them and to immediately know that they are enantiomers.

Enantiomers are stereoisomers that are not the same, but that are mirror images.

> *They have the same energy.*

Let's have a look at a reaction profile for this reaction.

We have one starting material, so this can only ever be one energy. We have two products, but they are enantiomers, with the same energy (by definition). As the reagent approaches the alkene from the top or from the bottom, the energy all along the reaction pathway is the same. Assuming a simple reaction profile with one transition state,[28] the transition states for attack from the top and from the bottom would have the same energy. This means that the activation energies are the same, so that the rates of reaction are the same.

---

[28] You don't have to make this assumption—all the arguments would still hold even if the mechanism was more complex.

SECTION 2 THE CHEMISTRY OF ALKENES, BENZENE RINGS AND ALKYNES

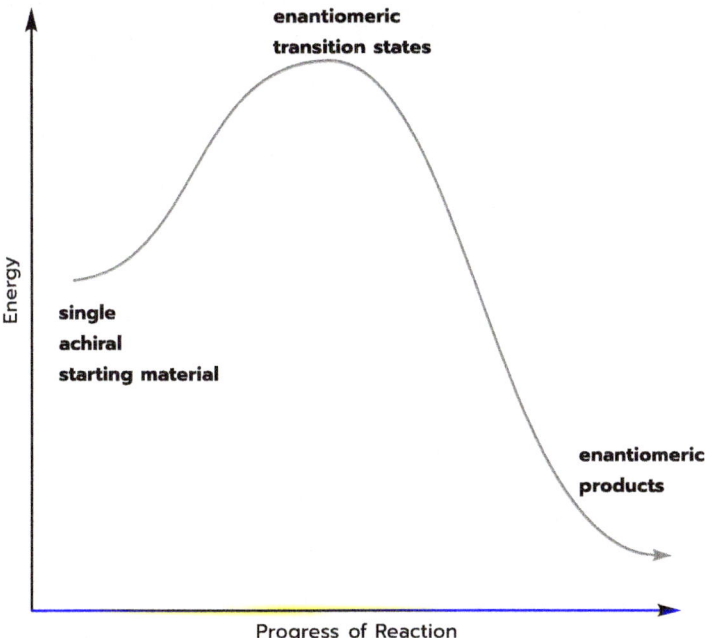

> It is important to be able to make these connections.

Once we get to this point, the outcome is inevitable—you would form a 1:1 mixture of the two enantiomers, a racemic mixture.

To put it another way, the reaction cannot be stereoselective.

> We will see how to make this reaction stereoselective in Stereochemistry 7.

DIASTEREOISOMERS

Diastereoisomers are stereoisomers that are not the same and are not mirror images. Here is a reaction that we are going to look at in some detail in Stereochemistry 6.

Once again, we have reaction of an alkene taking place either on the top face or on the bottom face. In this case, the two products are not the same, and they are also not mirror images.

> For them to be mirror images, the stereogenic centre on the left (with PhMe₂Si) would have to be different in the two products.

Diastereoisomers have different energy. Very often they are not *that* different, but they are not *exactly* the same.

125

SECTION 2 THE CHEMISTRY OF ALKENES, BENZENE RINGS AND ALKYNES

> *Compare this situation with what we saw for enantiomers above. In that case, the reaction absolutely had to give a 1:1 mixture of the two possible products, because the reaction profile for formation of the two products had the same energy all the way along.*

Here, we don't know how different, but the energies of the products will be different, and the energies of the transition states leading to those products will be different. Here is a simple reaction profile.

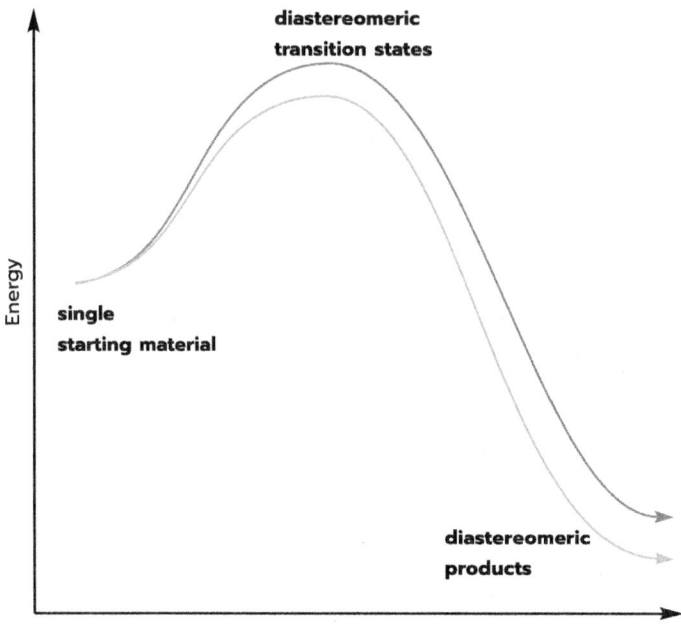

Of course, if the transition state energies are different, then the rates of formation of the two products will be different. Therefore, they will be formed in unequal amounts.

> *The reaction can be stereoselective!*

ADDITIONAL POINTS

In the reaction profile above, the lowest energy transition state leads to the most stable product. This does not have to be the case.

> *Don't assume you will always get more of the most stable product. It is entirely possible that the least stable product will be formed fastest.*

We will refine our thinking about this particular reaction in **Stereochemistry 6**, and in order to do this we will need to look at the energetics of conformational isomers (**Stereochemistry 5**).

SECTION 2 THE CHEMISTRY OF ALKENES, BENZENE RINGS AND ALKYNES

STEREOCHEMISTRY 3

Stereochemistry of Alkene Bromination

In **Reaction Detail 1**, we looked at bromination of alkenes. We found that the intermediate in this reaction is a bromonium ion—a compound with a three-membered ring. Towards the end of that chapter, we then looked at which carbon atom the nucleophilic bromine attacks. Here is the scheme again.

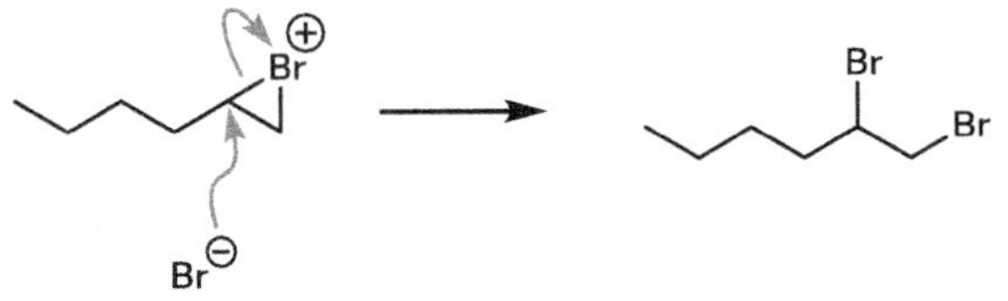

Everything you need to know about the stereochemistry of alkene bromination is contained in this scheme.

> The reaction shown above is an S_N2 substitution. It proceeds with inversion of configuration (HTSIOC **Reaction Detail 2**).

There are a couple of problems. The first is that in this particular example, there are no observable consequences. The second problem, in terms of learning this chemistry, is that you haven't yet developed the ability to glance at a structure and to spot the stereochemical possibilities instinctively.

The second problem is the one we need to focus on. Let's do this by considering the bromination of cyclohexene.

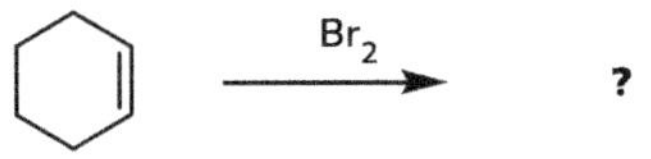

We already know (**Reaction Detail 1**) we will be adding one bromine to each of the alkene carbon atoms, with loss of the double bond. Now, we are considering the stereochemistry.

> First question, which face of the alkene will the Br_2 approach preferentially?

Have another look at **Stereochemistry 2**. We looked at a related epoxidation reaction there.

Yes, the top and bottom face are exactly equivalent in terms of energy, so we will get equal attack on both. We know that we will get a bromonium ion intermediate. Here is formation of the bromonium ion from attack on the top face.

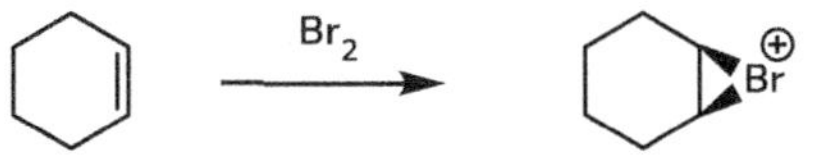

SECTION 2 THE CHEMISTRY OF ALKENES, BENZENE RINGS AND ALKYNES

If we take the structure on the right, and rotate it 180° about a horizontal axis, it will look like this.

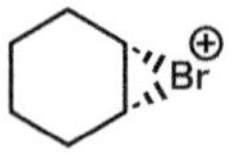

Of course, this is what we would have got had we had Br₂ approach the lower face of cyclohexene as originally drawn.

> It isn't just that attack on the top and bottom face have the same energy.

We would have this if these two scenarios gave products which were enantiomers.

> In fact, attack on the top and bottom face give **exactly the same** structure.

> Get your molecular models out and make sure you can see it. You might think you can see it, but it's still worth taking a little time to check.

We are taking our time with this one, but we now need to look at attack of bromide on the bromonium ion intermediate. At the start of this chapter, we saw that this was 'just' an S$_N$2 substitution reaction, so the nucleophile will attack from the opposite side to the Br.

> You will notice that I didn't use the more common wording that the S$_N$2 reaction takes place with inversion. That is a consequence of the more fundamental wording I used.

Here is one possibility.

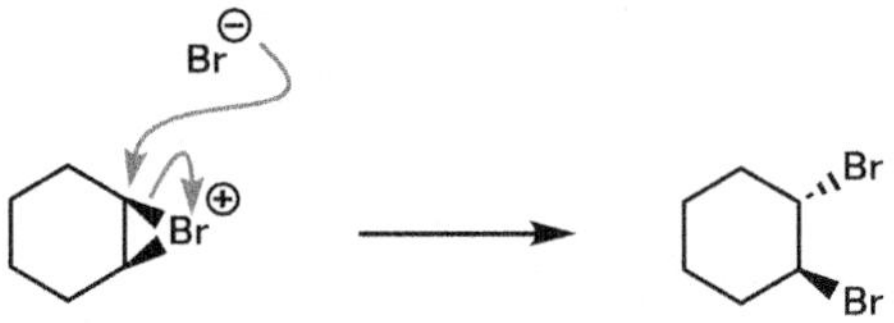

We have an absolute. The two Br atoms are on opposite sides of the cyclohexane ring.

What happens if we have attack of bromide at the other carbon atom? Here it is.

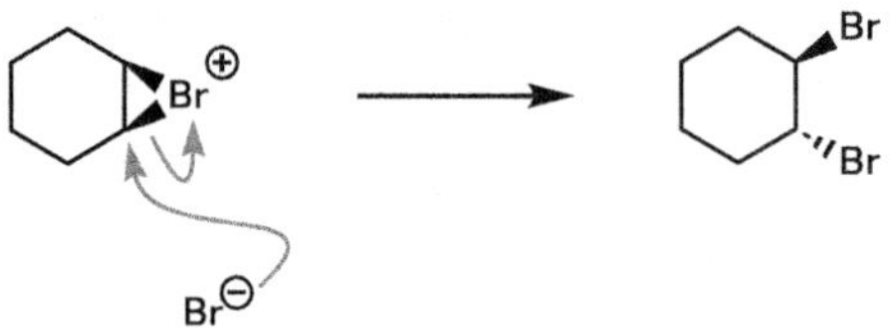

We also get a product with two Br atoms on opposite sides of a cyclohexane ring. But is it the same structure or a different structure?

Here are the two structures again. This time, I have rotated the one on the right.

128

SECTION 2 THE CHEMISTRY OF ALKENES, BENZENE RINGS AND ALKYNES

These look like mirror images, and in fact are not the same structure—they are enantiomers.

> How can you be sure they are not superimposable?

That's the big question! You make a model of each one, and you try to superimpose them. You rotate each structure and see if they are the same. It's a bit more complicated, because they are cyclohexanes, and you have two possible chair conformers (**Stereochemistry 1**) for each.

There is one more approach you could take—assign the R or S stereochemistry for each stereogenic centre using the Cahn-Ingold-Prelog rules (**HTSOIC Habit 6**). Until you can spot the stereochemical relationships at a glance, I would say this is the safer approach.

> Assign the stereochemistry for the two stereogenic centres in each of the two structures above. You might have to add the 'missing' hydrogen atoms on each structure first.

Here is the answer, but I'm not going to show you the working.

You might find this approach frustrating, but ultimately it is what will help you become good at this. If you didn't get it right, have another go. As further guidance, can you identify the highest and lowest priority substituent on each stereogenic centre? If so, you are then only trying to prioritize two substituents.

> And the most common error — you need to 'look' from the stereogenic centre along the bond to the lowest priority carbon atom.

Use your molecular models for this, and keep using your molecular models until you are absolutely confident that you can 'look' at the structure and see it in three dimensions. As with so many other aspects of organic chemistry, I wish there was a shortcut, but there isn't.

Okay, I think we've got to the end of the problem now. We have identified that we can get two products, and those products are enantiomers. We will inevitably get a 1:1 (racemic) mixture of the two enantiomers because there is no reason why one should be favoured over the other (**Stereochemistry 2**).

Perhaps, more importantly, there is a third possible structure for the 1,2-dibromide, which is as follows.

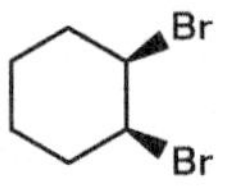

SECTION 2 THE CHEMISTRY OF ALKENES, BENZENE RINGS AND ALKYNES

You will never get this stereoisomer from a bromination reaction of cyclohexene. The reaction mechanism does not permit it.

> This is really important. We don't have to even think about which stereoisomer, or conformer, is more stable. We just 'follow the mechanism'.

I know I'm nagging! But it is all too common at first to look at the possible structures and make comparisons or assumptions. You might look at which product is more stable, and assume that is the one that will be formed. I need to convince you to start by drawing the mechanism, to add wedges and dashes whenever you form a stereogenic centre.

> If you don't make this a habit, you won't be able to spot the times when it matters.

ANOTHER EXAMPLE

We looked at cyclohexene for a reason. We can talk in absolute terms about whether the bromine atoms are on the same side of the ring or on opposite sides. In the next example, we cannot do this.

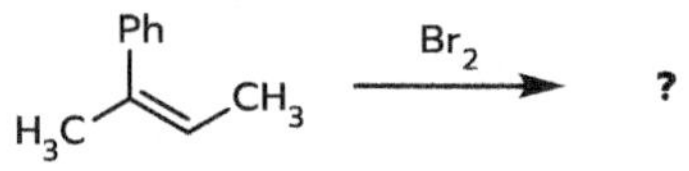

Of course, we follow the same process as above, and draw the intermediate and consider how it will react. First of all, here are the bromonium ions that are produced by attack on the top face and on the bottom face.

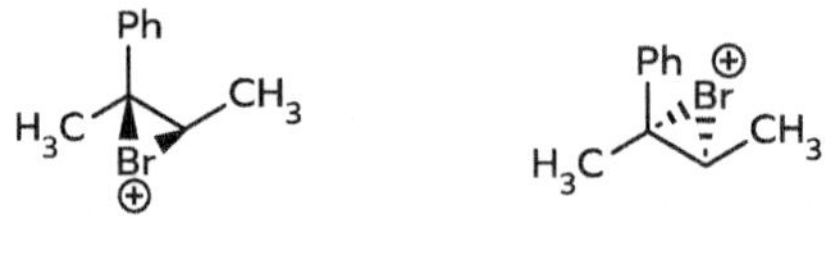

In this case, they are enantiomers—mirror images that are not the same and are not superimposable.

> Enantiomers have the same energy, so they will be produced in equal amounts.

We are going to do a little more work here, to show a particular problem. I'm going to take the structure on the left and draw it in two different ways.

> Which is correct?

They are actually the same, but there's a cognitive problem.

I added the hydrogen atom on the right-hand carbon and also added the substituent priorities for this carbon atom according to the Cahn-Ingold-Prelog rules. We use these to assign the stereochemistry.

$$\underset{\underset{\oplus\ H^4}{1\,Br}}{\overset{\overset{Ph}{|}2}{H_3C\underset{}{\overset{}{\diagup}}\!\!\!\diagdown\overset{3}{CH_3}}} \qquad \underset{H^4}{\overset{\overset{Ph\ 1\oplus}{\underset{|}{Br}\ 3}}{H_3C\underset{2}{\diagup}\diagdown CH_3}}$$

Look at the left-hand structure. Substituents 1→2→3 are arranged clockwise. In the right-hand structure, 1→2→3 are arranged anticlockwise.

There is a risk that if you assign the stereochemistry from these structures, you could get the wrong answer. Let's come back to our previous question regarding the representations.

> Which is correct?

The rather unsatisfying answer is that they are both correct, but neither of them look very good. What you have to do is *really* visualize the shape of the molecule, and convince yourself that this stereogenic centre is *R*.

> When you look down the C–H bond, 1→2→3 are arranged clockwise.

> Now assign the stereochemistry of the other stereogenic centre.

AN ADDITIONAL COMPLICATION

To finish this problem, we need to consider which carbon atom is attacked by the nucleophilic bromide anion. Here is the structure we have been working with, now with carbon atoms labelled 'a' and 'b'.

$$\underset{\underset{\oplus}{Br}}{\overset{\overset{Ph}{|}a}{H_3C\underset{}{\diagup}\!\!\!\diagdown\overset{b}{CH_3}}}$$

Carbon atom 'a' is better able to support a partial positive charge.

> It has an additional Ph (benzene ring) substituent. We saw in **Recap 2** that this can stabilize positive charge.

As a result of this, the C–Br bond to carbon 'a' will be longer than that to carbon 'b'. Carbon 'a' will have a larger amount of positive charge.

> We saw the same thing in **Reaction Detail 1**. The bromonium ion is not symmetrical. One bond is longer than the other.

We get attack of the nucleophilic bromide at carbon 'a'. This is the stereochemical outcome.

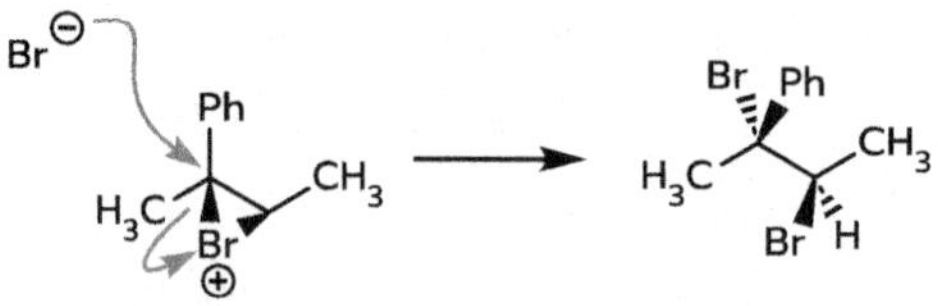

SECTION 2 THE CHEMISTRY OF ALKENES, BENZENE RINGS AND ALKYNES

This is another of those 'honest' moments. This is not at all easy to see. To check the outcome, I 'look down' the central C–C bond and try to visualize whether Br→Ph→CH$_3$ will appear clockwise or anticlockwise. I do this on the left-hand structure, because I know where the bromide attacks. I then make sure the orientation is the same in the right-hand structure.

> *This is how I do it, and how I can visualize it. It's not the only way, and you may find another approach that works better for you. All I can do is try to guide you through the steps until you find your own best way.*

FINISHING THE JOB

We are pretty much there now. I want to close this chapter with a couple of exercises for you, to reinforce the points we have made.

> Draw attack of bromide onto the other enantiomer of the bromonium ion. Remember that we would expect to form the two bromonium ions in a 1:1 ratio. What does this mean for the overall outcome of the reaction?
>
> Make a model of the product, and rotate around the central C–C bond, to convince yourself that it is impossible to talk in absolute terms about the Br atoms being on the same side or on the opposite side.
>
> Assign the stereochemistry (R or S) of all stereogenic centres in every structure in this chapter.
>
> Now do all the 'same stuff' with the following compound, which is the double bond isomer of the one we have been looking at. Convince yourself that it will give a diastereoisomer (**Stereochemistry 2**) of the compound at the top of this page.

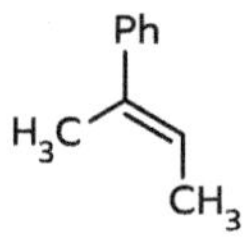

SECTION 2 THE CHEMISTRY OF ALKENES, BENZENE RINGS AND ALKYNES

STEREOCHEMISTRY 4

Diastereoselective Epoxidation of Cyclic Alkenes

WHY?

As before, different stereoisomers have different properties—not just their biological properties, but other properties as well. We need to understand when a reaction can be stereoselective, and we need to be able to develop a model to predict the outcome of a stereoselective reaction.

There is nothing 'special' about epoxidation. It's a neat reaction that we saw in Reaction Detail 6. Of course, epoxides are useful. We already looked at some of the reasons for this.

> But the more fundamental reason for the chemistry in this chapter is that it applies to all reactions of alkenes.

If we have an alkene reaction that can give two different stereochemical outcomes, and if we understand the reagent we are using, then we will be able to apply the principles in this chapter to almost anything.

> Understanding and prediction is better than memorization.

INTRODUCTION

This chapter is concerned with the diastereoselective epoxidation of cyclic alkenes. We looked at stereoisomers and energy in Stereochemistry 2. We looked at cyclohexane and cyclohexene conformers in Stereochemistry 1.

In this chapter we are going to look at some epoxidation reactions that can be stereoselective—they give more of one stereoisomer than the other.

> We can do this for epoxidation because we understand the mechanism of the reaction.

Let's start by looking at the reaction. Epoxidation of alkene **1** gives a mixture of epoxides **2** and **3**.

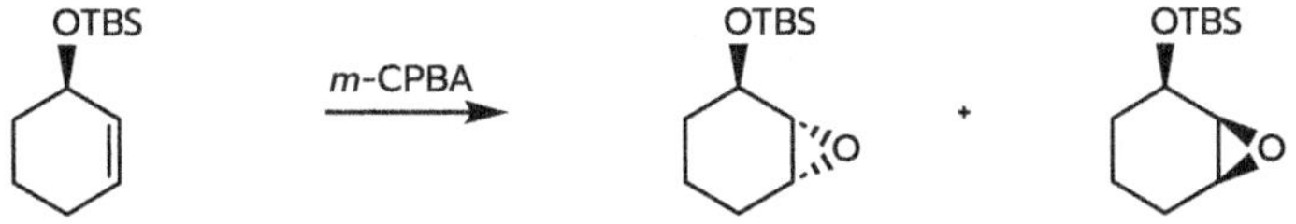

SECTION 2 THE CHEMISTRY OF ALKENES, BENZENE RINGS AND ALKYNES

These are diastereoisomers, having different energy. And the transition states that lead to them have different energy as well. Have another look at the reaction profile in **Stereochemistry 2**.

Of course, we could simply say that the top and bottom face of the alkene are different because there is a OTBS group on the top face.

> We could say this, and it's the same argument. If the two faces of the alkene were not different, the products would not be diastereoisomers.

Now let's get one really important thing out of the way. *In this particular case* we find that the OTBS group (we will get to what this is in a moment) blocks the top face of the alkene. But in the general sense, don't see a substituent with a wedge bond and automatically put the new group (the epoxide in this case) on a dashed bond.

> In a little while I'm going to make a prediction about a reaction that I don't think has ever been done. Maybe at some point we will find out if I am right!

For now, it's all about the method. We will work through this carefully.

EPOXIDATION OF CYCLIC ALKENES

Let's stick with the same example. First of all, we should identify the substituent. The abbreviation 'TBS' refers to '*t*-butyldimethylsilyl'. This is pretty bulky. Here is compound **1** when we draw it out fully.

1

Now let's look at the stereochemical outcome of the reaction. We get a lot more of compound **2** than compound **3**.

1 →(*m*-CPBA) **2** + **3**

7:1 ratio **2**:**3**

So, we do indeed have a very bulky group blocking the top face of the alkene bond, so that epoxidation takes place predominantly from the lower face.

We are going to consider the conformers of the cyclohexene. Do we need to do this? Probably not, in this case. But what about in other cases?

SECTION 2 THE CHEMISTRY OF ALKENES, BENZENE RINGS AND ALKYNES

> The simple fact is that the representation above is incomplete. It doesn't fully show the shape of the molecule. A representation that shows the shape of the molecule will always be better. With experience, you will find that you learn to recognize the reactions you need to draw out more fully. For now, let's do it with all of them.

We looked at conformations of cyclohexenes in **Stereochemistry 1**. Compound **1** will look something like this. We would expect the conformer on the right to be more favoured.

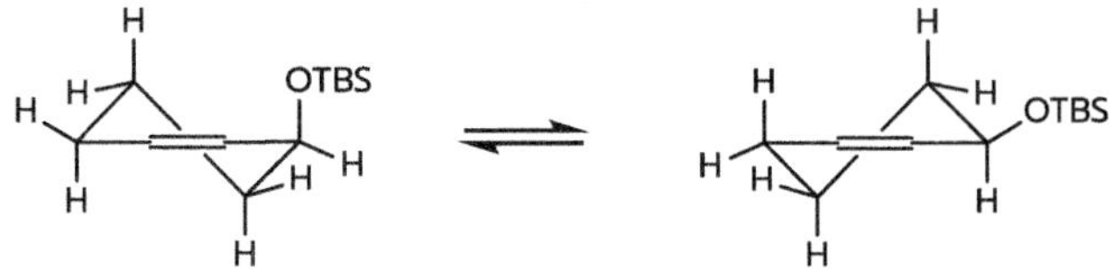

> Make a model of the compound and convince yourself that these conformers look okay.

In either conformer, there is more 'stuff' on the top face than on the bottom face. The OTBS group blocks the top face of the double bond.

> Did I draw the correct enantiomer of compound **1** in this form? Make sure you know how to check!

A PREDICTION

As far as I know, as of 2021, cyclohexene **4** has never been prepared. No-one knows for sure what happens if we react it with *m*-CPBA. Will epoxidation take place predominantly from the top face or from the bottom face?

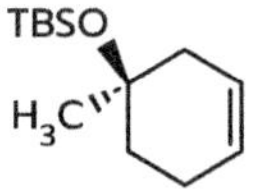

> You may think this isn't very exciting. But if you need one diastereoisomer of this epoxide for the synthesis of a blockbuster drug, you need to know how to make it stereoselectively. And if we can make a prediction about this reaction, we can make predictions about other reactions.

Consider *only* the structure above. The OTBS group is bigger than the CH₃ group. We might expect more epoxidation from the bottom face. But the OTBS group is further from the alkene in compound **4** than it is in compound **1**, so how much stereoselectivity do we expect?

> It isn't easy to put a number to it.

Now let's consider the conformers of compound **4**.

135

SECTION 2 THE CHEMISTRY OF ALKENES, BENZENE RINGS AND ALKYNES

In the conformer on the left, the OTBS group is *pseudo* equatorial. In the conformer on the right, it is *pseudo* axial. The conformer on the left is going to be more stable, which means that any individual molecule will spend more time in the conformer on the left.

In the conformer on the left, the methyl group is blocking the lower face. The OTBS group isn't really blocking the upper face.

> When we draw the structure more realistically, and carefully consider the orientation in space of the OTBS and CH₃ groups, we reach a different conclusion to the one we get from the 'flat' structure above.

The more realistic the representation, the better will be our ability to predict the outcome of a reaction.

> And that is the key point. Replace the simple 'top face' and 'bottom face' with a more rigorous description.

I would predict more epoxidation on the top face—the same face as the more bulky OTBS group. But I probably wouldn't expect very high levels of stereoselectivity.

ONE MORE EXAMPLE

Now we are going to look at a related example, this time without the TBS group.

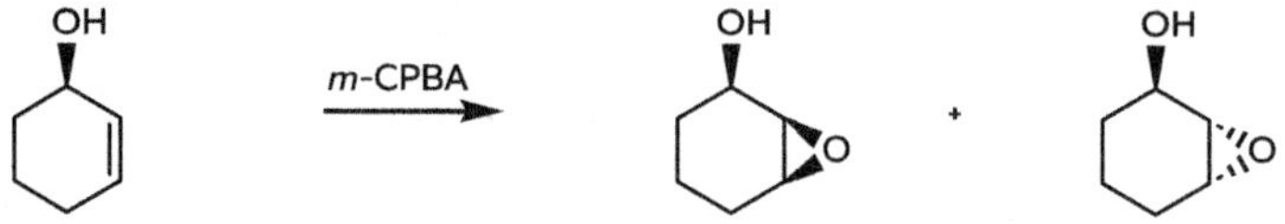

5 → **6** + **7**

m-CPBA

10:1 ratio **6:7**

Here, we have the opposite stereochemical outcome. Epoxidation is taking place from the same side as the OH group. We could take the approach we used above, and we could ask what is blocking the lower face. If we do this, we don't find a good answer. Maybe we need to find a better question to ask.

> What is leading to epoxidation on the same face as the OH group?

Perhaps there is some sort of 'attractive' interaction? Once we start along this line of thinking, there really is only one possibility.

> The OH group must be interacting in some way with m-CPBA. The 'in some way' can only really be a hydrogen bond.

Here is a plausible representation of a transition state for this process.

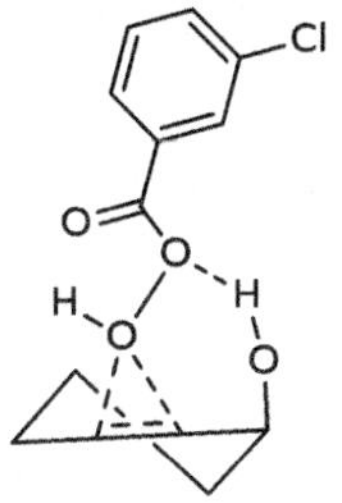

Of course, there are a few points to consider. First of all, this is basically a 'normal' transition state for epoxidation, but with one added interaction. Stereochemical effects rarely change the fundamental nature of a process. Note that compared to the structures above, I got rid of most of the hydrogen atoms on the cyclohexene ring. Leaving these hydrogen atoms in makes the structure look quite cluttered, and I'm not going to forget that they are there!

We will look at some more stereoselective epoxidation reactions in Stereochemistry 6. We will find we need to add to our methodology, but we won't need to make any dramatic changes.

A COMMENT ABOUT STEREOCHEMISTRY

We have just had four consecutive chapters focusing on aspects of stereochemistry. In Stereochemistry 1 we looked at the shape of cyclohexanes and cyclohexenes. In Stereochemistry 3 we looked at bromination of a cyclohexene, but we did not use the structures from Stereochemistry 1.

> Because we didn't have substituents on the cyclohexene, we didn't need to consider conformers.

In this chapter, we did have substituents, so we could **only** rationalize the stereochemical outcomes of reactions using these structures.

And, of course, irrespective of the representations we used, we considered how stereochemistry relates to energy (Stereochemistry 2).

Almost every reaction requires some consideration of stereochemistry. After all, reactivity is dictated by the shape of a molecule. You must build the habit of thinking about shape, even if it is only to reassure yourself that it doesn't matter 'this time'. But developing this habit will take time and effort. That's why I am starting early with this.

We are deliberately going to take a break from stereochemistry now, and consider some reactions that don't have (many) stereochemical consequences. Use some of that time to go back over these last four chapters, and make sure you can 'just see it'.

> For me, molecular models are the only way to go.

REACTION DETAIL 7

Hydrogenation of Alkenes, Alkynes and Aromatic Compounds

INTRODUCTION

Here are three reactions. They are all useful, and we will consider some of the reasons why in a minute. First of all, we will consider how we might accomplish these transformations.

$$R\text{-CH=CH-}R \xrightarrow{H_2} R\text{-CH}_2\text{-CH}_2\text{-}R$$

$$R\text{-C}\equiv\text{C-}R \xrightarrow{2\,H_2} R\text{-CH}_2\text{-CH}_2\text{-}R$$

$$C_6H_6 \xrightarrow{3\,H_2} C_6H_{12}$$

You can mix hydrogen and oxygen relatively safely, as long as you don't light a match. Hydrogen is unstable, but it does need quite a lot of energy to react. Another way to put this is that hydrogen is thermodynamically unstable, but kinetically stable.

I don't really like the term 'kinetically stable'. Kinetics is about rate of reaction, not about stability. What we mean by 'kinetically stable' is that its reactions have a high activation barrier.

> We could mix hydrogen with an alkene, an alkyne or an aromatic compound, and nothing would happen, despite these reactions being *exothermic*. We need to lower the barrier to reaction, and for this, we need a catalyst.[29]

We tend to use catalysts based on expensive transition metals, such as palladium or platinum. Molecular hydrogen is adsorbed onto the metal surface, and this facilitates cleavage of the H–H bond. The unsaturated organic compound approaches, and the two hydrogen atoms are transferred.

ALKENE HYDROGENATION

Alkene hydrogenation is the simplest case. Here is a more complete reaction scheme.

$$R\text{-CH=CH-}R \xrightarrow[\text{Pd/C}]{H_2} R\text{-CH}_2\text{-CH}_2\text{-}R$$

---

[29] I am not going to explain the principles of catalysis. That would just get in the way. If you need a reminder, I'm sure you can find a suitable resource.

The catalyst is normally a finely divided metal adsorbed onto a surface such as charcoal. Palladium is commonly used, and we tend to abbreviate this as Pd/C (palladium on carbon).

There is only one more point. Most of the time, it doesn't matter, but the two hydrogen atoms are always delivered to the **same face** of the double bond. Sometimes, it does matter, and you get a specific stereoisomer of the alkane product. We can see this easily with the following, cyclic, alkene.

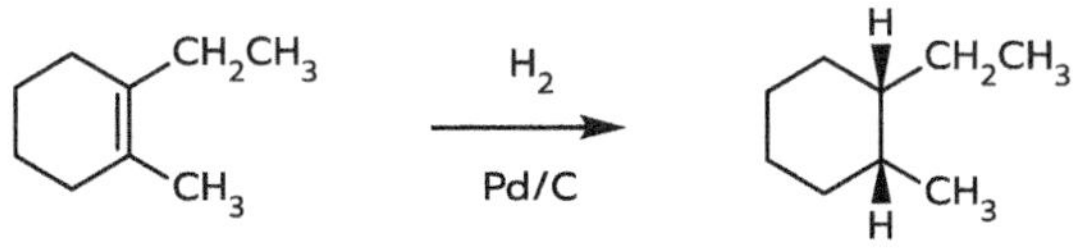

> I said we were taking a break from stereochemistry, but it's impossible to consider hydrogenation without briefly mentioning stereochemistry.

There is no reason why the two hydrogen atoms will only be delivered to the top face of the alkene as drawn. They could just as easily be delivered to the bottom face. Therefore, the product will be formed as a racemic mixture of the two enantiomers of product. We will come back to this aspect in **Stereochemistry 7**.

We would observe the same phenomenon with acyclic alkenes.

> Try to draw some acyclic alkene structures for which there would be observable consequences.

> We added two groups to opposite sides of a double bond in Stereochemistry 3. Different double bond isomers gave products with different stereochemistry. This is more of the same! You would use the same skill set to determine the stereochemical outcome of a hydrogenation reaction.

That's all I want to say about this reaction right now. We're going to see the ideas again with alkynes.

ALKYNE HYDROGENATION

There is one absolute with alkynes. They are more electron-rich than alkenes. They tend to react faster. This is true with hydrogenation reactions. If we treat an alkyne under the conditions shown above, we will form an alkane. The corresponding alkene is an intermediate in this reaction,[30] and of course it will be hydrogenated under the reaction conditions.[31]

---

[30] Although it will be the (Z)-alkene! Make sure you understand why! We've already covered the reasons in this chapter.

[31] You normally do these reactions under an atmosphere of hydrogen, although it is possible to vary the hydrogen pressure. It is possible to measure the amount of hydrogen consumed in a reaction, although this is not done routinely.

SECTION 2 THE CHEMISTRY OF ALKENES, BENZENE RINGS AND ALKYNES

$$R-\equiv-R \xrightarrow{H_2, \text{Pd/C}} R\frown R$$

With alkynes, it would be really useful to be able to reduce them to the alkene, and then stop. How could we do this?

Let's consider the two steps in this process.

$$\underset{1}{R-\equiv-R} \xrightarrow{H_2, \text{Pd/C}} \underset{2}{R\frown R} \xrightarrow{H_2, \text{Pd/C}} \underset{3}{R\frown R}$$

> We should ask a very simple question. Which reaction is faster, the hydrogenation of 1 to give 2, or the hydrogenation of 2 to give 3?

It turns out that the first hydrogenation is the fastest, for the reason mentioned above. Alkynes are more electron-rich than alkenes. To put this another way, the activation barrier for the first step is lower than that for the second step.

Now we have established that, we could, in principle, stop the reaction at the alkene **2**. All we need is a catalyst that is good, but not too good!

> I really like this approach. You establish what is possible, and what you need to make it happen. Then you can start looking for the solution, or invent it if it doesn't already exist.

This is such a useful transformation, that the catalyst already exists. The catalyst is referred to as a Lindlar catalyst, and consists of palladium supported on calcium carbonate, and 'poisoned' with lead acetate, lead oxide or quinoline (a heterocyclic base).

$$\underset{1}{R-\equiv-R} \xrightarrow[\text{Lindlar catalyst}]{H_2} \underset{2}{R\frown R}$$

ARENE HYDROGENATION

We have looked at aromaticity in Recap 6. We saw electrophilic substitution of aromatic compounds in Fundamental Reaction Type 3, and there is plenty more coming up. If we hydrogenate a benzene ring, the first hydrogenation will be slow, because we have to lose the aromatic stabilization. The second and third hydrogenations will be faster, because they do not have this additional energetic barrier.

SECTION 2 THE CHEMISTRY OF ALKENES, BENZENE RINGS AND ALKYNES

> There is no way we can stop this process after one hydrogenation.[32]

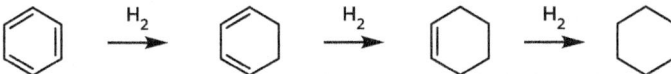

We tend to use a finely divided nickel catalyst, along with elevated temperatures and pressures for this transformation, although of course, researchers are constantly searching for alternative, better, catalysts. For now, it is the principles that matter.

WHY DO THIS?

Hydrogenation reactions get rid of functional groups (alkenes, alkynes) rather than introducing them. You might think this makes them inherently less useful than reactions that lead to the formation of functional groups. However, alkene hydrogenation leads to the formation of a carbon-carbon single bond that doesn't need to have any adjacent functional groups. These C–C bonds are not always easy to form. However, it is relatively easy to form a carbon-carbon double bond with no adjacent functionality. You can form the double bond, and the reduce it to leave the single bond that you wanted!

WHAT ARE THE ALTERNATIVES?

Why use hydrogen and a catalyst? The short answer is that it works. The longer answer is that there are not many viable alternatives. If we wanted to reduce an alkene using a more conventional 'polar' reagent, we would need a hydrogen electrophile (H⊕) followed by addition of a hydrogen nucleophile (H⊖). The problem is, we need to add the acid first, and a hydride reagent such as sodium borohydride would react rapidly with acid, so this won't work. It is okay when reducing a carbonyl group, because we add the borohydride first and then add acid at the end.

---

[32] Organic chemists are creative. Some chemists have devised clever ways to partially reduce benzene rings. We aren't going to cover these reactions in this book, but if you want to know more, look up the 'Birch reduction'.

SECTION 2 THE CHEMISTRY OF ALKENES, BENZENE RINGS AND ALKYNES

BASICS 4

Regioselectivity in Aromatic Chemistry

INTRODUCTION

Aromatic compounds are very commonly found in pharmaceuticals. When you have a benzene ring with multiple substituents, the defined bond angles mean that you know exactly where the substituents are.

> *If you can predict the shape of a potential pharmaceutical ingredient, you know where the interactions are with an enzyme.*

In this short chapter, we will look at the very basic idea of how one substituent on an aromatic ring can affect where the next substituent is introduced in an aromatic electrophilic substitution reaction.

Let's start with benzene. We know (**Fundamental Reaction Type 3**) that benzene undergoes substitution reactions. First we will carry out a substitution reaction with electrophile 'A'. For the purpose of this chapter, it doesn't matter what it is.

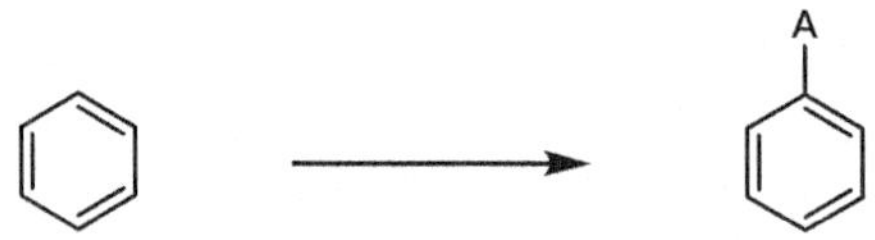

If we now carry out a second substitution reaction with electrophile 'B' (not an element symbol), we find there are three possible products, depending on where the second substituent goes. We refer to these as *ortho*, *meta*, and *para*.

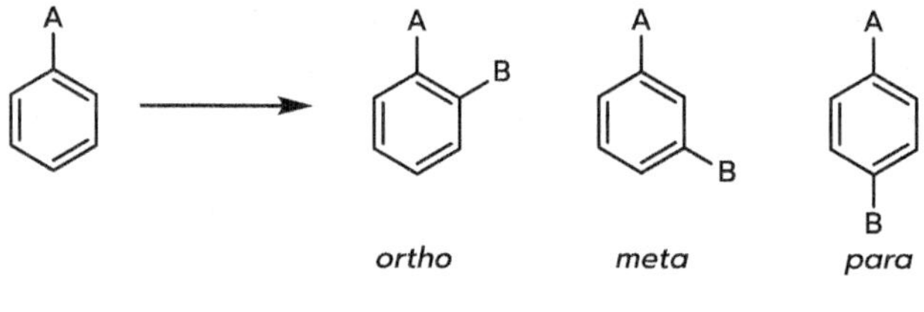

ortho   *meta*   *para*

CAUTION

You've probably encountered this reaction before university, so if I talk about a substituent 'A' being *meta* directing, you will identify it as a substituent that results in 'B' going to the *meta* position. Other substituents direct to the *ortho* and *para* positions.

> *This often leads to a problem.*

Let's use an example. Here is methoxybenzene, also known as anisole.

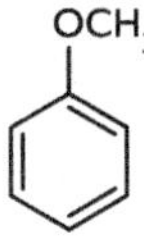

SECTION 2 THE CHEMISTRY OF ALKENES, BENZENE RINGS AND ALKYNES

The methoxy group directs the next substituent (we are going to use NO$_2$) to the *ortho* and *para* positions. We get a mixture of the two products as shown below.

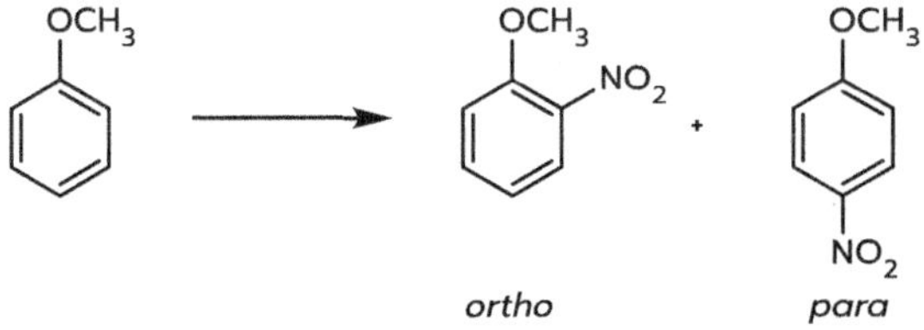

ortho   *para*

Because you have learned this reaction before studying at university, if I ask you why we get these products, you might tell me it is "because methoxy is *ortho/para* directing".

> On one level, this is absolutely correct. But what I am really wanting is an understanding of *why* methoxy is *ortho/para* directing.

If you give the above answer, what will you do if the substituent is one you haven't seen before?

> Hopefully you will compare it to substituents you know about, and make a decision based on the substituent that the new one is most like!

But it would always be easier to really understand **why** methoxy is *ortho/para* directing, and to be able to apply the same reasoning to any new substituent.

THE MOST COMMON ERROR

In **Reaction Detail 8**, we will work through a range of substituted benzene derivatives, and determine from first principles which isomers are formed in a given electrophilic substitution reaction. In order to do that, we will need to draw good curly arrows for the addition of the electrophile, and good curly arrows for the resonance forms of the intermediate.

There is a mistake that is quite common for students to make while they are learning this reaction. I want to present the mistake to you, in isolation, so you can examine it carefully and understand why it is wrong.

Let's look at the ***incorrect*** scheme and see what processes you can apply to establish why this is wrong.

REMEMBER, THIS IS WRONG!

143

SECTION 2 THE CHEMISTRY OF ALKENES, BENZENE RINGS AND ALKYNES

> *Let's cut to the chase. We have the electrophile and the positive charge on the same carbon atom. This is wrong!*

How can we establish that this is wrong? I would add the hydrogen atoms on the carbon atoms.

THIS IS WHY IT'S WRONG

In the carbocation, carbon atom 'a' has four bonds *and* a positive charge. This cannot be right. Carbon atom 'b' has only three bonds and no charge.

> *Carbon atom 'b' is harder to spot, because we are used to seeing structures drawn without all the hydrogen atoms.*

If we move the positive charge to where it should be, we get the following.

Carbon atom 'a' has four bonds and no charge—that makes me happy! Carbon atom 'b' has three bonds and a positive charge—I'm still happy!

> *Don't move on from this point until you are confident that you have got this sorted. Draw the structures and the curly arrows a few times, to build 'muscle memory'. In a little while, you will be drawing these structures and curly arrows quickly. You don't want to be drawing them quickly and wrong!*

THE POINT!

In the next chapter, we are going to look at this reaction in some detail. Here is key point to get in up front.

> *Forget about the directing effects you have memorized.*

We need to replace this superficial description with a reaction mechanism where you can consider the stability of intermediates and relate this to the likely activation energy for each possible outcome.

Benzene is not the only aromatic compound. We will encounter a selection of heterocycles in **Basics 7** and in **Applications 9**. If we know the underlying mechanisms and principles, we will be able to determine where any given heterocyclic compound will undergo substitution.

SECTION 2 THE CHEMISTRY OF ALKENES, BENZENE RINGS AND ALKYNES

> *And only then, when you are familiar with all the different substituent types, benzene rings and heterocycles, will you be able to glance at **any** aromatic compound structure and know automatically where it will undergo electrophilic substitution.*

And at that point, you will see a methoxy group on a benzene ring, and you will say, "that's *ortho/para* directing". When you get to that point, it will be fine, as you are remembering it easily because you understand it.

SECTION 2 THE CHEMISTRY OF ALKENES, BENZENE RINGS AND ALKYNES

REACTION DETAIL 8

Aromatic Electrophilic Substitution

GETTING STARTED!

This is a chapter with quite a few examples of the same reaction type with different outcomes for very good reasons.

We have already established the essentials in **Fundamental Reaction Type 3**, and we defined the terminology and one common problem in **Basics 4**.

At first, we will look at nitration reactions. In much of the discussion so far, we have simply called the electrophile 'E⊕'. Now we need to recognize that in a nitration reaction, the electrophile is $NO_2^⊕$, and we need to know how to generate it. We use a mixture of nitric and sulfuric acids, and we will simply consider the sulfuric acid to be a source of protons.

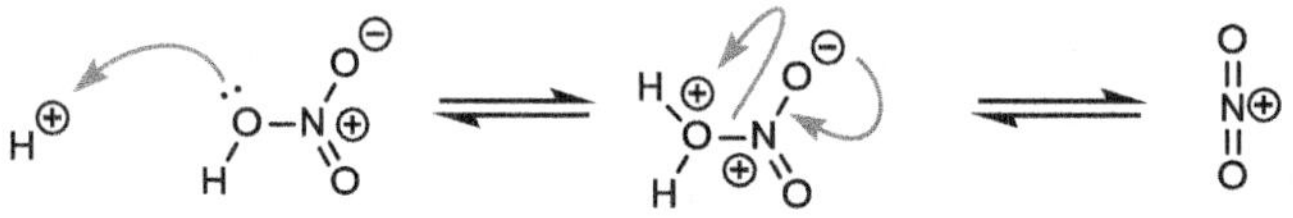

First, we protonate nitric acid. This may seem strange—after all, nitric acid is an acid, not a base! But we are using a very strong acid, sulfuric acid, and there is a lone pair that can be protonated. Follow (and draw!) the curly arrows and make sure you can understand why they lead to the formation of the nitronium ion **3**.

Right, now we have our electrophile, it's over to you. Here is the reaction.

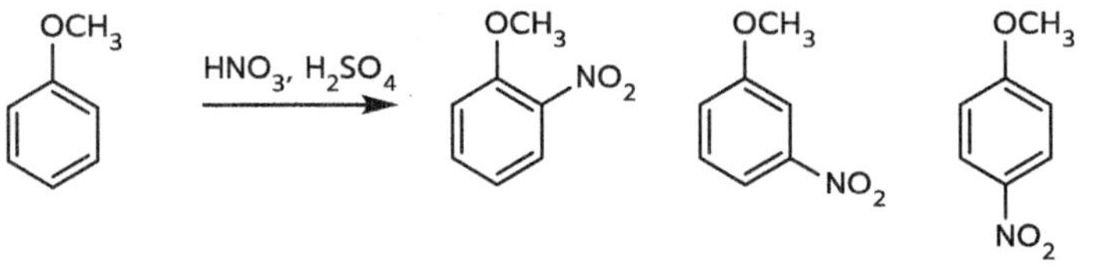

> Draw a curly arrow mechanism for the reaction, and by consideration of the intermediate, predict which of the products (**5, 6, 7**) will be formed preferentially. Consider *everything*.

Don't you just hate instructions like 'consider everything'? What is 'everything' and how do you know if you've missed something? Don't worry. We are going to work through the problem.

NITRATION OF METHOXYBENZENE

First of all, let's consider a reaction, and a reaction energy profile. We have an aromatic compound, reacting with an electrophile to generate an intermediate.

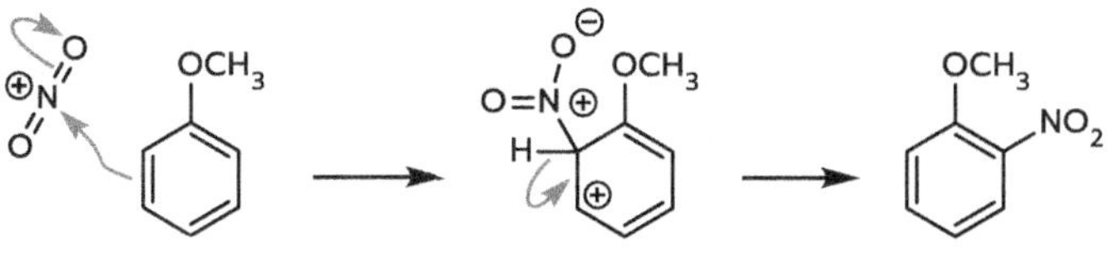

This intermediate then loses a hydrogen atom (as a proton) **from the same carbon atom** as shown above.

> It has to be lost from the same carbon atom, or it would not be substitution!

The carbocation intermediate is stabilized by resonance.

> We are going to spend a lot of time looking at this!

But it is not aromatic. Formation of this intermediate will be significantly *endothermic*. It then loses a proton to regain aromaticity. We will assume that the overall process will be *exothermic*. Here is a reaction profile, straight out of **Recap 2**.

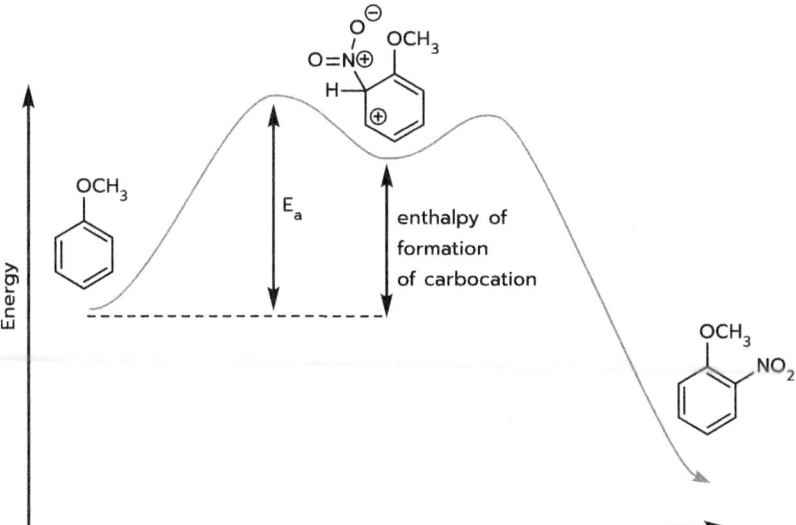

Once again, we can invoke the Hammond Postulate. The rate of reaction is determined by the activation energy. The transition state for carbocation formation is close in energy to that of the carbocation itself. Although we covered it in **Recap 2**, I'm going to give you the wording of the Hammond Postulate again.

> *"If two states, as, for example, a transition state and an unstable intermediate, occur consecutively during a reaction process and have nearly the same energy content, their interconversion will involve only a small reorganization of the molecular structures."*

SECTION 2 THE CHEMISTRY OF ALKENES, BENZENE RINGS AND ALKYNES

Remember what this really means. If there is only a small reorganization of the molecular structures, this means that something that stabilizes the carbocation intermediate will also stabilize the transition state.

What this, in turn, means is that when we compare three different carbocation intermediates, we can be confident that the most stable will be formed fastest. Therefore, the product distribution will depend on the intermediate distribution. Whenever we need to consider the relative amounts of things with differing energies, we could use the Boltzmann distribution.

> We aren't going to do this in this book, but I wanted to connect organic reaction mechanism with your basic physical chemistry.

Right, we are taking this slowly, but we have got to the point where we know that we need to look at the carbocation intermediate for substitution at each position, and to work out which intermediate is most stable.

We will need to consider inductive effects, mesomeric effects (resonance forms!) and steric effects.

> I think that's 'everything'!

Here is the first one (*ortho*).

> Draw the curly arrows for formation of resonance form **A** from methoxybenzene and the nitronium ion. Check your working against the answer on the previous page.

The positive charge is delocalized (not localized!) between three of the benzene ring carbon atoms, but also onto the oxygen. Let's have a look at resonance form **C**. Oxygen is electronegative, so that an inductive electron-withdrawing effect will destabilize the positive charge. But the electron-donating mesomeric effect shown by the curly arrow (and leading to resonance form **D**) is much more significant. We quantified this in **Recap 2** for a simpler carbocation.

> In summary, the methoxy group stabilizes the intermediate.

Let's move on to the next one (*meta*).

> Again, draw the curly arrows for formation of resonance form **E**. If you need to draw all the hydrogen atoms in, that's fine. It is important to keep track.

SECTION 2 THE CHEMISTRY OF ALKENES, BENZENE RINGS AND ALKYNES

[Resonance structures E, F, G showing intermediate with OCH₃ and O₂N groups]

We only have three resonance forms this time. We do not have a resonance form in which the positive charge is particularly *destabilized* by the inductive effect of the electronegative oxygen atom. Nor do we have the mesomeric electron-donating effect from the oxygen.

> *The methoxy group has little or no effect on the stability of the intermediate in which nitration is meta to the methoxy group.*

Now the last one (*para*).

> And again, draw the curly arrows for the formation of the intermediate.

[Resonance structures H, I, J, K showing intermediate for para substitution]

Here, we have resonance form **I**, which will be destabilized by the inductive effect of the electronegative oxygen. However, once again, the oxygen has a mesomeric electron-donating effect which will be bigger. This is shown in resonance form **K**.

Let's take stock of where we are. We have predicted, based only on electronic effects, that *ortho* and *para* substitution will be favoured. There is absolutely no situation in which a mesomeric effect can favour *ortho* substitution and not favour *para* substitution.

> *Methoxy is an ortho/para directing group.*

We aren't quite done yet. We should still consider the ratio of *ortho:para* substitution products. Let's ignore steric effects.

> *In the example we are using, we have two ortho positions and one para position. If we only consider this 'statistical' effect, we would expect to obtain a 2:1 ratio of ortho and para products.*

Now let's look again at one resonance form for the intermediate in each case. We've compared electronic effects. Now let's look at steric effects. Structure **A** (it doesn't matter which of the resonance forms we consider) is more crowded than structure **H**. If we assume that resonance stabilization means that structures **A** and **H** are 'equal', then the steric effects will disfavour structure **A**.

149

SECTION 2 THE CHEMISTRY OF ALKENES, BENZENE RINGS AND ALKYNES

A (structure with O₂N, H, OCH₃) **H** (structure with OCH₃, O₂N, H)

> There are still two *ortho* positions. We cannot accurately predict the ratio of products, but when we consider steric and electronic effects together, we would expect more than 33% of the *para* product and less than 67% of the *ortho* product.

Here is the actual outcome. We do actually get a **lot** more *para* product than *ortho* product.

OCH₃ + HNO₃, H₂SO₄, acetic acid, 45 °C → ortho-NO₂ + meta-NO₂ + para-NO₂

ratio 31:2:67

This reaction has been carried out numerous times, with a wide range of conditions. The above outcome is typical. In fact, although attack *meta* to the methoxy group is disfavoured, we still get some of this product.

> I'm going to show you a few product ratios. Try to understand the general trends, but don't feel that you should have been able to predict these.

RELATIVE RATES OF REACTION

This is something we need to consider. Perhaps we have a compound with more than one benzene ring, so that not only are we trying to carry out substitution at one particular position on a benzene ring. We could also be trying to react one benzene ring selectively in the presence of another.

> Can we compare the reactivity of methoxybenzene with that of benzene itself?

(OCH₃-benzene and benzene structures)

We have actually done all the hard work. We drew resonance forms **A – K**, and we saw that resonance forms **D** and **K** particularly stabilize the intermediates for *ortho* and *para* substitution.

> Let's look at this wording carefully. I'm not saying that these resonance forms are more stable than the other ones. That would have no meaning.

SECTION 2 THE CHEMISTRY OF ALKENES, BENZENE RINGS AND ALKYNES

I am saying that as a result of the contribution from the oxygen lone pairs, these intermediates are more stable than they would be without the oxygen atoms. To put this another way, they are more stable than those derived from the reaction of benzene itself.

We often talk about methoxy being *ortho/para* directing and **activating**. That is, the net electron-donating effect of the substituent makes the ring more reactive towards electrophiles.

> *The terminology is good, but perhaps not necessary. We can draw resonance forms and consider their stability, which gives us a better ability to make predictions without relying on recall.*

The relative reactivity of methoxybenzene and benzene itself depends on the reaction we are talking about, but in general, methoxybenzene reacts with electrophiles >10,000 times more rapidly than benzene itself does.

TWO DIFFERENT ELECTRON-WITHDRAWING GROUPS

There is another way to express the chemistry in the previous section. If a benzene ring has a substituent which has a **net** electron-donating effect, it will be more reactive than benzene in electrophilic substitution reactions.

The counterpoint to this is that if a benzene ring has a substituent with a **net** electron-withdrawing effect, it will be less reactive than benzene itself. We are going to look at two such substituents.

First of all, we will consider nitration of chlorobenzene. I have not drawn the curly arrows for addition of the nitro group.

> You should do this! You can look back at the previous example to check your work.

Here is the intermediate, with its resonance forms, for *ortho* attack.

Of course, this is exactly the same as with the methoxy group, so we need to focus on what the differences are. Chlorine and oxygen have virtually the same electronegativity.

> *The inductive effects will be the same.*

The curly arrow on resonance structure **C** shows overlap of a chlorine lone pair with the empty p-orbital on carbon. But chlorine is in row three of the periodic table and

SECTION 2 THE CHEMISTRY OF ALKENES, BENZENE RINGS AND ALKYNES

carbon is in row two. This overlap is not as effective as it was with the methoxy group. To put this another way, resonance form **D** does not contribute as much.

> It does still contribute. We do get some stabilization of the intermediate for ortho attack. Just not as much.

Now let's look at substitution *meta* to chlorine.

[structures E, F, G showing meta substitution intermediate resonance forms with Cl and O₂N-H groups]

As with the methoxy group, the lone pair on chlorine does not stabilize the positive charge.

What about *para* substitution?

> This is one for you to draw. Give it a go!

To check your working, compare it with the corresponding structures with a methoxy group. They should basically be the same.

> Our conclusion is that meta substitution is less favoured than ortho or para substitution. Chlorine is an ortho/para directing group.

We can also consider the relative reactivity. Methoxybenzene is >10,000 times as reactive as benzene itself. Chlorobenzene is 50 times **less** reactive than benzene.

> The inductive electron-withdrawing effect destabilizes all the intermediates. This is mitigated, **slightly**, by electron-donation from the chlorine lone pair.

Here is the reaction outcome.

[Reaction: chlorobenzene + HNO₃/H₂SO₄ → ortho-, meta-, and para-nitrochlorobenzene]

ratio 35:1:64

As before, we would expect more *ortho* nitration than *para* statistically, and again, we can account for the observation of more *para* substitution by invoking steric hindrance to *ortho* nitration.

> In fact, there is no meaningful difference compared to methoxybenzene, apart from the rate of reaction. I have to be honest—I would have expected a slightly higher proportion of the *meta* product based on the above argument.

SECTION 2 THE CHEMISTRY OF ALKENES, BENZENE RINGS AND ALKYNES

Now let's look at a nitration reaction of nitrobenzene. Again, we have a process to follow. We draw the curly arrows for the attack of $NO_2^{\oplus}$ at each possible position on the benzene ring.

> This is something for you to do. You can check your products against the resonance forms we are drawing next.

Then we draw the resonance forms for each intermediate. Here they are for *ortho* attack.

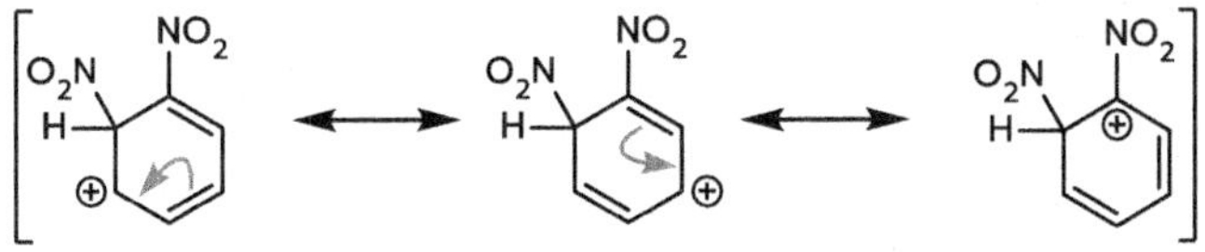

Make sure you are certain that you know which nitro group was the 'original' nitrobenzene one, and which is the electrophile. Of course, I could have helped you by highlighting them.

> But long term, that wouldn't really help you, so I didn't!

The nitro group is strongly electron-withdrawing. The resonance form on the right is not going to contribute much to the overall structure. A positive charge next to an electron-withdrawing group cannot be good.

There is another way to think about this, if we draw the nitro group out 'fully' as shown below.

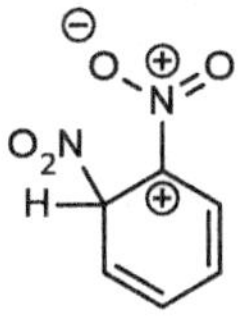

Here, we have a positive charge next to a positive charge. This cannot be good.

Let's compare this with the situation for *meta* substitution. Here, we have three resonance forms, and none of them have a positive charge next to the nitro group.

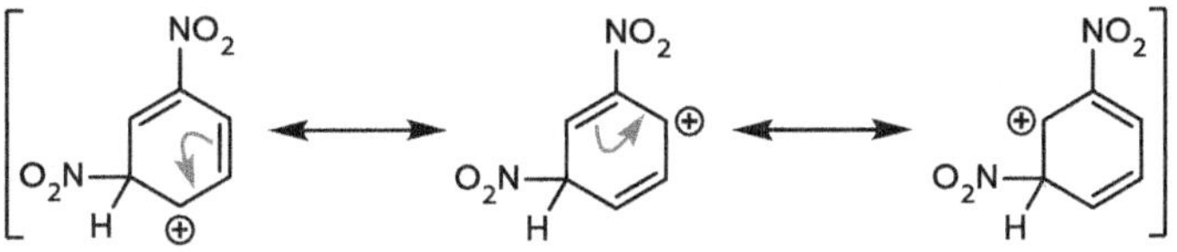

Meta substitution is better than *ortho* substitution.

> 'Better' is a relative term. Perhaps it would be more correct to say it is 'less bad'.

Here is the outcome of the reaction.

153

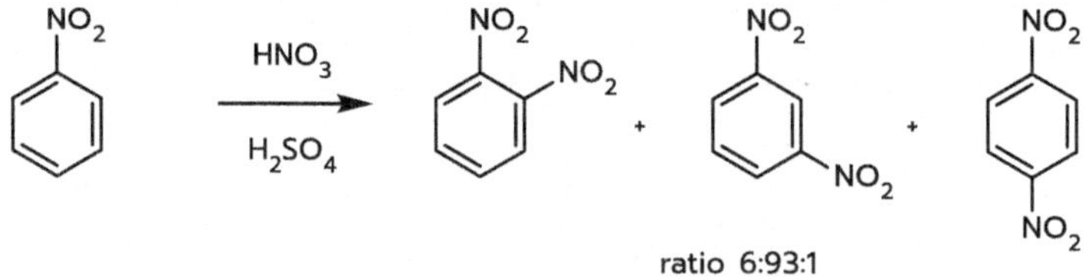

SECTION 2 THE CHEMISTRY OF ALKENES, BENZENE RINGS AND ALKYNES

ratio 6:93:1

The nitration of nitrobenzene is about 17 million times slower than the nitration of benzene itself. All the intermediates are destabilized, but the intermediate leading to nitration at the *meta* position is not destabilized quite as much.

> Did you notice that we didn't draw the intermediate for *para* substitution? Draw it, and then work out why we were able to take this shortcut.

Oh, and why do we get six times as much *ortho* product compared to *para* product? I wonder if there might be some hydrogen bonding involving the acid catalyst, but I honestly don't know!

AMINES, AMIDES AND REACTION CONDITIONS

Of course, we always need to look at the bigger picture. Nitration of methoxybenzene (anisole) and nitration of aminobenzene (aniline) look the same.

> Draw the resonance forms for both processes. We already drew all the resonance forms for nitration of methoxybenzene. Draw them again, this time for aniline!

But there's a key difference. Aniline is not very basic, but it is basic!

> What would happen if you put aniline in a mixture of nitric and sulfuric acids?

Hopefully you drew something like this.

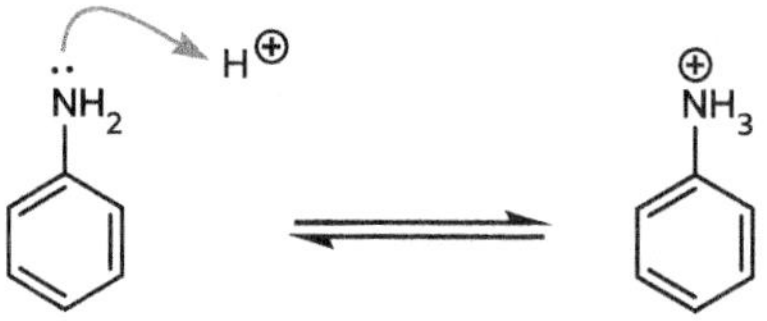

Under the reaction conditions, we don't have an amino group with a lone pair 'directing' the attack. We have an ammonium ion with a positive charge.

> Draw the resonance forms again, this time with the ammonium ion.

This should lead you to a different conclusion. You should predict that *meta* substitution is more favoured (or less disfavoured). You should also predict that aniline will react more slowly than benzene.

SECTION 2 THE CHEMISTRY OF ALKENES, BENZENE RINGS AND ALKYNES

> I've given you enough information to finish the job. I'm not going to give you all the structures this time.

Now let's propose a problem. Suppose you need to put a substituent *para* to an amino group on a benzene ring.[33] You need the amino group to be electron-donating. Therefore, it needs to have its lone pair 'available'.

> We have just seen that if the lone pair is *too available*, it will simply be protonated under the reaction conditions.

We can react the amine with acetic anhydride to make an amide.

We will see the mechanism of this reaction in **Reaction Detail 16**.

> Once you've read that chapter, come back here and draw the mechanism of the reaction above. It's a slight variation, and variety is good.

We still have a lone pair on nitrogen, but this one won't be protonated significantly under the conditions used for the nitration reaction we are using as our example.

But when we draw the intermediates for nitration, and consider the resonance forms, we will still use the nitrogen lone pair. Here it is for *ortho* nitration.

Well, I say 'here it is'. What I really mean is that here is just enough to show that you will get stabilization of the intermediate for *ortho* nitration.

> Once again, it's over to you to finish the job. Draw the curly arrows for the nitration step at each position. Draw all the resonance forms in each case. Draw the final loss of a proton to give the product.

I have two reasons for getting you to do the work here. First of all, it is how you will get better at doing it. If I draw them for you, you will be confident that you understand everything. Secondly, if you draw the structures and resonance forms,

---

[33] We accept that we will also get some *ortho* substituted product as well, as there can never be a substituent that is *para* directing but not *ortho* directing. Does this answer the question towards the top of the previous page?

155

SECTION 2 THE CHEMISTRY OF ALKENES, BENZENE RINGS AND ALKYNES

you will predict the correct outcome for the reaction, and then you will be confident that you can do this for **any** aromatic electrophilic substitution reaction.

HALOGENATION OF AROMATIC COMPOUNDS

In **Reaction Detail 10**, we will see a reaction in which a halogen on an aromatic ring can be replaced by a carbon substituent. The fact that we have such reactions at our disposal means that introduction of halides onto aromatic rings is a really useful transformation.

In **Reaction Detail 1** we looked at the bromination of an alkene. In **Recap 6** we saw how much stability is imparted by aromaticity.

> *Alkenes react with bromine. The corresponding reaction with benzene would be much slower.*

We need to find a way to make this reaction faster. The slow (rate-determining) step in bromination is attack of the electrophile. Recall that when we simply use bromine, the electrons in the double bond attack the Br–Br bond. We now need to add a little more detail. When an alkene approaches Br_2, a dipole is induced in the Br–Br bond as follows.

This 'gets the reaction started'. But if we have a less reactive substrate, we need to polarize the Br–Br bond even more. Let's add a Lewis acid such as $FeBr_3$. This is what happens.

$$FeBr_3 + Br_2 \longrightarrow Br-Br-FeBr_3$$

Now, when the benzene ring attacks, we kick out $FeBr_4^{\ominus}$. This is more favourable.

> *And then we lose a proton to regain aromaticity!*

If we want to carry out a chlorination reaction, we use Cl_2 and $AlCl_3$. But it's the same!

> *This, of course, is the point.*

We could spend more pages working through the halogenation of various substituted benzenes, and we would find that everything we have said about nitration applies in exactly the same way. All we have here is a different electrophile.

SECTION 2 THE CHEMISTRY OF ALKENES, BENZENE RINGS AND ALKYNES

It is different in two ways. First of all, it is different from the nitro group we have been using as our benchmark so far. Secondly, it is different from the bromine electrophile we used when we wanted to react it with an alkene.

> The key point is to understand what is different and why it needs to be different.

I worry that by covering this so quickly (because there is nothing new!) it might not make an impact on you.

> Pick one of the substituted benzene compounds we have looked at so far. Draw the bromination mechanism out fully at each position on the benzene ring. Draw the intermediates and their resonance forms. Rationalize their stability. Then go back and compare with the nitration reactions we have covered fully.

SULFONATION OF NAPHTHALENE

Here is a really instructive example. Remember, that's the point. It's not about remembering this example. It's about understanding the process we are following.

If we sulfonate naphthalene at 80 °C, we get predominantly naphthalene-1-sulfonic acid. If we carry out the same reaction at 160 °C, we get predominantly naphthalene-2-sulfonic acid.

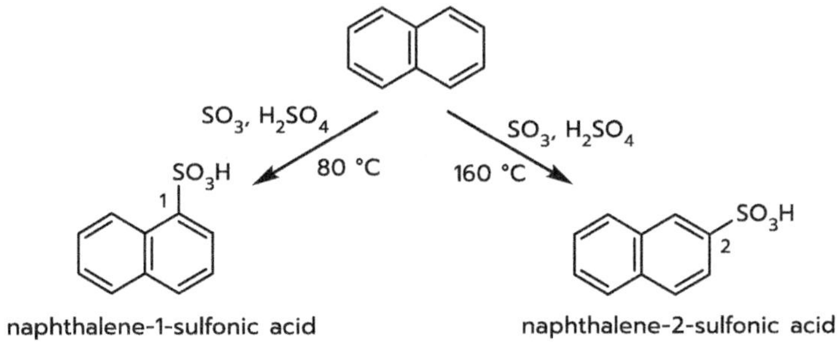

This is an example of a reaction in which one product is formed faster, and the other product is more stable. We will see another example of such kinetic and thermodynamic control in Applications 5.

Here is the structure of naphthalene.

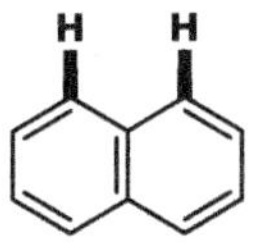

I have highlighted two hydrogen atoms. They are quite close to one another, and this causes torsional and steric strain (we will define these terms in Stereochemistry 5). It is rather like a 1,3-diaxial interaction in a cyclohexane (Stereochemistry 1).

> It's always good to relate it back to something we know.

SECTION 2 THE CHEMISTRY OF ALKENES, BENZENE RINGS AND ALKYNES

On this basis, it looks reasonable that naphthalene-2-sulfonic acid is formed at higher temperatures—it is more stable.

> This makes a key assumption—that naphthalene-1-sulfonic acid is formed reversibly.

If naphthalene-1-sulfonic acid is formed quickly and **irreversibly**, then it will simply be formed, and that is that. The formation of the intermediate in each case is shown below.

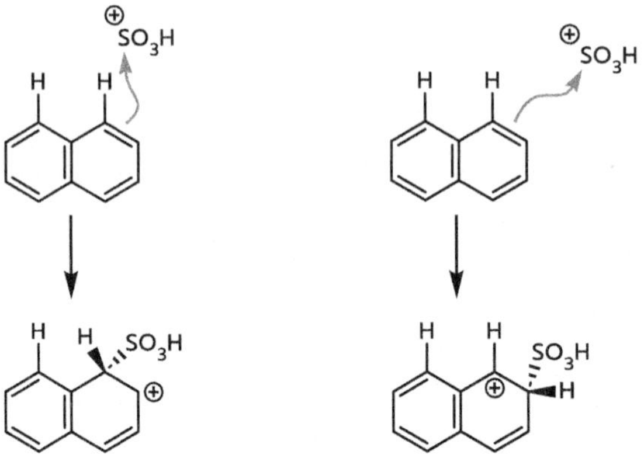

On the left, we lose some of the torsional/steric strain upon addition of the electrophile. On the right, we do not.

> On steric grounds, it seems reasonable to expect that we will get faster sulfonation at the 1- position.

> Draw a reaction profile for this. Remember that formation of the intermediate shown is going to be *endothermic*, and the intermediate is preceded by a transition state. Remind yourself of the Hammond Postulate which we saw earlier in this chapter.

Are we done yet? It's tempting to find a good reason why a particular reaction will be favoured, and to stop there. But we have only considered one aspect of the stability of the intermediate. What about resonance stabilization?

> Draw resonance forms for the stabilization of the two intermediates drawn above. How many of the resonance forms in each case still have an intact benzene ring?

You should have drawn something like this, to explain why the left-hand intermediate is more stable.

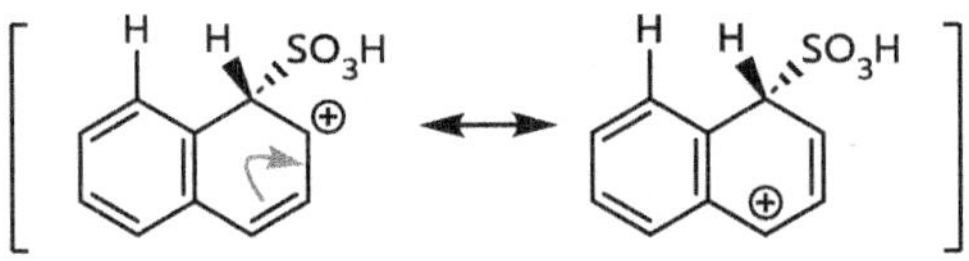

SECTION 2 THE CHEMISTRY OF ALKENES, BENZENE RINGS AND ALKYNES

Okay, so sulfonation of the 1-position should be faster on steric grounds, and it should also be faster on electronic grounds.

> Now we are done!

Well, almost! We know why we get naphthalene-1-sulfonic acid at 'lower' temperature, but we also need to know how we can get conversion of naphthalene-1-sulfonic acid into naphthalene-2-sulfonic acid.

> Don't over-think this. The reaction conditions are acidic. Protonate the product. Then lose the electrophile we added above, and we get naphthalene back. And then it gets sulfonated again.

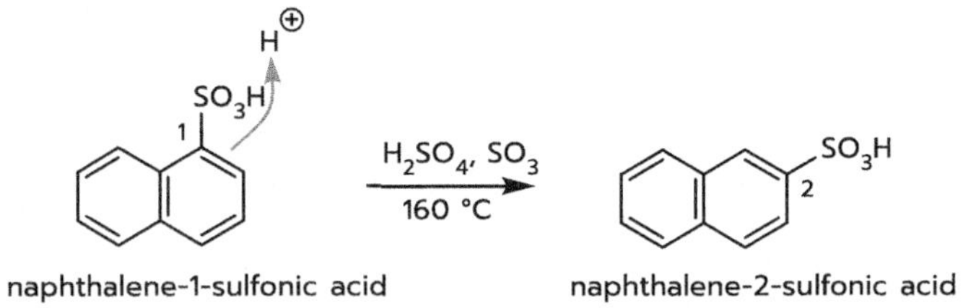

naphthalene-1-sulfonic acid → (H$_2$SO$_4$, SO$_3$, 160 °C) → naphthalene-2-sulfonic acid

> Draw out the mechanism. I've made a start!

DEALING WITH A COMMON ERROR

In **Basics 4** we looked at one common error when drawing aromatic electrophilic substitution mechanisms. There is another common error, but I deliberately left discussing it until you had drawn some mechanisms of your own.

The problem is losing a proton from the wrong carbon atom to re-form the aromatic ring. First of all, here is the correct curly arrow.

In the product, we need each carbon atom to have one thing attached—either Br or H. We are adding Br to an atom that already had a hydrogen atom attached.

> This is the one we need to lose.

> Redraw this step with **every** hydrogen atom on the structure.

Now here is the 'wrong' curly arrow.

SECTION 2 THE CHEMISTRY OF ALKENES, BENZENE RINGS AND ALKYNES

We have drawn in one hydrogen atom, and it is certainly there. But by drawing the wrong hydrogen atom, we draw an incorrect curly arrow. In the product, that hydrogen atom (which I have highlighted) still needs to be there.

> Now go back to the mechanisms you have drawn while you were working through this chapter. Did you make this mistake? If so, redraw the last step.

Don't worry too much if you made this mistake. I wouldn't be highlighting it if I didn't see it a lot. Just make sure you understand why it is a mistake so you can avoid it moving forward.

IN SUMMARY

We have looked at a range of aromatic electrophilic substitution reactions, and we have identified a process by which we can analyse any of these reactions and work out approximately how fast the reaction will be and where a new substituent will be introduced relative to an existing substituent.

An experienced organic chemist will simply glance at a structure and 'know' where the electrophile will add. To some extent they are doing this by the sort of pattern recognition that you learn prior to university study (*this is an ortho/para directing substituent*).

> *The difference is that the pattern has become easy to remember because they have drawn the structures and resonance forms out so many times that they fully understand them.*

Short term, this is more effort. Long term, it leads to better results.

And then, when you see a 'new electrophilic substitution' reaction with a different electrophile, then only thing you will have to learn is what the electrophile is and how it is generated.

SECTION 2 THE CHEMISTRY OF ALKENES, BENZENE RINGS AND ALKYNES

REACTION DETAIL 9

Friedel-Crafts Alkylation

INTRODUCTION

There is nothing new in this chapter. Friedel-Crafts reactions are aromatic electrophilic substitution reactions (**Reaction Detail 8**) with carbon electrophiles.

> This means that they are used to form carbon-carbon bonds, which we like!

There are two types of Friedel-Crafts reactions, and a thorough understanding of the difference between them is essential.

> To put it another way, if you really understand the mechanisms, you will be able to predict the difference between them.

FRIEDEL-CRAFTS ALKYLATION

Let's see what the reaction is, and then we can see what the problems are. Take 1-chlorobutane and add aluminium chloride. This will polarize the C–Cl bond, increasing the amount of positive charge on carbon.

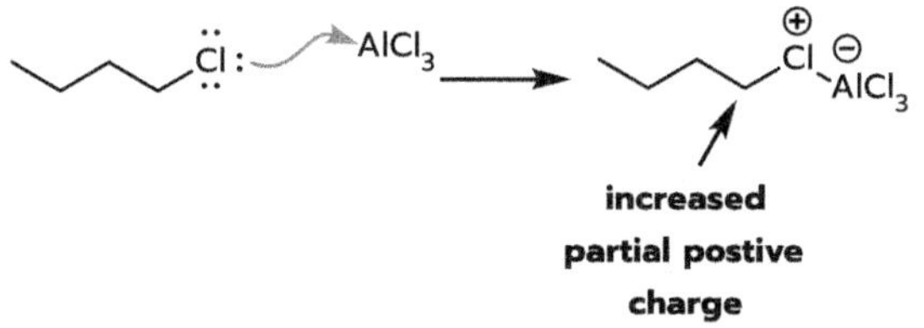

increased partial postive charge

Let's see how this intermediate will react with benzene.

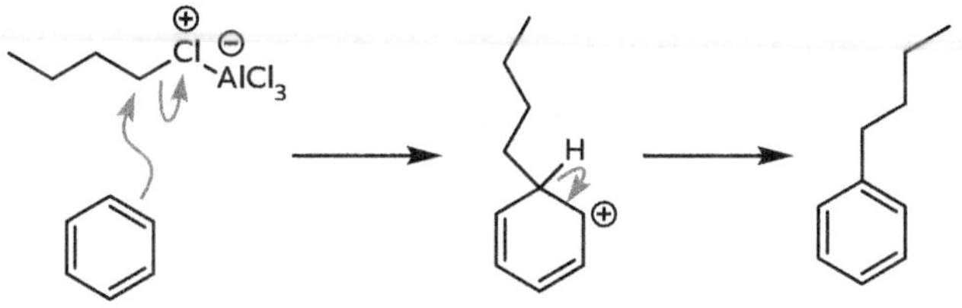

> We saw exactly the same thing with Br_2/Cl_2 with Lewis acids in **Reaction Detail 8**.

A CLOSER LOOK AT THE MECHANISM

We looked at S_N1 and S_N2 substitution in great detail in **HTSIOC**. Benzene is a nucleophile. In this case, we have a *primary* alkyl halide, so this reaction is S_N2 as far as the alkyl halide is concerned.

> As far as the benzene is concerned, it is an aromatic electrophilic substitution.

SECTION 2 THE CHEMISTRY OF ALKENES, BENZENE RINGS AND ALKYNES

If we used a *tertiary* alkyl halide, the mechanism (as far as the alkyl halide is concerned) would be S_N1.

> It would be S_N1 because a *tertiary* alkyl carbocation is 'stable enough'. Have another look at **Recap 2**.

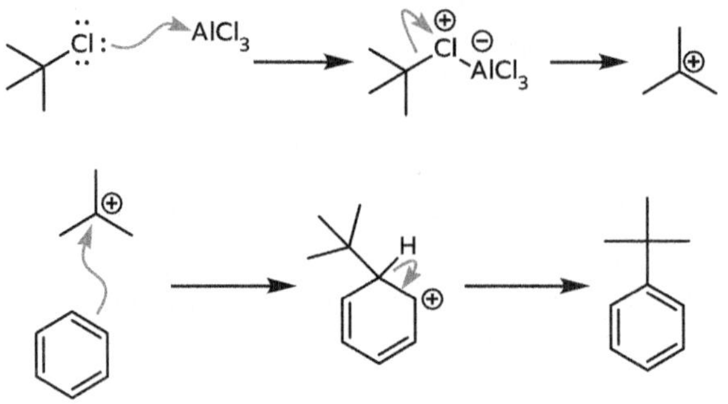

As always, when there's a change of mechanism, there's a good reason for it.

THERE'S A PROBLEM

We rarely, if ever, get useful amounts of products in which there is only a single alkylation of the benzene ring. Let's see why this is.

After 10% of the benzene has reacted, we will have 90% of benzene and 10% of butylbenzene in the flask. We have an electrophile.

> Which of the two will react faster with the remaining electrophile?

As always, we have a process to follow. We look at the energies of the starting materials and the intermediates.

> We invoke the Hammond Postulate (**Recap 2/Reaction Detail 8**) to justify talking about the energy of the intermediate rather than that of the transition state that precedes it.

We saw in **Recap 3** that more substituted alkenes are more stable, but they also tend to be more reactive.

> I'd suggest re-reading that chapter before carrying on with this one.

With aromatic compounds, it is common to simply focus on the fact that an alkyl group is 'electron-donating' and makes the benzene ring more electron-rich. While this is true, we can dig a bit deeper and find the same situation as we saw with alkenes.

Thermodynamically, butylbenzene is more stable than benzene. But it's a very small effect. The mechanism of stabilization is the same as we saw with alkenes—donation from a σ bond orbital into a π* orbital.

SECTION 2 THE CHEMISTRY OF ALKENES, BENZENE RINGS AND ALKYNES

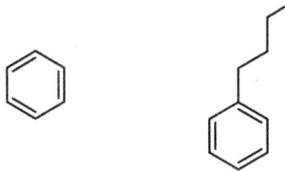

But what we really want to consider is the rate of reaction of each, and to do this we need to consider the activation energy. In order to do that, we will look at the intermediate that would be formed if each of these were to react with 1-chlorobutane as above.

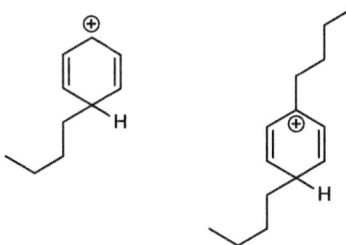

I have done something really important here. On the right, I drew the one resonance form in which it is easy to see the stabilization of the positive charge by the butyl group.

> How do you get to the point where you know which structures to draw to best demonstrate your point? This is a key question, and the answer is simple but unsatisfying. Draw all of the resonance forms out until you 'just know'.

There is something else I did. I drew *para* substitution. By recognizing that an alkyl group will stabilize a cationic intermediate, I implicitly recognized that it would be *ortho/para* directing.

> Draw the second attack of a butyl group onto butyl benzene at each of the three positions. Draw the resonance forms and consider their relative stability.

You are not doing anything that you did not already do in **Reaction Detail 8**.

Right, let's get back to the point. The amount of stabilization that a butyl group provides to the intermediate (and hence to the transition state that precedes it) will be far more than the butyl group provides to butylbenzene itself.

> Butylbenzene will undergo electrophilic substitution quite a lot faster than benzene itself.

If you react benzene with chlorobutane in the presence of aluminium chloride, you will get a mixture of alkylation products. This is unlikely to be a synthetically useful process.

> And that's only the start of the complications!

SECTION 2 THE CHEMISTRY OF ALKENES, BENZENE RINGS AND ALKYNES

THERE'S ANOTHER PROBLEM!

Here is a general outcome of the reaction we have been looking at. This is a reaction that has been carried out a great many times, and under a range of conditions. It gives a mixture of three products as shown below.

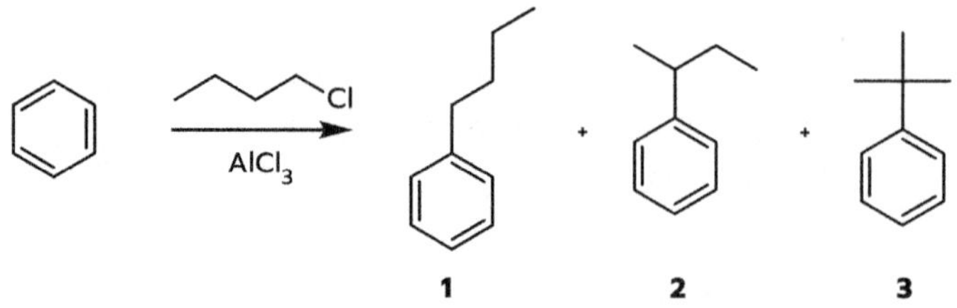

> Logically, there are only two ways this can happen.

Either we form compound **1**, and then it is converted into compounds **2** and **3** under the reaction conditions.

> This would be very easy to check, and it doesn't happen.

Or we somehow form the following two carbocations under the reaction conditions, and these also act as electrophiles.

> This is what is happening!

If we only consider products in which there has been a single alkylation, compound **2** is the major product.

Let's start with a simple reaction scheme to show what is happening. If we assume that we start with a *primary* carbocation, it can be converted into a *secondary* carbocation.

This is a rearrangement reaction. These are the subject of Section 5. We will see the curly arrow mechanism for this reaction in Fundamental Reaction Type 5.

There is no doubt that the *secondary* carbocation formed would be more stable than the *primary* carbocation drawn.

We have already seen that we won't actually form the *primary* carbocation, but we can still get the *secondary* carbocation from the *primary* alkyl chloride as shown below. I don't want to show you how we draw the curly arrows for this process just yet. They are a bit different, and we need to build patterns before we look at something that doesn't quite fit the pattern.

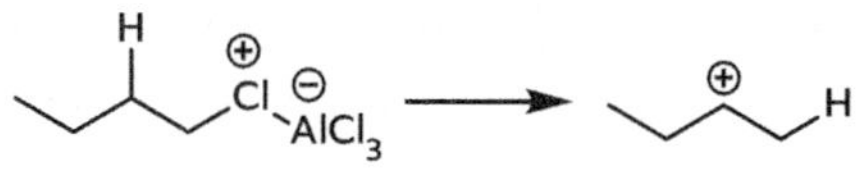

SECTION 2 THE CHEMISTRY OF ALKENES, BENZENE RINGS AND ALKYNES

This carbocation then reacts with benzene as shown below.

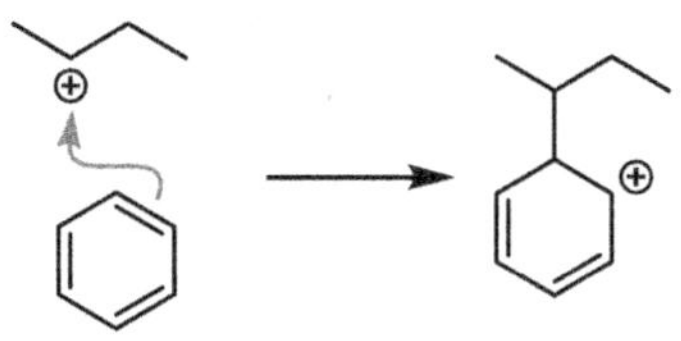

> Now draw the curly arrow for loss of a proton to give the aromatic product. Make sure you are losing the proton from the correct carbon atom.

Under some reaction conditions (and I don't want to give you too much detail to remember) you can get more of product **3** than product **2**. Here are the three possible butyl carbocations.

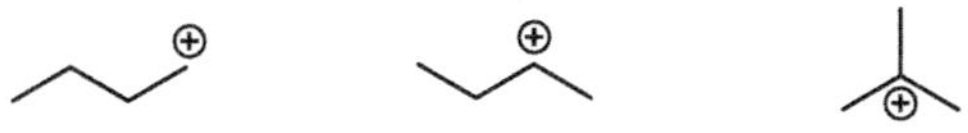

n-butyl   *sec*-butyl   *t*-butyl

For now, I want to keep things simple. We know (**Recap 2**) that the *tertiary*-butyl carbocation is more stable than the *secondary*-butyl carbocation. This further rearrangement reaction is not easy, but it can happen. We will come back to this in **Fundamental Reaction Type 5**.

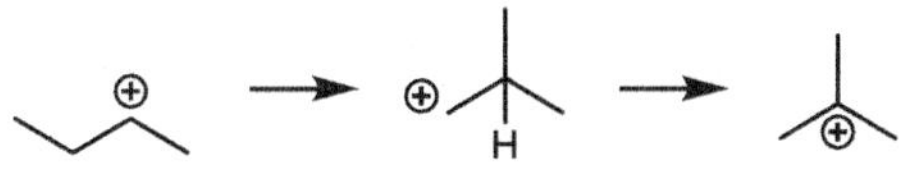

> Why doesn't this happen in S_N2 substitution reactions? After all, this is basically (as far as the alkyl halide is concerned) an S_N2 substitution.

The difference in this case is the bond polarization caused by the Lewis acid. In a 'normal' S_N2 substitution reaction, the alkyl halide is already attacking by the time we start to break the C–Cl bond. In that case, we are not building up more positive charge on the carbon atom, which starts the rearrangement.

> As usual, it's good to check when we have an apparent inconsistency.

SUMMARY OF THE PROBLEMS

In the previous section, we had quite a lot of detail, and quite a lot of application of principles. It is easy to lose sight of the overall message.

If we try to alkylate pretty much any aromatic compound, we will find it difficult to stop at one alkylation reaction. Even if we did, there is the possibility that we will obtain products in which the carbon electrophile has undergone rearrangement.

SECTION 2 THE CHEMISTRY OF ALKENES, BENZENE RINGS AND ALKYNES

> *This doesn't sound like a synthetically useful process.*

Don't lose sight of the bigger picture. You are studying organic chemistry because making molecules is important. As natural resources become scarcer, it is increasingly important that we can carry out a synthesis in the fewest possible number of steps, and with the highest levels of selectivity.

> *We want every molecule of starting material to be converted into the desired product.*

MOVING FORWARD

Forming a C–C bond to a benzene ring is a really useful process. You can bet that someone has worked out a way around these problems.

> *They have!*

The criteria are clear. We need a reagent/intermediate that will not undergo rearrangement, and we need the product to be less reactive than the starting material.

> *We will look at how this can be done once we have seen a bit more carbonyl chemistry. You'll have to wait for* **Reaction Detail 18**.

In fact, there are other ways to form C–C bonds to aromatic rings, and these are very widely used nowadays. We are going to look at a few examples of these in **Reaction Detail 10**.

For now, make sure you understand why the reaction in this chapter works, but make sure you also understand why it will give an horrendous mixture of products.

REACTION DETAIL 10

Cross-Coupling

MANAGING EXPECTATIONS

The major focus of this book is to help you develop your skills in drawing good reaction mechanisms, and to learn how to predict which of various possibilities will be the preferred pathway.

This chapter is an exception. We won't be able to draw a conventional curly arrow mechanism for the reactions, and because of that, you won't be able to predict which of the reactions is likely to work best.

There is a further consequence of this. As you try to learn these reactions, you will be relying on memory rather than understanding, and this will make your task harder.

> *If you increase cognitive load in one aspect, you must try to reduce cognitive load in other areas to compensate. The best way to do this is to better understand the chemistry we have covered so far.*

OVERVIEW—PALLADIUM CATALYSED CROSS-COUPLING

Transition metals are increasingly important as catalysts in organic chemistry. These reactions were developed because they allow us to do things that would otherwise be very difficult. They can give us selectivity (chemoselectivity, regioselectivity, stereoselectivity) that would be difficult to achieve in 'standard' organic transformations.

Recall that in aromatic electrophilic substitution reactions, an existing substituent determines where the next substituent goes (Reaction Detail 8). We may find that we cannot stop at one reaction (Reaction Detail 9). We may find the reaction is very slow indeed. And of course, the major limitation is that the thing we are reacting with *must* be an electrophile.

Palladium catalysed cross-coupling provides an alternative that does not have these limitations. Let's look at a simple example.

PhB(OH)$_2$ + PhI $\xrightarrow{\text{Pd(PPh}_3\text{)}_4\text{, Na}_2\text{CO}_3}{\text{DMF, H}_2\text{O}}$ Ph–Ph

99%

SECTION 2 THE CHEMISTRY OF ALKENES, BENZENE RINGS AND ALKYNES

This is a reaction known as a Suzuki coupling, named after its pioneer, Akira Suzuki. We have two reacting partners, benzeneboronic acid and iodobenzene. These are 'clipped together' in a catalytic reaction. A palladium catalyst is used.

We can make several points about this reaction. First, we have no regioselectivity issues. The carbon atoms bearing the substituents are joined together directly. Second, the yield is very high—almost perfect in fact. This is a particularly good example, but high yields are often obtained.

The third point is that this is a transformation that we would not have been able to do by aromatic electrophilic substitution.

> DMF is a solvent, dimethylformamide. I want you to get used to looking up structures when you encounter something you are not familiar with.

Here is an example with quite a bit more functionality. The aryl halide is a heterocycle (pyridine, Applications 9). It is also a bromide rather than an iodide. It still works. We have an ester and a nitro group present, and neither of these react.

$$MeO_2C\text{-pyridine-Br} + HO\text{-B(OH)-Ar-}NO_2 \xrightarrow[\text{Na}_2\text{CO}_3,\ H_2O\ \text{benzene}]{Pd(PPh_3)_4} MeO_2C\text{-pyridine-Ar-}NO_2$$

73%

In fact, we can couple together pretty much any pair of sp² hybridized carbon atoms. Couplings involving sp³ carbon atoms are less common, but they are now widely developed. The potential of the Suzuki coupling reaction is immense, so it is not surprising that Suzuki received a share of the 2010 Nobel Prize in Chemistry for this work.

> I keep mentioning Nobel Prizes, because they are the highest accolade, and so much of the organic chemistry you learn about was ground-breaking at first. The fact that it has become 'routine' tells us how important it is.

These reactions are not limited to boronic acids. Many organometallic compounds can be used as reacting partners, and there are almost as many 'named' cross-coupling reactions as there are metals.

Here is one more example, using tin. This is known as a Stille coupling, after Professor John K. Stille.

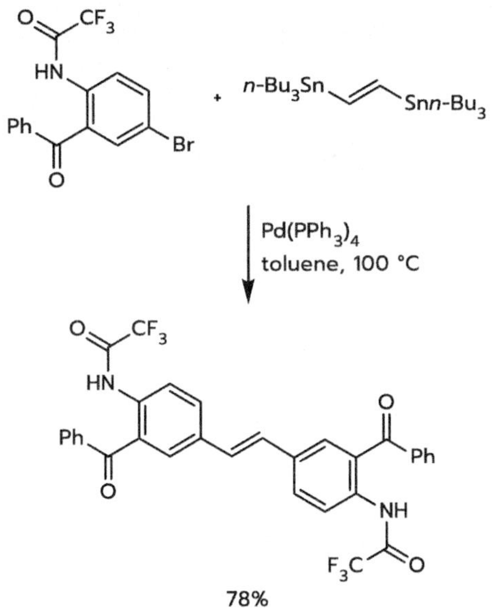

SECTION 2 THE CHEMISTRY OF ALKENES, BENZENE RINGS AND ALKYNES

In fact, in this example we have two Stille coupling reactions. You need to heat it a bit, but it works well.

SUMMARY

Remember why you are studying organic chemistry. Organic compounds are really important to life and society. We need to learn how to make them efficiently. If there are reactions that can be used to form chemical bonds in organic molecules, we need to know these reactions.

In this chapter, we have reactions that we do not draw curly arrow mechanisms for. When you are taught these reactions 'properly', you will see a catalytic cycle used. Only a tiny amount of the palladium catalyst is needed.

It would be difficult to imagine organic chemistry without transition-metal catalysed coupling reactions. Some transition-metal catalysed reactions allow us to do the same thing as 'normal' reactions, but 'better'. We will see this in **Stereochemistry 7** and **Stereochemistry 8**. In other cases, like this, we can accomplish transformations that are not possible using 'more traditional' reactions.

Don't try to learn this chemistry yet. Just be aware that it is there, and it is useful.

REACTION DETAIL 11

Structure and Addition Reactions of Alkynes

INTRODUCTION

An alkyne contains a carbon–carbon triple bond. We saw in **Recap 1** that the carbon atoms are sp hybridized, with a bond angle of 180°. The bond length is typically 1.2Å. This is shorter than a C–C double bond.

> *Always draw alkynes as linear! Don't make excuses or convince yourself that even though you have drawn them bent, you really actually do know that they are linear.*

A typical bond dissociation energy for a carbon–carbon triple bond is 954 kJ mol^{-1}. However, it is not just the C–C bond that is stronger. In particular, for terminal alkynes (those with a hydrogen directly attached to the triple bond), the C–H bond dissociation energy is in the region of 549 kJ mol^{-1}, which is much stronger than an sp^3 hybridized C–H bond (420 kJ mol^{-1}).

We will look at the chemistry of alkyne C–H bonds in more detail in **Reaction Detail 13**. For now, we will focus on the chemistry of the triple bond.

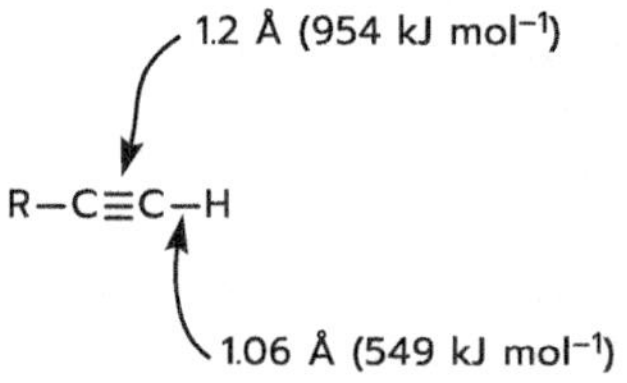

$$R-C\equiv C-H$$

1.2 Å (954 kJ mol^{-1})

1.06 Å (549 kJ mol^{-1})

The triple bond involves a σ-bond made up of the sp hybrid orbitals, leaving two p orbitals on each carbon to form the π bonds. The π bonds are therefore at 90° to one another as shown below.

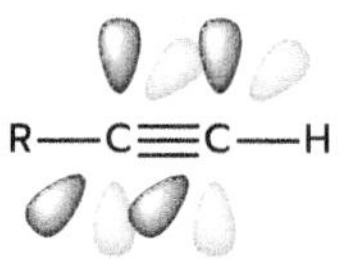

$$R-C\equiv C-H$$

This results in a cylinder of electron-density all around the bond. Basically, alkynes react very similarly to alkenes. They are very electron-rich, and addition of an electrophile can take place at either end of the bond. The stability of the cation that is left behind will determine the regiochemical outcome. This is exactly the same as reactions of alkene bonds, which makes it easier to learn, since there is only one thing to understand.

SECTION 2 THE CHEMISTRY OF ALKENES, BENZENE RINGS AND ALKYNES

> We are going to look at many of the same addition reactions that we saw for alkenes. The focus is to identify where the outcomes are the same, or where they are sometimes different but for good reasons.

ADDITION OF HYDROGEN HALIDES TO THE ALKYNE TRIPLE BOND

Let's look at the addition of HBr to an alkyne. As with the corresponding alkene, the bromine is added at the more substituted position and the hydrogen at the less substituted end.

The reasons are the same—we get a more substituted (more stable) carbocation intermediate.

> Note that in this case, the double bond does not stabilize the carbocation intermediate. The carbocation is not *allylic*. We saw allylic carbocations, and why they are stabilized, in **Recap 2**.

This is important stuff! You know that double bonds can stabilize carbocations. You see a double bond. You make a connection, but in this case it would be the wrong connection. Let's see what you can do about this.

> Draw the carbocation structure above, and try to draw resonance forms in which the electrons in the π bond stabilize the positive charge. You should not be able to do this!
>
> Now draw the molecular orbitals for the carbocation and the π bond. Assume the carbocation is represented by an empty p orbital, which would make the corresponding carbon atom sp hybridized. Can you draw any orbital overlap that is expected to lead to stabilization of the carbocation?

Now we've got that out of the way, there are two possible carbocations we could form. We form the one show above because it is more stable. Whenever we talk about 'more substituted = more stable', we always need to think about what is stabilizing the positive charge. In this case, it is a butyl group.

HYDRATION OF THE ALKYNE TRIPLE BOND

What if we add water across an alkyne bond? Well, we have seen this with alkenes, and we have looked at three distinct methods:

1. Direct addition of water catalysed by acid.
2. Oxymercuration.
3. Hydroboration.

SECTION 2 THE CHEMISTRY OF ALKENES, BENZENE RINGS AND ALKYNES

> Draw a mechanism for the 'simple' hydration of an alkyne with acid and water. Don't over-think this. And don't have a negative charge on any atom at any point, since this is being done under acidic conditions.

As we saw with alkenes, this doesn't tend to get used very often. The organic compound isn't very soluble in water. We can use co-solvents to help with the solubility, or we can consider alternatives such as oxymercuration.

There is one important difference between oxymercuration of alkenes and of alkynes. In the case of alkynes, we do not need to use a reducing agent to cleave the C–Hg bond.

> *It's okay to remember that we don't need a reducing agent. But the reasons for this should be apparent when we draw the mechanism.*

There is an additional useful consequence to this. We do not need to use mercury salts as stoichiometric reagents. We can use them catalytically. Don't worry about the reasons for this up front. Just start drawing something sensible.

Initially, we form a mercurinium ion **1** from the alkyne. This is the same as we saw in **Reaction Detail 5** for alkene oxymercuration, but it has one additional double bond. We saw the resonance forms of bromonium ions in **Reaction Detail 1**, and the ring-opening of mercurinium ions in **Reaction Detail 5**. We can now combine these, and we expect attack of water on intermediate **1/2**[34] to give structure **3** as shown.

> *I don't really like the way I have drawn resonance form 2, with two positive charges. The reality is that if we use mercury acetate as the mercury salt we will have acetate groups around. All we are doing here is keeping track of the overall charge.*

From this point, we can lose a proton from oxygen. This looks rather like an enol— because it is! It just happens to have mercury attached to the double bond. We

---

[34] Of course, since they are resonance forms, they both represent the same structure. We should not worry about which structure is being attacked.

know that enols are not as stable as the corresponding keto form, and we know that they can be converted into the keto form by tautomerization (**Fundamental Reaction Type 4**).

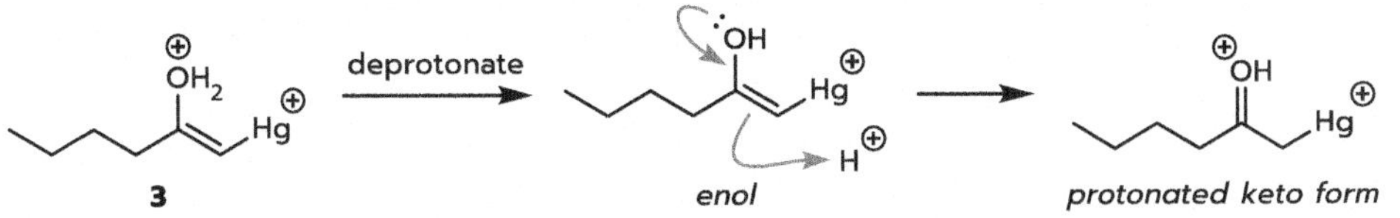

> There is just one small adjustment we will make. When we get to the protonated keto form, we can now lose the mercury as Hg^{2+}. This is what we started with, and the mercury (if we started with mercury acetate, we will regenerate mercury acetate—minor details!) is free to react with another molecule of the alkyne.

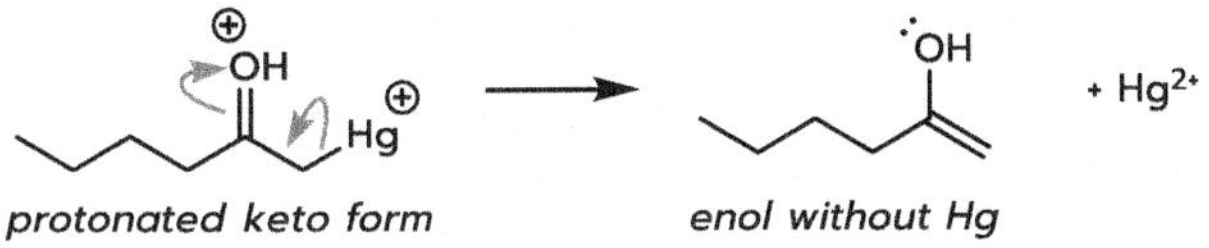

This gives us the enol, but without mercury. Once again, it can tautomerize to give the ketone.

In **Fundamental Reaction Type 4** we saw the formation of an enol from a keto form.

> Now draw the mechanism for the formation of the keto form from the enol. Make sure you do not have separation of charge (a ⊕ and a ⊖ in the same structure) or an overall negative charge at any point in your mechanism.

When we looked at oxymercuration of alkenes, we found that you add a hydrogen to one end of the alkene and an OH to the other end. With alkynes you add two hydrogen atoms to one end and an oxygen to the other end. This is the reality, but the difference is a consequence of the mechanism rather than being any more fundamental.

> It starts out the same, and then tautomerizes.

We also looked at hydroboration reactions of alkenes (**Reaction Detail 5**), and we saw that these proceed with the opposite regioselectivity to oxymercuration reactions. Since this is a result of an alkyl group stabilizing an intermediate or transition state, perhaps we should expect the same with alkynes.

Let's start with the mechanism, and then look at reasons for the differences compared to the reaction with alkenes.

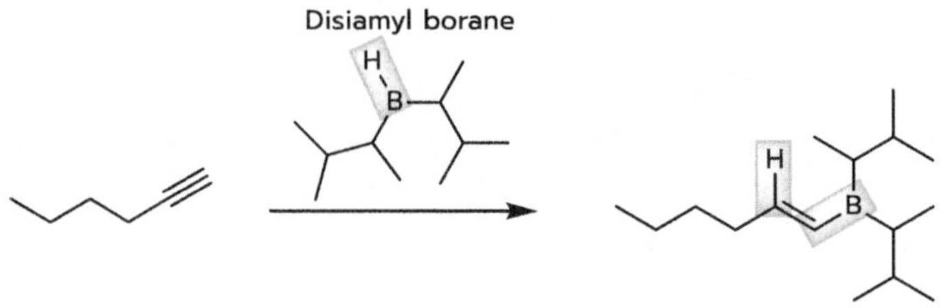

SECTION 2 THE CHEMISTRY OF ALKENES, BENZENE RINGS AND ALKYNES

The first step is the same, apart from the fact that we are left with a double bond. The other key difference is that we use a dialkylborane. This borane is quite bulky, and it is used for one very simple reason. It can hydroborate the alkyne, but not the resulting vinylborane.

> It only reacts once!

This is actually a very common tactic in synthesis, but it doesn't change the fundamental reaction. The priority is to understand the reaction. Then you can worry about the details, such as which reagent will give the higher yield.

We will eventually see the mechanism of oxidation of boranes in Reaction Detail 33. For now, as we did in Reaction Detail 5, accept that we replace the boron with an OH group.

> With an alkyne, this gives us an enol, and we know what enols do.

> Draw the keto-enol tautomerization mechanism out for this compound.

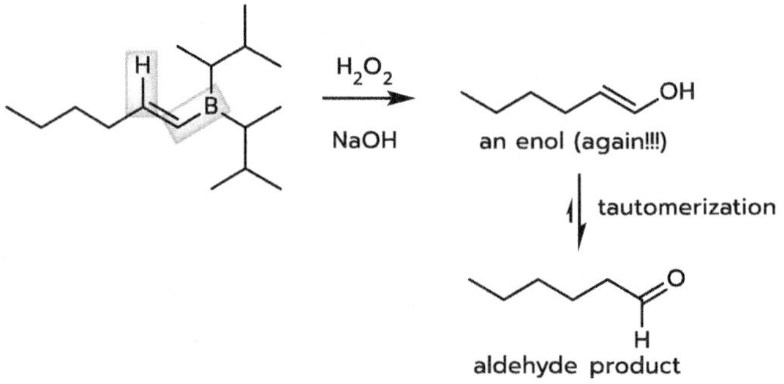

I know you've just done it for a different compound, but you should always take an opportunity to draw a mechanism and reinforce your learning.

> We have two hydration reactions of alkynes which closely parallel those of alkenes. You get the different regiochemical outcomes because of the different intermediates/transition states. The only difference between alkenes and alkynes is the oxidation state (Basics 3).

SECTION 2 THE CHEMISTRY OF ALKENES, BENZENE RINGS AND ALKYNES

STEREOCHEMISTRY 5

Conformers—Rotation of Bonds in Acyclic Systems

Here, we will recap material covered in HTSIOC Basics 30, but we will also add to it in preparation for Stereochemistry 6.

Recall that conformational isomers (conformers) are a special type of stereoisomer related by rotation around single bonds. They might be 'special' but the only organic compound I can think of without conformational isomers is methane—conformational isomers are everywhere!

CONFORMATIONAL ISOMERS OF BUTANE

Here, we are not going as far back as consideration of ethane. Let's start with butane and work from there. Here are the key conformers of butane.

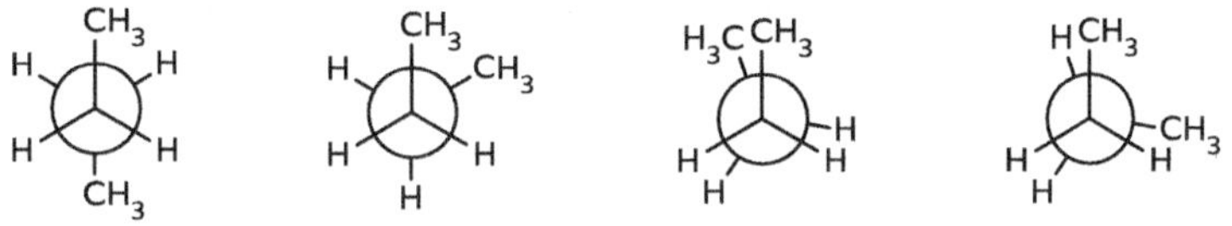

anti	gauche	Me-Me eclipsed	Me-H eclipsed
(0 kJ mol⁻¹)	(+3.8 kJ mol⁻¹)	(+19 kJ mol⁻¹)	(+16 kJ mol⁻¹)

These are Newman projections. Here is the *anti* conformer with the front atom shown in **bold**.

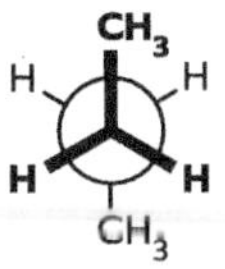

Newman projections are really useful. You need to practise using them until you are familiar with the connection.

For the two conformers I have indicated as 'eclipsed', the bonds line up exactly. It would be a very messy Newman projection drawn like that, so we accept the drawing above as an approximation.

The important point about conformers is that they have different energies.

> *To put it another way, as you rotate the central C–C bond in butane, the energy of the molecule changes.*

Here is a diagram showing the energy changes.

175

SECTION 2 THE CHEMISTRY OF ALKENES, BENZENE RINGS AND ALKYNES

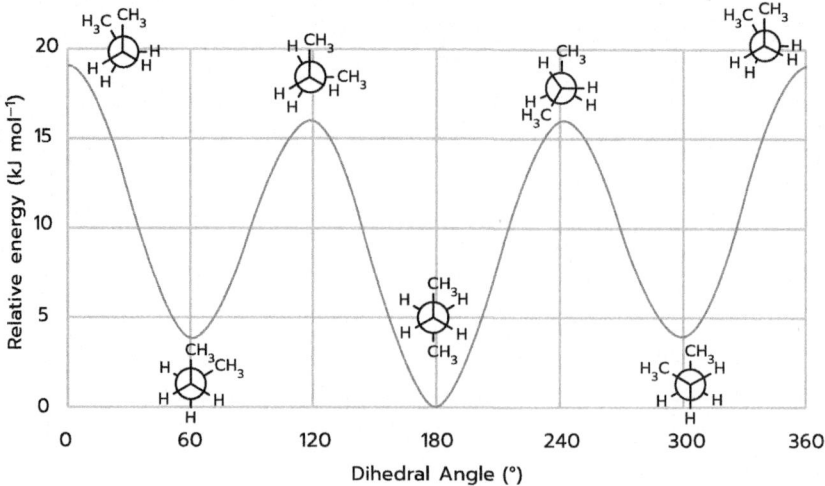

The **dihedral angle** I have indicated is the angle between substituents on **adjacent** carbon atoms—in this case between the methyl groups.

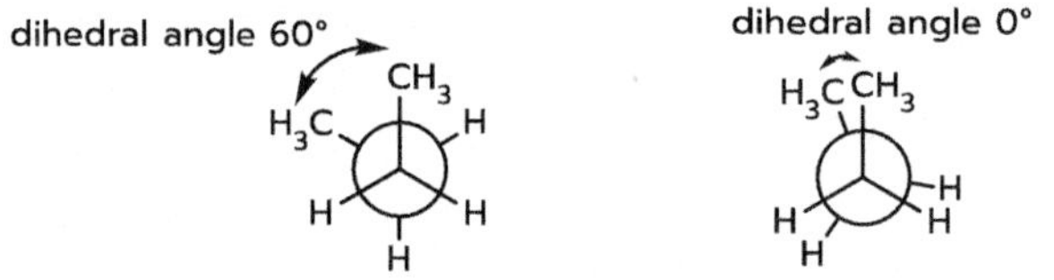

Organic chemists like to have names for everything. We describe the two staggered (not eclipsed) conformers as **anti** and **gauche**.

Overall, the *anti* isomer is the most stable conformer. On the diagram above, we define this as having an energy of 0 kJ mol^{-1}. The *gauche* conformer is approximately 4 kJ mol^{-1} less stable.

> These two conformers do not have any bonds that are eclipsed.

The remaining two conformers **do** have eclipsed bonds. We have a conformer in which the methyl group is eclipsed with a hydrogen atom. This conformer is 16 kJ mol^{-1} less stable than the *anti* conformer. Finally, we have a conformer in which the two methyl groups are eclipsed. This conformer has a relative energy of +19 kJ mol^{-1}. As the bonds can rotate by any arbitrary amount, there are an infinite number of conformers in between.

Quite simply, the methyl-methyl eclipsed conformer of butane is unstable because there is a steric 'clash' between the two methyl groups.

> Now you can make a prediction. If we had bigger groups, the energy difference would be bigger.

Where the eclipsing atoms are hydrogen (small!) the significant interaction is electrostatic repulsion between the electrons in the bonds. We call this interaction **'torsional strain'**.

SECTION 2 THE CHEMISTRY OF ALKENES, BENZENE RINGS AND ALKYNES

> *If the worst mistake you ever make in organic chemistry is considering the hydrogen-hydrogen eclipsing to be a steric effect, you will still be fine!*

A molecule will spend more time in a lower energy conformer, so that the lowest energy conformer is likely to be the one that reacts.

CONFORMERS IN ALLYLIC SYSTEMS—A^{1,2} AND A^{1,3} STRAIN

In order to get ready for Stereochemistry 6, let's have a look at the following compound.

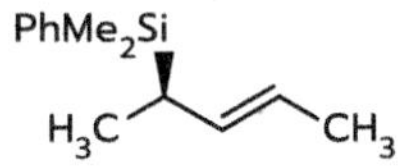

> *The key question is 'when we epoxidize this alkene with m-CPBA, do we epoxidize the top face or the bottom face?'*

The reasoning commonly given is that 'the PhMe$_2$Si group is big[35] and is blocking the top face of the alkene, so we will get epoxidation on the bottom face.[36]

The problem, in the most general sense, is that as soon as you start rotating around the bond highlighted below, the PhMe$_2$Si group is not necessarily on the top face any more.

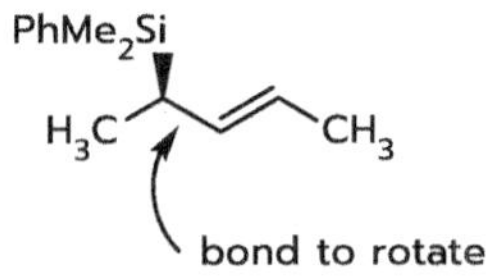

> *This would be a good time to get your molecular model kit out and prove it!*

What we need to do is look at all possible conformers, and determine which is the most stable (or least unstable!). Of course, the key difference between this compound and butane is that we have an sp^2 carbon on one end of the bond.

We will deal fully with this compound in Stereochemistry 6. What I want to do here is look at two 'simpler' examples so that we can introduce the terminology and look at the important conformers. Here is the first structure. Again, I have indicated the bond we are looking at.

---

[35] This is true!

[36] This is, unfortunately, also true. I say 'unfortunately' because it promotes an incomplete solution to the problem.

SECTION 2 THE CHEMISTRY OF ALKENES, BENZENE RINGS AND ALKYNES

We can calculate the energy changes as we rotate the bond, exactly as we have just done for butane. Here is the profile. We have a total of four **different** conformers to discuss, these being the maxima and minima.

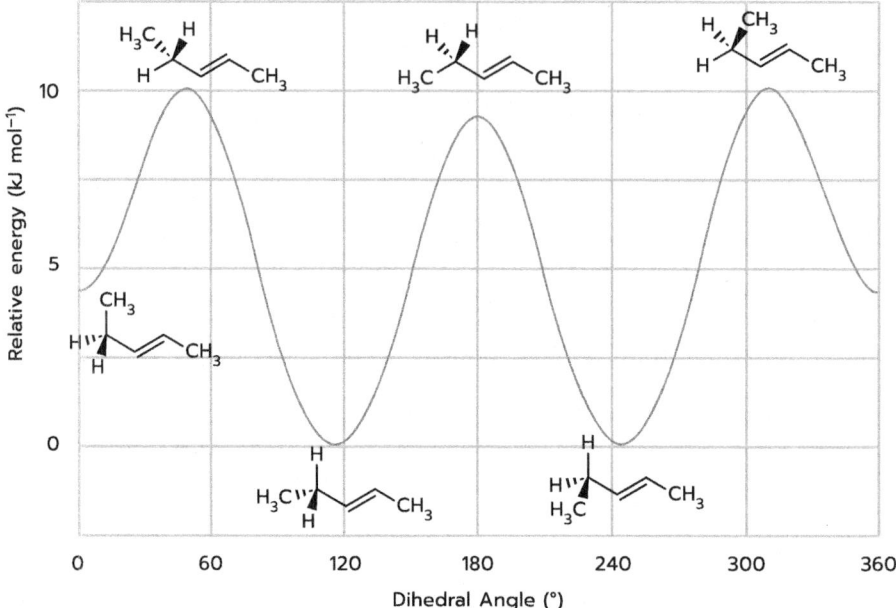

Here is the lowest energy conformer, **A**. I have highlighted a key interaction between two hydrogen atoms that are quite close together.

A

This conformer has a hydrogen atom on the allylic carbon eclipsed with the double bond, and its associated hydrogen atom.

> 'Allylic' simply means 'the carbon next to the double bond'. We saw allyl cations in Recap 2.

We refer to this interaction between the two hydrogen atoms as 1,3-allylic strain or $A^{1,3}$ strain. Let's reiterate, the lowest energy conformer **A** has the smallest substituent on the allylic carbon eclipsed with the double bond.

SECTION 2 THE CHEMISTRY OF ALKENES, BENZENE RINGS AND ALKYNES

> *This is what we need to look for! In this case, the dihedral angle between the methyl group and the alkene bond is 120°. Because the allylic carbon has two hydrogen atoms attached, we have an identical energy conformer at 240°.*

Here is the structure again with a different two hydrogen atoms indicated.

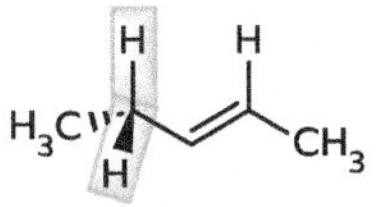

A

> What is the stereochemical relationship between these two hydrogen atoms? You will need to read ahead to **Stereochemistry 11**.

The next conformer is **B**. It is about 4 kJ mol^{-1} less stable than **D**. Here, we have a methyl group eclipsed with the double bond, and close to the hydrogen atom.

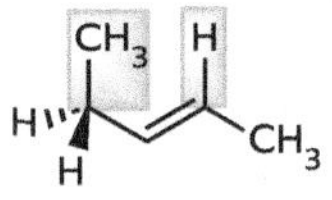

B

We then have two 'maxima' on the energy profile, conformers **C** and **D**. I've highlighted a different interaction now.

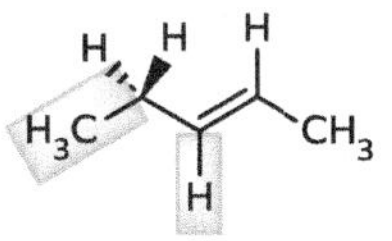

C 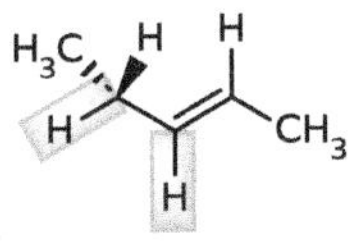 **D**

This interaction represents eclipsing between the substituent (H or CH$_3$) on the allylic carbon and the C–H bond on the adjacent carbon atom. We call this 1,2-allylic strain or A^{1,2} strain. You might be surprised that conformer **D** is less stable than conformer **C**.

> *Don't read too much into this—the difference in energy between the two conformers is tiny.*

Having said that, in conformer **D**, the CH$_3$ group is getting quite close to the other H on the alkene. There is an energetic penalty associated with this.

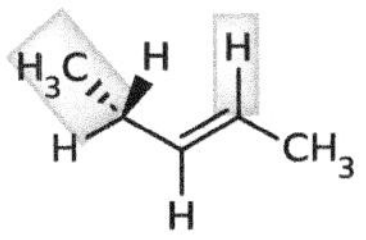

D

SECTION 2 THE CHEMISTRY OF ALKENES, BENZENE RINGS AND ALKYNES

INTERLUDE

We have defined $A^{1,3}$ and $A^{1,2}$ strain. It definitely helps to have a 'label' for a type of interaction.

We have then identified that the lowest energy conformers tend to have eclipsing between the allylic bonds and the double bond.

> *We have the $A^{1,3}$ interaction. When you try to find the lowest energy conformer, you should look for the conformer where this interaction is smallest.*

Hopefully this gives you a plan.

ANOTHER EXAMPLE

I'm going to give you another example, which shows that the key point in the above interlude is true, but the energy profile can change. Here is a compound in which the double bond geometry has been changed to *cis*. The previous example was *trans*.

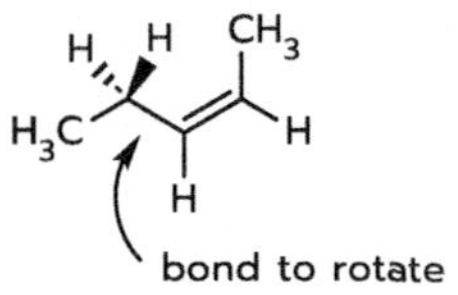

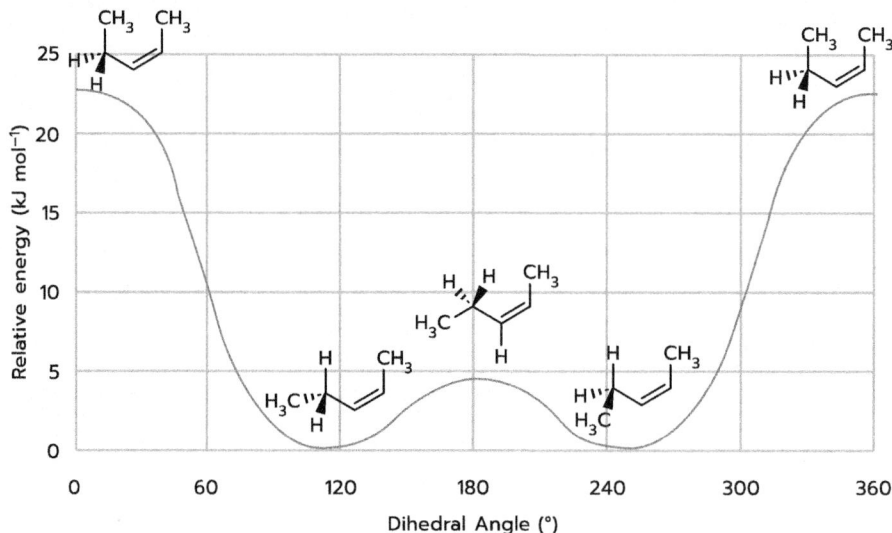

As before, I have calculated the energy of a range of conformers of this compound. Here is a graph, with several 'extreme' conformers drawn on.

Let's get the first point out of the way. Conformer **E**, which corresponds to conformer **A** in the previous example, is the lowest in energy. As before, because we have two hydrogen atoms on the allylic carbon, there are two of these conformers (approximately 120° and 240° on the above diagram).

180

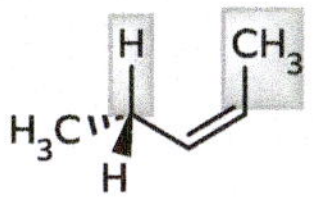

SECTION 2 THE CHEMISTRY OF ALKENES, BENZENE RINGS AND ALKYNES

E

In the middle, we have a local maximum, conformer **F**. In this case, the CH_3 on the allylic carbon eclipses a H on the adjacent carbon of the alkene—an $A^{1,2}$ interaction.

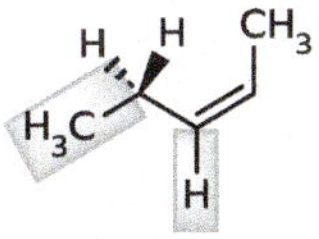

F

There is only a very small energy difference between conformers **E** and **F**. Conformer **F** corresponds to conformer **C** in the previous example. Conformer **E** is destabilized by the $A^{1,3}$ interaction, reducing the energy difference between conformers **E** and **F** compared to **A** and **C**.

The final two conformers we can consider are **G** and **H**, corresponding to conformers **B** and **D** in the previous example.

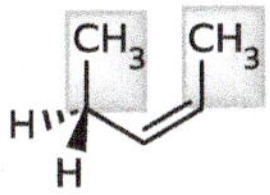

G 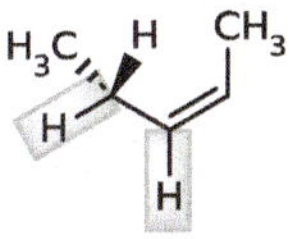 **H**

Conformer **G** is now the maximum energy, whereas conformer **B** in the previous example was a minimum (but not the global minimum). Here, the $A^{1,3}$ clash of two methyl groups is energetically very significant, and corresponds to about 22 kJ mol^{-1} compared to the most stable conformer, **E**.

It then turns out that conformer **H** is neither a maximum nor a minimum.

> How can you tell, with absolute confidence, whether any conformer you draw will be a minimum or a maximum?

Honestly, you cannot always tell for sure. What you can tell is that conformer **G** will be significantly higher in energy than conformer **E**. There *might* be another conformer that is even higher in energy, but the methyl-methyl interaction here does look like the 'worst case scenario'.

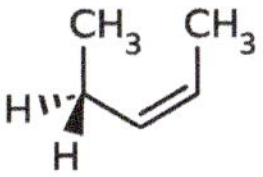

G 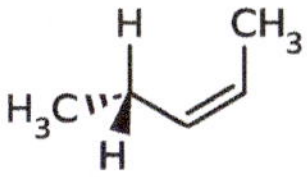 **E**

It is not unreasonable to speculate that conformer **G** could be a maximum, but what we really need to be able to do is identify the minimum energy conformer.

181

As we have seen, this was something we could be confident about.

THE POINT OF ALL THIS

The key point is than in many cases, you must identify the lowest energy conformer of a given structure. On average, the compound will spend more time in this conformer.

We are going to see, in **Stereochemistry 6**, that if we can identify the lowest energy conformer of a compound of this type, we can predict the stereochemical outcome of epoxidation reactions.

In this chapter, we have spent some time 'looking at structures'. By doing this, you will get better at looking at structures and working out which interactions are favourable/unfavourable. We started by looking at the following alkene.

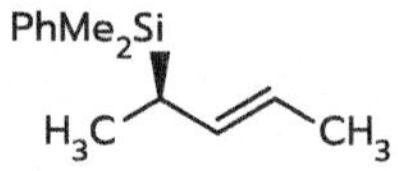

$$PhMe_2Si-CH(CH_3)-CH=CH-CH_3$$

Now, instead of looking at this structure *as drawn* and assuming that epoxidation will take place from the face opposite the bulky silyl substituent, you will rotate bonds, and try to identify the most stable conformer. We are going to do this in the next chapter.

This will take time and practice. All I can try to do is convince you to work towards this. It will be difficult at first, and then it will get easier. Eventually, it will become your default way of thinking about a problem, and you will do it quickly and correctly every time.

SECTION 2 THE CHEMISTRY OF ALKENES, BENZENE RINGS AND ALKYNES

STEREOCHEMISTRY 6

Diastereoselective Epoxidation of Acyclic Alkenes

LOOKING AT A REACTION

Here is the reaction we started looking at in the last chapter. In fact, we are going to start with the *cis* double bond isomer **1**. This is because epoxidation of this compound is highly stereoselective for attack on the lower face of the alkene.[37] It's always good to look at a process with high selectivity first.

>95:<5 ratio **2**:**3**

What we will do now is reiterate the key points from the previous chapter, and we will add a little refinement. In **Stereochemistry 5** we looked at the conformers of a simplified compound. Now let's look at compound **1**. Before we do this, let's make it really clear **why** we are doing this. If a compound can exist in two different conformers, it will spend more time in the lower energy conformer. In many cases (we will see exceptions to this in **Basics 8**), this means that the lowest energy conformer is the one that leads to the product.

> So, if we can work out which face of the alkene the PhMe₂Si group is blocking *in the lowest energy conformer*, we will know which face the *m*-CPBA will attack.

Remember, we are looking for the conformer with the lowest $A^{1,3}$ strain. Recall that $A^{1,3}$ strain is a steric interaction between a substituent on the stereogenic centre and the methyl group on the double bond. We can identify three conformers, **A**, **B**, and **C**.

If you refer back to the comparable structures in **Stereochemistry 5**, you could safely conclude that conformer **A** is the best. Conformer **B** looks pretty bad. Conformer **C** looks horrific.

---

[37] By this, I mean "the lower face, as it is currently drawn"—there is no absolute lower face.

183

SECTION 2 THE CHEMISTRY OF ALKENES, BENZENE RINGS AND ALKYNES

A BRIEF DIGRESSION

We have just dismissed conformer **C**. This is because we have a 1,3 steric clash between a PhMe₂Si group and a methyl group. To give you a bit more context, it would be like having the following cyclohexane.

This cyclohexane has two substituents on the same side of the molecule in an axial position. It would be massively unstable.

> If you find it difficult to assess the stability of conformer **C**, find something similar that you do know.

The cyclohexane chair would flip to put both substituents equatorial.

> Reach for the molecular models and try it! Then draw it! While you are doing this, assign the stereochemistry (R or S) of both stereogenic centres, and make sure you have the same absolute stereochemistry after you have flipped the chair form.

A C–C bond in conformer **C** would rotate to move the PhMe₂Si group out of the way. It is basically doing the same thing for the same reasons.

BACK TO THE STORY

Conformer **A** has the bulky silyl group blocking the top face. Epoxidation takes place preferentially from the bottom face. We are probably done!

But just for completeness, let's check whether there might be any other conformers that could be of low enough energy to be important. Consider conformers **D**, **E**, and **F**.

In each case, there is a substituent (indicated) eclipsed with a C–H bond on an adjacent carbon atom. Recall that this is $A^{1,2}$ strain. In **Stereochemistry 5**, we had a conformer corresponding to **D** which was quite stable.

But in this case, conformer **D** has the silyl group close to the methyl group on the alkene.[38]

---

[38] I want to make another point about conformer **D**. It looks as though the PhMe₂Si group in conformer **D** is closer to the methyl group (on the double bond) than it is in

184

> That cannot be good.

Conformer **F** is almost certainly worse. This leaves conformer **E** with the silyl group eclipsing the adjacent C–H bond.

I will be honest. I would struggle to put conformers **D**, **E**, and **F** into rank order.

> They are all really bad.

We looked at them. We considered them. We dismissed them as not having any meaningful contribution.

> It looks like we were indeed done as soon as we looked at conformer A, but it didn't take too long to check.

MAKING A PREDICTION!

Which epoxide will be the major diastereoisomer formed from compound **4**. Can we predict what level of stereoselectivity will be obtained in the epoxidation of this compound?

PhMe₂Si / H₃C — CH=CH — CH₃
4

> It's not going to be 'better' than compound 1, but it might be as good as. Or we might predict lower stereoselectivity, or even a different preferred epoxide stereoisomer. Let's find out!

Once again, we will assume that we get more product formed from the most stable conformer, and that epoxidation will preferentially take place on the face opposite the PhMe₂Si group in this conformer. We have reduced the problem to 'identify the lowest energy conformer of compound **4**'.

The only difference between compounds **1** and **4** is double bond geometry. In **Stereochemistry 5**, we looked at the following compound, now drawn in (one of) its lowest energy conformers.

H₃C — CH(H)(H) — CH=CH — CH₃

---

conformer **F**. So close, in fact, that I had to abbreviate it to SiR₃ to fit it in. This is only the case because of how the structure is drawn. In fact, the PhMe₂Si group is the same distance from the methyl group in conformer **D** and conformer **F**. Check this with your molecular models!

SECTION 2 THE CHEMISTRY OF ALKENES, BENZENE RINGS AND ALKYNES

Clearly, moving the CH₃ group on the alkene into the *trans* position considerably reduces the A^{1,3} strain in **all** conformers, and we can now apply what we learned from this example to the compound at hand. Here are the conformers.

G   H   I

Here, I have drawn the H on the double bond explicitly.

> We need to do this, because we are looking for A^{1,3} strain. Deciding which bits of a molecule to draw at any given time is an important key skill.

Based on the discussion in **Stereochemistry 5**, we expect conformer **G** to be the lowest energy. It's pretty safe to say that anything with PhMe₂Si eclipsed (conformer **I**) will be **bad**!

So where does that leave us with conformer **H**? Here are two conformers from **Stereochemistry 5**. Their calculated energy difference was only 4 kJ mol⁻¹.

Conformer **H** looks a little worse than the conformer on the right, as there is the extra silyl group. On the other hand, conformer **G** looks worse than the conformer on the left as the CH₃ and silyl groups are close to the hydrogen on the adjacent carbon atom of the alkene.

On balance, it's looking pretty close. You would expect conformer **G** to be more stable, and undergo epoxidation of the lower face, and you would expect conformer **H** to be marginally less stable and to undergo epoxidation from the upper face.

> So, we have a prediction about which epoxide stereoisomer will be the major one, but we don't expect to get a very high level of stereoselectivity.

Here is the actual reaction.

4 →(*m*-CPBA) 5 + 6

61:39 ratio **5**:**6**

A 61:39 ratio is not very selective at all. Our prediction was good.

DIRECTED EPOXIDATION

In **Stereochemistry 4**, we looked at an epoxidation reaction in which hydrogen bonding between an OH group and the reagent determined the stereochemical

SECTION 2 THE CHEMISTRY OF ALKENES, BENZENE RINGS AND ALKYNES

course of the reaction. This was for a cyclic alkene, but there is no reason not to apply it to acyclic alkenes.

Let's have a look at epoxidation of compound **7**.

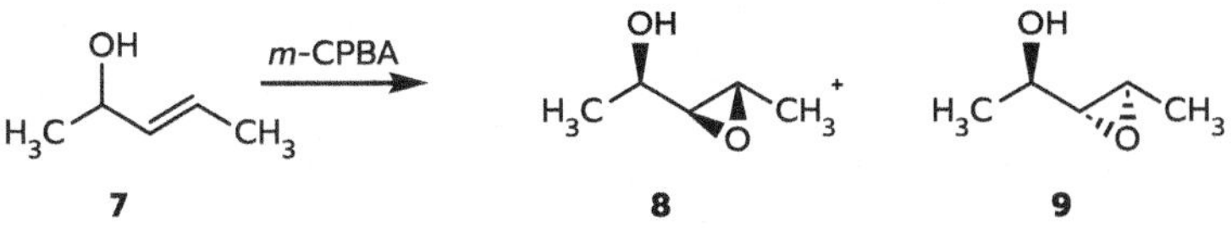

64:36 ratio **8:9**

It doesn't matter that we have the OH group. We start the analysis in the same way. We draw three conformers.

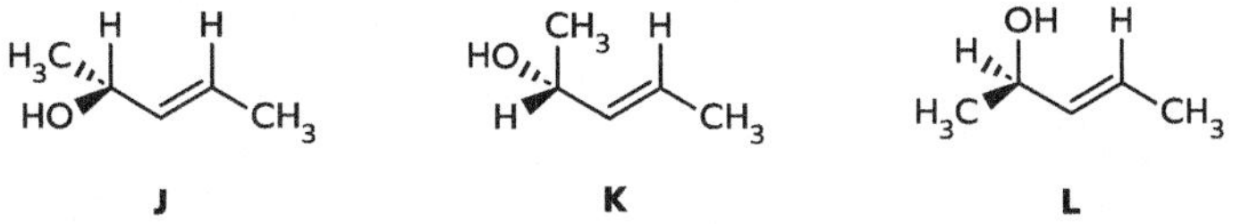

Let's address one question first.

> Which conformer will predominate between K and L?

I have two answers to this question. Firstly, if we are considering hydrogen bonding of *m*-CPBA to the OH group directing the epoxidation, hydrogen bonding anything to the OH group will make it bigger. Also, and more simply, in conformer **L**, the OH group is not on the top or on the bottom, so it doesn't really matter.

> We can stop worrying about conformer L.

Conformer **J** is more stable. Hydrogen bonding of *m*-CPBA will lead to epoxidation of the top face.[39] This should predominate, and it does!

> The 64:36 ratio here is very similar to the 61:39 ratio from compound 4. We would expect the ratio of conformers G:H and that of conformers J:K to be broadly similar. This further supports the hypothesis that the ratio of products is largely determined by the ratio of conformers.

ONE MORE DETAIL

Strictly speaking, we shouldn't be comparing the energies of conformers of the alkene starting material. We get the rate of reaction from the activation energy, so we should really be comparing the energies of the different transition states for reaction. Here is a schematic transition state for the epoxidation of compound **1**.

---

[39] As drawn! It doesn't hurt to ~~nag~~ remind you.

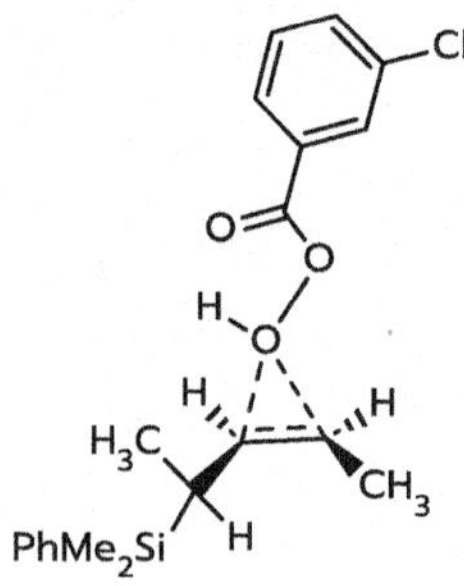

SECTION 2 THE CHEMISTRY OF ALKENES, BENZENE RINGS AND ALKYNES

It turns out that we could rotate the same bond in the transition state, and we would reach the same conclusions.

> It is simpler to do this in the alkene starting material, and it gives us broadly the same information.

As an added insight, it is possible to calculate the energy of the transition states for epoxidation of each face for each conformer. From this, we could then calculate how much of each stereoisomer we expect to form. It can be done, but it's not trivial.

CONCLUSION

We have looked at epoxidation of various alkenes with *m*-CPBA, and we have identified minimization of allylic 1,3-strain as a strong driving force for determining the stereoselectivity.

Bear in mind that there are many reagents that can be used for carrying out epoxidation reactions. When you encounter a different reagent, you can apply the processes in this chapter to determine the stereochemical outcome of a given reaction. You must identify what interactions are possible for the reagent.

Similarly, the principles in this chapter can be applied to a whole range of different reactions, and we will see some of them over the next few chapters. Make sure you fully understand this chapter before you proceed.

SECTION 2 THE CHEMISTRY OF ALKENES, BENZENE RINGS AND ALKYNES

STEREOCHEMISTRY 7

Catalytic Asymmetric Alkene Epoxidation

INTRODUCTION

An enantioselective reaction is one that produces two enantiomeric products, but produces more of one of them than the other.

> *The most important thing is to understand when a reaction can be enantioselective, and when it cannot.*

In this chapter, we are going to cover the fundamental principles, and we will then apply them to one reaction—the enantioselective epoxidation of alkenes.

THE FUNDAMENTAL PRINCIPLES

Have another look at Stereochemistry 2. There, we saw that if a reaction produces two enantiomeric products, with equal energy, it will inevitably produce these enantiomers as a racemic (1:1) mixture of the two enantiomers.

> *That's all well and good, but the two enantiomers will behave differently in an inherently chiral biological system. We need to have a way to make one enantiomer selectively over the other enantiomer.*

We need to see how this might happen. Let's stick with the previous example of enantioselective epoxidation. We are going to choose a different example, as it will lead into the next section, but the principles won't change. Here is the example.

This epoxide product has two stereogenic centres. They both have *R* stereochemistry.

> Don't just trust me. Check! Take every opportunity to practice.

Strictly speaking, we should refer to this as the (*R*,*R*) epoxide. Epoxidation of the lower face would give the (*S*,*S*) epoxide. We could never get the (*R*,*S*) or (*S*,*R*) epoxides from this particular *trans* alkene, so let's keep it simple and just use one stereochemical descriptor – *R*.

Here is the reaction profile from Stereochemistry 2 again. Remember that the *R* epoxide and the *S* epoxide have the same energy. If we use *m*-CPBA as the reagent, the transition states for production of the two epoxides will also be mirror images with the same energy.

SECTION 2 THE CHEMISTRY OF ALKENES, BENZENE RINGS AND ALKYNES

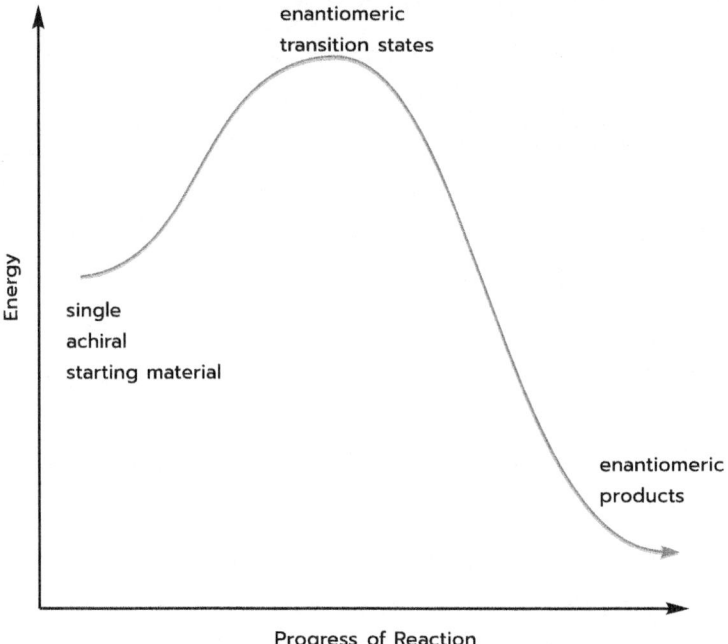

> To put this another way, attack on the top face of the alkene and the bottom face of the alkene will give enantiomeric pathways along the entire reaction coordinate. This can never be enantioselective.

Whatever else happens, we have a single achiral starting material with a single energy. We have two possible enantiomeric products with the same energy. There is only one thing that we could possibly change. To make this reaction selective, we must make the rate of formation of the two enantiomers different, which means making the energies of the transition states leading to those enantiomers different.

> The reaction profile needs to look like this!

SECTION 2 THE CHEMISTRY OF ALKENES, BENZENE RINGS AND ALKYNES

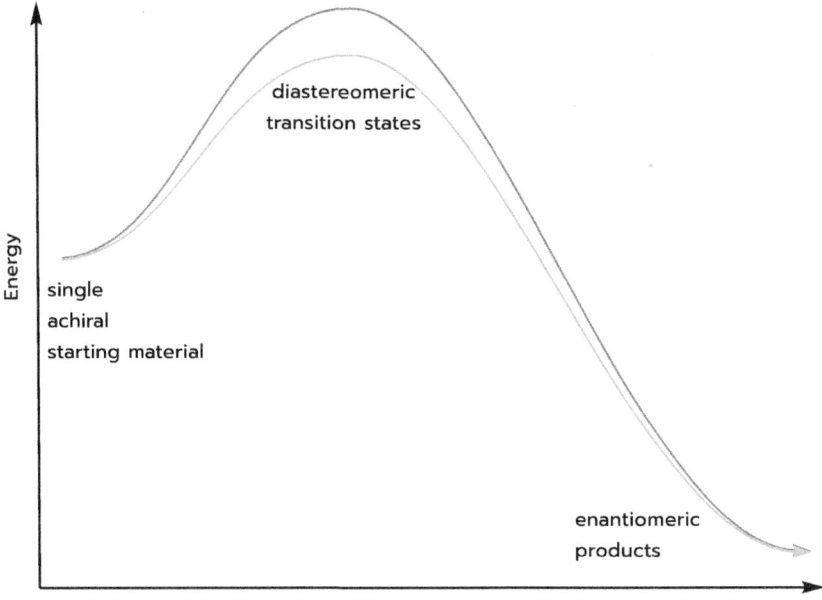

> How do we make the transition state energies different?

I've already given the game away on the profile. We only have one type of stereoisomer with different energy. The transition states need to be diastereomeric. In order to have this situation, we need to add some external chirality. Let's look at this schematically, and then we will move on to look at how this reaction is actually carried out.

We add a catalyst and an oxidizing agent. The catalyst has two purposes. It creates a chiral environment, and it delivers an oxygen atom to the substrate.

Let's assume that for this transformation we need a catalyst with S stereochemistry. As the catalyst delivers the oxygen, we have two possible transition states. Transition state **2** leads to the R epoxide (once again, we are simplifying the stereochemical description). We could describe this as the SR transition state.

SECTION 2 THE CHEMISTRY OF ALKENES, BENZENE RINGS AND ALKYNES

Transition state **3** leads to the *S* epoxide. We could describe this as the *SS* transition state.

> *The SR transition state and the SS transition state are diastereoisomers. They have different energy. We have a rational basis on which to base an enantioselective reaction.*

This will give us the reaction profile on the previous page.

Finally, the catalyst departs, leaving the oxygen atom behind. It can then take part in another epoxidation reaction, which is why it is classed as a catalyst.

> *Catalytic enantioselective reactions are wonderful, because they lead to the production of large amounts of enantiomerically-enriched products from small amounts of chiral catalyst.*

The scheme shown above is very general. We could draw something similar for any of the enantioselective reactions we are going to look at over the next couple of chapters.

INTERLUDE 1—'UNITS' OF STEREOCHEMISTRY

In the above discussion, we talked about the epoxide as a single stereochemical unit with one stereochemical descriptor. This makes the discussion simpler. If we start with the *trans* alkene, we are only ever going to get the *trans* epoxide shown below, or its mirror image.

We cannot possibly get the *cis* epoxide.

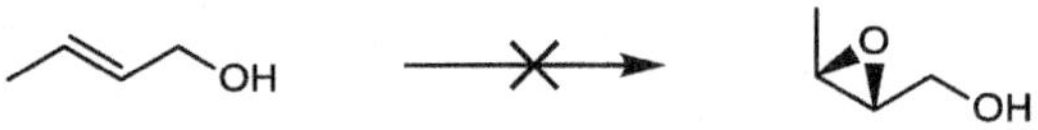

It would be possible to talk about both stereogenic centres in the epoxide, and all stereogenic centres in the chiral catalyst, but it won't add anything to the discussion.

> *It's absolutely fine to think in terms of the two enantiomers of the product that could be formed, and in terms of the catalyst, as a single 'unit' of stereochemistry.*

That is, although the epoxide has two stereogenic centres, we do not need to consider them independently, because if we know the stereochemistry of one centre, we automatically know the stereochemistry of the other. You can do the same with chiral catalysts that have multiple stereogenic centres. As we will see, we only really care about which enantiomer we have, so we don't need to 'micromanage' each individual stereogenic centre.

> *I haven't seen the term 'unit of stereochemistry' used elsewhere, but I find this approach convenient, because it allows me to focus on the principles.*

SECTION 2 THE CHEMISTRY OF ALKENES, BENZENE RINGS AND ALKYNES

INTERLUDE 2—ENANTIOMERIC EXCESS

We need a way to 'keep score'. The scientists who invent enantioselective reactions are constantly striving for better and better levels of stereoselectivity.

> *You could simply report a yield, but that wouldn't give much information. You could report a selectivity in the form "x:1". That would be okay, and stereoselectivity has often been reported in this way.*

There is no absolute standard method for reporting the outcome of stereoselective reactions, but "enantiomeric excess" is as near to a standard as we have. Let's see how this works.

Suppose we have a reaction that produces an 85:15 mixture of enantiomers.

> *The ratio in this case is reported as a percentage.*

There is 70% more of one enantiomer than the other (85 − 15 = 70). Therefore, an 85:15 ratio corresponds to 70% enantiomeric excess. The measure is useful. As we have seen in **HTSIOC Basics 29**, a single enantiomer of a chiral compound can rotate plane polarized light. A racemic mixture (50:50 ratio of enantiomers) does not rotate plane polarized light. The two enantiomers 'cancel each other out'.

> *A 50:50 mixture has 0% enantiomeric excess.*

A single enantiomer (100:0 ratio, 100% enantiomeric excess) has the maximum possible rotation of plane polarized light (optical rotation) for the compound. In most cases, the optical rotation is proportional to the enantiomeric excess, so that measurement of optical rotation allows us to directly calculate the enantiomeric excess.[40]

> *We tend to abbreviate enantiomeric excess as e.e.*

There is an additional term, diastereomeric excess (d.e.) that we can apply to diastereoselective reactions. In this case, it cannot be related to a physical property such as optical rotation. However, it does give us a consistent approach to reporting the stereochemical outcome of all stereoselective reactions.

It's now time to look at an enantioselective reaction. We often refer to these as 'asymmetric'.

SHARPLESS ASYMMETRIC EPOXIDATION

This is the reaction we have been (indirectly) looking at. Sharpless received a share in the 2001 Nobel Prize in Chemistry for his work on catalytic enantioselective

---

[40] There are exceptions to the proportional relationship, and measurement of optical rotation is not that precise. In practice, the ratio of enantiomers would normally be determined by other methods.

SECTION 2 THE CHEMISTRY OF ALKENES, BENZENE RINGS AND ALKYNES

oxidation reactions. In 2022, Sharpless received a share in a second Nobel Prize in Chemistry!

Here is the full reaction.

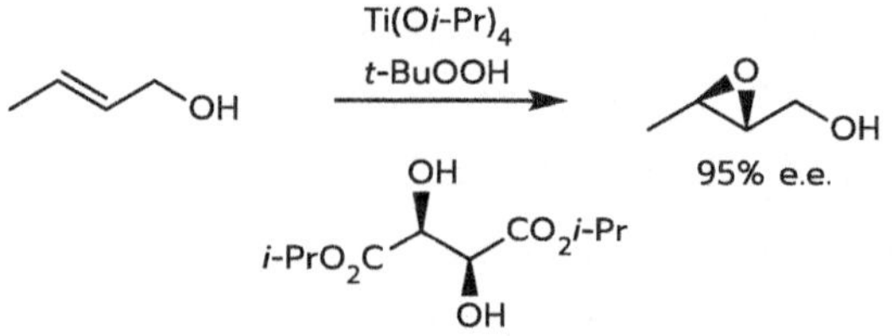

The 'chiral influence' is provided by the diol, diisopropyl tartrate. In this case, it is the (S,S) enantiomer. The only thing used in stoichiometric amounts in the above reaction is the oxidizing agent, t-butyl hydroperoxide.

| This oxidizing agent is used instead of m-CPBA because it works! |

Okay, there's a bit more to it. You wouldn't want to use a 'good' oxidizing agent that would epoxidize the alkene without needing the catalyst.

The following structure is a representation of the transition state for this reaction.

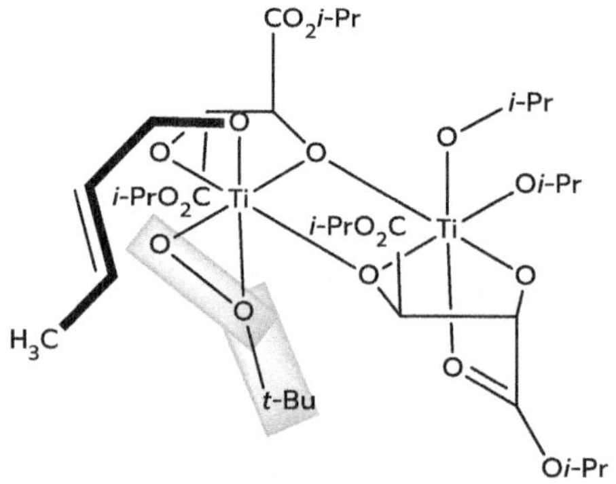

There is a lot of information in this diagram.

| You don't need to remember all of it. I will highlight a few points. |

Firstly, this reaction is well known because it is highly stereoselective for a range of allylic alcohols. This is because it proceeds via a highly organized transition state in which only one face of the alkene (bold) is exposed to the oxidizing agent (highlighted).

This reaction only works for allylic alcohols. We need the OH group to bind to titanium. This is really useful, as we can chemoselectively epoxidize an alkene that is part of an allylic alcohol in the presence of other alkenes.[41]

---

[41] Remind yourself what chemoselective (Recap 4) and allylic (Recap 2) mean.

SECTION 2 THE CHEMISTRY OF ALKENES, BENZENE RINGS AND ALKYNES

There are two titanium atoms and two diisopropyl tartrate ligands in the transition state. It turns out that this is the way it is, and there are no real profound consequences in this particular instance.

At this level, you should be aware that highly stereoselective reactions tend to have very well-defined transition states. Don't try to memorize the structure of the transition state. It would be a waste of time!

JACOBSEN-KATSUKI EPOXIDATION

As we have seen, many reactions are associated with the names of their discoverers. This reaction is often known as the Jacobsen-Katsuki epoxidation, giving credit to both of the pioneers in this area. Jacobsen and Katsuki both trained in the Sharpless group!

Here is a nice example. This particular chiral epoxide is used to make indinavir, a HIV protease inhibitor.

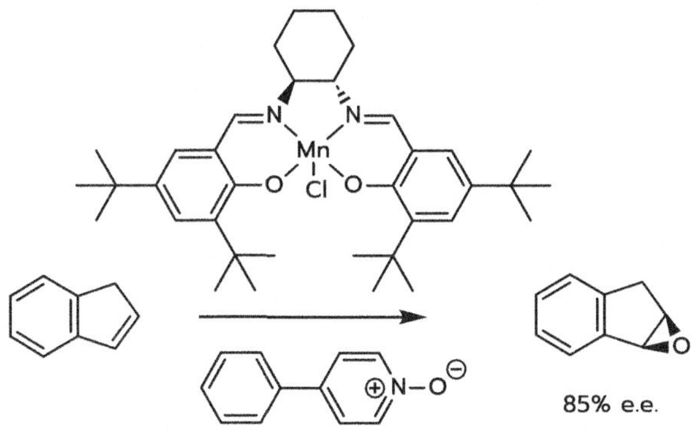

In this case, the chiral influence is provided by a manganese complex with a chiral ligand. There has been much debate in the chemical literature about the precise mechanism and the approach of the alkene to the catalyst. It isn't always straightforward. The oxygen atom is ultimately delivered from an *N*-oxide, but it is still the principles that matter!

SUMMARY

What we have here are two of the most important methods for the enantioselective epoxidation of alkenes. They are not the only methods. The key point is that they are different methods, but they share a principle. They use a single enantiomer of a catalytic species.

SECTION 2 THE CHEMISTRY OF ALKENES, BENZENE RINGS AND ALKYNES

STEREOCHEMISTRY 8

Asymmetric Dihydroxylation of Alkenes

Sharpless didn't stop with one highly stereoselective reaction. We encountered the dihydroxylation of alkenes with osmium tetroxide in **Reaction Detail 6**. Whenever we have a reaction involving a metal, we can consider complexing a chiral ligand to the metal to impart stereoselectivity.

It is a huge challenge to find the best chiral ligand for a given process. Of course, all we ever know is what is the best that has been discovered.

In this case, the 'standard' system is shown below.

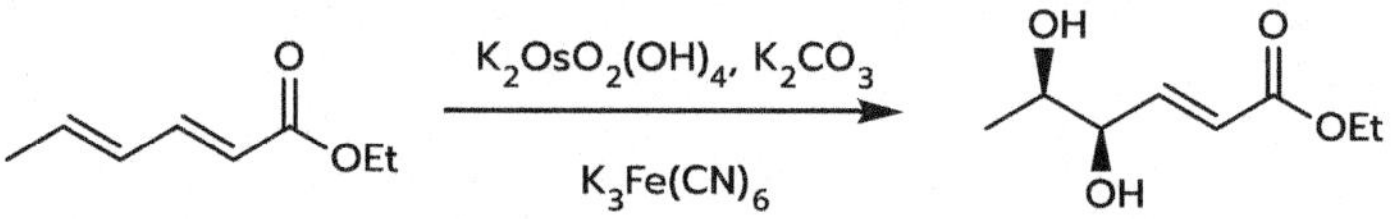

[Reaction scheme: diene ester + $K_2OsO_2(OH)_4$, K_2CO_3, $K_3Fe(CN)_6$ → diol ester, 92% e.e.]

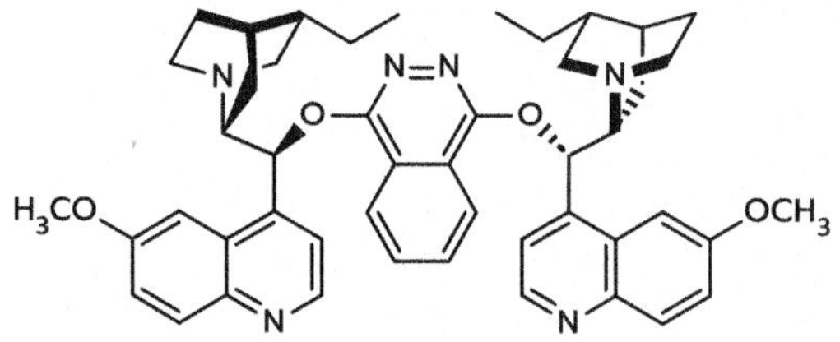

[Structure of (DHQD)₂PHAL ligand]

The ligand in this case is abbreviated as (DHQD)₂PHAL. The 'DHQD' part refers to dihydroquinidine, an alkaloid. We only need a catalytic amount of the osmium and the ligand.

THE OTHER ENANTIOMER

Whenever we have an enantioselective reaction, we want to be able to make both enantiomers of the product. Usually, we make the other enantiomer of the product by using the other enantiomer of the ligand.

> *In this case quinidine is a natural product, and the enantiomer (opposite stereochemistry at all stereogenic centres) is not available.*

Instead, we can use a ligand shown below, based on a related natural product, dihydroquinine. This has the opposite stereochemistry at all stereocentres except the ethyl group.

> *It turns out that this doesn't matter. The ligand is a diastereoisomer of that shown above, but it functions as if it were an enantiomer.*

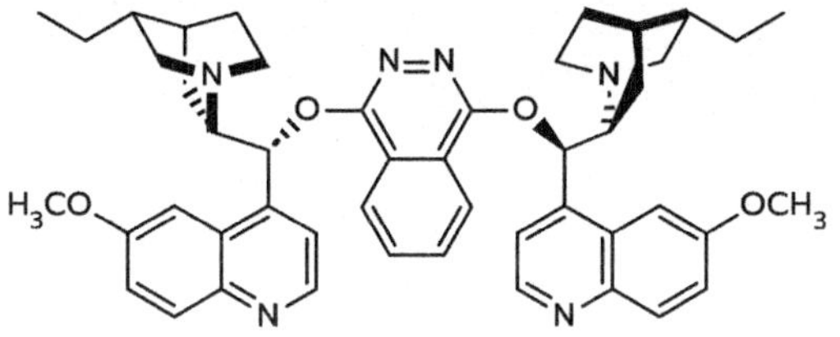

SECTION 2 THE CHEMISTRY OF ALKENES, BENZENE RINGS AND ALKYNES

(DHQ)₂PHAL

> Assign all stereogenic centres in (DHQ)₂PHAL and (DHQD)₂PHAL using the Cahn-Ingold-Prelog rules. There are eight stereogenic centres in each case, although you can take some shortcuts.

SUMMARY

Once again, we have a catalytic enantioselective reaction, and we need a single enantiomer of the chiral ligand in order to achieve diastereomeric transition states.

> *This shouldn't surprise you. It's the only way this can happen!*

The product of alkene dihydroxylation is a 1,2-diol. In **Stereochemistry 7** we saw that a 1,2-diol was the ligand in another catalytic enantioselective reaction. That diol also had two carboxylic ester groups, and we saw they were essential for the mechanism.

STEREOCHEMISTRY 9

Asymmetric Hydrogenation

We looked at hydrogenation of alkenes in Reaction Detail 7. We saw that this reaction generally takes place at a metal surface, and the two added hydrogen atoms are delivered to the same face of the alkene.

> Since the catalyst and substrate are in different phases (solid and solution) we refer to this as a **heterogeneous** reaction.

We can also carry out this reaction using a **homogeneous** metal complex, and in this case we can add a chiral ligand in order to make the process enantioselective.

> Everything is in solution.

Here is an example.

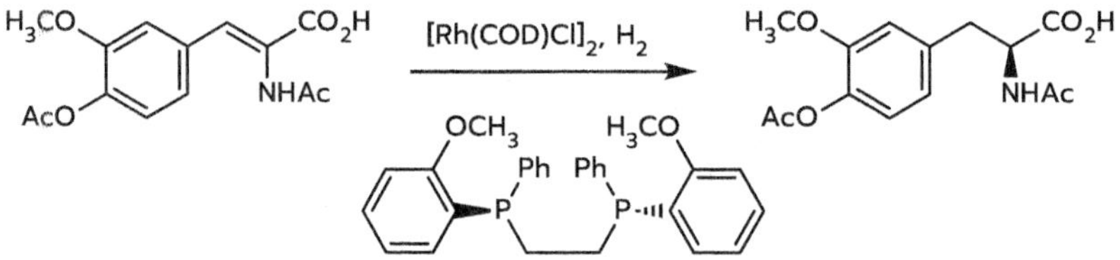

There isn't much more to say about this example. It follows the same principles as the reactions in the previous two chapters.

> Perhaps you can't see the chirality in the ligand, despite the wedge and dash. Phosphorus has three bonds, and a lone pair. The phosphorus atom is a stereogenic centre, and it is configurationally-stable.[42]

The product of this reaction is structurally-related to L-DOPA, a compound used in the treatment of Parkinson's disease.

> It's easy to get bogged down with learning reactions, and to lose sight of the bigger picture. We make organic compounds because they are important to life and to society. We need to make chiral compounds as single enantiomers, because the two enantiomers of a chiral compound can have dramatically different biological properties.

It isn't just alkene double bonds that can undergo asymmetric hydrogenation. Carbonyl bonds can be hydrogenated as well.

Here is an example.

---

[42] The same would not be true of a nitrogen atom. While amines can be chiral in principle, they undergo rapid racemization.

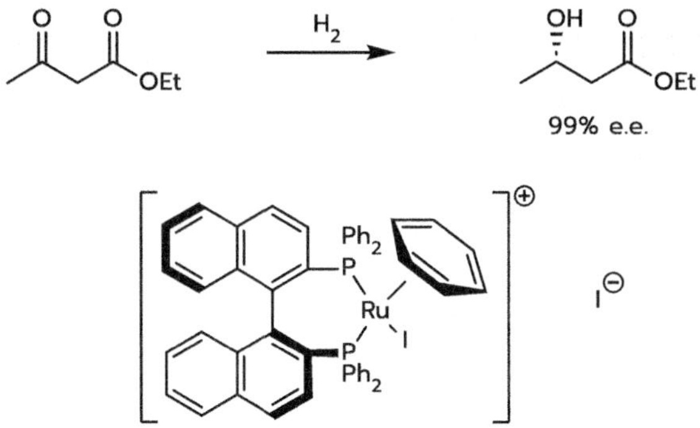

SECTION 2 THE CHEMISTRY OF ALKENES, BENZENE RINGS AND ALKYNES

> *In this case, the chiral ligand has **axial chirality**. We haven't talked about this. The ligand is not superimposable with its mirror image, despite not having a 'traditional' stereogenic centre. This is why the fundamental definition is more important than looking for one structural feature.*

SECTION 3

NUCLEOPHILIC ADDITION TO THE CARBONYL GROUP

SECTION 3 NUCLEOPHILIC ADDITION TO THE CARBONYL GROUP

INTRODUCTION

The carbonyl group is definitely **the** most important functional group in organic chemistry. There are a vast number of reactions associated with the carbonyl group. Many of these are forever associated with the names of the chemists who played a part in their discovery and development.

> *Very soon we will encounter Grignard reagents, Claisen condensations, the wonderful Knoevenagel condensation, and many more!*

This tends to lead to the perception that organic chemistry is 'just a list' of reactions that have to be **memorized**.

The truth is that you do have a lot to learn, but if you can build a framework for your knowledge, you will be able to add new reactions to it with ease.

The carbonyl group only does two things.[43] This section covers one of them—addition of a nucleophile to the carbonyl carbon atom.

The other reaction type is covered in Section 4.

There are many possible carbonyl compounds, and many possible nucleophiles. The permutations are almost limitless, and yet the curly arrows for these reactions all look the same.

> *There are variations, depending on 'which electrons' the nucleophile uses, but that's about it.*

The goal of this section is to show you that, no matter how complex an individual reaction might look, the carbonyl group only does two things!

> *The curly arrows always look the same, so you can focus on building your habits so that you can draw good clear curly arrows with confidence.*

---

[43] This is a bit of a simplification, but the exceptions are much less common, so we won't worry about them for now.

SECTION 3 NUCLEOPHILIC ADDITION TO THE CARBONYL GROUP

BASICS 5

Linking Carbanions and Organometallic Reagents

INTRODUCTION

Before we talk about some specific reactions of carbonyl compounds, it turns out that we will need to learn about a very specific class of compound—the organometallic reagent.

Let's start with a couple of important points. First of all, carbon-carbon bonds are really important. We can only make a bigger compound from smaller compounds by being able to form new carbon-carbon bonds.

We have seen that we can add nucleophiles to carbonyl compounds. Having a carbon nucleophile would give us a way to form a carbon-carbon bond.

RECAP—STABILITY OF CARBANIONS

Here is the trend for the stability of 'simple'[44] carbanions (**Recap 5**).

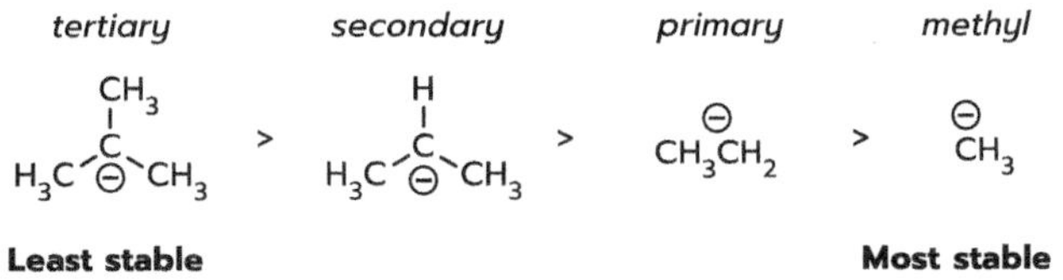

tertiary > secondary > primary > methyl

Least stable   **Most stable**

There is a fundamental problem here. None of these are stable in any meaningful sense. The methyl carbanion is the 'most stable', and yet methane has a pK_a of 48. Remember what this means.

$$CH_4 \rightleftharpoons {}^{\ominus}CH_3 + H^{\oplus} \qquad K_a = 10^{-48}$$

An equilibrium constant of 10^{-48} is vanishingly small. The equilibrium is so far to the left that none of the carbanion is ever formed.

> To put it another way, you cannot simply buy a bottle of "$CH_3^{\ominus}$".

HOW DOES THIS RELATE TO ORGANOMETALLIC REAGENTS?

An organometallic species is something with a metal-carbon bond. They are very useful in synthesis. When we considered electronegativity (**HTSIOC Basics 5**), we saw

---

[44] Just carbon and hydrogen, and a negative charge.

203

SECTION 3 NUCLEOPHILIC ADDITION TO THE CARBONYL GROUP

that the carbon-lithium bond in a compound such as methyllithium is polarized as follows.

$$\overset{\delta-}{H_3C}-\overset{\delta+}{Li}$$

Therefore, methyllithium can be considered to be equivalent to "$CH_3^{\ominus}$". This is useful, as we will see in **Reaction Detail 12**. Now let's consider the stability of a range of 'organolithium' reagents.

We can apply the same trend to the stability of organolithium reagents.

$$(CH_3)_3\overset{|}{\underset{}{C}}-Li \quad > \quad (CH_3)_2\overset{}{\underset{}{CH}}-Li \quad > \quad CH_3CH_2-Li \quad > \quad CH_3-Li$$

Least stable   **Most stable**

Now we can see why *t*-BuLi (the one on the left—*t*-Bu is an abbreviation for $(CH_3)_3C$, *tertiary* butyl) catches fire when exposed to moisture in the air, while *n*-BuLi usually does not. The *tertiary* butyl anion, with three destabilizing methyl groups, is far less stable/more reactive.

You can also think of Grignard reagents, RMgBr, as carbanions. These are generally formed by insertion of magnesium metal into a carbon-halogen bond. Although the bond in these compounds is significantly covalent, it reacts as if it were a carbon with a negative charge.

> *There are as many types of organometallic reagent as there are metals. The wonderful thing is that they all (as a first approximation) do the same thing. Some are more reactive than others. Some are more selective than others. But the principles governing the types of reaction they undergo can all be understood following from a reasonable understanding of organolithium or organomagnesium reagents.*

We are now in a position to discuss the addition of organometallics to carbonyl groups.

SECTION 3 NUCLEOPHILIC ADDITION TO THE CARBONYL GROUP

REACTION DETAIL 12

Addition of Carbon Nucleophiles to Carbonyls

WHY?

If we add a carbon nucleophile to a carbonyl group, we form a C–C bond. The carbon nucleophiles in this chapter are readily available from alkyl halides, which means we can make lots of different useful compounds.

INTRODUCTION

We looked at addition of nucleophiles to carbonyl groups in **Fundamental Reaction Type 2**. We looked at the addition of a simple carbon nucleophile, cyanide, in **Reaction Detail 2**. However, there are far more useful carbon nucleophiles than cyanide. We looked at organometallic reagents in **Basics 5**. It is now time to look at addition of organometallic reagents to carbonyl groups.

ADDITION OF GRIGNARD REAGENTS TO CARBONYL COMPOUNDS

In **Basics 5**, we made a link between carbanions and organometallic reagents. A Grignard reagent contains a C–Mg bond. Because magnesium is more electropositive than carbon, the carbon atom has a partial negative charge.

Grignard reagents are generally prepared from alkyl halides as shown below. In the alkyl halide, the halide has the partial negative charge, and the carbon atom is δ+. So, formation of the Grignard reagent reverses the charge on carbon.[45]

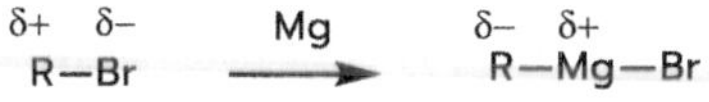

$$\overset{\delta+\ \ \delta-}{R-Br} \quad \xrightarrow{Mg} \quad \overset{\delta-\ \ \delta+}{R-Mg-Br}$$

As the carbon is now partially negative, a Grignard reagent behaves as if it were a source of "R$^\ominus$", a carbon nucleophile. The reaction with an aldehyde is shown below.

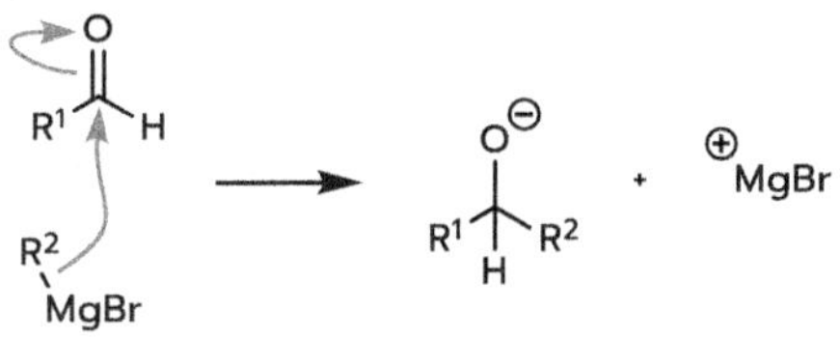

> We use "R" rather than "Nu" for the nucleophile, because we want to specify that it is a 'very general' carbon nucleophile. We wouldn't use this for a 'specific' carbon nucleophile such as cyanide, but in that case we wouldn't need to use a Grignard reagent.

---

[45] In German, there is a wonderful word for this reversal of polarity—umpolung!

205

SECTION 3 NUCLEOPHILIC ADDITION TO THE CARBONYL GROUP

As drawn, this gives an alkoxide product, and a positively charged magnesium bromide by-product. We will look at the fate of this in a moment. First, we have a couple of 'housekeeping' points to consider.

> *The curly arrow starts at the C–Mg bond. That is simply because this is where the electrons are, and curly arrows represent the flow of electrons.*
>
> *I have represented the different 'R groups' as R^1 and R^2. These numbers must be superscripts. If they were subscripts, R_2 would mean two 'R' groups, and R_1 would simply be confusing!*

Now let's revisit the by-product. In reality, you would expect the $O^\ominus$ and the $Mg^\oplus$ to be associated. It would probably look more like the following.

> Draw a curly arrow mechanism for the formation of this product.

$$\underset{H}{\underset{R^1 \quad R^2}{\diagup\!\!\!\diagdown}} \!\!-\!\! O\!-\!MgBr$$

Now we need to consider the eventual fate of this compound. If we add an acid, we can protonate the oxygen, and remove the magnesium. Most of the time when we draw a mechanism, we would simply consider the alkoxide to be the intermediate, and just protonate it as follows.

The protonation step is most definitely separate. If we added a Grignard reagent and an acid together to a carbonyl compound, the Grignard reagent would simply react with the acid.

> *Remember, we looked at this point when we discussed acid catalysis (**Basics 2**).*

This reactivity dictates that we must **first** add the Grignard reagent, and **then** the acid. If this is what happens when we do the reaction, the mechanism we draw should reflect this. Now we have a couple of strategic points to consider.

> 1. This reaction makes a new C–C bond. This is really useful.
>
> 2. Many different Grignard reagents can be formed, with R being a simple alkyl group, or a more complex functionalized alkyl group (if you can make the alkyl halide, you can generally make the Grignard reagent!) or aryl (a benzene ring or similar).
>
> 3. The reaction produces an alcohol which can be used in subsequent reactions.

SECTION 3 NUCLEOPHILIC ADDITION TO THE CARBONYL GROUP

The reaction is not limited to aldehydes (or ketones). We will now look at the reaction of a Grignard reagent with an ester, and use this example to work through the question of 'what else can the O$^\ominus$ do?'

IF YOU DON'T PROTONATE THE O$^\ominus$, WHAT ELSE COULD HAPPEN?

Here's an ester, and a Grignard reagent. We can draw the first step as above. There is absolutely no reason to look for something different.

Because we do not have a source of protons in the reaction at this point, perhaps this initial product can do something else?

Consider the following equilibrium.

In terms of enthalpy, we have lost a C–O σ-bond and gained a C–O π-bond. If we calculate the enthalpy change, and assume that the two O$^\ominus$ species have similar stability, then we have lost two C–O single bonds and gained a C–O double bond.

> Use the data in **Fundamental Reaction Type 2** to convince yourself that this is predicted to be *exothermic* by 32 kJ mol^{-1}.

In addition, we have two species on the right and only one on the left, so that we should also consider the entropy change. After all, we need to consider whether the reaction is *exergonic*, not whether it is *exothermic*.

> We predicted that the reaction would be *exothermic*, and entropy can only help.

This is a lot to think about. There is no way you can memorize these details for every reaction. However, you can work to understand the principles, and systematically apply them to every reaction you see until you do this naturally. Now would be a good time to start making a checklist of things to look for. You can keep adding to it. I could give you a checklist, but then it wouldn't be your own, and you wouldn't *really* own it!

Finally, we should think about what can happen under the reaction conditions. Even if we only form a very small amount of the ketone in this equilibrium, it will react with more of the Grignard reagent as follows.

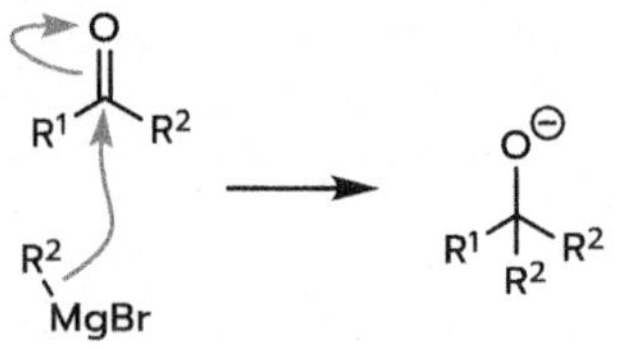

SECTION 3 NUCLEOPHILIC ADDITION TO THE CARBONYL GROUP

Putting this together, we get the following mechanism, so that you get an alcohol product containing two of the Grignard reagent "R" groups.

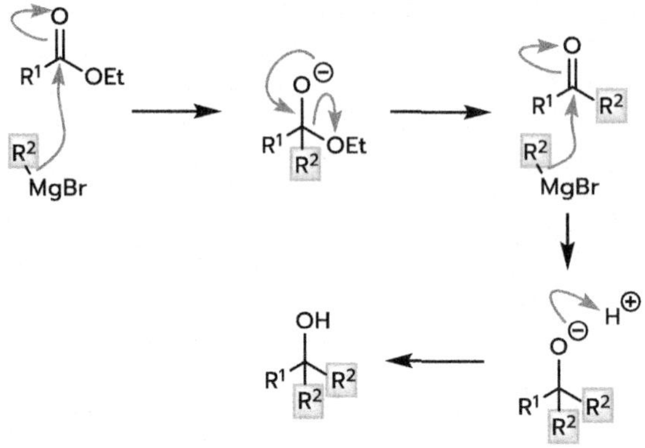

> Try drawing a reaction profile for this overall process, with the various intermediates and transition states shown.

To get you started, the addition of each Grignard reagent forms an intermediate. This will be preceded by a transition state in which the new C–C bond is partly formed.

SECTION 3 NUCLEOPHILIC ADDITION TO THE CARBONYL GROUP

REACTION DETAIL 13

Acetylide Chemistry

INTRODUCTION

Alkanes are sp³ hybridized, alkenes are sp² hybridized, and alkynes are sp hybridized.

When we looked at the implications of hybridization (**Recap 1**), we saw that a C–H bond to an alkene carbon atom is shorter and stronger than an alkane C–H bond. An alkyne C–H bond is even shorter and even stronger. Here are the numbers that we saw in **Recap 1**.

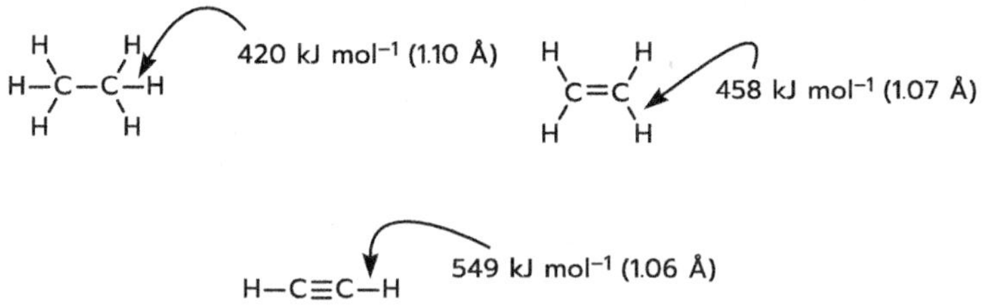

ethane: 420 kJ mol⁻¹ (1.10 Å)
ethene: 458 kJ mol⁻¹ (1.07 Å)
ethyne (H–C≡C–H): 549 kJ mol⁻¹ (1.06 Å)

> *You might expect that since an alkyne C–H bond is very strong, it would be very difficult to break. In **HTSIOC Basics 12** we saw that bond-dissociation energies are for homolytic dissociation to form (in this case) a carbon radical and a hydrogen radical. It would be much harder to break an alkyne C–H bond homolytically, but this isn't what we are going to do.*

In **Recap 7** we saw that the anion we would form by deprotonation of an alkyne is relatively stable. Another way to express this is to say that this hydrogen is acidic. The hydrogen atom has a pK_a of approximately 25.

The pK_a values of the corresponding hydrogen atoms are summarized below.

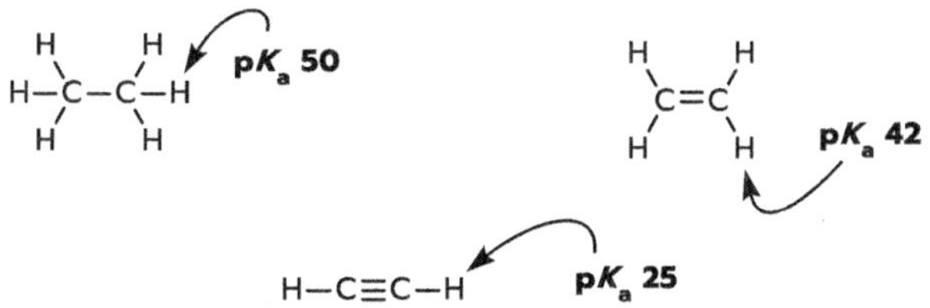

ethane pK_a 50
ethene pK_a 42
H–C≡C–H pK_a 25

To make another connection, we encountered the cyanide anion in **Reaction Detail 2**. Hydrogen cyanide, HCN, has a pK_a of 9.2. The nitrogen atom makes the hydrogen atom much more acidic, but the sp hybridization is a massive factor.

209

SECTION 3 NUCLEOPHILIC ADDITION TO THE CARBONYL GROUP

FORMATION OF ACETYLIDE ANIONS

Let's put this into context. A hydrogen atom α- to an ester has a pK_a of 25 (**Recap 7**). We will see in the next section that it is quite easy to form carbanions with this level of stability, and they have some very useful reactivity.

The anion that we form by deprotonation of an alkyne is known as an **acetylide anion**. It can be formed by treatment of a terminal alkyne with a strong base. Bases that are commonly used to achieve this include sodium amide (NaNH$_2$), alkyllithium reagents and Grignard reagents.

> Hopefully, by now, the method is kicking in. You won't assume that *every* alkyne is acidic. Just the ones that have a hydrogen atom on an alkyne carbon atom.

We have already seen that alkyllithium reagents and Grignard reagents are good nucleophiles. Bases and nucleophiles are essentially the same thing. A nucleophile becomes a base when the electrophilic site it attacks is a hydrogen atom!

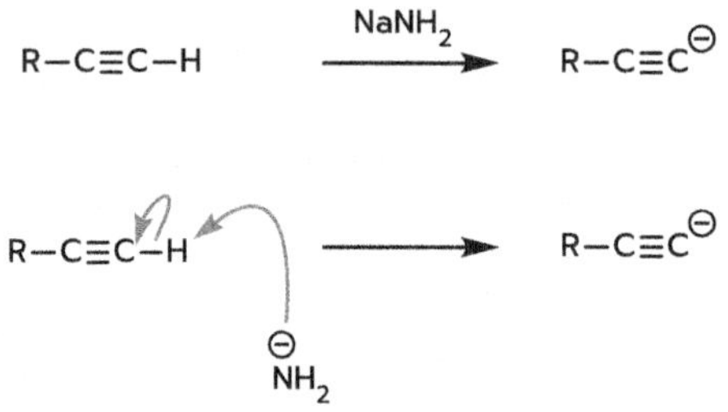

$$R-C\equiv C-H \xrightarrow{NaNH_2} R-C\equiv C^{\ominus}$$

$$R-C\equiv C-H \quad + \quad {}^{\ominus}NH_2 \longrightarrow R-C\equiv C^{\ominus}$$

R'-Li or R'-MgX also used

REACTIONS OF ACETYLIDE ANIONS

An acetylide anion is a species of moderate stability, with a negative charge on carbon atom. It is a nucleophile.[46]

Once we have formed the acetylide anion, it can react with a range of electrophiles. We can represent this in the most general sense as shown below.

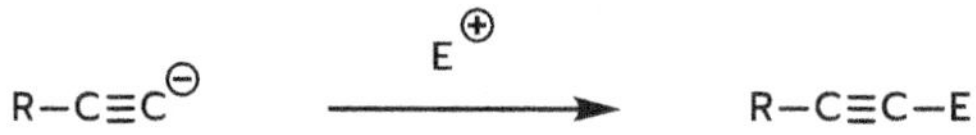

$$R-C\equiv C^{\ominus} \xrightarrow{E^{\oplus}} R-C\equiv C-E$$

General Reaction – $E^{\oplus}$ is a generic electrophile

Now let's think about what sort of electrophiles we might use. We know that alkyl halides (**HTSIOC Fundamental Reaction Type 1**) and carbonyl compounds (**Fundamental Reaction Type 2**) react with nucleophiles. There is nothing special about the reactions shown below. They are just another example of the corresponding fundamental reaction types!

---

[46] Or a base! You wouldn't try to do any of the following reactions in water!

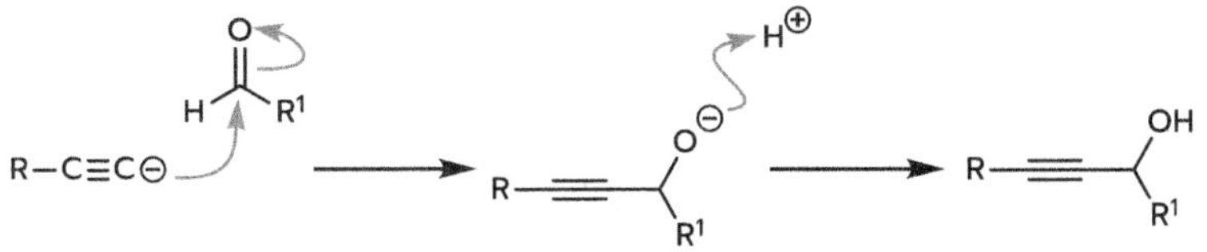

SECTION 3 NUCLEOPHILIC ADDITION TO THE CARBONYL GROUP

In the latter case, we need to add a proton after the nucleophilic attack.[47] This is the point where you add water at the end of the reaction before extracting into an organic solvent.

Don't confuse 'nothing special' with 'not useful'. These reactions are extremely useful. They lead to the formation of C–C bonds with additional functionality (at least the carbon-carbon triple bond). The products of these reactions can be used in many other transformations.

Here is one more example, opening of an epoxide.

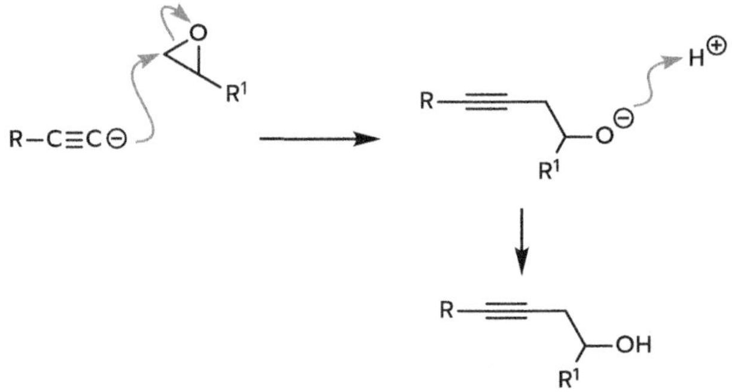

We looked at epoxide-opening reactions in **HTSIOC Worked Problem 5**. We looked at epoxidation reactions in **Reaction Detail 6**. We looked at stereoselective epoxidation reactions in **Stereochemistry 4** and in **Stereochemistry 6**. We are seeing how to build some moderately complex molecules.

---

[47] Unless you want the alkoxide anion to do something else, of course!

SECTION 3 NUCLEOPHILIC ADDITION TO THE CARBONYL GROUP

REACTION DETAIL 14

Addition of Hydrogen Nucleophiles to Carbonyls

TERMINOLOGY—NUCLEOPHILIC ADDITION VERSUS REDUCTION

In **Basics 3**, we explored the concept of oxidation states in organic compounds. Addition of a hydrogen nucleophile to a carbonyl group corresponds to a reduction.

> In this case, because the nucleophile is hydride, nucleophilic addition and reduction are exactly the same thing.

REDUCTION OF CARBONYL COMPOUNDS

This is the overall transformation, again using an aldehyde as an example. There are two hydrogen atoms added, both highlighted below.

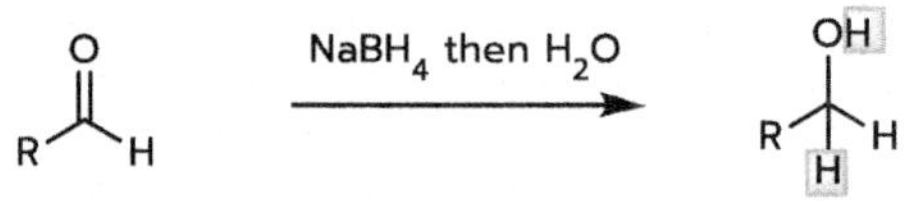

Sodium borohydride, $NaBH_4$,[48] is a commonly used reagent for the delivery of a hydride nucleophile. We also need to add a proton as well. This makes sense, because if we are adding two hydrogen atoms and not changing the charge on the compound, then if we add one as $H^{\ominus}$ we must add the other as $H^{\oplus}$.

Sodium borohydride, $NaBH_4$, and the corresponding reagent lithium aluminium hydride ($LiAlH_4$) both contain a tetrahedral borate (or aluminate) anion, and deliver hydride as shown in the curly arrows below.

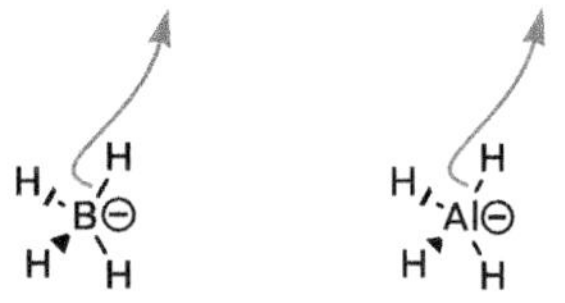

Here is a ***really*** important point. The negative charge on boron (or aluminium) is a result of the atom having a full octet of electrons.

> This is not a pair of available electrons.

---

[48] Sodium hydride, NaH, contains a hydride anion. However, it turns out that this hydride acts as a base (quite a strong one) and not as a nucleophile.

SECTION 3 NUCLEOPHILIC ADDITION TO THE CARBONYL GROUP

It is a very common mistake to draw the curly arrow starting from the negative charge. Of course, this is because you get into the habit of starting a curly arrow from a negative charge.

> This is a good habit to get into!

But not so much that you draw a curly arrow from a negative charge that doesn't represent a pair of electrons. In this case, the correct habit to get into is always starting the curly arrow where the relevant electrons are. In this case, they are in the B–H bond.

Lithium aluminium hydride is much more reactive than sodium borohydride. However, they both basically do the same thing. The reduction of an aldehyde using borohydride is shown below.

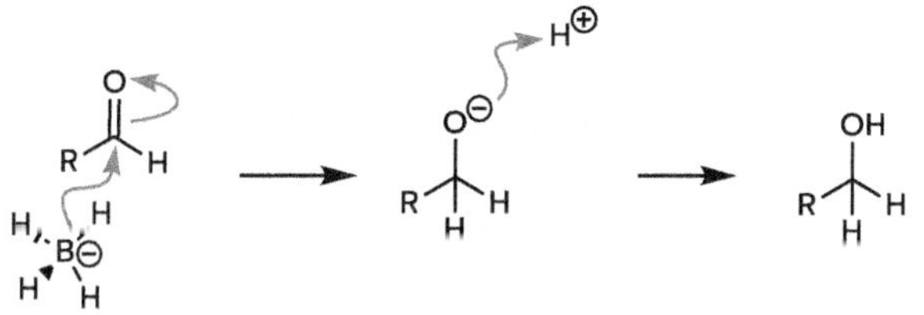

If you try to balance this equation, you will find that you have produced borane, BH_3. You may remember from your inorganic chemistry that the boron has only six outer-shell electrons, and BH_3 can therefore accept a pair of electrons from the alkoxide intermediate.

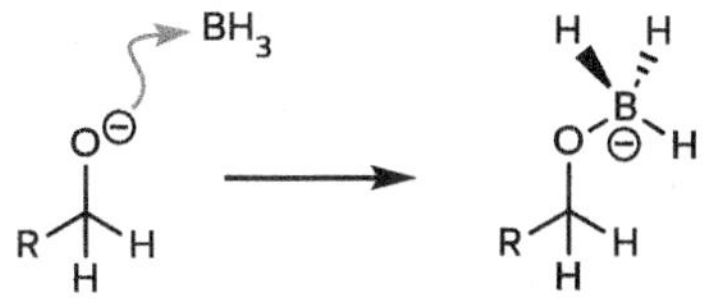

The product shown above can be hydrolysed to give the alcohol that we saw above. This is good justification for drawing the simplified mechanism. Alternatively, we could add a little more complexity, by recognizing that this species could deliver a second hydride to another molecule of aldehyde—it can, and it does!

From a practical point of view, sodium borohydride tends to be used in **protic** solvents[49] (methanol, ethanol). It will react with these solvents, but it will react more rapidly with an aldehyde or a ketone.

Lithium aluminium hydride must be used in **aprotic** solvents (without acidic hydrogens—for example diethyl ether). It would react rapidly with a protic solvent (and generally catches fire!).

---

[49] A **protic** solvent is one that can act as a source of protons.

SECTION 3 NUCLEOPHILIC ADDITION TO THE CARBONYL GROUP

STEREOCHEMISTRY 10

Diastereoselective Addition of Nucleophiles to Carbonyl Compounds

INTRODUCTION

Have another quick read of **Stereochemistry 2**. The key point we need here is that diastereoisomers are different molecules with different energies, and a reaction that leads to the formation of diastereoisomers can produce them in unequal amounts.

> *We can have a stereoselective reaction.*

A carbonyl group is a double bond. It has two faces, and they are not necessarily equivalent.

> *If a nucleophile adding to one face or the other leads to the production of two different diastereoisomers, this establishes the difference in the two faces. If the two faces were equivalent, addition of a nucleophile to either of them would lead to the same outcome.*

We will look at two distinct scenarios, and we will see that the fundamental principles apply—for a reaction to be stereoselective, we need the products and/or transition states to be diastereomeric.

If the products are diastereomeric, the transition states that lead to them will also be diastereomeric. This is inevitable, and it is the scenario we will consider in this chapter.

THE BÜRGI-DUNITZ TRAJECTORY

If we are going to fully rationalize the best side of the carbonyl for the nucleophile to attack, we need to know which direction it approaches from.

> *It turns out that, in general, nucleophiles attack at an angle of about 107° from the carbonyl oxygen atom. This is known as the Bürgi-Dunitz angle.*

We can relate this back to the orbital interactions involved. Recall (**Fundamental Reaction Type 2**) that the nucleophile donates into the π* orbital of the carbonyl group.

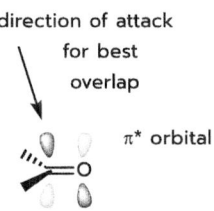

214

SECTION 3 NUCLEOPHILIC ADDITION TO THE CARBONYL GROUP

ADDITION OF NUCLEOPHILES TO CYCLIC CARBONYL COMPOUNDS

First, we will consider cyclohexanones. Once again, there is nothing special about a six-membered ring, apart from the convenience of it having a very well-defined shape. The inclusion of one sp^2 hybridized carbon atom into the ring doesn't change that.

> Caution! I occasionally see the carbonyl bond drawn as either axial or equatorial. It is neither. The carbonyl carbon, the two carbon atoms it is attached to, and the oxygen atom are all in a plane. I have tried to draw this realistically in all the structures below.

We are going to consider 4-*t*-butylcyclohexanone. We know that the *t*-butyl group will be equatorial, so we only need consider whether a given nucleophile attacks to give an axial alcohol product or an equatorial alcohol product.

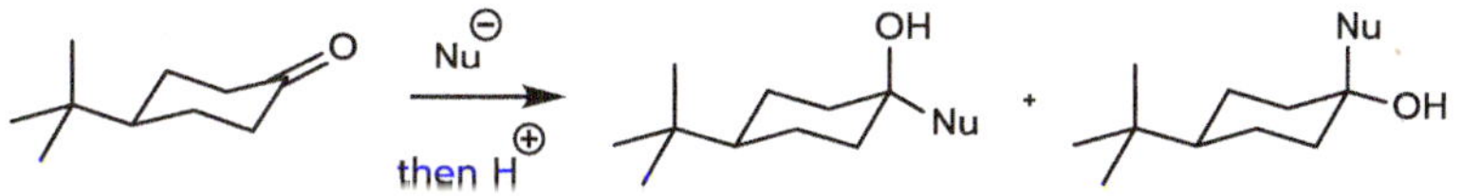

The reactions of this type that produce defined stereochemical outcomes tend to be irreversible. The outcome is kinetically controlled. We get the product that is formed fastest. This being the case, we **only** need to consider the direction of attack of the nucleophile, and what steric interactions are at play.

> Let's make one point really clear. This chemistry is actually very complicated. You don't have to change the reaction conditions much in order to have a big change in the outcome. We can identify some general trends, but we should be cautious about making detailed predictions without checking the results of a range of reactions similar to the one that interests us.

In terms of learning, the most important thing is that you understand the methodology we are using, and that you will be able to apply the general principles and analytical skills to a range of other reactions when you need to do so.

EXPLAINING THE STEREOSELECTIVITY

We are going to keep this simple. A nucleophile attacking axially has to get past two axial hydrogen atoms on the β-carbon atoms. This is a simple steric effect.

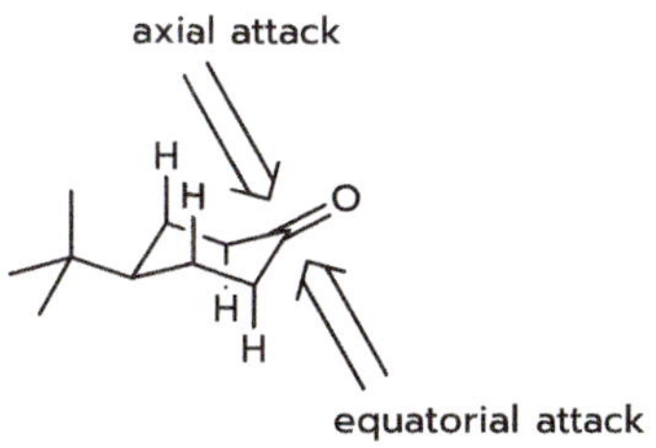

215

SECTION 3 NUCLEOPHILIC ADDITION TO THE CARBONYL GROUP

A nucleophile attacking equatorially will enter in a trajectory that eclipses it with two axial hydrogen atoms on the α-carbon atoms. We describe this as a **torsional interaction** (Stereochemistry 5). Both effects are significant, but the steric effect is more significant with larger nucleophiles.

> Did you look at the structure above and work out which carbon is α- and which carbon is β-?

This leads us to a generalization. Small nucleophiles attack axially, larger nucleophiles attack equatorially.

> This sounds like a 'fact to remember'. I suppose it is, but it's easier when you understand the reasoning behind it.

SOME EXAMPLES

Let's start with the most clear-cut example. Addition of *t*-butylmagnesium bromide gives exclusively equatorial attack.

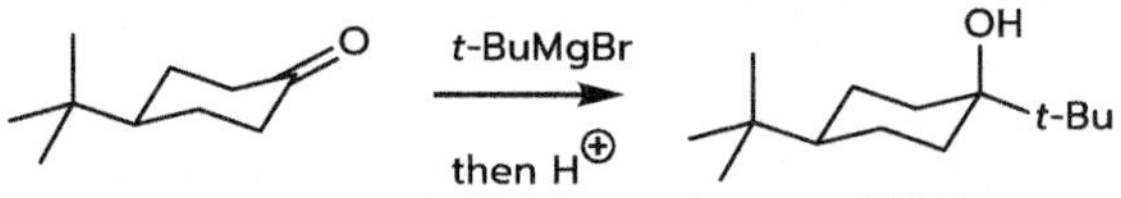

100:0
stereoselectivity

In Stereochemistry 1 we said that a *t*-butyl group will never be axial. Here, we are saying that you won't get axial attack of a *t*-butyl nucleophile.

> They two statements are not quite the same, although they rely on the same underlying principles.

I've done something 'annoying' here. I've drawn the original *t*-butyl group and the new *t*-butyl group differently. Of course, they are exactly the same thing! It's important that you get used to seeing different ways that structural features are commonly drawn.

Now let's look at a smaller nucleophile, methylmagnesium bromide. This gives a 60:40 mixture of equatorial:axial attack (resulting in a 60:40 mixture of the axial:equatorial alcohol products).

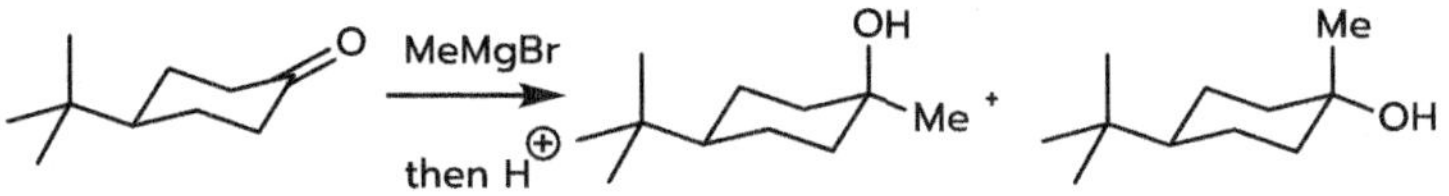

60:40
stereoselectivity

You could draw a reaction profile for the two reactions we have seen. You would hopefully draw the reactions being *exothermic/exergonic* in both cases. You should draw two products in each case, even though we only obtain one product from the first reaction. Hopefully you drew a much larger energy difference between the two

SECTION 3 NUCLEOPHILIC ADDITION TO THE CARBONYL GROUP

possible products in the first reaction. Then, you should have also drawn a larger energy difference for the transition states for the first reaction.

> Did you do all this? I've done the hard work, but by drawing the profile, you will be reinforcing your understanding!

If we use a smaller nucleophile, such as acetylide (**Reaction Detail 13**), you get a 12:88 mixture of equatorial:axial attack. Yes, more of the axial attack! Here, we have been more specific about where the hydrogen atom is coming from—water rather than just $H^{\oplus}$.

12:88 stereoselectivity

> *Since we saw a methyl nucleophile used above, I bet you didn't think we would be able to find a smaller nucleophile.[50] For anyone who can remember very old adverts, acetylide is sometimes described as a 'Heineken nucleophile'. It reaches the places that other nucleophiles can't reach![51]*

I just want to show you one more cyclohexanone derivative, to show you how substituents next to the carbonyl can affect things. First of all, let's see the reaction in the flying wedge projection.

84:16 stereoselectivity

You would be tempted to suggest that more of the product on the left is formed because the methyl group is blocking the top face, so that the new methyl group (highlighted) attacks the bottom face preferentially.

> And you'd be right, although you would miss some detail.

Let's apply the same reasoning to the first reaction we saw. In this case, the equivalent argument is that we get attack from the lower face because the *t*-butyl group already attached to the ring is blocking the upper face. Therefore, the new *t*-butyl group (highlighted) approaches the lower face.

---

[50] Apart from hydride, but this would be delivered from a bulkier reagent such as borohydride, so perhaps it isn't smaller after all.

[51] Just Google it! The advert can still be found online!

217

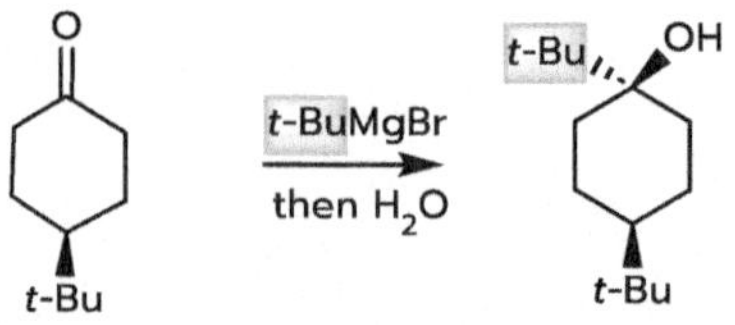

SECTION 3 NUCLEOPHILIC ADDITION TO THE CARBONYL GROUP

> We have seen that while this describes the outcome, the reasoning is not correct.

The *t*-butyl group is too far away to exert any direct steric influence on the approach of the nucleophile. It exerts its effect by locking the conformation of the ring.

Let's have a look at the new reaction again, this time with the cyclohexane chairs. I want to make a couple of points.

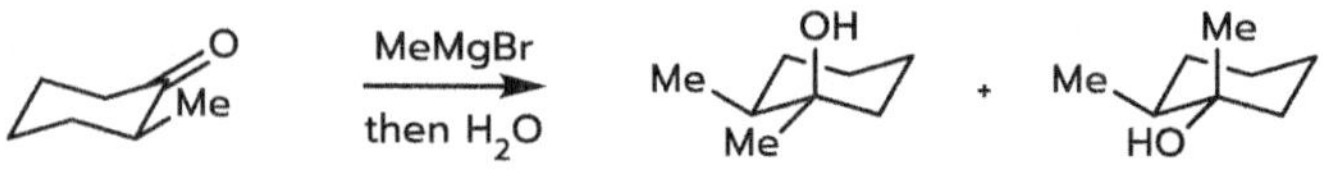

84:16 stereoselectivity

First of all, I have done something horrible with this diagram.

> The products are shown in a very different orientation to the starting material.

This is because when I draw the products in the same orientation, the equatorial substituent (Me or OH) overlaps with the existing Me group. I had to tidy it up. Sometimes this is necessary, but hopefully by now this did not confuse you!

> If it does confuse you, make a model of the structure and check it.

More importantly, we can start to look at exactly how the methyl group influences the course of reaction. In the above diagram, it doesn't look as though the methyl group is really affecting either the top or the bottom faces.

> We have not yet considered everything!

When we draw the methyl group out 'fully', we can see a little more.

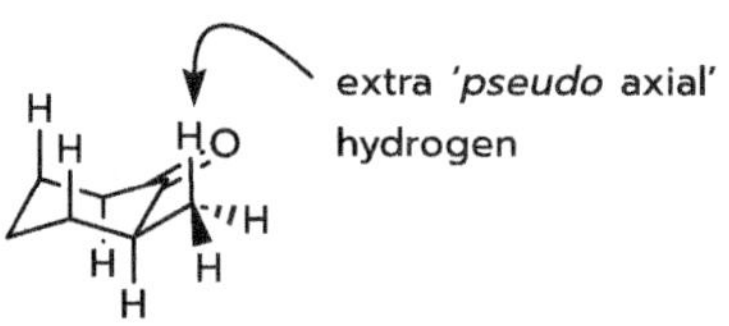

extra *'pseudo* axial' hydrogen

One of the hydrogen atoms on the methyl group will always be *pseudo* axial. Make a model of the compound and look at the conformers! It really is as if the compound has gained an additional axial hydrogen atom.

> How would you know that you needed to do this?

SECTION 3 NUCLEOPHILIC ADDITION TO THE CARBONYL GROUP

That, of course, is the million-dollar question! And that is the reason for including this example. If you try to remember the outcome of this particular reaction, you probably won't gain much. However, if you work to understand the process, then the next time you encounter a related problem, you will apply what we have done here.

ADDITION OF NUCLEOPHILES TO ACYCLIC CHIRAL CARBONYL COMPOUNDS

We looked at cyclic ketones first, because we have an absolute. Whenever there is a substituent, we can talk about which face of the carbonyl bond it is affecting.

> *Of course, there is no 'absolute' top and bottom face—all we have to do to swap them is turn the entire molecule upside-down.*

With acyclic carbonyl compounds, we are not limited to consideration of ketones—most of the examples we will look at are aldehydes—and we can rotate a bond in order to change the face of the carbonyl that any given substituent appears to affect.

Have another look at **Stereochemistry 6**. We considered the epoxidation of acyclic alkenes, and we found that we had to identify the correct reacting conformer (**Stereochemistry 5**).

If we want to rationalize the outcome of diastereoselective addition of a nucleophile to a carbonyl compound in which bond rotation is possible, we need to do the same thing.

> *That's good. It would be surprising if we had to follow a fundamentally different process.*

Let's look at a reaction.

First of all, we are going to take the 'simple' approach, and we will see why it doesn't work. Here is the structure again. I have added the hydrogen atom on the stereogenic centre.

Let's look at the 'simple' (**WRONG!**) logic. The phenyl group is in the same plane as the carbonyl. It isn't affecting either face of the bond. The methyl group is on the

SECTION 3 NUCLEOPHILIC ADDITION TO THE CARBONYL GROUP

front, and the hydrogen atom is on the back. The methyl group is bigger, so we would expect more attack from the back.

> Every statement in the above paragraph is correct—for the structure as drawn. But have a look at structure 2. The new phenyl group has actually attacked from the front.[52] So our prediction was incorrect. Before we consider which face of the carbonyl is 'most blocked' we need to identify the correct conformer in the reaction.

You won't be surprised to find that this is the lowest energy conformer, although you don't have enough experience yet to know which conformer this will be.

This is a bit of a history lesson. In order to understand why and how you need to change your thinking, it is instructive to see how others have done the same. The chemistry we are going to look at was developed between the early 1950s and mid 1970s, initially by Professor Donald Cram (1987 Nobel Prize in Chemistry—although not for this work), based at the University of California, Los Angeles.

Cram and co-workers established which stereoisomers were being produced, and in what ratio, for a range of carbonyl addition reactions. This was not trivial! As a result of their research, they were able to propose a working model to explain the outcome of the reactions they investigated. Here is the reaction again.

1   2, 80%   3, 20%

We have already established that the Grignard reagent prefers to attack the front face of the carbonyl bond. Cram rationalized this as being due to a preferred conformer in which the larger group (phenyl) eclipses the aldehyde hydrogen.

> As with diastereoselective epoxidation in **Stereochemistry 6**, we identify the preferred (most stable) conformer and only then decide which face of the double bond is attacked.

See the Newman projection (**Stereochemistry 5**) below.

preferential attack of nucleophile on same side as H

> Newman projections really are the best representation for this sort of thing.

---

[52] Make a model of the compound and ensure you can see that this is the case. It doesn't help that a hydrogen atom is not drawn, but you can work out where it *has to be*. This is why we often don't draw it.

SECTION 3 NUCLEOPHILIC ADDITION TO THE CARBONYL GROUP

The nucleophile can attack from either the same side as the small hydrogen atom, or the same side as the larger methyl group. You won't be surprised that the Cram model predicts attack from the same side as the hydrogen atom.

> Of course, this was the experimental result (in terms of the actual stereochemical outcome) so naturally Cram proposed a model that would rationalize this nicely!

This model worked well for a number of years, but increasingly, chemists found reactions that did not give the predicted outcome.

> When this happens, you need to determine whether the model is fine, but the new reactions are exceptions, or perhaps there is a new model that can explain all of the new reactions and the 'old' ones.

It turns out that the conformer above is *not* the reacting conformer. There *is* a better model. The new model is much closer to the reality of the reaction.

The new model was proposed Felkin and was supported by elegant computational work by Anh. Instead of placing the large substituent eclipsing the aldehyde hydrogen atom, they oriented it at 90°. There are two possible conformations that will fit this model.

In either of these conformers, we could get attack of the nucleophile from the left or from the right. Let's get two possibilities out of the way quickly.

In the first conformer, attack from the right is not good. The nucleophile would almost eclipse the bulky phenyl group. The same applies to attack from the left in the second conformer.

> This is good. We have something we can be confident of.

Next, let's consider attack from the opposite side of each conformer.

In the left-hand conformer, attack from the left brings the nucleophile close to the hydrogen atom. In the right-hand conformer, attack from the right brings the nucleophile close to the methyl group.

SECTION 3 NUCLEOPHILIC ADDITION TO THE CARBONYL GROUP

I would say this is pretty clear cut. We have two situations that are bad. We get rid of those. We then have one that is better than the other.

Remember, this is the same compound, and the same stereoisomer, that we saw for the 'Cram model' above. We predicted attack from the left there, and we predict the same with the Felkin-Anh model.

> There is a conceptual difference between the Cram and Felkin-Anh models. In the former case, we have only one conformer, and we have attack from either side determining the outcome. In the latter case, we have two conformers to consider. In principle, either could lead to the major observed diastereoisomer, but in all likelihood, each diastereoisomer is produced by reaction of one conformer.

In all of this, the process is more important than recall of facts. You only need to remember one fact—where to put the largest substituent. After that, you consider all of the options for attack, and choose the best one.

SIZE DOESN'T ALWAYS MATTER!

In the above discussion, we focused primarily on where we put the largest substituent. We get some interesting results with electronegative elements as substituents, and this requires refinement of the models.

7:3 stereoselectivity

We explain this by invoking the following conformer. This time, we place the electronegative element perpendicular to the carbonyl group, not the largest group.

Don't over-think this. In fact, we are not actually saying that the carbonyl compound adopts this conformer preferentially, and then the nucleophile attacks. There is no doubt that attack on the carbonyl group takes place at the Bürgi-Dunitz angle. As we approach the transition state, dipole repulsion between the incoming nucleophile and the electronegative element favours this conformer.

We saw a similar situation in **Stereochemistry 6**. In general, we need to compare the energies of the transition states for the reaction. In that case, we saw that the same steric factors ($A^{1,3}$ strain) were present in the transition state. Here, we essentially use the conformer of the carbonyl compound as a 'mnemonic' for predicting the stereochemical outcome.

SECTION 3 NUCLEOPHILIC ADDITION TO THE CARBONYL GROUP

CONTROLLING THE CONFORMATIONS—THE CRAM CHELATION MODEL

There is one thing that you can guarantee with Grignard reagents—magnesium will complex to the carbonyl oxygen atom. You can bet that if you put another 'donor atom' nearby, it will get in on the act. Chelation is a very strong thermodynamic driving force.

In Cram's early work, he realized that the stereochemical outcome of some reactions is best rationalized by assuming that the metal (e.g. magnesium) complexes to the carbonyl group and to another atom. Here is a typical reaction.

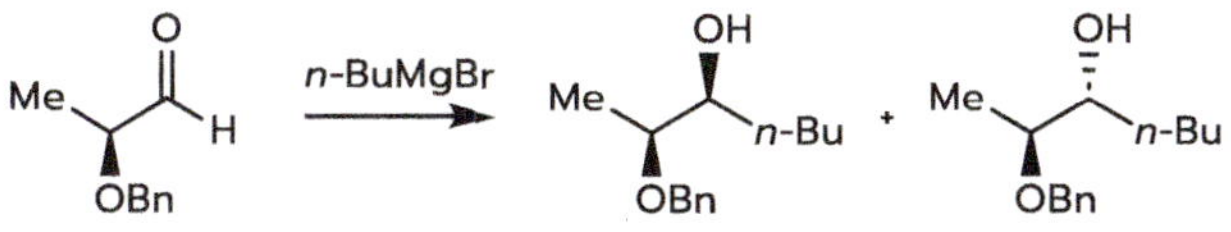

11:1 stereoselectivity

In this case, the Newman projection looks quite cluttered, so I will draw a flying wedge projection. It's all part of the plan! You need to decide which representation suits your needs.

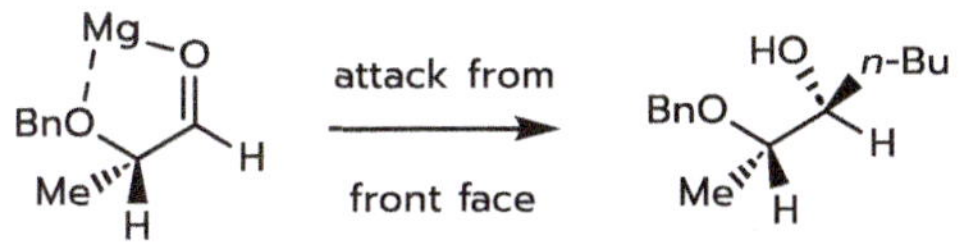

> Check that the product drawn here is the same as the major product shown in the first scheme. It isn't trivial, but it will get easier with practice!

> What do you do if the reagent isn't good enough at chelating? Add something else that will do the job! In the above case, addition of ZnBr₂ increased the stereoselectivity to 71:1.

In the above schemes, Bn is benzyl, PhCH₂. It isn't big enough to prevent chelation, but you can bet there are things you could put on the oxygen atom that would prevent this.

> Oh, and don't forget that Ph is phenyl, not phenol!

HOW GENERAL IS THIS?

Once we have considered chelation and non-chelation models, there isn't much more to it. It doesn't matter whether the nucleophile is hydride, a Grignard reagent or an enolate. The same principles apply, so the same outcome is observed.

> If you add methylmagnesium bromide to an aldehyde, or sodium borohydride to the corresponding methyl ketone, what will happen?

223

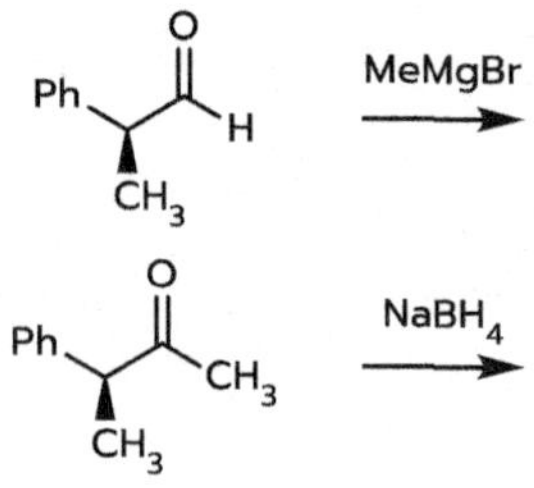

SECTION 3 NUCLEOPHILIC ADDITION TO THE CARBONYL GROUP

Ph–CH(CH₃)–CHO + MeMgBr →

Ph–CH(CH₃)–C(O)–CH₃ + NaBH₄ →

In this case 'what will happen?' means 'what will the stereochemical outcome be?'

I'm not going to give you the answer. Follow the process, draw the conformers, directions of attack and work out which is favoured, and which is not. If you've got the right answer, you will know why I included this as a question.

CONSIDERATION OF REACTION PROFILES

Now here is an exercise for you. It is really important that you do not lose sight of the fundamentals as things become more complicated. For this exercise, take the reasonable assumption that addition of a highly reactive organometallic nucleophile is an *exothermic* reaction with a transition state but no intermediate.

> Draw a reaction profile for this reaction.

> Explain why the transition state resembles the starting material rather than the product. Which idea do you need to use?

> Following on from this, explain why it is appropriate to consider the conformational preferences of the starting material as a good model for the transition state.

SECTION 3 NUCLEOPHILIC ADDITION TO THE CARBONYL GROUP

STEREOCHEMISTRY 11

Some Important Stereochemical Definitions

A lot of your success with stereochemistry will depend on you becoming confident using the terminology. We have a few definitions to introduce now.

DIASTEREOTOPIC

Here is a relatively simple structure with a stereogenic centre. I have highlighted two hydrogen atoms.

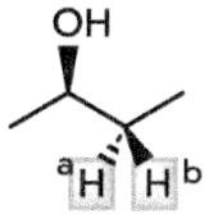

It is tempting to assume that the two hydrogen atoms are identical.

> After all, they are attached to the same carbon atom!

In fact, they are not the same. They are **diastereotopic**. Let's see how we can define this.

Suppose we independently replace H^a and H^b with a different group X. We get the following two structures.

These two structures do not represent the same stereoisomer. They are also not mirror images.

> On this basis, they must be diastereoisomers.

We established in Stereochemistry 2 that diastereoisomers are different compounds with different energies. If we can replace the hydrogen atoms to give different compounds, then the hydrogen atoms must be different.

> That's one definition of diastereotopic. We will see another one in a moment.

We aren't covering NMR spectroscopy in this book, but because diastereotopic hydrogen atoms are different, they can have different chemical shifts. There are some additional implications, but we won't worry about them here.

ENANTIOTOPIC

Now let's take away the OH group.

225

SECTION 3 NUCLEOPHILIC ADDITION TO THE CARBONYL GROUP

What happens if we replace the hydrogen atoms with X now?

The two structures I have drawn are enantiomers—non superimposable mirror images.

If replacement of the two hydrogen atoms results in the formation of enantiomers, then we describe the hydrogen atoms as **enantiotopic**.

HOMOTOPIC

We have one final definition. Here is another structure, with two hydrogen atoms highlighted.

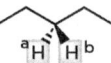

Let's do the same thing. Here are the two structures we get.

In this case, the structures are actually identical. The hydrogen atoms are described as **homotopic**.

SCOPE OF THE DEFINITIONS

These definitions don't just apply to hydrogen atoms. Any two identical groups on a carbon atom (or indeed on any other atom) can be diastereotopic, enantiotopic or homotopic.

There is one further subtlety. You can apply the terms to the faces of π-bonds.

> Let's see how this works.

Here is reaction of an aldehyde with methylmagnesium bromide.

The product on the left results from addition of the Grignard reagent to the front face of the carbonyl group. The product on the right results from addition to the back face.

> Check that you can see this. It isn't trivial.

We have two faces of a double bond. Addition of a nucleophile to one face or the other results in the formation of a pair of enantiomers. The faces of the double bond are **enantiotopic**.

SECTION 3 NUCLEOPHILIC ADDITION TO THE CARBONYL GROUP

We need to be a little careful with this definition. If we had used sodium borohydride in the reaction above, we would have got the same product from attack on either face.

> The problem isn't the definition. It's the reaction we apply to 'test' the definition.

We could also classify the faces as diastereotopic or homotopic. The faces of formaldehyde are homotopic. It doesn't matter what nucleophile we add—we will get the same product from attack on either face. The faces of the carbonyl bond in 2-hydroxypropanal are diastereotopic. In this case, addition of a (non hydride!) nucleophile to the front face or to the back face will give different diastereoisomer products.

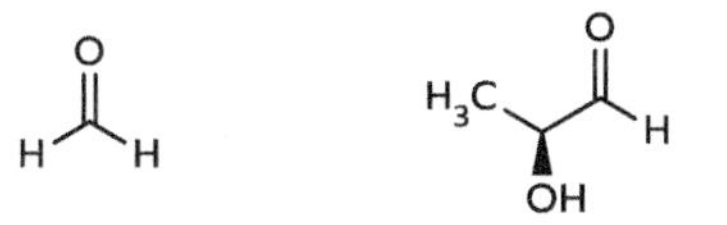

formaldehyde   2-hydroxypropanal

I drew a single enantiomer of 2-hydroxypropanal.

> Which one?

But the definition would still apply to the racemic compound.

ASSIGNING STEREOCHEMISTRY TO FACES OF DOUBLE BONDS

We won't use these definitions in the book, but it is important that you are aware of this idea. Let's look at acetaldehyde. As we have seen above, if we add a nucleophile, we can create a new stereogenic centre. The two faces of the double bond are enantiotopic. There is another term that is used to indicate the potential for formation of a stereogenic centre. The faces of the double bond are said to be **prochiral**.

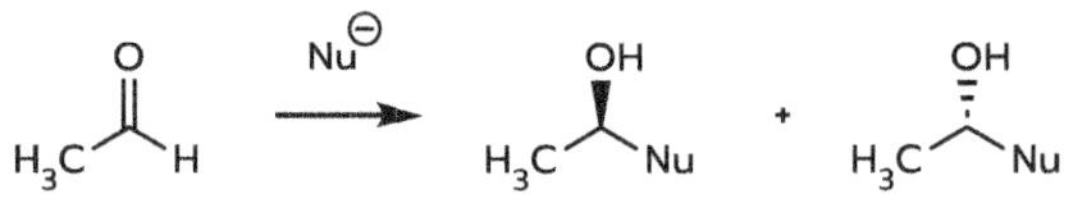

When we have this situation, we can refer to the faces by stereochemical descriptors *Re* and *Si*. These are like *R* and *S* for flat molecules.

The application of these terms is shown by the following diagram. We assign the Cahn-Ingold-Prelog priorities to the substituents on the carbon atom. If they are arranged clockwise, then we are looking at the *Re* face. If they are arranged anticlockwise, we are looking at the *Si* face.

227

SECTION 3 NUCLEOPHILIC ADDITION TO THE CARBONYL GROUP

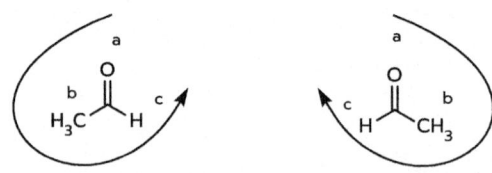

> There is absolutely no connection between Re and Si and the stereochemistry (R/S) that you would get by addition of a nucleophile to the carbonyl group. This would depend on the nucleophile added and its priority according to the Cahn-Ingold-Prelog rules.

Now let's look at an enolate and see an important limitation with this terminology.

The question 'which face of the enolate are we looking at?' is meaningless. If we were looking at carbon atom 'a', we would call this the Re face, but for carbon atom 'b' it is the Si face.

When we get to Section 4, we will see that enolates react at carbon atom 'a'. If I needed to apply a stereochemical descriptor to this double bond, I would relate it to carbon 'a'.

> But I would still be cautious because the ambiguity is still there.

SECTION 3 NUCLEOPHILIC ADDITION TO THE CARBONYL GROUP

STEREOCHEMISTRY 12

The Origin of Chirality in Life

Okay, it is *just possible* that I am over-selling this one!

Nature uses lots of single-enantiomer compounds—sugars, aminoacids, alkaloids—and no-one really knows for sure how it came to be that life evolved to use only one enantiomer of these compounds. There are lots of theories, but they are remarkably hard to test experimentally.

What Professor Kenso Soai and his group at Tokyo University of Science did is truly amazing. And yet it starts with 'just another stereoselective reaction'! Here is the reaction.

Ph-CHO + Me$_2$Zn →(DAIB) Ph-CH(OH)-Me, 95% e.e.

We have a reagent, dimethylzinc. It is another '$CH_3^{\ominus}$' (**Basics 5**). It just happens that this one doesn't react with the aldehyde until we add a catalyst. We encountered enantiomeric excess (e.e.) in **Stereochemistry 7**.

> I did say that some organometallic reagents are less reactive/more selective than others.

The best catalysts tend to be amino-alcohols, such as DAIB. The structure of this compound is shown on the scheme above. It turns out that we only need a tiny amount of this compound, typically 2 mol% (or 0.02 equivalents).

> So far, so good. DAIB is chiral, and a single enantiomer. As we have seen before, this results in the transition states for production of the two product enantiomers being diastereomeric. We form one enantiomer considerably faster than the other.

If this was 'it', it would still be significant. We would be able to make a broad range of chiral *secondary* alcohols with high levels of enantiomeric excess.

ASYMMETRIC AMPLIFICATION

If we mix dimethylzinc with DAIB, we get a dimeric structure in equilibrium with a monomeric structure.

SECTION 3 NUCLEOPHILIC ADDITION TO THE CARBONYL GROUP

If we have a single enantiomer of DAIB, then the structure on the right can only exist in one stereochemical form.

Now it is time for a hypothetical experiment. What happens if we take a 55:45 mixture of enantiomers of DAIB? Let's simplify the discussion by referring to them as R and S, even though DAIB has four stereogenic centres.

> We considered multiple stereogenic centres to be a 'unit of stereochemistry' in Stereochemistry 7. I personally find this approach to be useful, but it isn't 'standard' so you probably won't see it in other books.

The dimeric structure could be RR, RS (which is the same as SR) or SS.

> The RR and SS dimers are enantiomers. They must have the same energy.

In contrast, the RR and RS dimers will not necessarily have the same energy. They are diastereoisomers!

It turns out that the RS dimer is pretty stable. So stable, in fact, that it doesn't do anything.

> In not doing anything, it does something quite amazing.

$$RR + SS \rightleftharpoons 2R + 2S \rightleftharpoons 2RS$$

The 45% of the minor enantiomer forms a dimer with 45% of the major enantiomer. That is 90% of the compound that does nothing. This leaves 10% of the major enantiomer to get on with the catalysis.

> We can consider this to be a smaller amount of a single enantiomer of catalyst.

This excess 10% of the chiral ligand leads to formation of the product with a much higher level of enantiomeric excess than that of the original ligand.

SOME PROFOUND OBSERVATIONS!

Living organisms are self-replicating. Natural selection favours those genes/traits that lead to survival of the organism/species. The self-replicating nature of life stems from the self-replicating behaviour of some of the chemical processes involved.

> Let's start with a big question. How could life have evolved to use only one enantiomer of aminoacids in enzymes?

SECTION 3 NUCLEOPHILIC ADDITION TO THE CARBONYL GROUP

At one level the answer could be very simple. If life used a stereochemical mess of enzymes, each stereoisomer would be a different compound, and most of them would not catalyse the reaction of concern.

> So had life not evolved in this way, I would not be here to ask the question.

This is something called the anthropic principle, which is a neat way of avoiding the questions 'how' and 'why'!

But since the precursors to the biological molecules that we take for granted **had to be** racemic at some point in prehistory, how did the primordial soup start to become enantiomerically enriched?

What Soai was able to show is that some reactions can lead to enhancement of enantiomeric excess. Look at the reaction of aldehyde **1** with diisopropylzinc.

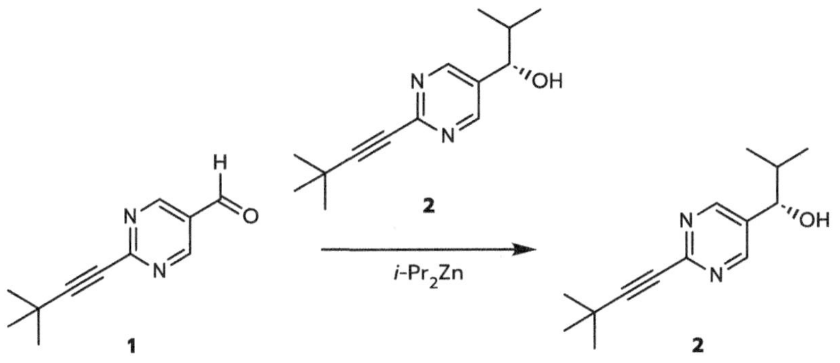

The chiral ligand **2** also happens to be the product of the reaction. If we start with a small amount of a single enantiomer of **2**, we will make much more of the same compound with higher enantiomeric excess.

In one truly awesome experiment, a sample of compound **2** with only 0.00005% e.e. was used as catalyst.

> We would have trouble measuring such a small enantiomeric excess. This is virtually racemic.

The reaction was run, and a sample of the product was isolated. It had a higher level of enantiomeric excess. A small (catalytic) amount of this compound was then used in a second reaction, and so on. After six iterations of this process, the final product had >99.5% e.e.

> We go from a situation where we can barely measure the excess, to a situation where we can barely detect the minor enantiomer. This level of enantiomeric enrichment is remarkable!

SECTION 3 NUCLEOPHILIC ADDITION TO THE CARBONYL GROUP

Of course, prebiotic life did not begin with the addition of dialkylzinc reagents to aldehydes. Nevertheless, there are many other autocatalytic reactions, including some which could lead to amplification of stereochemistry. While this doesn't answer all of the questions, it does establish that with suitable reactions, it is at least theoretically possible that a small initial excess of one enantiomer, coupled with an autocatalytic reaction *might* lead to the evolution of life based only on one enantiomer of the key building blocks.

> *And Nature had more than six iterations, and a **very** large reaction flask to work with. Even the tiniest stereochemical fluctuation, amplified over millions of years, could potentially lead to organisms with single enantiomers of compounds.*

We will probably never know for sure how it happened, but we know it did!

SECTION 3 NUCLEOPHILIC ADDITION TO THE CARBONYL GROUP

EXERCISE 2

Bond Dissociation Energies and Predicting Outcome of Reactions

In **Exercise 1** we looked at the thermodynamics of hydrate and ketal formation. We are now going to add a few more related reactions to get us ready for the discussion in **Reaction Detail 15** and **Applications 2**.

Once again, we will apply the method described in **HTSIOC Basics 13** (and extended in **HTSIOC Basics 35**) to predict/rationalize the outcome of these reactions.

Aldehydes and ketones react with alcohols or amines under acidic conditions. Here are several possible outcomes for these reactions. We will stick with similar, simple, examples to illustrate the point. This doesn't mean that any of these are particularly good reactions.

Reaction 1 – Ketal Formation

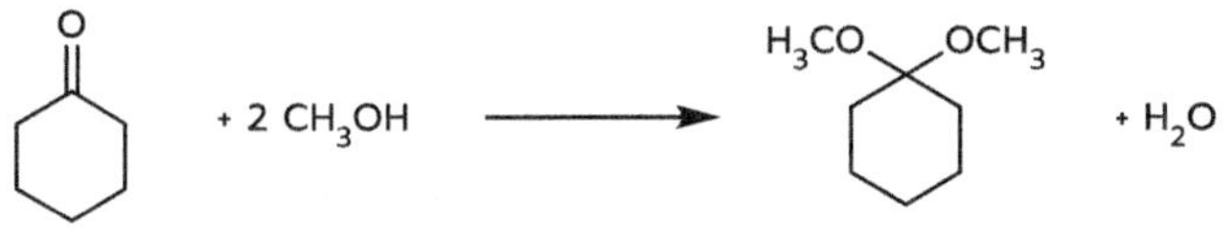

cyclohexanone + 2 CH$_3$OH ⟶ 1,1-dimethoxycyclohexane + H$_2$O

Reaction 2 – Enol Ether Formation

cyclohexanone + CH$_3$OH ⟶ 1-methoxycyclohexene + H$_2$O

Reaction 3 – Imine Formation

cyclohexanone + CH$_3$NH$_2$ ⟶ N-methylcyclohexanimine + H$_2$O

Reaction 4 – Enamine Formation

cyclohexanone + CH$_3$NH$_2$ ⟶ N-methyl-1-cyclohexenylamine + H$_2$O

233

SECTION 3 NUCLEOPHILIC ADDITION TO THE CARBONYL GROUP

Reaction 5 – Aminal[53] Formation

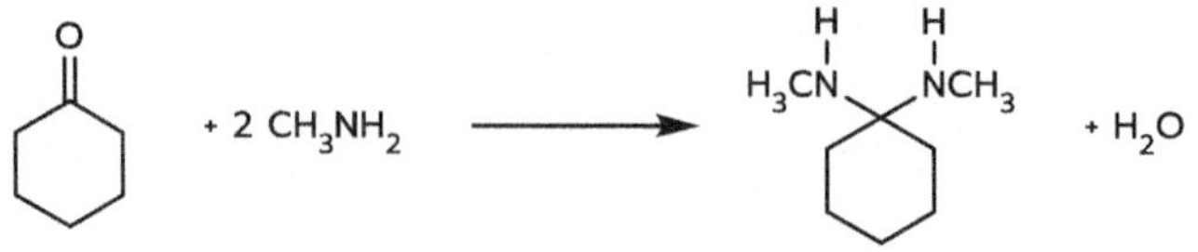

cyclohexanone + 2 CH$_3$NH$_2$ ⟶ cyclohexane with H$_3$CN(H) and NCH$_3$(H) substituents + H$_2$O

Calculate the enthalpy of reaction for all five of these using data in the following Table.

I've deliberately kept the Table simple, and just given you what you need. Work them out before looking at the answers below.

Bond Dissociation Energy / kJ mol^{-1}			
C–C	350	C=C	611
C–H	410	C=O	732
C–O	350	C≡N	615
C–N	300		
O–H	460		
N–H	390		

REACTION 1—KETAL FORMATION

Here, we break two O–H bonds, but we also form two O–H bonds. We can assume that these cancel out. In addition, we break a C=O bond, and form two C–O bonds.

$$\Delta H = 732 - (2 \times 350) = +32 \text{ kJ mol}^{-1}$$

This reaction is predicted to be **endothermic**.

REACTION 2—ENOL ETHER FORMATION

In this case, we have broken an O–H bond and a C–H bond, and formed two O–H bonds. We will allow one of the O–H bonds to cancel the other. We break a C=O bond, and a C–C bond, and we form a C–O bond and a C=C bond.

$$\Delta H = (410 + 732 + 350) - (460 + 350 + 611) = +71 \text{ kJ mol}^{-1}$$

This reaction is predicted to be significantly **endothermic**. So, we expect a ketone to react with an alcohol to give a ketal rather than an enol ether.

---

[53] This has to be one of the coolest names for a functional group. It's not an animal!!!

SECTION 3 NUCLEOPHILIC ADDITION TO THE CARBONYL GROUP

> This makes two assumptions. First of all, there has to be a plausible mechanism which allows us to get to either product. Secondly, the reaction has to be run under thermodynamic conditions—it has to be allowed to reach equilibrium.

REACTION 3—IMINE FORMATION

Here, we break two N–H bonds and form two O–H bonds. We have to include these in the calculation, obviously. In addition, we break a C=O bond and form a C=N bond.

$$\Delta H = (390 + 390 + 732) - (460 + 460 + 615) = -23 \text{ kJ mol}^{-1}$$

This reaction is predicted to be **exothermic** even though we are replacing a stronger C=O bond with a weaker C=N bond.

REACTION 4—ENAMINE FORMATION

Here, we break a N–H bond and a C–H bond and form two O–H bonds. In addition, we break a C=O bond and a C–C bond, and we form a C=C bond and a C–N bond.

$$\Delta H = (390 + 410 + 732 + 350) - (460 + 460 + 611 + 300) = +51 \text{ kJ mol}^{-1}$$

This reaction is calculated to be **endothermic**. Given the choice, a *primary* amine will react with a ketone to form an imine, not an enamine.

REACTION 5—AMINAL FORMATION

In this case, we break two N–H bonds, and form two O–H bonds. We also break a C=O bond and form two C–N bonds.

$$\Delta H = (390 + 390 + 732) - (460 + 460 + 300 + 300) = -8 \text{ kJ mol}^{-1}$$

This isn't as **exothermic** as imine formation, so we still expect imine formation to be preferred.

YOU CANNOT ALWAYS FORM AN IMINE

In order to form an imine, you need a *primary* amine. Nitrogen can only form three bonds. If you start with a *secondary* amine, you would either get an enamine or an aminal.

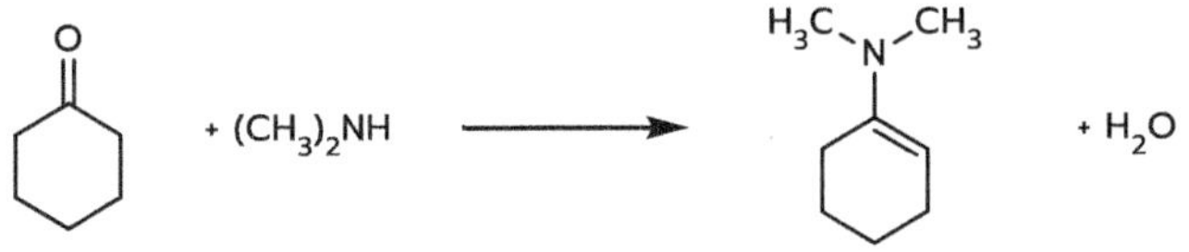

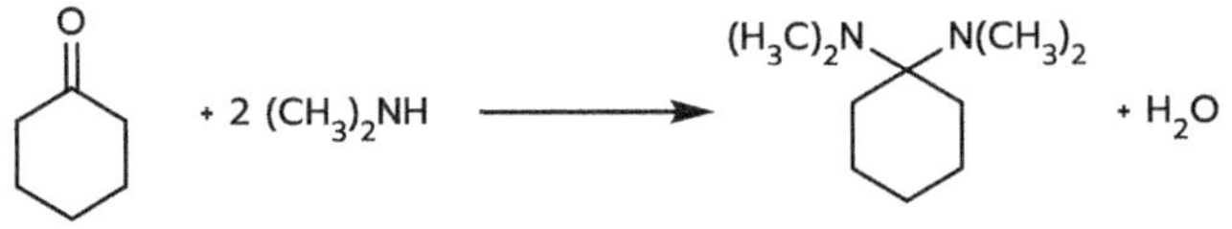

SECTION 3 NUCLEOPHILIC ADDITION TO THE CARBONYL GROUP

It turns out that, despite the above numbers, enamine formation is preferred. The reasons for this are not difficult to understand, but it isn't easy to work out what to do about the problem!

Remember that we are using *average* values for bond-dissociation energies. A C–N bond in an enamine is not the same as a C–N bond in an amine. We would expect, on the basis of hybridization alone, that this bond would be quite a bit stronger (Recap 1). When we add resonance stabilization (overlap of the nitrogen lone pair with the double bond) we get a bit more stabilization.

> Draw the resonance forms!

Furthermore, the numbers take no account of steric factors—the aminal is very crowded, much more so than the enamine.

> *In order to do this sort of calculation in a meaningful way, the bond dissociation energies you use must be representative of the type of bond, in the type of functional group, that you are forming.*

In addition to these inaccuracies, none of these calculations consider entropy—these are enthalpies of reaction, not free-energies. When you consider entropy, the formation of a ketal from a ketone and two equivalents of an alcohol becomes a lot less favourable.

> *These calculations are a useful guide, but be aware of their limitations.*

WHERE NEXT?

Now we will look at the formation of each of these compounds, and consider how to draw plausible mechanisms, and how we might identify the more likely mechanism in cases where there is more than one possibility.

SECTION 3 NUCLEOPHILIC ADDITION TO THE CARBONYL GROUP

REACTION DETAIL 15
Formation of Acetals and Ketals

INTRODUCTION

In **Fundamental Reaction Type 2** we considered attack of nucleophiles onto carbonyl groups in the general sense. In **Reaction Detail 12** and **Reaction Detail 14** we considered the attack of carbon and hydrogen nucleophiles respectively. There isn't really a whole lot more we need to say about those reactions (until we get to some other carbon nucleophiles in **Reaction Detail 23**). Now let's consider oxygen nucleophiles in detail. In **Applications 2** we will then consider the attack of nitrogen nucleophiles.

> *All these reactions can be used to form useful and important products. In addition, comparison of the detailed mechanisms will help us learn. We will see carbonyl groups doing fundamentally the same thing, but with subtle differences for good reasons.*

FORMATION OF HYDRATES

We already looked at hydrates in **Reaction Detail 3**. Go back and have another look, in preparation for the rest of this chapter.

ACETAL AND KETAL FORMATION

The general structures of acetals and ketals are shown below. The only difference between these two functional groups is the presence or absence of a hydrogen atom (shown).

$$R^1O \quad OR^1 \qquad R^1O \quad OR^1$$
$$R \quad H \qquad\qquad R \quad R$$

acetal   ketal

> *An acetal is derived from an aldehyde, while a ketal is derived from a ketone. Apart from that, they are exactly the same, and should not be considered as two different functional groups.*

There is a lot of similarity between hydrate formation and acetal/ketal formation. They are all functional groups with two C–O single bonds to the same carbon atom, so you would expect similarities.

If we consider the addition of an alcohol to an aldehyde under acidic conditions, we would expect to see exactly the same as we saw for hydrate formation, at least in the first instance.

Let's see the overall reaction for acetal formation. Instead of using two molecules of an alcohol, we are going to use one molecule of a diol.

237

SECTION 3 NUCLEOPHILIC ADDITION TO THE CARBONYL GROUP

> Think about why this might be—we will come back to this point at the end.

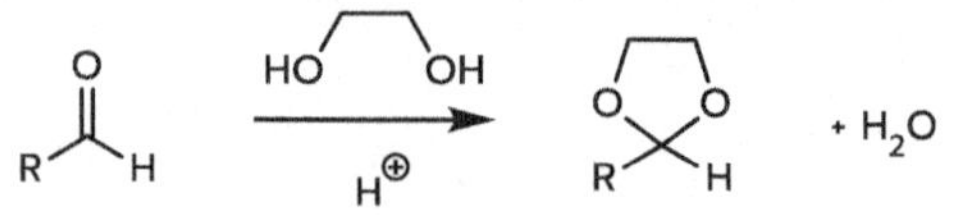

Here's another of those fundamental questions.

> Which oxygen atom will become the water molecule?

We should answer this by drawing plausible steps, and considering the energetics of alternatives. We should not try to guess or pre-judge this.

In the case of hydrate formation, we saw that we protonated the carbonyl oxygen atom in **1** and then added water as a nucleophile to intermediate **2**. There is no obvious reason to do anything different here. Here is the start of a plausible mechanism. The only difference is that we are using one of the diol OH groups as the nucleophile. We get intermediate **3**, which can lose a proton to give compound **4**.

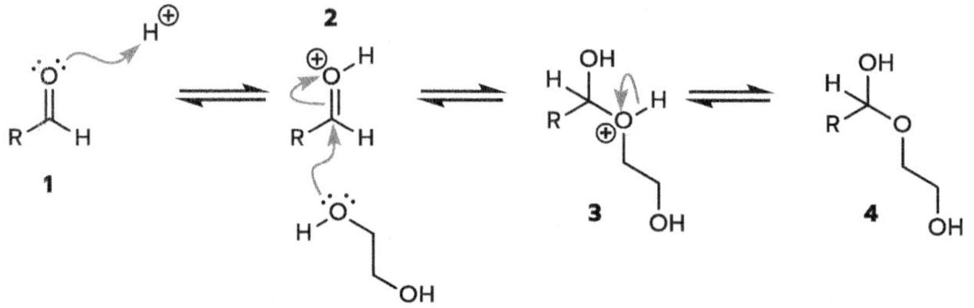

At this point, it is probably clear to you that there is one oxygen atom that won't be lost. It's the one that is bonded to two carbon atoms. Either of the other two seem fair game though. Let's try to distinguish between the possibilities.

Here is option 1. We protonate compound **4** on the *primary* alcohol. The other oxygen atom then attacks the carbon to kick out water, giving the protonated product **6**.

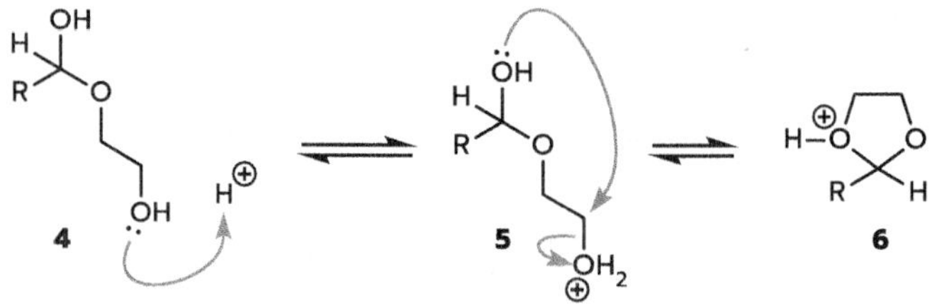

This is a substitution reaction. It would have to be S_N2, as this is a *primary* alcohol. However, even water isn't a great leaving group in S_N2 substitution reactions. We should have reservations about this.

> More formally, we would expect this reaction to have quite a large activation barrier.

Here is option 2.

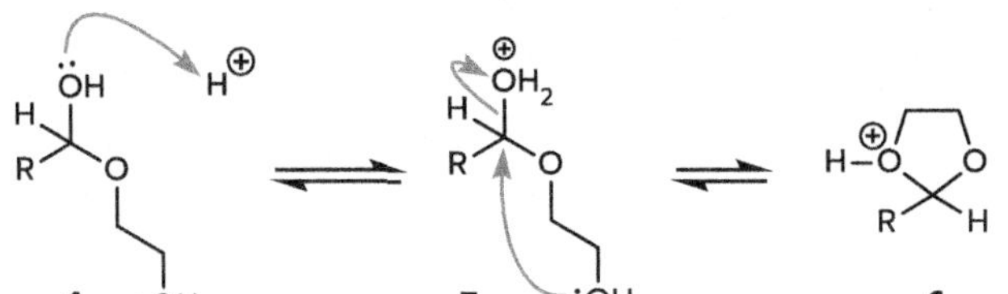

SECTION 3 NUCLEOPHILIC ADDITION TO THE CARBONYL GROUP

We still have a substitution reaction. However, this is at a *tertiary* centre. This is more crowded, so it will disfavour S_N2 substitution (which is the mechanism I have drawn). The electronic effect of each of the substituents on S_N2 substitution isn't quite so clear, but it is easier to see for S_N1 substitution.

Now to option 3. If we lose water from intermediate **7**, we get a pretty stable carbocation. In fact, it is so stable that I would not draw it as a carbocation at all. Here is a mechanism that shows this.

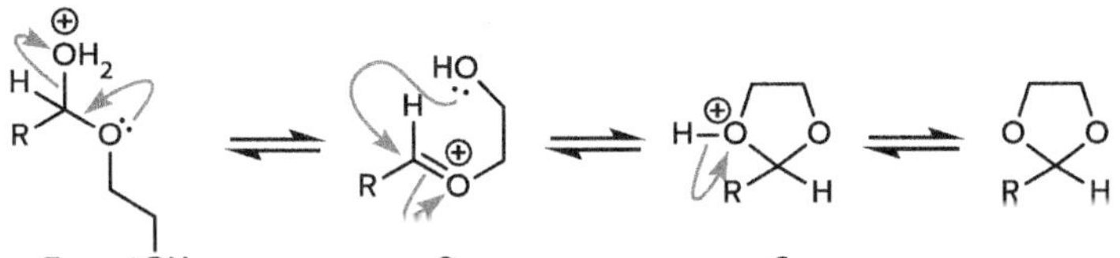

> At least to me, this looks much better. I guess it looks much better to me because this is what I am used to seeing.

If you carry out this reaction using an ^{18}O labelled aldehyde, the ^{18}O label ends up in the water, not in the acetal product. We know, therefore, that the mechanism must be either option 2 or option 3. Option 1 would not give the observed outcome. We cannot directly distinguish between option 2 and 3 on the basis of the data, but option 3 is much more reasonable.

NOW YOUR TURN

That mechanism was a bit fragmented.

> Draw it all out in a single reaction scheme. Then draw the same mechanism with two equivalents of methanol, instead of the 1,2-diol.

Overall, it should look the same.

ACETAL AND KETAL HYDROLYSIS—ANOTHER EXERCISE

You will notice that all steps in the acetal formation above are equilibria—they are reversible. If we start with an acetal and add acid and water, we will get an aldehyde. If we start with the aldehyde and add an acid and an alcohol, we will get an acetal.

239

SECTION 3 NUCLEOPHILIC ADDITION TO THE CARBONYL GROUP

> Based on the above mechanism for acetal formation, draw the corresponding mechanism for acetal hydrolysis.

If you find that you have different intermediates, you have made a mistake. All you need to do is draw the same intermediates, but draw the curly arrows for the reverse reaction.

> *Are you getting tempted to leave these activities for later? You know what I think about that! This is where you will reinforce your learning, and make mistakes with no consequences.*

WHY ARE ACETALS AND KETALS IMPORTANT?

There are two fundamental reactivity types for carbonyl compounds, as shown on the structure below.

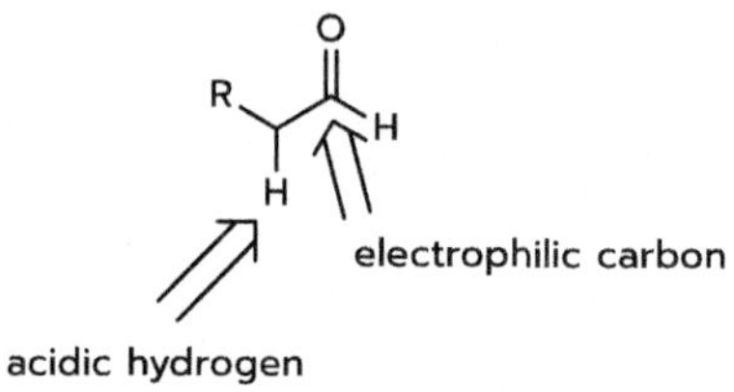

electrophilic carbon
acidic hydrogen

If we compare the corresponding acetal, we see that the acetal carbon atom (which was the carbonyl carbon atom) is not really electrophilic. Sure, it is δ+, so it could (in principle) be attacked by a nucleophile. But the acetal oxygen atoms are not very good leaving groups. That's why we need acid catalysis for the hydrolysis of an acetal. Similarly, the α-hydrogen atoms are no longer acidic.

> In making the acetal, we have 'shut down' all of the typical carbonyl reactivity.

If you have a carbonyl group in a molecule, and you don't want it to react in a given step (one in which it would normally react), but you want to preserve it for later, you can convert it into the corresponding acetal, do your other chemistry, and then regenerate the carbonyl group from the acetal by hydrolysis.

> The carbonyl group is 'protected' from reagents it would normally react with.

Protecting groups are central to the success of complex target synthesis.[54]

HEMIACETALS AND SUGARS

Let's look at structure **4** again. This is the half-way stage between an aldehyde and an acetal. We call this a 'hemiacetal'.

---

[54] I have several books on 'Protecting Groups'. These books were expensive, so protecting groups must be important to organic chemists. They are beyond the scope of this book, but it is useful to see why they work, and why they are needed.

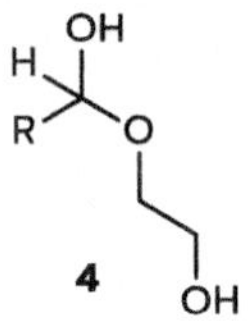

SECTION 3 NUCLEOPHILIC ADDITION TO THE CARBONYL GROUP

Are hemiacetals important? Here is the structure of glucose.

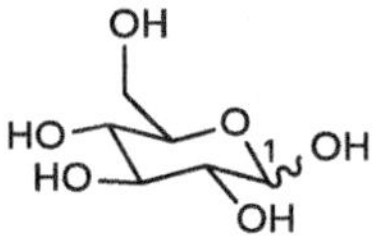

glucose

Carbon atom 1 is a hemiacetal carbon. Most sugars can exist in a hemiacetal or hemiketal (bet you can work out what this is!) form.

> I'd say hemiacetals are important!

ANOTHER QUICK EXERCISE

Glucose can be drawn in an open chain form, with an aldehyde functional group.

> Using the mechanisms in this chapter, identify the open-chain form and draw mechanisms for the hemiacetal form of glucose being converted into the open chain form, and back again.[55]

THE ROLE OF ENTROPY

Let's look at two reactions side by side. Which is most favourable?

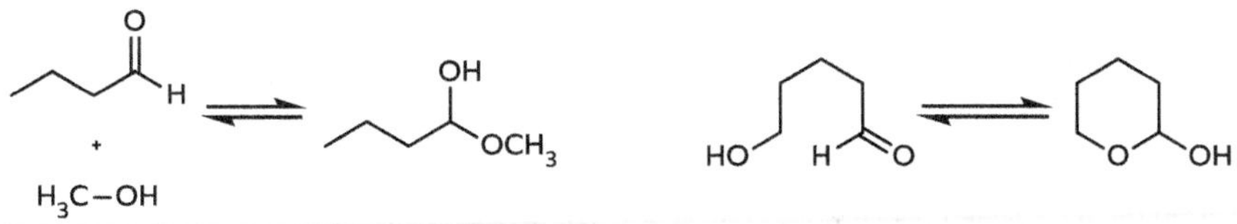

The reaction on the left is **intermolecular**—the OH group and the aldehyde are in different molecules. The reaction on the right is **intramolecular**—they are in the same molecule.

We could analyse this reaction in terms of bonds broken and bonds formed. If we are using average bond dissociation energies, we would get the same answer (**Exercise 2**).

The difference is in the *entropy*, not the *enthalpy*. The reaction on the right has one molecule going to one molecule. The entropy change, ΔS, will be close to zero. The

---

[55] Did you draw the open-chain form with stereochemistry? The correct stereochemistry? If not, why not? This needs to become your habit!

241

SECTION 3 NUCLEOPHILIC ADDITION TO THE CARBONYL GROUP

reaction on the left has two molecules coming together to give one molecule. The entropy change, ΔS, will be negative.

> If we start from our premise that these reactions will be *endergonic*, the reaction on the left will be **more** *endergonic* than the one on the right.

There's another point to make here. There usually is! For the intramolecular example, I chose one with a six-membered ring. Have another look at Stereochemistry 1. In a cyclohexane, all the bonds are staggered and all bond angles approximately 109°. If we chose a smaller ring size, there would be strain involved, which would affect the enthalpy of reaction.

> Of course, this is another change that would not be reflected in any calculation we do using average bond dissociation energies.

We will look at the stability/strain associated with different size rings in Basics 6. It is one more thing to understand, but it can be applied consistently to all processes.

Now we can come back to the acetal formation we looked at. We deliberately used a 1,2-diol. Again, let's compare two reactions. Here is the reaction we looked at before.

There is only a very small entropy change here. Two molecules are reacting to give two molecules. The alternative, below, is worse.

Here, we have three molecules reacting together to give two molecules.

HOW DO WE GET UNFAVOURABLE REACTIONS TO WORK?

Here is our acetal formation equilibrium again.

We know that there is a perfectly good mechanism to get us from the aldehyde to the acetal. But we also know that this process is *endothermic/endergonic*. We chose a cyclic acetal, because, as we have just seen, the reaction will be a bit less endergonic.

> But that doesn't solve the fundamental problem.

We can add various things to reactions that will absorb water molecules. That will help drive the reaction irreversibly to products. Or there is equipment that we can

SECTION 3 NUCLEOPHILIC ADDITION TO THE CARBONYL GROUP

use in the lab that will remove the water from a reaction. We don't need to go into details here. For now, you need to recognize that there is a problem, and understand that there is a single conceptual solution.

> We can only drive the equilibrium to the right by either removing product (not easy) or water (easy). This is an application of Le Chatelier's principle, which you've known about for ages!

SECTION 3 NUCLEOPHILIC ADDITION TO THE CARBONYL GROUP

APPLICATIONS 2

Formation of Imines and Enamines

We have just (Reaction Detail 15) looked at addition of oxygen nucleophiles to aldehydes and ketones. Now we are going to do the same for nitrogen nucleophiles. We have seen (Exercise 2) that thermodynamically we might expect different outcomes in these reactions. However, we should not expect the fundamentals, such as the reaction types or curly arrows, to be any different. Carbonyl groups only do two things. Addition of a nucleophile to the carbonyl carbon atom is one of them.

The previous chapter was a 'Reaction Detail'. This one is 'Applications'. We will take what we learned in the previous chapter and apply it to a slightly different system.

AMINES AND ACIDS

In Reaction Detail 3, we established that a carbonyl group is protonated more readily than water. However, both will be protonated to some extent in a bulk solution of both, so we don't really need to use this fact to explain why we can protonate the carbonyl group and add water as a nucleophile.

With amines, the situation is different, yet the same principles apply. We think of amines as bases (because they are!). We don't think of carbonyl groups as bases. It is considerably easier to protonate an amine than it is to protonate a carbonyl group. Protonation of an amine removes the lone pair.

Here are two competing equilibria.

The bottom one wins!

> So, how do we add a protonated amine to an unprotonated carbonyl group?

We don't! This would be impossible. We still need to add an unprotonated amine to a protonated carbonyl group. Fortunately, there is still enough unprotonated amine and protonated carbonyl group in an equilibrium mixture, and when they bump into one another, they react.

> Find the pK_a values for the protonated amine and carbonyl group, draw the equilibria and try to work out the equilibrium constants.

SECTION 3 NUCLEOPHILIC ADDITION TO THE CARBONYL GROUP

IMINE FORMATION

Now let's start to work through the mechanism of the reaction. I am going to look at various aspects of individual steps, and then show you the complete mechanism at the end.

We are going to start in the same way that we did for the acetal/ketal mechanism. We will use a *primary* amine. You'll see why in a minute. We protonate the carbonyl group (acid catalysis, Basics 2), and then add the amine as a nucleophile. This gives compound **3**. We can lose a proton from compound **3** as shown, to give 'aminol' **4**. Don't get worried by the name of yet another functional group. This is quite an obscure one.

Now we can protonate the oxygen atom in compound **4**. Sure, compound **3** is more stable than compound **5**—nitrogen is more basic than oxygen—but we will get a small amount of compound **5** at equilibrium, and this can lose water as shown. Formally we could think of this as carbocation formation (only one curly arrow needed) but the carbocation we would form is stabilized by the nitrogen atom.

> It is better to show this.

In the final step, we can lose a proton from the nitrogen atom to give the imine, compound **7**.

> Now you can see why we needed to use a *primary* amine. We lose one of the hydrogen atoms to form compound **4** and the second hydrogen atom to form compound **7**.

COMPARING ACETAL AND IMINE FORMATION

When we discussed hydrate formation, we were able to justify protonation of the carbonyl group based on our understanding of bonding.

> That's the reason I did it—to reinforce your understanding of bonding.

It turns out that it didn't actually matter. These are equilibria, and a product can be formed from an intermediate that is only ever present in small concentrations.

245

SECTION 3 NUCLEOPHILIC ADDITION TO THE CARBONYL GROUP

The other point we need to make is why these reactions do different things.

> Clearly one reaction has O and the other has N, but even so, the products look different.

Here is structure **6**. Alongside it is the comparable structure, **8**, from the acetal formation reaction in the previous chapter.

In structure **6**, we could either lose the H, or add a nucleophile to the carbon atom. In structure **8**, the only option is to add a nucleophile to the carbon atom.

Of course, in **Exercise 2** we established that imine formation was the most favourable, or the least unfavourable, process.

AVOIDING A COMMON ERROR

We are going to look at what would happen if we only had one hydrogen atom (a *secondary* amine) in a moment. I just want to have a look at a common error encountered when drawing mechanisms of this type.

Let's look at the formation of compound **5** from compound **3** again.

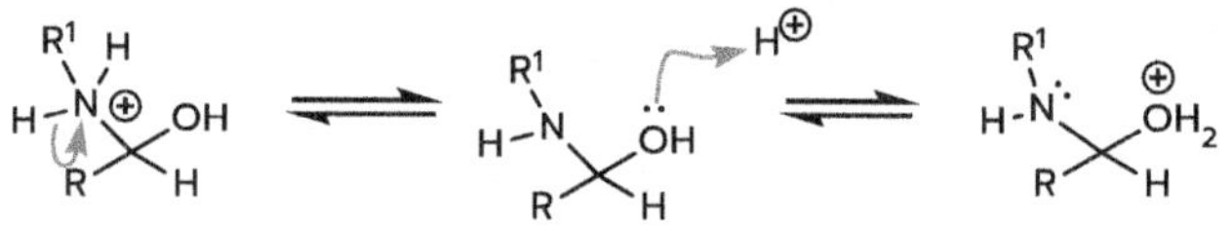

We could potentially abbreviate this as follows.

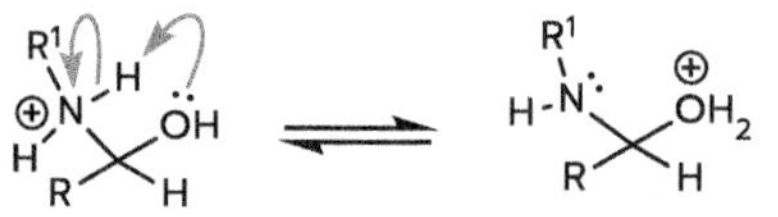

> It would be wrong to do this. Let's look at why.

We are breaking the N–H bond. In terms of the curly arrows, this looks a lot like an S_N2 reaction at hydrogen rather than at carbon. We know that in an S_N2 substitution reaction, the nucleophile attacks from the opposite side to the leaving group because this allows orbital overlap (**HTSIOC Reaction Detail 2**).

Look carefully at the structure **3** above. We have a four-membered ring transition state (N, C, O, H). The O–H–N bond angle in the transition state would be close to 90°—nowhere near the 180° that would be needed.

246

SECTION 3 NUCLEOPHILIC ADDITION TO THE CARBONYL GROUP

Basically, intramolecular proton transfer cannot happen in this way. But that's okay, because the proton can drop off N (to solvent?) and add onto O (from solvent?).

> Let's rephrase this in energetic terms. A direct intramolecular proton transfer has a high activation energy. A two-step process has a lower activation energy.

Now let's add a comment. An experienced organic chemist will draw the intramolecular proton transfer **because it's quicker to draw!** They know it's wrong, and they don't mean to imply that it happens in a single step. They are just taking a shortcut.

> It's okay to draw a single step, as long as you know why it's not strictly correct!

Now we can look at the complete mechanism.

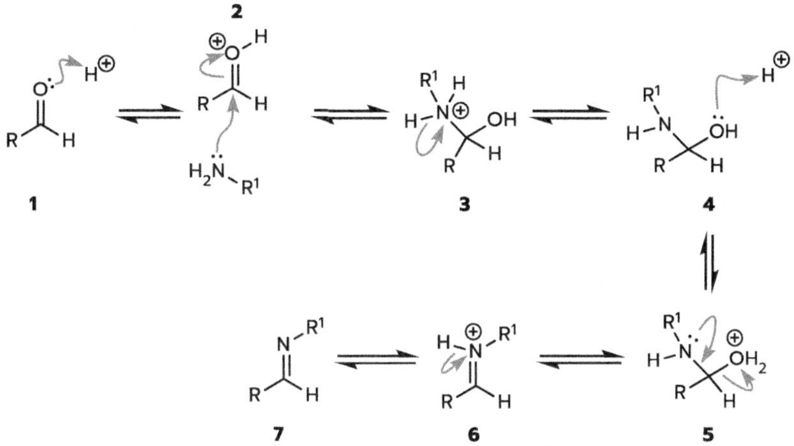

ENAMINE FORMATION

Now, let's look at what would happen if you start with a *secondary* amine. Here is structure **8**, the equivalent of structure **6**, but derived from a *secondary* amine. It cannot simply lose a proton from nitrogen in order to get back to being neutral. Alongside it, I've drawn structure **9**, which we drew above as an intermediate in acetal formation.

There are two ways to get rid of a positive charge. Either remove something positive (always a proton!) or add something negative. During acetal formation, we added an oxygen nucleophile to the carbon atom, and moved the electrons from the π-bond back to the oxygen atom. In **Exercise 2** we saw why—it gives the more stable product.

SECTION 3 NUCLEOPHILIC ADDITION TO THE CARBONYL GROUP

In the case of compound **8**, we don't add another nitrogen nucleophile, and we cannot lose a proton directly from the nitrogen atom, so we need to find an alternative. I'm going to draw a new structure below, with an additional CH₂ group, so we can see what is possible.

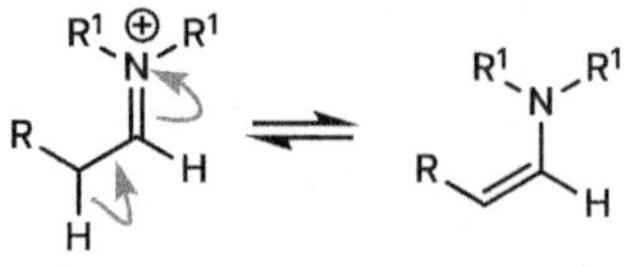

10 ⇌ **11**

We can lose a proton from the α-carbon. Iminium ion **10** is very similar to a carbonyl compound, and we have seen in **Recap 7** and **Fundamental Reaction Type 4** that the hydrogen atoms α- to a carbonyl are acidic. Hydrogen atoms α- to an iminium ion are even more acidic, because we get a neutral compound after loss of a proton.

We don't need to labour the point too much now. Here is the complete mechanism. Again, all the way to intermediate **10**, the pathway is the same. The only difference is how it can get to become a stable neutral product. We can't lose a hydrogen atom from nitrogen, so we must lose it from the α-carbon atom.

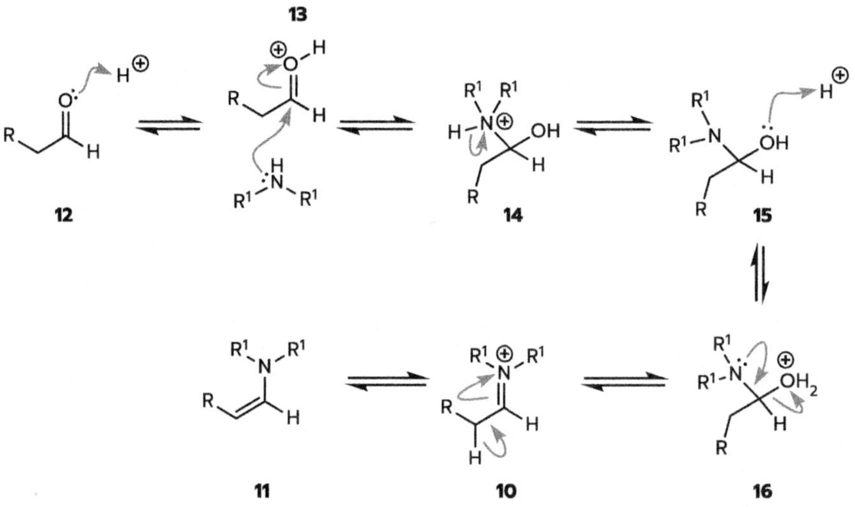

12 ⇌ **13** ⇌ **14** ⇌ **15**

11 ⇌ **10** ⇌ **16**

WHERE IS THIS GOING?

Imines and enamines are important compounds. They have a range of versatile reactivity. Compounds with nitrogen are really important, and reacting an amine of some sort with a carbonyl compound of some sort is a good way to get nitrogen into an organic molecule.

We will look at the reactions of amines with carboxylic acid derivatives in **Reaction Detail 16**. There are similarities and differences as a result of the different oxidation states.

Furthermore, many heterocyclic compounds contain imines and enamines. The structures of pyridine and pyrrole are shown below.

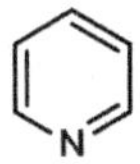

pyridine 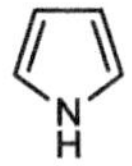   pyrrole

These structural motifs are often found in pharmaceuticals, and many methods for their formation involve the reactions of carbonyl compounds with amines. We will encounter some of these reactions in Applications 9.

REACTION DETAIL 16

Synthesis and Hydrolysis of Esters and Amides

Is this chapter 'Reaction Detail' or 'Applications'? There is nothing here that you don't already know. However, we need to emphasize *how* to apply what you know. We are reaching a point in the book where you shouldn't expect many more new ideas, and everything should be an application of ideas you have already seen. We are now using these examples to help you internalize the concepts. Of course, knowing the reactions is still important in its own right.

INTRODUCTION

This chapter is a one-trick pony. It is all about pK_a. If you understand the stability of a leaving group, or the reactivity of a nucleophile, you can predict which reactions will work. For the ones that you expect to not work well, you can work out a strategy to make them better.

To get us started, let's summarize the reactions we are talking about on a single scheme.

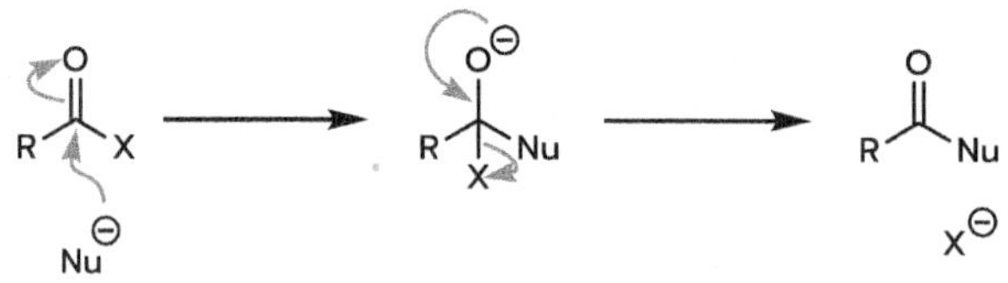

> We can always think in terms of stability. If RCOX is more stable than RCONu then you might expect the process to be disfavoured.[56] Of course, if $X^\ominus$ is much more stable than $Nu^\ominus$ then we might be able to over-ride this inherent preference.

The problem is, if RCOX is more stable, this would generally mean that $X^\ominus$ would be less stable.

CALCULATING ENTHALPY OF REACTION

We saw how to calculate the enthalpy change for a given reaction in **HTSIOC Basics 13**. We added to this by considering how to include a consideration of pK_a in **HTSIOC Basics 35**. So, we ought to be ready to calculate the enthalpy change for any given process represented by the above reaction.

> Be careful!

---

[56] I am drawing these out in shorthand form so you get used to seeing this!

SECTION 3 NUCLEOPHILIC ADDITION TO THE CARBONYL GROUP

You *can* do this calculation, but you need to pick your data carefully. A C–N bond in an amide is very different to a C–N bond in an amine. If you use a table of 'average' bond dissociation energies, you might not be using a value that is appropriate for the functional group you have in your molecule. By now we need to be thinking about this in a more refined way. If your compound is an amide, find a typical bond dissociation energy for an amide.

We aren't going to include these calculations in this chapter. However, you are strongly encouraged to find the values for pK_a and bond dissociation energy and try the calculations yourself. You know what I think about the value of practice!

ESTERS AND AMIDES—CONTEXT

Esters and amides are really important industrially, and are manufactured by the tonne. They are present in many synthetic fibres (nylon, polyester!). They are also incredibly important functional groups in pharmaceutical products and as part of biological systems (for example enzymes).

Esters are formed from carboxylic acids and alcohols by a formal dehydration (loss of water) process. Amides are formally made by dehydration of a carboxylic acid and an amine, although as we will see, the reaction is generally carried out in a different way.

It turns out that we sometimes carry out these reactions under acidic conditions, and in other cases under basic conditions. The curly arrows always look the same, apart from the number and timing of protonation and deprotonation steps.

> Make sure you have another look at Basics 2 before you carry on reading. Acid catalysis is at the heart of this discussion.

WHY DON'T YOU FORM ESTERS UNDER NEUTRAL CONDITIONS?

Here is a hypothetical equilibrium.

$$\underset{R}{\overset{O}{\|}}\text{–OH} + \text{HO–R}^1 \rightleftharpoons \underset{R}{\overset{O}{\|}}\text{–O}^{\ominus} + \text{H}_2\overset{\oplus}{\text{O}}\text{–R}^1$$

A carboxylic acid is an acid. It's all in the name! However, it's not a strong acid. A pK_a in the region of 5 is typical. The protonated alcohol on the right has a pK_a in the region of −2. This is a strong acid! You would probably need to look up the exact value, but losing a proton from a positively charged species will generally be more favourable than losing a proton from a neutral species. The equilibrium will definitely be towards the left.

Even if this does happen, we would then need the following to occur.

251

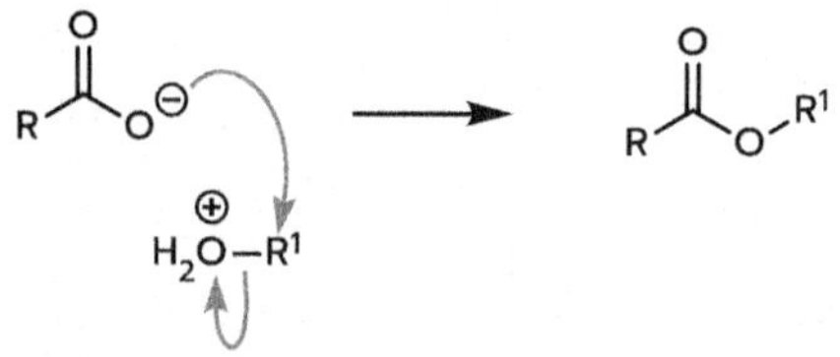

SECTION 3 NUCLEOPHILIC ADDITION TO THE CARBONYL GROUP

The carboxylate anion is a poor nucleophile (it is too stable!) and there won't be much of the protonated alcohol anyway. Remember that this is just an S_N2 reaction, and you have seen those before.

> We have a disfavoured acid-base equilibrium and a poor S_N2 substitution reaction. It isn't looking good!

ESTER FORMATION UNDER ACIDIC CONDITIONS

It turns out that the best way to form an ester directly from a carboxylic acid and an alcohol is using an acid catalyst. We have a carboxylic acid with two oxygen atoms, and an alcohol with one oxygen atom.

> Before you read further, try to work out the best mechanism based on which steps are favoured following protonation at each of the possible sites, followed by attack of a nucleophile.

Have you done it? Reading ahead without drawing the possibilities is the easy option.

Let's get the simple one out of the way first. We will protonate the alcohol and use the carboxylic acid as the nucleophile. I haven't drawn the actual protonation step.

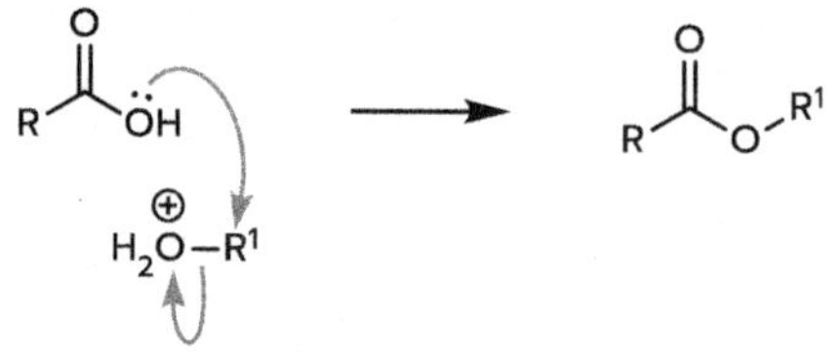

> We saw above that when this was a carboxylate anion as the nucleophile, it wasn't good. Here, we have made it worse. A carboxylic acid is an appallingly diabolically poor nucleophile. This reaction is simply not going to happen!

Let's consider option two. We protonate the carboxylic acid OH group, and use the alcohol as a nucleophile.

Here we have one step shown explicitly, and a second step hinted at by the curly arrows on the right-hand structure. Protonation of the OH group does not help the

SECTION 3 NUCLEOPHILIC ADDITION TO THE CARBONYL GROUP

first step. We are always going to end up with a product with a positive charge, but the additional separation of charge cannot be a good thing. We already know from **Basics 2** how to use acid catalysis to facilitate nucleophilic attack onto a carbonyl group.

> *The second step would be very much favoured, as we have an unstable negatively charged oxygen and a good leaving group. But it will never get to this point, so this won't matter!*

This all leads us to the third option—protonation of the carbonyl group. I am going to draw this out fully, because it is the 'right answer' and it needs reinforcing.

> *A protonated carboxylic acid has a pK_a of around −8. You need a strong acid to protonate a carboxylic acid. It is easier to protonate the alcohol, but this doesn't get us anywhere, as we have already established.*

What have we achieved by doing this? We have made the carbonyl carbon more susceptible to nucleophilic attack (**Basics 2**). The alcohol can attack now.

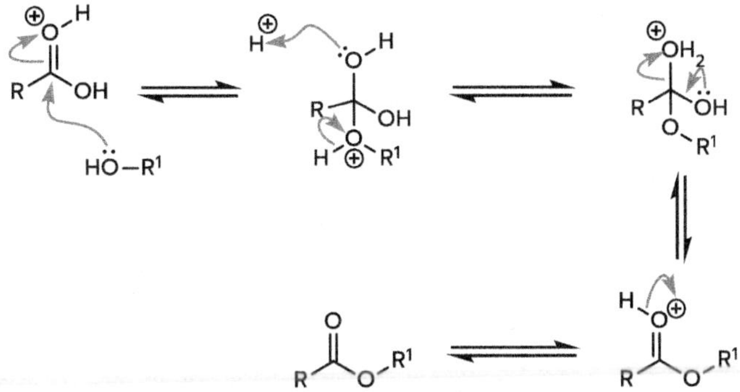

After the initial nucleophilic attack, we need a proton transfer step to make the OH group into a better leaving group (compare this with the acetal formation mechanism in **Reaction Detail 15**).[57]

> *You will notice that in the structure in the middle, we have a protonation and a deprotonation step. We do this to save drawing an extra structure. We don't really mean that they happen at exactly the same time.*

---

[57] It would probably have been easier for you to follow this scheme if I had put numbers on structures. I'm not being lazy. I want you to have to work for this. That way, you won't just glance at this and move on.

SECTION 3 NUCLEOPHILIC ADDITION TO THE CARBONYL GROUP

ESTER FORMATION UNDER BASIC CONDITIONS

What would happen if we tried to do the same thing under basic conditions? If we treat a mixture of a carboxylic acid and an alcohol with a base, there is no doubt that we will deprotonate the carboxylic acid.

Now we can see the problem. To form the ester, we would need the carboxylate anion to attack the alcohol with loss of hydroxide. The carboxylate anion is so much more stable than the hydroxide anion that this is never going to happen.

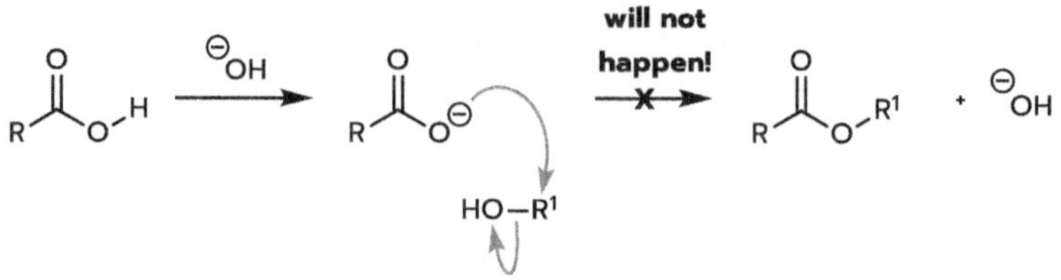

> We made the same argument above under acidic conditions. If the carboxylate isn't a good enough nucleophile to displace the protonated alcohol, it certainly won't displace it when it isn't protonated.

How could we improve this situation? We might use a better leaving group than hydroxide. We cannot protonate the alcohol at the same time as we deprotonate the carboxylic acid (we saw this above!). Instead, we might use iodide as the leaving group. Carboxylate anions will react with iodomethane to give the corresponding esters. However, even this is not a very favourable reaction. It is an S_N2 substitution, and the carboxylate is a terrible nucleophile.

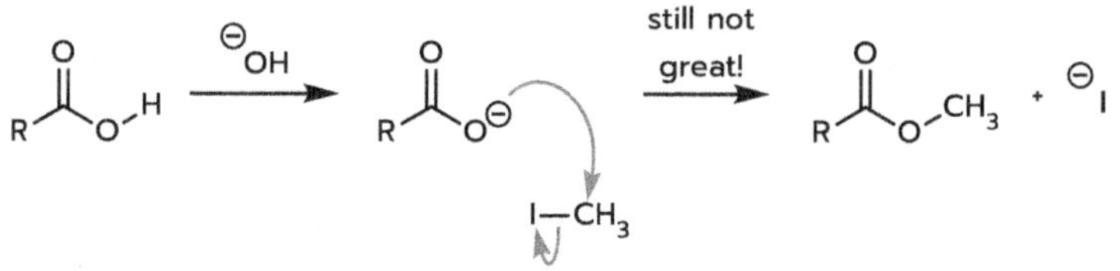

> Look back at S_N2 substitution in your preferred textbook. We might get away with this approach for making methyl or benzyl esters, but it's unlikely to be a good general method. Make sure that you are making connections, not just learning reactions.

Now, let's move on to looking at the reaction in the other direction—hydrolysis of esters.

ESTER HYDROLYSIS UNDER ACIDIC CONDITIONS

Esters can be hydrolysed under acidic or basic conditions. The hydrolysis under acidic conditions requires quite a strong acid, and the mechanism is exactly the reverse of the formation, shown above.

Here it is!

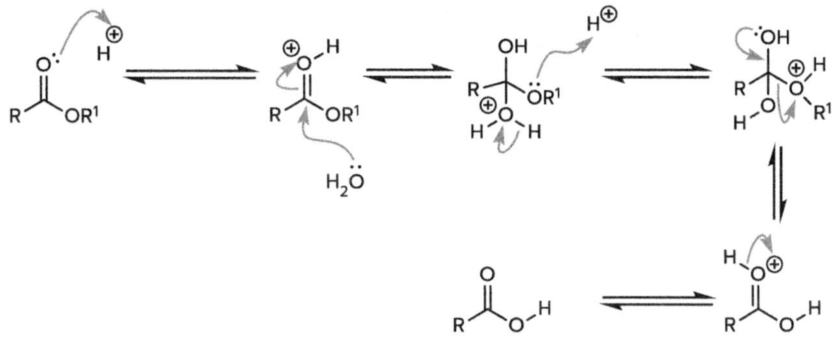

SECTION 3 NUCLEOPHILIC ADDITION TO THE CARBONYL GROUP

I am genuinely struggling to think of something new to say about this reaction. We have already seen all the explanation for this in the ester formation. The only difference is that here the nucleophile is water rather than an alcohol.

> Look through this mechanism, and then try to draw it without looking at the scheme. Make sure that after the initial protonation, and before the final deprotonation, every intermediate only has one positive charge.

From a practical point of view, if you dissolve a carboxylic acid in an alcohol solvent, and add acid, you will make an ester. If you dissolve an ester in water and add acid, you will hydrolyse an ester.[58]

ESTER HYDROLYSIS UNDER BASIC CONDITIONS

In this case, the heading is a little misleading.

> The base is not acting as a base.

This reaction needs hydroxide, not just any base. The mechanism actually starts with the base (hydroxide) acting as a nucleophile and attacking the carbonyl carbon. The negative charge on oxygen can go back to form the C=O bond, but this time kick out the alkoxide anion, to give the carboxylic acid.

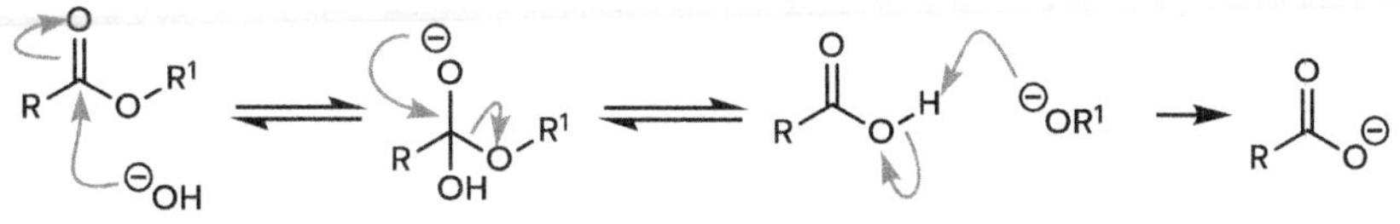

> The hydroxide anion is the conjugate base of water. The alkoxide anion is the conjugate base of an alcohol. As a first approximation, they are of similar stability.

Now, we have generated an acid under basic conditions. The alkoxide anion which we kicked out can come back and remove the proton from the carboxylic acid group.

---

[58] Apart from the minor complication that most carboxylic esters are not very soluble in water. This is a practical complication that doesn't change the principles.

255

SECTION 3 NUCLEOPHILIC ADDITION TO THE CARBONYL GROUP

This step is essentially irreversible, so that the reaction is driven to completion.

> Under acidic conditions, we need to dissolve the ester in a poor solvent (water). Under basic (well, nucleophilic!) conditions, we can use an organic solvent (just not an alcohol or an ester!) and make the reaction irreversible. Knowing the mechanism of the reaction can guide you towards the best way to do it.

In summary, the best way to form an ester from an alcohol and a carboxylic acid is under acidic conditions. The best way to hydrolyse an ester to give a carboxylic acid and an alcohol is under basic conditions.

Now let's look at amides. This time we will look at hydrolysis first.

AMIDE HYDROLYSIS

We have just established that we can carry out ester hydrolysis under acidic or basic conditions. What about amide hydrolysis?

> Before you carry on, redraw the reaction schemes for ester hydrolysis under acidic and basic conditions, but change the substrate from an ester to an amide. Consider each step in turn, and decide if the change from O to N makes some steps better or worse.[59]

Hopefully you came to the conclusion that acidic conditions will be better than basic (nucleophilic!) conditions.

Why won't the reaction work under basic conditions? Here is the reaction mechanism.

[Reaction mechanism showing structures 1, 2, 3]

Amides are pretty stable.

> Draw the resonance forms to show this stability. Have a look back at **Applications 1** if you need a hint.

If the amide is stable, then the activation barrier for the first step will be high. This is always going to be the case (we will lose the resonance stabilization under acidic or basic conditions) but we already know that acidic conditions will stabilize **2**.

> Perhaps you should draw a reaction profile to compare esters with amides!

---

[59] You would think that by now I would have stopped nagging you. Think again! If you find that drawing these schemes out is tedious, then it is exactly what you need to do. It will get easier. If it already has got easier, then it wasn't such a pain anyway.

SECTION 3 NUCLEOPHILIC ADDITION TO THE CARBONYL GROUP

The more pressing problem is the conversion of **2** into **3**. This requires the loss of an $NH_2^\ominus$, whereas the hydrolysis of an ester under basic conditions required the loss of an alkoxide leaving group. Negative charges on oxygen are relatively stable (pK_a around 15). However, a similar leaving group with a negative charge on N would be **much** less stable ($pK_a > 30$). Even if the hydroxide attacks the amide, to form **2**, it won't go any further.

> As always, don't just learn this. Make sure you understand the reasons. Go back and identify the chapters where we covered the basic principles.

Now let's think about what happens when we use acidic conditions instead. This mechanism is almost identical to the mechanism of ester hydrolysis under acidic conditions. This is no coincidence. It does the same things for the same reasons.

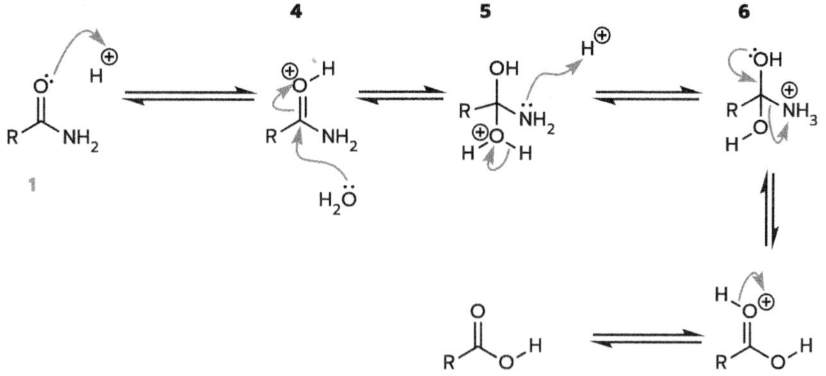

The mechanism, shown above, is for the hydrolysis of a *primary* (NH_2) amide, but it works the same for *secondary* and *tertiary* amides as well. Protonate the carbonyl group, not the nitrogen atom (the amide nitrogen is much less basic than an amine nitrogen) and water can attack the carbonyl group. The proton transfer steps occur to give you a neutral nitrogen leaving group (ammonia in this case).

Now we need to analyse a couple of the steps in more detail. We will directly compare intermediates in the ester and amide hydrolysis mechanisms.

First of all, comparing the ester **7** and the amide **1**, we have seen that the amide is significantly more stable (**Applications 1**). Protonation of the carbonyl oxygen atom will not disrupt the resonance stabilization. Draw the resonance forms to convince yourself that this is the case.

Now we should compare **5** and **8**. In both cases the resonance stabilization no-longer exists, because the double bond has been lost. Of course, there will be some difference in stability, but it won't be as much as the difference between **7** and **1**.

SECTION 3 NUCLEOPHILIC ADDITION TO THE CARBONYL GROUP

$$\underset{8}{\overset{OH}{\underset{H\overset{\oplus}{O}H}{R-\overset{|}{\underset{|}{C}}-\ddot{O}R^1}}} \qquad \underset{5}{\overset{OH}{\underset{H\overset{\oplus}{O}H}{R-\overset{|}{\underset{|}{C}}-\ddot{N}H_2}}}$$

Now it's over to you to finish the job. We know the relative stability of **1** and **7**, and of **5** and **8**.

> You can draw these on a reaction profile.

In each case you can safely assume that the formation of the intermediate will be endothermic. You can then apply the Hammond Postulate (**Recap 2**) to draw conclusions about the relative sizes of the activation barriers for the formation of **5** and **8**.

> As always, it's a lot to do the first few times you do it. However, taking this methodical approach pays dividends in the long run. You will develop the neural pathways that allow you to take these steps subconsciously. You will know the answers without really working through the problem.[60]

You should have reached the conclusion that amide hydrolysis will be considerably slower than ester hydrolysis.

AMIDE FORMATION

It would be really nice and simple if we could simply react an amine with a carboxylic acid to give an amide. We have already looked at why you don't form esters from carboxylic acids and alcohols under neutral or basic conditions. The situation is far worse for amide formation, since the amine itself is a base.

> If we make the amine less basic, we also make it less nucleophilic. We always need the amine to act as a nucleophile, so we cannot get around the problem by reducing the basicity of the amine.

Here is the equilibrium we saw with a carboxylic acid and an alcohol. We established that the position of the equilibrium is very much to the left.

$$R-\overset{O}{\underset{}{C}}-OH \; + \; HO-R^1 \; \rightleftharpoons \; R-\overset{O}{\underset{}{C}}-O^\ominus \; + \; H_2\overset{\oplus}{O}-R^1$$

Here it is with a carboxylic acid and ammonia. Ammonia (and any amine) is basic, so we would expect the equilibrium to be shifted towards the right. This would not help us make an amide. The ammonium ion is not a nucleophile.

---

[60] Sorry to nag (not sorry!) but it's worth reminding you that while this isn't 'elegant learning', it is how you got good at most of the things you do routinely. Just keep practising until you get good at it!

SECTION 3 NUCLEOPHILIC ADDITION TO THE CARBONYL GROUP

When we have a problem like this, we should always take it back to basics. Here is the scheme we saw at the beginning of this chapter.

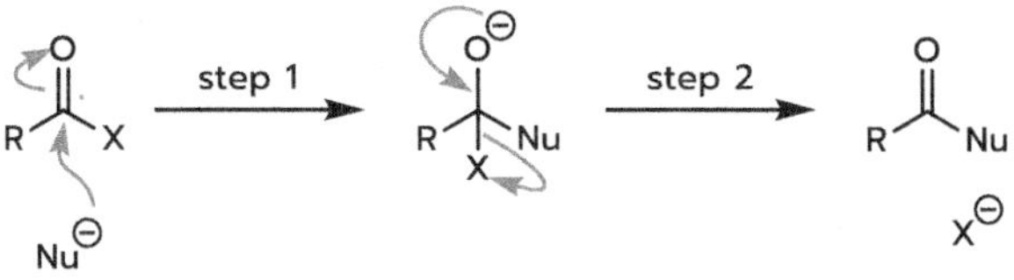

We definitely need a nitrogen nucleophile. After all, we need to form a C–N bond. The 'X' group cannot be an OH. We don't want anything with an acidic hydrogen. We want 'X' to make the carbonyl susceptible to nucleophilic attack (make step 1 faster) and to be a good leaving group (make step 2 faster).

It turns out that one of the best ways to accomplish this transformation is to use an acid chloride.[61] Chloride is a pretty good leaving group, and highly electronegative, so if we have a carbonyl with a chloride attached, this will be a very reactive carbonyl group (Applications 1), so it will be readily attacked by a nucleophile. From this point, I have drawn proton loss, and then the negative charge on oxygen moving back in and kicking out the chloride.

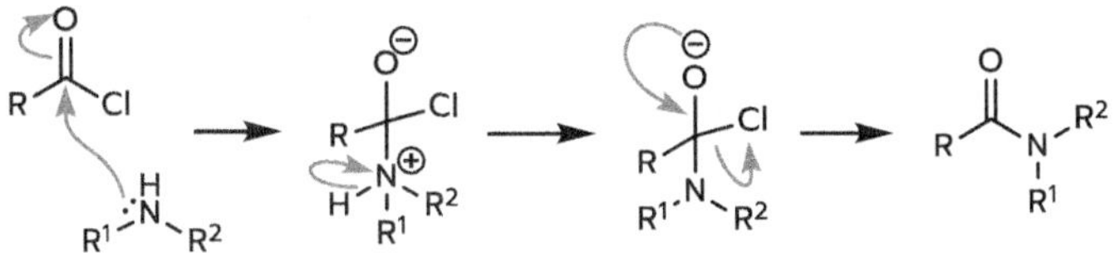

This is a very general method for the formation of amides. Acid anhydrides are also commonly used for this transformation. Have a look at the structure below, and see if you can work out why the use of an anhydride might be less attractive.

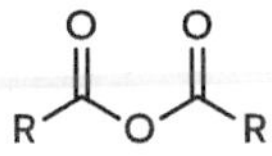

It requires an extra step to make an acid chloride from a carboxylic acid. To be fair, it also requires an extra step to make an acid anhydride from a carboxylic acid. If the acid anhydride you need is commercially available and very cheap, it might be better to use the anhydride. Otherwise, it is generally better to make the acid chloride.

We will look at how we make acid chlorides in Reaction Detail 17. In some ways, this will be an application, because we have already seen all the ideas. Of course, there will be some differences, so there is still stuff to learn.

---

[61] There are many other methods, but they all apply the same principles. If you can understand this, you will understand the other related methods.

SECTION 3 NUCLEOPHILIC ADDITION TO THE CARBONYL GROUP

APPLICATIONS 3

Lithium Aluminium Hydride Reduction of Esters and Amides

In **Reaction Detail 16**, we looked at the formation and hydrolysis of esters and amides. One of the things we saw was that a nitrogen atom with a negative charge is so much less stable than an otherwise similar oxygen atom with a negative charge.

Whenever you have two functional groups that are similar, and yet do things slightly differently, it is always good to focus on the most important differences.

> In this case, it is the lower electronegativity of nitrogen compared to oxygen.

In **Reaction Detail 14**, we looked at hydride reduction of aldehydes and ketones. Now, we will look at hydride reduction of esters and amides. We will apply the same fundamental principles as we did above, but this time you will do all the work (with a little guidance!).

HERE'S THE PROBLEM:

Lithium aluminium hydride can be used to reduce a range of other carbonyl compounds. For example, reduction of an ester gives an alcohol. Reduction of an amide tends to give the amine.

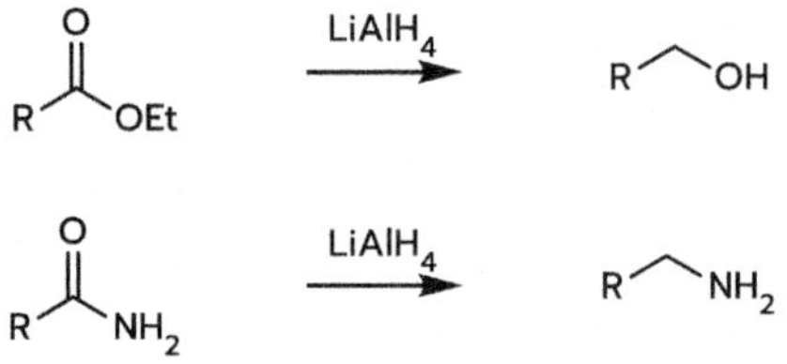

> Why do you get a different outcome in these cases?

GUIDANCE

You saw the reduction of aldehydes in **Reaction Detail 14**.

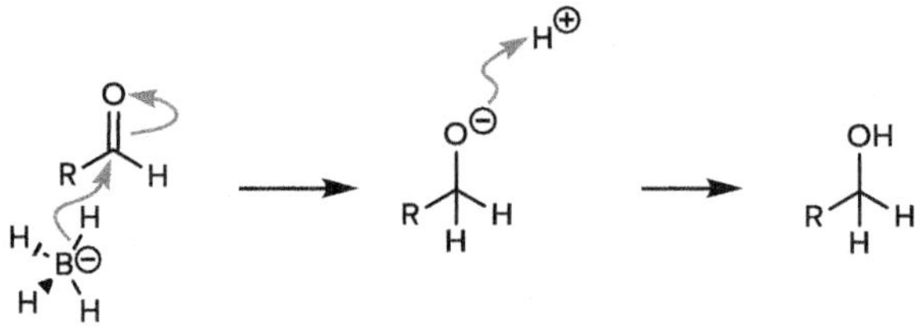

You also saw the reaction of Grignard reagents with esters in **Reaction Detail 12**.

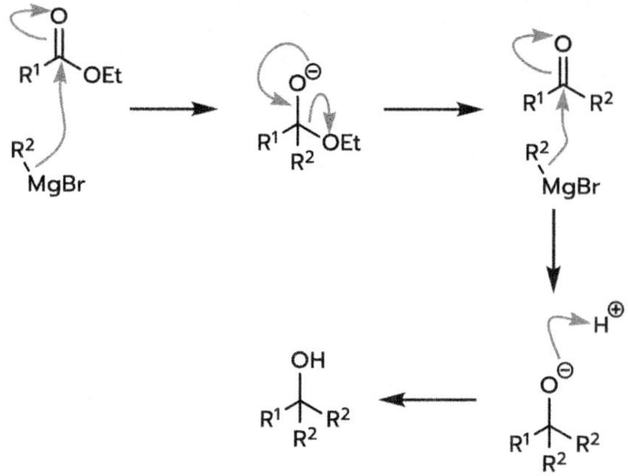

SECTION 3 NUCLEOPHILIC ADDITION TO THE CARBONYL GROUP

You also saw the reaction of the intermediate alkoxide with the borane produced from sodium borohydride.

> This will help the oxygen act as a leaving group, if needed.

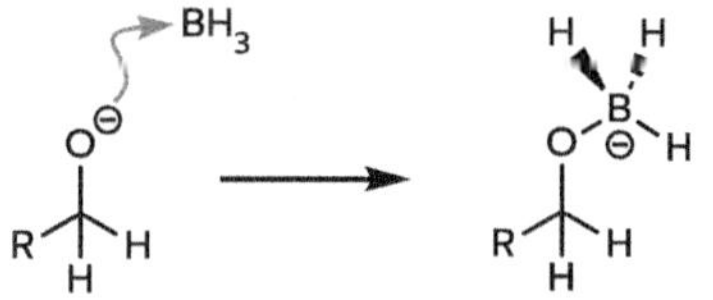

WHAT ELSE DO YOU NEED TO CONSIDER?

Make sure you know, and understand, the pK_a of alcohols and amines. Draw out the equilibria that these numbers correspond to.

> This will allow you to relate pK_a to leaving group ability.

Remember that there are no protons present until you add water at the end of the reaction. Don't protonate an intermediate and then use LiAlH₄ to deliver another hydride afterwards—there wouldn't be any left around!

Fundamentally, just draw the same curly arrows we have been drawing so far in all the chapters I have mentioned. As long as you don't draw anything 'silly', you should be fine.

> And if you aren't fine the first time, look at what you did wrong, learn from it, and have another go!

SECTION 3 NUCLEOPHILIC ADDITION TO THE CARBONYL GROUP

REACTION DETAIL 17

Formation of Acid Chlorides

INTRODUCTION

We encountered acid chlorides in **Reaction Detail 16**. Now we will look at how they are made.

There is one thing we need to **really** consider.

> Acid chlorides are very reactive (= unstable) carbonyl compounds. If we want to go from a carboxylic acid to an acid chloride, we are going uphill in energetic terms. We cannot get this for free. Something else *must* come downhill.

To put this in more scientific terms, acid chloride formation must be *exergonic*, or it would not happen.

> Don't get into the habit of learning and drawing curly arrow mechanisms (even very good ones) without considering what they mean.

THE REACTION MECHANISM

The mechanism for acid chloride formation looks complex, but (like every other mechanism we will see) it follows some basic principles.

We start with a carboxylic acid and thionyl chloride (SOCl$_2$). Doesn't thionyl chloride look like an acid chloride? It reacts like one as well. In this case the carboxylic acid oxygen atom attacks the sulfur atom, to kick out chloride. This would leave us with a protonated intermediate, but we can just lose the proton.

I have done something a little different with the curly arrows here. Look at the curly arrow along the S–O double bond. It has an arrowhead at each end.

This is a form of shorthand to indicate that the electrons move out onto the oxygen to form a tetrahedral intermediate (as we saw for carbonyl groups in **Reaction Detail 16**) and then back in to kick out the chloride. It just saves a little time to draw it this way.

> This looks okay in energetic terms. Thionyl chloride is reactive, and we are forming something that looks a bit like an anhydride (we saw these at the end of **Reaction Detail 16**)—this is no worse than an acid chloride.

SECTION 3 NUCLEOPHILIC ADDITION TO THE CARBONYL GROUP

Then the chloride can kick back in and attack the carbonyl to kick out the oxygen. This oxygen is a really good leaving group. This gives us the acid chloride.

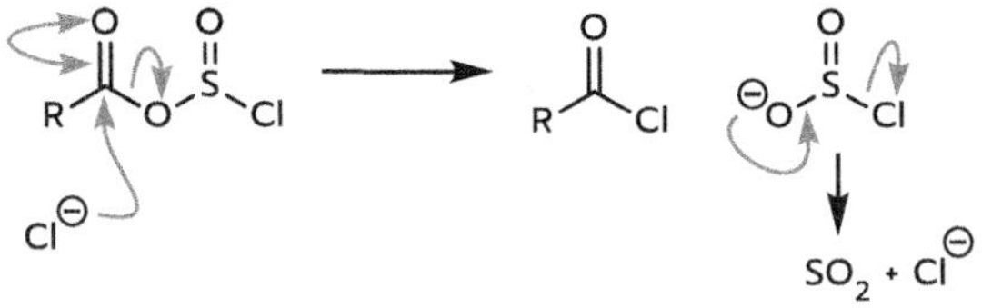

The payback in this reaction is the by-product. This can break down as shown to give SO_2 (a gas—entropy) and HCl (also a gas).

> Hang on! You said it gives HCl, but the scheme shows it losing a chloride anion. Yes, but remember the proton we lost in the first scheme? You must keep track of the overall equation!

OTHER REACTIONS OF ACID CHLORIDES

Now we have seen acid chlorides, we may as well look a little harder at them.

If you remember **Reaction Detail 12**, you will recall that we looked at the addition of organometallic reagents to aldehydes, ketones and esters. In particular, if you add a Grignard reagent to an ester, it tends to add twice. It is hard to stop this reaction at the ketone stage because ketones are more reactive than esters.

However, acid chlorides are much more reactive than ketones, so if we can find a less reactive organometallic reagent—something that won't react with a ketone—we can stop this reaction at the ketone stage.

The most common reagents for this transformation involve copper, and in many cases lithium as well. In the example shown below, we have a lithium dialkylcuprate.

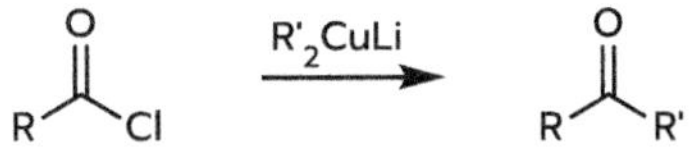

This is really useful, since we have seen how many things we can do with ketones. Now we have a simple way of making them in two steps from carboxylic acids.

> Once again, everyone reading this book is able to understand that if you have a more reactive substrate, you can use a less reactive reagent. Once you see it in these terms, it is easier to put organocopper reagents and acid chlorides into context.

We will see cuprates taking part in conjugate addition reactions in **Reaction Detail 28**.

SECTION 3 NUCLEOPHILIC ADDITION TO THE CARBONYL GROUP

REACTION DETAIL 18
Friedel-Crafts Acylation

INTRODUCTION

This chapter features a really important reaction, but there won't actually be much new chemistry. As always, it's about making connections.

In **Reaction Detail 9** we encountered the Friedel-Crafts **alkylation** reaction, and we saw that this is a really neat way to form a C–C bond to an aromatic ring, as long as you are prepared to accept a mixture of products.

> We certainly don't want that!

In **Applications 1**, we saw that acid chlorides are among the most reactive types of carbonyl compound. In **Reaction Detail 16** we saw that you can add a neutral nucleophile (an amine) to an acid chloride, and you then lose HCl to get a neutral product.

> We are going to do the same thing here, but with a different nucleophile.

In **Recap 6** we saw that benzene is about 150 kJ mol^{-1} more stable "than it should be".

> When a benzene ring undergoes electrophilic substitution, we lose the aromaticity upon formation of the intermediate.

The activation energy will be relatively high. While acid chlorides are very reactive carbonyl compounds, we need to make them even more reactive (less stable!).

> We know how to do this from **Basics 2**.

All we need now is to add the details.

DRAWING THE MECHANISM

Let's go with the 'simple' example of benzene reacting with acetyl chloride. We will use aluminium trichloride as a catalyst. Here is what happens when we add the catalyst.

We have coordinated the aluminium chloride to the chlorine atom of the acid chloride. This makes sense because aluminium and chlorine are in the same row of the periodic table. You would expect this interaction to be favourable.

SECTION 3 NUCLEOPHILIC ADDITION TO THE CARBONYL GROUP

We refer to the highlighted intermediate as an acylium ion. This is more reactive than an acid chloride, and it reacts with benzene as follows.

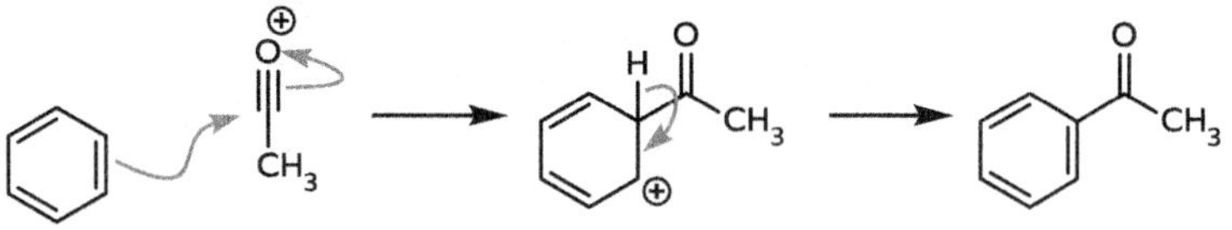

HOW DO WE KNOW?

What I mean by this heading is 'how do we know to draw this?' It's a fair question. Every other time we have added an acid to a carbonyl compound, we have protonated the oxygen atom. Just because this is a Lewis acid, why should it be different?

> Why don't we try drawing this 'alternative' mechanism?

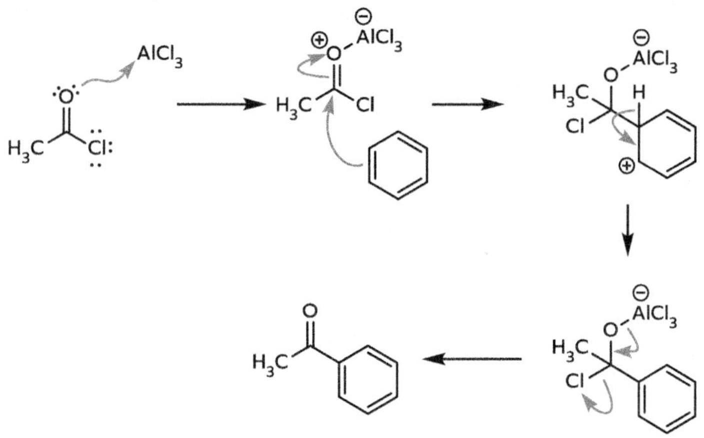

Every step in this mechanism looks perfectly sensible.

> Draw it out for yourself. Check that you understand why we have the charges as drawn.

So, we have two mechanisms that both appear to be plausible. How do we know which mechanism operates?

> We justified coordination of $AlCl_3$ to the chlorine of acetyl chloride based on our knowledge of the periodic table.

This is sensible, but this doesn't make it right, especially since we have another sensible mechanism. I would say the short answer is 'it probably doesn't matter'. Both mechanisms give the same product.

It turns out that there is experimental evidence for both mechanisms operating in various Friedel-Crafts acylation reactions. At this level, I am not going to present the evidence or get bogged down in detail. The key point is that you might draw the second mechanism and worry about having drawn 'the wrong mechanism' and wonder whether you should have known it was wrong.

> *Some very good chemists, about 60 years ago, did the same thing.*

They then wondered which acid halides (not always chloride) might react with which aromatic compounds by either mechanism, and they devised some nice experiments to test their theories.

The first mechanism is 'the textbook mechanism'. This is because it is more common. Draw it a few times. It will 'sink in' with experience, but if you are asked to draw a mechanism, it is much more important that you draw a sensible mechanism with good curly arrows and all reagents doing 'the right sort of thing'.

> *This approach is much better than trying to remember the mechanism without understanding it. If you take this approach, you might find you draw something that does look broadly like the textbook mechanism, but that has fundamental errors.*

THE POINT

We started in **Reaction Detail 8**, looking at reactions in which we form bonds to benzene rings. Knowing that C–C bond formation has a very special place in organic chemistry, we saw, in **Reaction Detail 9**, that we can take the same general approach to C–C bond formation.

> *But it had major problems.*

We have now seen a good way to resolve these problems.

QUESTIONS

In this reaction, we have added a carbonyl group to a benzene ring.

> Will the product be more reactive than the starting material?

Deep down, you already know the answer to this question. But draw out a second Friedel-Crafts acylation, this time starting with the product of the above reaction.

> Is the acetyl group activating or deactivating?

Another way to put this is 'does the carbonyl group stabilize or destabilize the intermediate?'

> Where on the benzene ring will the second acetyl group attack?

The point is to work this out by considering all possible alternatives, and deciding which is favoured. By doing all this, you will gain confidence that you can do the same for **any** aromatic electrophilic substitution reaction.

SECTION 3 NUCLEOPHILIC ADDITION TO THE CARBONYL GROUP

> If you are struggling, work through **Reaction Detail 8** again. You might also need to look at **Recap 2** for the stability (or not!) of positive charges adjacent to carbonyl groups.

ONE MORE REACTION—VILSMEIER-HAACK

I've mentioned named reactions before, particularly the cognitive load associated with learning which name relates to which reaction.

> *Don't try to remember which is which. Focus on the structures and drawing sensible mechanisms when given the reagents.*

There are many hundreds of common named reactions, and in a book focusing on teaching you mechanistic principles, most of them won't make the cut. This one did!

Let's look at the following transformation.

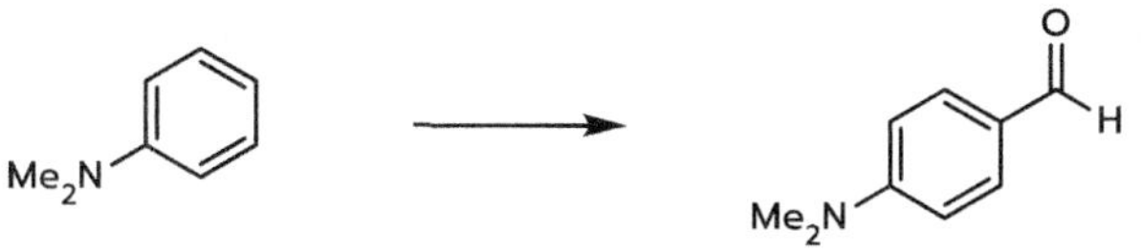

In principle, this looks like a potential Friedel-Crafts reaction, but it turns out that it doesn't work very well.

> *You wouldn't know that!*

There is a better way to accomplish this transformation.

> *This is something else that you would not know without being told.*

Let's make a start. We are going to take dimethylformamide and add phosphoryl chloride.

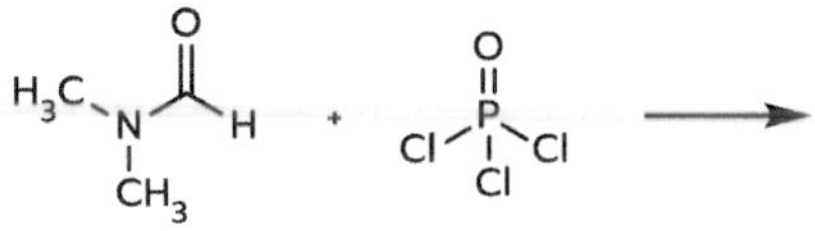

We saw thionyl chloride in **Reaction Detail 17**. Phosphorus forms strong bonds to oxygen (DNA has lots of P–O bonds). The nitrogen atom of dimethylformamide is not very nucleophilic, but the oxygen atom will react with phosphoryl chloride.

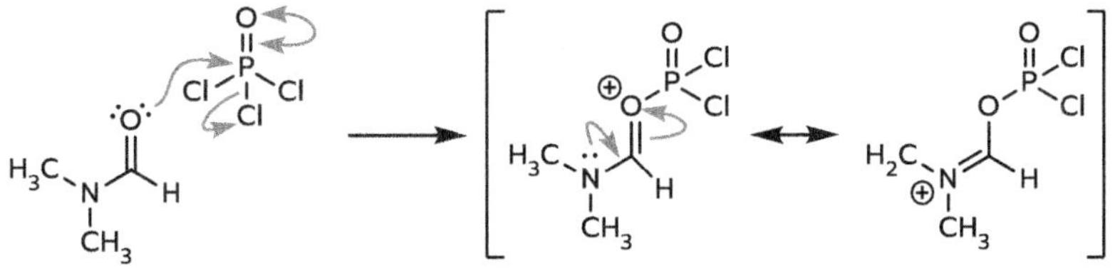

The product is a resonance-stabilized iminium ion (**Applications 2**). What we have done here is to take a nice stable amide and turn it into something that is a lot more reactive. But a phosphate is a good leaving group.

SECTION 3 NUCLEOPHILIC ADDITION TO THE CARBONYL GROUP

> Why is phosphate a good leaving group? Think about pK_a.

It turns out that the chloride that we lost in the first step attacks as follows.

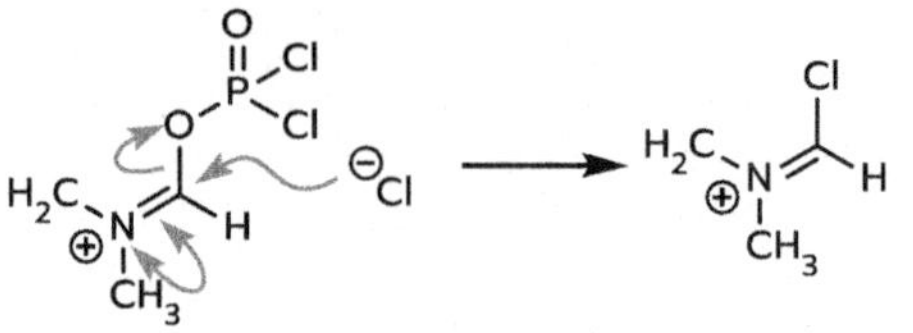

Let's take a close look at the structure on the right. It's a lot like an acid chloride, except that we have nitrogen instead of oxygen. But the nitrogen has a positive charge. It is reasonable to 'guess' that this species (we call it the Vilsmeier reagent) is more reactive than an acid chloride.

> By identifying what this looks like, and how reactive it will be, we can identify the curly arrows to draw.

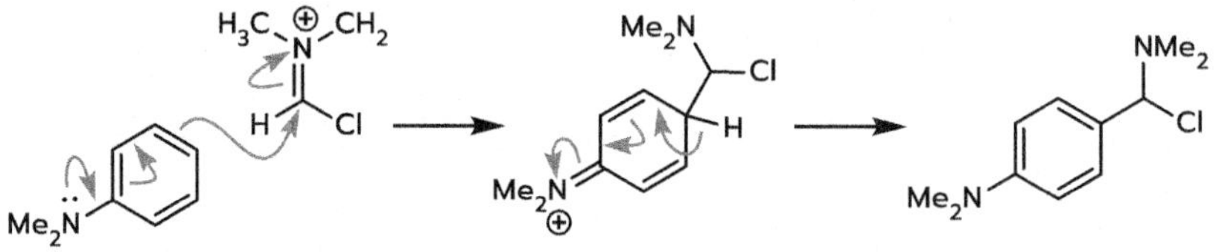

Have a very careful look at how I have drawn this. Now you have seen aromatic electrophilic substitution (**Reaction Detail 8**) drawn a few times, I made a change. Instead of 'just' drawing the curly arrow from the benzene ring, I recognized that the nitrogen lone pair would stabilize the intermediate, so I took a short-cut and drew all the curly arrows I needed to show this directly.

> I'm not saying you should have been able to do this yourself yet. But it is something to be aware of, and to work towards.

> Which other product would you expect to be formed in this reaction? Back to **Reaction Detail 8** if you aren't sure what I am asking.

We are almost done now. We add water and acid, and this happens.

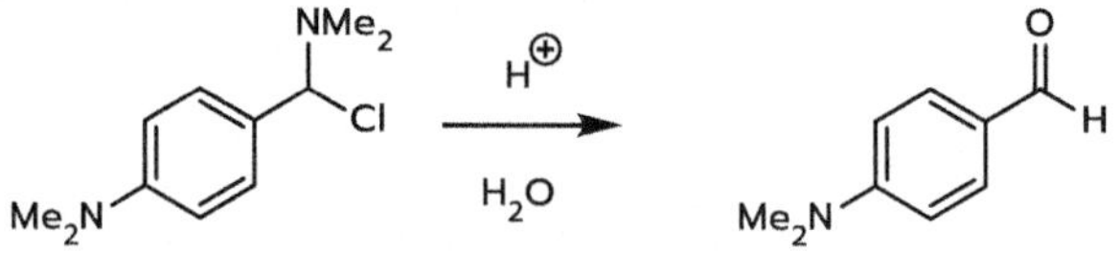

> Draw a mechanism for this reaction!

> First of all, the aldehyde hydrogen atom hasn't been added—it was already there.

268

SECTION 3 NUCLEOPHILIC ADDITION TO THE CARBONYL GROUP

If you are struggling, look at **Reaction Detail 15**. An acetal has an sp³ carbon with two oxygen atoms attached. Here we have an sp³ carbon with two electronegative elements attached.

> *There are far more similarities than differences.*

I deliberately want to leave this final part entirely to you, so that you can judge whether what you have drawn makes sense. You may need to do a little digging for similar mechanisms within the book, but this will only help you develop your understanding.

A PHILOSOPHICAL QUESTION

Have we been looking at nucleophilic addition to a carbonyl group in this chapter, or have we been looking at aromatic electrophilic substitution?

This is an entirely fair question to ask. I would have preferred to keep this chemistry with the rest of the aromatic chemistry, but you needed to see some carbonyl chemistry first.

> *Of course, covering a topic, and then coming back to it with another example is always a good way to reinforce the principles.*

Ultimately the Friedel-Crafts and Vilsmeier-Haack chemistry fits firmly into both camps.

> *Whenever we have an electrophile, we need a nucleophile.*

SECTION 3 NUCLEOPHILIC ADDITION TO THE CARBONYL GROUP

APPLICATIONS 4

The Wolff-Kishner Reduction

Here is another named reaction that made the cut. This is partly due to the mechanism, and partly because it allows us to see a nice connection with the Friedel-Crafts chemistry.

In Reaction Detail 9, we saw that Friedel-Crafts alkylation reactions are plagued with problems.

> Remind yourself what these are.

In Reaction Detail 18 we have seen that Friedel-Crafts acylation provides a way around some of these problems.

> But you get a product with a carbonyl group that you might not actually want.

Have a look at the following transformation. This would get us from the Friedel-Crafts acylation product to the Friedel-Crafts alkylation product. I wanted to deliberately show the hydrogen atoms in this case, so we know we are adding them.

This is a reduction.

> Work out the oxidation state (Basics 3) of the appropriate carbon atom.

In Reaction Detail 14, we saw reduction of ketones to give alcohol products. The reagents that we saw there will not reduce the alcohol further, but that is what we need to happen here.

I don't think you could readily predict the Wolff-Kishner reduction, so I am simply going to talk you through the mechanism.

The important reagent is hydrazine, H_2N-NH_2. It's an amine, with another amine attached. Let's have a closer look at the structure.

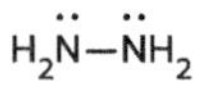

$$H_2\ddot{N}-\ddot{N}H_2$$

Each nitrogen atom has a lone pair. That is a lot of electron-density, so you can expect there to be some repulsion between the lone pairs.

> This makes hydrazine more reactive (more nucleophilic) than a typical amine.

SECTION 3 NUCLEOPHILIC ADDITION TO THE CARBONYL GROUP

We need to react the hydrazine with the ketone. You have seen nitrogen nucleophiles reacting with aldehydes (the mechanism with a ketone is the same) in **Applications 2**.

> Draw a mechanism for the following reaction.

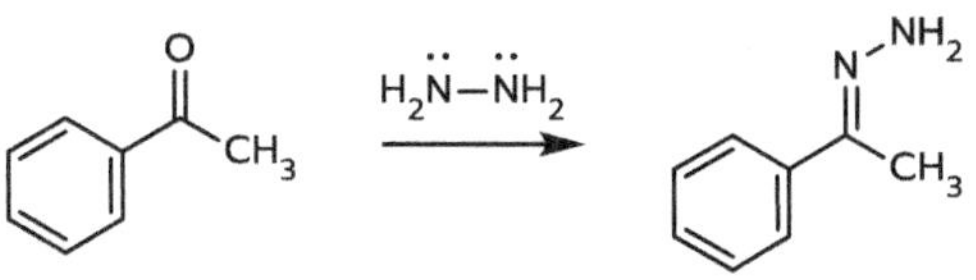

Now we move on to the 'interesting' bit. The Wolff-Kishner reduction is carried out under basic conditions (often potassium hydroxide) and in an alcohol solvent (ethane-1,2-diol is commonly used) at temperatures around 150 – 200 °C.

> *At such a high temperature, we can accept steps in the mechanism that have a higher activation energy.*

Let's see what happens next. We have a base, so deprotonation is a reasonable thing to draw. We will deprotonate one of the N–H hydrogen atoms.

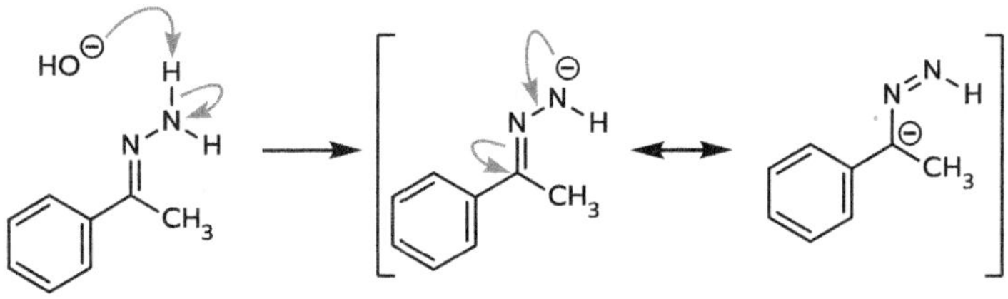

The anion we form is stabilized by resonance.

> *So, we draw the resonance forms! Don't worry, you will eventually be so confident drawing resonance forms that you will know when you need to do it and when you don't need to. Until then, draw them every time.*

Very early on, we established that we were adding two hydrogen atoms to the carbonyl carbon atom. It's time to add one of these now. In **Reaction Detail 14**, we reduced a carbonyl group by adding 'hydride'. Here, we do not have any hydride source. We must add a proton.

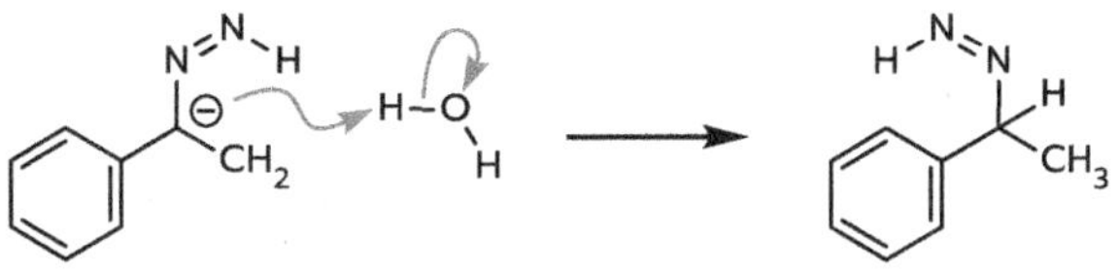

Most of the time, when we need to do this, we have just used H⊕. I've even told you not to worry too much about where the proton is coming from.

> *Very often, at the end of a reaction we add water and acid. We have plenty of H⊕.*

SECTION 3 NUCLEOPHILIC ADDITION TO THE CARBONYL GROUP

This one is different. We are not done yet, and we are working under strongly basic conditions. We should think about where the proton is coming from. We generated water in the deprotonation step, so we can confidently use that water in the protonation step.

I want to look at what we have just done. Here is the overall process. Have you seen anything like this before?

We have moved a hydrogen atom from nitrogen to carbon. Have another look at keto-enol tautomerization (**Fundamental Reaction Type 4**). Okay, we have nitrogen (two of them) instead of oxygen, but there are more similarities than differences. We could then start questioning which of the two structures is favoured. Even if the equilibrium favours the compound on the left, the one on the right can do something, and this will drive the reaction to completion. We are about to see an irreversible step.

It's time to make a couple more connections. A hydrogen atom attached to an sp carbon is more acidic than a hydrogen attached to an sp^2 carbon, which in turn is more acidic than a hydrogen attached to an sp^3 carbon. We saw this in **Reaction Detail 13**.

> But a hydrogen attached to an sp^2 carbon is not very acidic.

A hydrogen attached to O is more acidic than a hydrogen attached to N, which in turn is more acidic than a hydrogen attached to C.

> Right, so a hydrogen attached to sp^2 nitrogen might be relatively easily removed, even if the resulting anion is not resonance stabilized.

We are now coming to a step that I don't really like.

> Another way to put this is that it looks like it will have a high activation energy.

We know we need to break the C–N bond. I'm going to draw the curly arrows, and then discuss it.

SECTION 3 NUCLEOPHILIC ADDITION TO THE CARBONYL GROUP

We break the C–N bond as shown to give a carbanion. All the way back in Recap 5, I said I couldn't think of any reactions with a non-stabilized carbanion as an intermediate. Well, perhaps this is one.

> If this is an exception, we need to understand why it is an exception.

> What else is being produced in the first step?

Follow the curly arrows. We are forming a nitrogen-nitrogen triple bond. And that's it! We are forming nitrogen gas.

The triple bond in nitrogen gas has a bond dissociation energy of 945 kJ mol^{-1}.

> This is a really strong bond.

And nitrogen gas is a gas. This process is favoured on entropic grounds as well. This is the 'payback' for generating something that is unstable.

> Remember, we used a similar argument for acid chloride formation in Reaction Detail 17.

There is no doubt that the carbanion intermediate will be protonated very quickly indeed. I have drawn it being protonated by ethane-1,2-diol, which is commonly used as solvent for this reaction.

It may even be that the protonation begins as the negative charge is being produced. That is to say, we don't 'fully form' the carbanion. We actually used a similar argument in HTSIOC Perspective 4, when we considered mechanisms as a continuum rather than as discrete entities. We saw that there are substitution reactions in which we start to break a bond to form a carbocation, and then the nucleophile 'nips in' quickly at the end. In that case, we had what appeared to be an S$_N$2 substitution but with considerable S$_N$1 character.

Here, we may be building considerable carbanion character without fully forming a carbanion.

A COMMENT

This isn't the *only* way to completely reduce an aldehyde or a ketone. As you see more reactions, you'll find other ways of doing the same thing. But they will be easier to learn because you have something to relate them to.

SECTION 3 NUCLEOPHILIC ADDITION TO THE CARBONYL GROUP

REACTION DETAIL 19

Hydride is Never a Leaving Group—Except When It is!

A COMMENT

This is a chapter with an apparent 'exception to the rule'. It took me a long time to work out where to put this chapter, and even now I'm not totally happy with it. But the reactions *do* feature addition of a nucleophile to a carbonyl group, so it's in the right section. There's an obvious connection with the chemistry in Reaction Detail 14, but I thought these reactions might be too confusing if I put them straight after that chapter. So, here it is, in this section, but tucked nicely out of the way.

INTRODUCTION

When drawing reaction mechanisms, there is a really common problem in the early stages. In many reaction mechanisms there is a step where a hydrogen atom is lost.

In Applications 2, we had a lot of intermediates with protonated nitrogen atoms.

> Look carefully at the wording there—protonated!

But when you are talking about drawing the mechanism, you might use the expression 'get rid of the hydrogen'. This often leads to drawing something like the following.

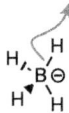

There's a big problem with this—it implies the loss of a hydride anion. And it would give a nitrogen atom with a double positive charge.

We did see a curly arrow very much like this in Reaction Detail 14. But there, we were losing the hydrogen atom from boron. More specifically, we were losing it from a borate anion.

There, we got a neutral boron from a negatively charged boron, which makes things better. That's important, but it's not the most important thing. Here's a more complete reaction scheme.

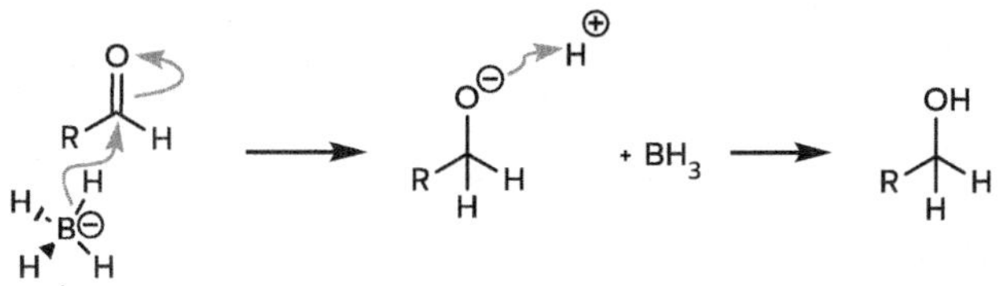

SECTION 3 NUCLEOPHILIC ADDITION TO THE CARBONYL GROUP

> Yes, the hydride anion has somewhere to go!

The key message here is that a proton is pretty stable. It can be solvated by many common solvents. You can lose a proton in most mechanisms without worrying exactly where it is going. On the other hand, a hydride anion is very unstable. You can't just lose it. It absolutely needs to go somewhere.

Here's a rule. If you draw a mechanism in which a hydride anion is lost, you've just drawn a **bad mechanism**.

Here are a few reactions that are the exceptions to the rule. We will look at the reason why these reactions work.

MEERWEIN-PONDORF-VERLEY REDUCTION

Here is a reaction scheme for the overall process.

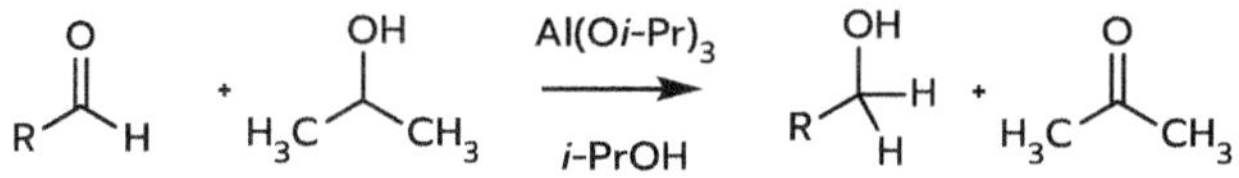

We are transferring two hydrogen atoms from isopropanol (propan-2-ol) to the aldehyde.

> Work out the oxidation states (**Basics 3**) of starting materials and products.

Yes, isopropanol is acting as a reducing agent.

We know that ketones are more stable than aldehydes (**Applications 1**). But the difference in stability alone would not account for good conversion of aldehyde to alcohol in this reaction.

> We use a *lot* of isopropanol in this reaction to drive the equilibrium to the right. Isopropanol is the solvent. That's why I drew it on the left of the reaction arrow, and again underneath it.

Here is a mechanism. The aldehyde complexes to the aluminium triisopropoxide. This is then followed by a hydride transfer reaction.

275

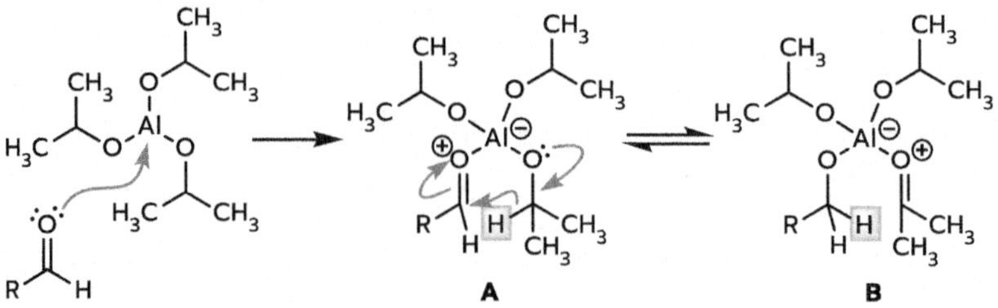

SECTION 3 NUCLEOPHILIC ADDITION TO THE CARBONYL GROUP

> The hydride *does* have somewhere to go!

Structures **A** and **B** are remarkably similar. The only difference is which is the (coordinated) alcohol and which is the coordinated carbonyl. You would expect them to be similar in energy.

> However, as we saw above, ketones are more stable than aldehydes. Structure **B** should be favoured, although possibly not by much.

Along the reaction coordinate, between **A** and **B**, we have a transition state, not an intermediate. The transition state corresponds to the hydrogen atom being 'half-way' between the two carbon atoms.

> The reaction takes place because the activation energy is sufficiently low.

The coordinated ketone can drop off. Note that the curly arrow starts at the Al–O bond, not at the negative charge on aluminium. We have a negative charge on aluminium because it has 'too many electrons'. These electrons are in the bond.

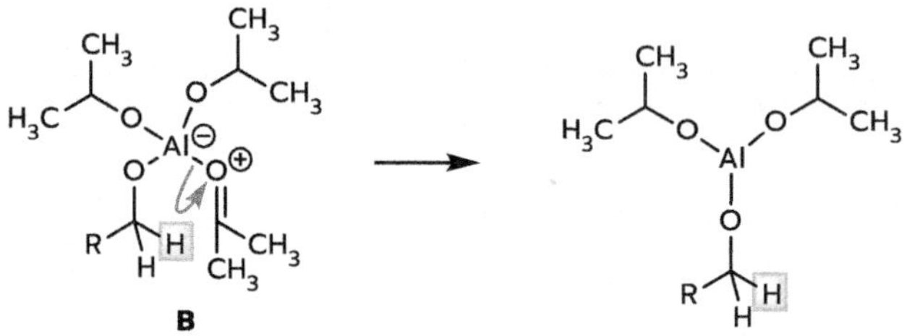

We have seen one reason why the reaction proceeds in the direction it does. The aldehyde is less stable than the ketone (acetone) by-product. There is another reason. We have lots of isopropanol (three coordinated to each aluminium, plus even more as solvent!) around.

> This drives the equilibrium towards the product.

From this point, I am going to leave it as an exercise to you. We need to get the product off the aluminium, without drawing anything 'silly'.[62]

---

[62] 'Silly' = high energy!

SECTION 3 NUCLEOPHILIC ADDITION TO THE CARBONYL GROUP

> Draw a mechanism in which another molecule of isopropanol attacks aluminium. Lose the isopropanol hydrogen atom (with the curly arrow going the right way!), then protonate the oxygen atom of the product, and draw one more curly arrow to liberate the alcohol product.

Getting you to draw this yourself will help you to build your own confidence. It might take a couple of attempts, but you'll know when you've got it right.

OPPENHAUER OXIDATION

There are two ways to look at the Meerwein-Pondorf-Verley reduction reaction. We have seen one of them. Here is an alternative.

> *We are using the aldehyde to oxidize isopropanol to acetone.*

We saw that a major driving force in the Meerwein-Pondorf-Verley (MPV) reduction was having enough isopropanol to reduce the aldehyde by driving the equilibrium. If we started with one equivalent of an alcohol, and a large excess of a ketone, we would drive the equilibrium in the opposite direction.

The Oppenhauer oxidation is **exactly the same reaction** as the Meerwein-Pondorf-Verley reduction, but in the opposite direction.

> Draw a full curly arrow mechanism for the oxidation of ethanol to ethanal (acetaldehyde). Use propanone (acetone) as the oxidizing agent and aluminium triisopropoxide (same as for the MPV reduction) as the catalyst.

Yes, once again I am getting you to do the work. It's how you will learn.

> *This isn't a trivial exercise. You have to be careful with each step. But take your time and you will do it. You can always look up the mechanism to check your working.*

Let's keep some perspective here. These are not widely used reactions. They are of interest because the mechanism features a 'wrong' step, and we need to understand why it isn't wrong in this instance. If you want to reduce an aldehyde, use sodium borohydride (**Reaction Detail 14**). If you want to oxidize an alcohol, use one of the reagents coming up in **Reaction Detail 20**.

THE CANNIZZARO REACTION

Although transfer of hydride is not a common step in a mechanism, there are some other important examples. Here is the Cannizzaro reaction.

$$2 \text{ PhCHO} \xrightarrow{{}^{\ominus}\text{OH}} \text{PhCH}_2\text{OH} + \text{PhCO}_2\text{H}$$

What we have is two molecules of benzaldehyde (it works for other aldehydes) reacting to give one molecule of benzyl alcohol and one molecule of benzoic acid.

SECTION 3 NUCLEOPHILIC ADDITION TO THE CARBONYL GROUP

> One molecule of benzaldehyde is being oxidized. The other is being reduced. We call this a disproportionation reaction.

Let's see how this works. The reagent is hydroxide—a nucleophile or a base. There is nothing that can be deprotonated. It must be acting as a nucleophile.

> This gives us a starting point.

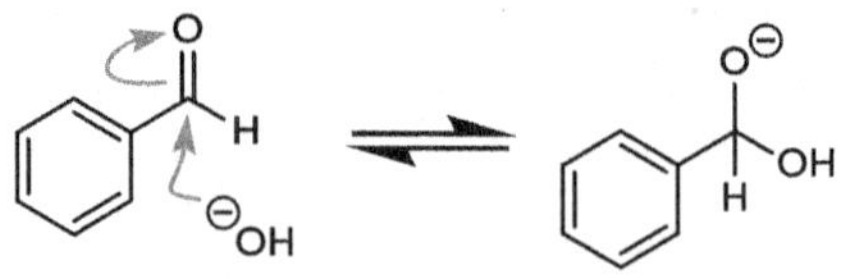

This is a mechanism that has been the subject of kinetic investigations. We know the order of reaction, so any mechanism we propose must be consistent with the known kinetics. Apparently, the next step is a further deprotonation to give the dianion.

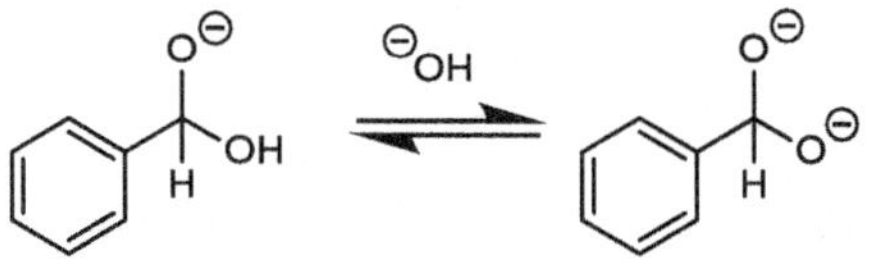

On the face of it, this looks really bad. We have two species, each with one negative charge. We have a product with two negative charges.

> This must be unfavourable!

It certainly is, and presumably this makes the next step 'less bad'. The negative charge on oxygen 'pushes back in', and it delivers a hydride to a second molecule of benzaldehyde. It's easy to see why this is favourable on thermodynamic grounds.

> Not only have we 'distributed' the negative charges between two molecules. One of them is particularly stable.

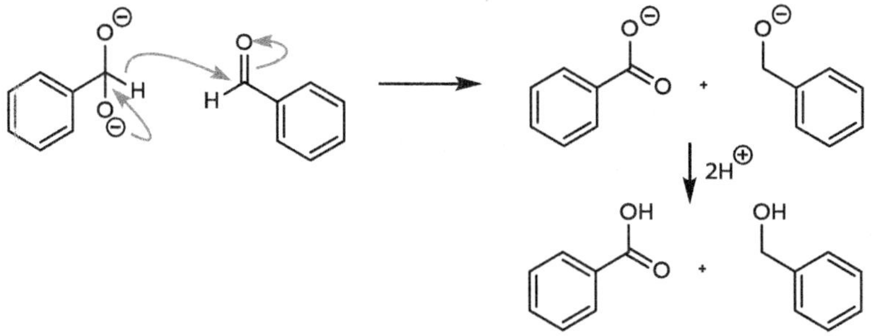

Finally, as shown above, we acidify the reaction to protonate both negative charges.

The Cannizzaro reaction is rarely, if ever, a synthetically useful reaction. Generally, it's an inconvenience. But it's an inconvenience that is helpful to understand.

SECTION 3 NUCLEOPHILIC ADDITION TO THE CARBONYL GROUP

> Usually, when you dig out an old bottle of benzaldehyde (a liquid) from the chemical store, it will have a solid in the bottom of the bottle. This solid is benzoic acid. You'll need to purify this benzaldehyde before you use it!

NATURE'S SODIUM BOROHYDRIDE

Whatever reactions we want to do in the laboratory, you can bet that Nature has already found a way to do them. The problem is that some of the reagents we use in the laboratory are totally incompatible with biological systems.

If Nature wants to carry out a hydride reduction reaction, it cannot use sodium borohydride. This reagent would not survive in the human (or any other!) body.

> The human body would probably not survive long either!

Here is a partial structure of NADH, nicotinamide adenine dinucleotide. For our purposes, we don't need to worry about what 'R' is.

The curly arrows show that NADH can deliver a hydride.

> Of course, it isn't just 'floating off'. It is being delivered to something. We will see one example in a moment.

The product of this reaction is NAD$^+$. It has an aromatic ring. In this case it is a pyridine ring, and it has a positively charged nitrogen atom. We will see some heterocycles in **Basics 7** and in **Applications 9**. Recall (**Recap 6**) that aromaticity 'gives us' about 150 kJ mol^{-1}, which provides a thermodynamic driving force for the reaction.

> We should always look for the driving force and the basic principles.

Now let's look at an example, the reduction of pyruvic acid to give lactic acid. This is part of a really important metabolic pathway.

> This is an enzyme-catalysed reaction. The NADH is described as a cofactor. The enzyme is lactate dehydrogenase.

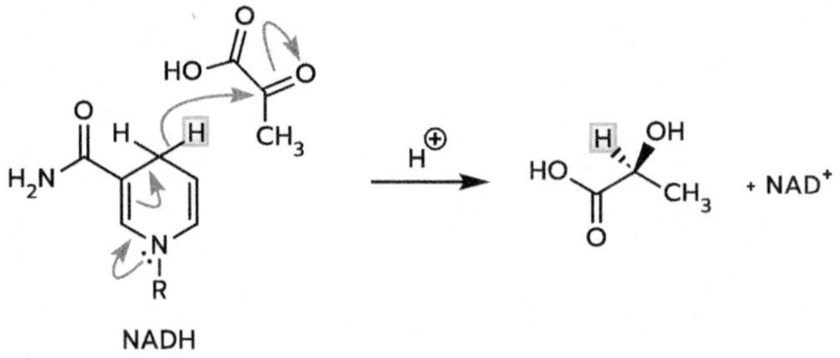

SECTION 3 NUCLEOPHILIC ADDITION TO THE CARBONYL GROUP

To finish this story, we need to add a couple of details. Lactic acid is chiral, and is formed as a single enantiomer.

> If we were to draw the NADH fully, we would see that it is chiral, and is a single enantiomer. This can be a stereoselective reaction (Stereochemistry 2).[63]

The reaction is reversible. NAD⁺ is an oxidizing agent.

> We have just seen something similar with the Meerwein-Pondorf-Verley reduction and the Oppenhauer oxidation.

Finally, just because this is a reaction that takes place in the body, all our existing principles apply. Sure, Nature has enzymes which lower the barriers of reactions that would otherwise be very slow. But this doesn't change the fundamental rules.

IN CLOSING

Remember why this chapter is included. Drawing a curly arrow mechanism in which a hydride is lost generally means you have a curly arrow going the wrong way. By understanding the handful of reactions where this *actually* happens, we reinforce the message of how unusual this is.

---

[63] In fact, as there is an enzyme involved, and this is also chiral, we don't actually need NADH to be chiral.

SECTION 3 NUCLEOPHILIC ADDITION TO THE CARBONYL GROUP

REACTION DETAIL 20

A Special Elimination—Oxidation of Alcohols

CONTEXT—OXIDATION OF ALCOHOLS INTO ALDEHYDES

When we looked at reduction of aldehydes and ketones using sodium borohydride (**Reaction Detail 14**) we found that we were adding two hydrogen atoms, one as $H^\ominus$ and one as $H^\oplus$.

Oxidation is the reverse of reduction, so to oxidize an alcohol into an aldehyde, we need to remove two hydrogen atoms. The starting material is neutral, and the product is neutral. Therefore, if we remove one of the hydrogen atoms as $H^\oplus$, we will need to remove the other one as $H^\ominus$.

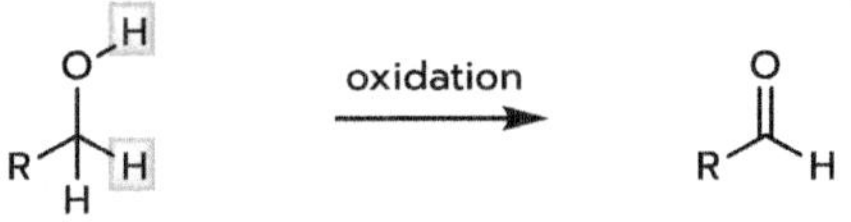

In **Reaction Detail 19**, we saw some reactions in which hydride appeared to be a leaving group. It turns out that this only works if the hydride has 'somewhere to go'. It never leaves as hydride, because that is too unstable.

For most oxidation reactions of alcohols, we need to employ a different strategy. Let's start with a fundamental principle.

> In the reaction above, work out the oxidation state (**Basics 3**) of the alcohol and aldehyde carbon atoms.

> *If we are oxidizing the organic compound, something must be reduced.*

That would be the oxidizing agent. Let's call it 'X'. If we could replace the acidic OH hydrogen atom with an 'X' group that could accept a pair of electrons (that is, it could leave as $X^\ominus$), then we would have the following scenario.

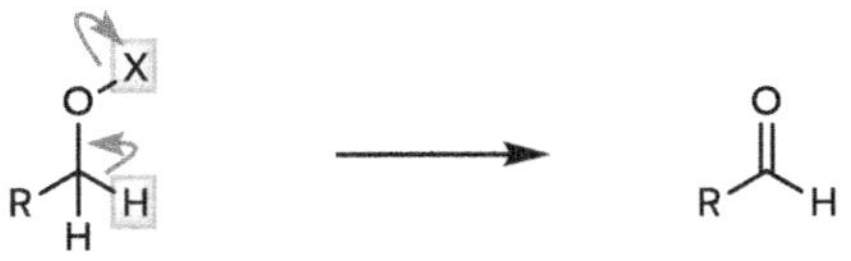

Let's expand this schematic approach, before we look at a real reagent. We will start with a positively charged reagent '$X^\oplus$'. We add this to the oxygen atom, and we then lose the proton.

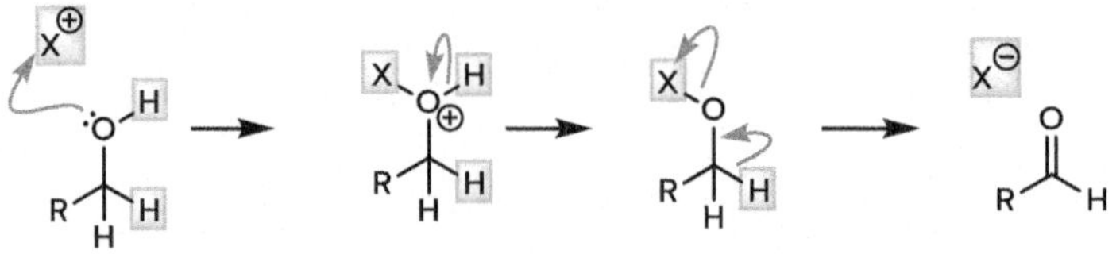

SECTION 3 NUCLEOPHILIC ADDITION TO THE CARBONYL GROUP

We then have our 'elimination' reaction, losing another proton, and $X^\ominus$. The 'X' group is being reduced, and along the way, we oxidize the alcohol.

> Now we have established the fundamental principles of what needs to happen, we can look at specific reagents and mechanisms.

We will focus on two mechanisms for the oxidation of alcohols. In the examples covered, both are shown with *primary* alcohols to give aldehydes. However, they could just as easily give ketones if the alcohol had one more alkyl group.

PYRIDINIUM CHLOROCHROMATE

This is the structure of pyridinium chlorochromate. From a mechanistic perspective, the fact that it is a pyridinium cation doesn't matter too much.

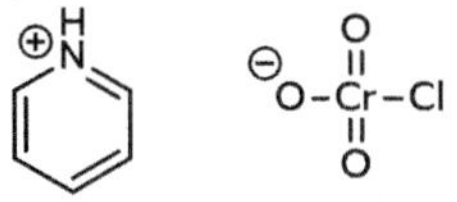

The chromium is in oxidation state +6, but it can easily be reduced.

> Another way to express that is 'it is an oxidizing agent'.

We talked about the formation of esters in **Reaction Detail 16**. We talked about the chemistry of acid chlorides in the same chapter. The chlorochromate anion looks a bit like an acid chloride. It reacts like one too!

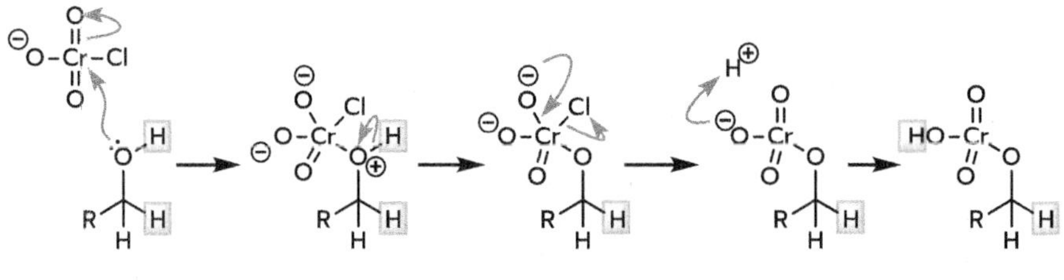

So far so good. We have made a chromate ester. I added the proton onto the chromate oxygen atom at the end to keep things tidy. It isn't always clear where a given proton is at any stage of a mechanism—unless this has been investigated and proven of course!

Now for the key step. Because the chromium can be reduced, this means it can accept more electrons.

SECTION 3 NUCLEOPHILIC ADDITION TO THE CARBONYL GROUP

These curly arrows might look a little strange. We are pushing an arrow directly to the chromium atom, and then taking away a pair of electrons from a Cr=O double bond. I am drawing it like this so that we end up with four bonds to Cr (oxidation state +4 now) and the negative charge on the most electronegative element.

The main point is that you understand how this mechanism relates to other mechanisms you have seen—ester formation, elimination and the general oxidation mechanism on the previous page.

THE SWERN OXIDATION

This oxidation reaction was developed by Daniel Swern in the 1970s. Generations of PhD students will hate him for this, as the mechanism is quite complicated, and is a very common question in PhD examinations.

There is no doubt that in the first instance this is a mechanism that you simply have to learn. However, there are effective ways of learning it. Make sure you understand that the actual oxidation step corresponds directly to the general oxidation mechanism on the previous page. In order to help you do this, here is the key step up front! We will worry about what removes the proton shortly.

The key point is how we get to this stage—how do we make the species on the left?

The reagents for the overall transformation are dimethylsulfoxide, oxalyl chloride and triethylamine. We will see each of these in turn. First of all, dimethylsulfoxide acts as a nucleophile, reacting with an acid chloride, oxalyl chloride. This is basically another ester formation.

I've drawn the resonance form of dimethyl sulfoxide in which there is charge separation rather than a double bond between S and O. Some people prefer one representation or the other. We saw both representations in **Reaction Detail 6**.

SECTION 3 NUCLEOPHILIC ADDITION TO THE CARBONYL GROUP

> *Once again, we are using the shorthand 'double-headed curly arrow' which we first saw in Reaction Detail 17. Instead of drawing the 'normal' curly arrow for the electrons going out onto the carbonyl oxygen atom, and a separate step having the arrow come back in and kick out the chloride, we just do it all in one step.*

From this point, something a little strange happens. The chloride anion that we just lost attacks the sulfur atom. When it does, it kicks out the oxygen, and then the whole leaving group 'falls apart' as shown below. When it does this, it forms one molecule of carbon dioxide, one of carbon monoxide and a chloride anion. That is, two molecules of a gas, so that entropy is definitely working in its favour!

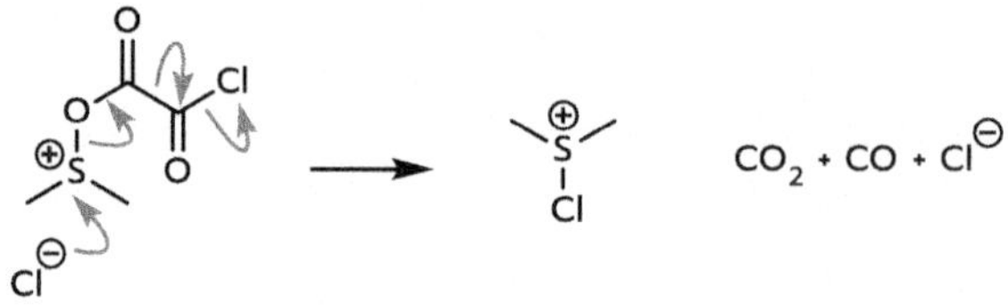

$$\text{structure} \longrightarrow \text{structure} \quad CO_2 + CO + Cl^{\ominus}$$

There are two curly arrows in the above scheme that look a little different. We have a curly arrow giving two electrons to carbon. We generally only see this when we are making a carbanion.

> *In this case, it would finish there to leave the carbon with a share in eight outer electrons.*

Then, we have a curly arrow taking two electrons away, leaving the carbon with only two bonds and six outer electrons. Carbon monoxide is an unusual 'organic' molecule, as it does only have two bonds to carbon.

> *When thinking about this step, remember that it is giving off two molecules of gas. ΔS is very positive, so that the process is likely to be exergonic even if it is a bit unusual.*

Okay, we are getting close now. We have formed a chlorosulfonium ion. This is the oxidant. The alcohol can attack this species as shown below, and we then lose the proton. We are replacing something that can only be lost as a positively charged species with something that can leave taking electrons with it.

The final reagent, triethylamine, is used to facilitate the elimination reaction.

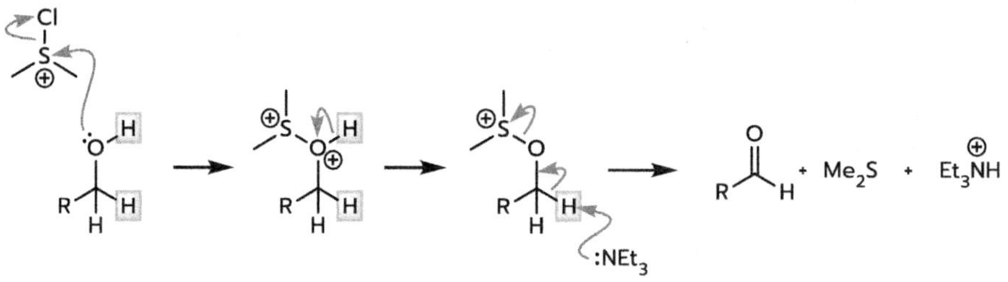

$$\text{structure} \longrightarrow \text{structure} \longrightarrow \text{structure} \longrightarrow R\text{-CHO} + Me_2S + Et_3NH^{\oplus}$$

We are using dimethyl sulfoxide as the oxidizing agent. Without any of the other reagents, we would have to think of the overall equation as follows.

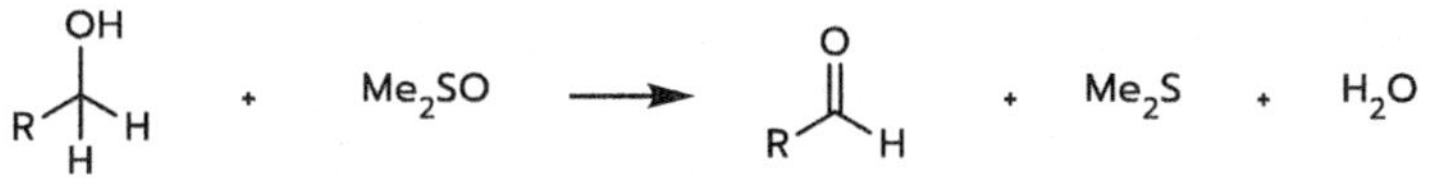

SECTION 3 NUCLEOPHILIC ADDITION TO THE CARBONYL GROUP

$$R\text{-CH(OH)-H} + Me_2SO \longrightarrow R\text{-CHO} + Me_2S + H_2O$$

doesn't work!

Of course, this doesn't work. Even if the reaction is *exothermic* (calculate this using the method shown in **HTSIOC Basics 13**!), there isn't a viable mechanism. We need oxalyl chloride to 'activate' the dimethyl sulfoxide. This acid chloride is 'special' because of the fragmentation reaction it can undergo.

By understanding why we need this complex sequence of steps, it becomes possible to devise alternative reagents and transformations.

This isn't an easy mechanism to learn.

> *For now, don't expect to remember it. But work through it and make sure you understand each individual step.*

The key point is to focus on the 'elimination' part of the mechanism.

MOVING FORWARD

There are lots of organic functional groups that can be oxidized. There are many different oxidizing agents that can be used, and some work better than others in particular instances. If you do enough organic chemistry, you will learn to recognize which work best for particular substrate types. This is just filling in the pieces of the jigsaw. For now, the important thing is to understand the general principles of the process.

SECTION 4

CARBONYL COMPOUNDS AS NUCLEOPHILES

SECTION 4 CARBONYL COMPOUNDS AS NUCLEOPHILES

INTRODUCTION

Carbonyl groups do two things. You can add a nucleophile to the carbonyl carbon. That's what **Section 3** was all about, and we saw lots of different nucleophiles and lots of different carbonyl compounds, all doing the same thing.

> *Well, the curly arrows looked the same, which is the important thing.*

The other thing that carbonyl compounds do is allow you to remove a proton from a carbon atom adjacent to the carbonyl group.

We saw this in **Fundamental Reaction Type 4**. We either form an enol, or an enolate.

> *They are both nucleophiles.*

What do we add nucleophiles to?

> *Mainly, carbonyl groups!*

The focus of this section is reactions of enols or enolates with some sort of electrophile. We will start by looking at alkyl halides as electrophiles.

> *This will allow you to learn the 'new' reactivity.*

Then we will use carbonyl compounds as the electrophiles as well. At that point, you have two carbonyl reagents (either the same or different) to contend with. As always, I want to present the reactions in such a way that you can believe that if you go slowly and think carefully, you might have predicted the outcome of the reactions.

> *We will focus on the underlying principles, so that if two compounds that look similar do different things, we see why this happens.*

But because a lot of these reactions look quite similar, that makes them harder to learn. And that will be the case right up until the point where you 'see it', and then you will wonder why you didn't see it sooner.

That is the nature of learning. I was not stupid as a student, and I found this chemistry difficult. You will also find it difficult. When you find it difficult, just draw the reaction mechanisms out, over and over, and this will help you make the connections.

> *I suspect you won't like that approach, but you should. It gives you something to do, rather than wondering why these reactions aren't 'sinking in'. Good learning takes practice.*

SECTION 4 CARBONYL COMPOUNDS AS NUCLEOPHILES

REACTION DETAIL 21

Alkylation of Enolates

INTRODUCTION

In **Recap 7** we encountered anions α- to carbonyl groups. We saw that the carbonyl group stabilizes the negative charge.

In **Fundamental Reaction Type 4** we saw, at least in principle, that you can deprotonate α- to a carbonyl group, and then react the resulting enolate with an alkyl halide.

> This is all true, although sometimes these reactions don't work all that well.

We are going to be looking at quite a lot of enol and enolate reactivity, so it is worth consolidating this reaction type with a few examples. We can look at why some work better than others.

We will start with cyclohexanone (**1**) and use lithium diisopropylamide (LDA) as the base. We saw this in **Fundamental Reaction Type 4**. The resulting enolate then reacts with alkyl halide **2**. The product is formed in 58% yield.

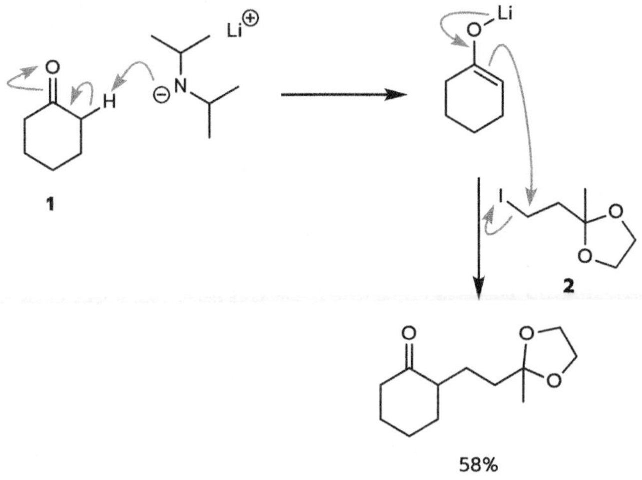

A 58% yield is not bad. More than half of the starting material is converted into product. Then again, 42% of the starting material is *not* converted into product, so it isn't great!

I've also missed something out on the scheme. The reaction is done in a solvent, tetrahydrofuran, and we have an additive, HMPA. Here are the structures.

SECTION 4 CARBONYL COMPOUNDS AS NUCLEOPHILES

tetrahydrofuran (THF)   HMPA

You have to use a solvent that won't react with anything you are using. Tetrahydrofuran is pretty unreactive.

> It is also a liquid between −100 °C and +66 °C. Being a liquid is an important property for a solvent!

I am not being facetious. This particular reaction is carried out at −78 °C, and finished off at room temperature. Some reactions need to be heated. When you think of the properties you need in a solvent, don't miss the obvious one!

The second compound, HMPA, solvates lithium really well. Although I have drawn the lithium associated with the enolate above, we might consider the enolate to be 'simply' an anion, and it will therefore be more reactive. If we don't have the HMPA in there, the reaction doesn't work as well.

INTERLUDE

In this one, relatively simple, reaction, we have LDA, THF and HMPA. We are getting into acronym overload.

> How are you supposed to remember all this stuff?

The short answer is that I very much doubt that the people marking your exams will be awarding many (any?) marks for recall of every acronym. You don't necessarily need to remember what LDA is, although you probably should recognize that the formation of the enolate requires removal of a proton, and a pretty strong base is needed. It would be a good idea to understand that an enolate is not a great nucleophile in an S_N2 substitution reaction,[64] but there is something we can add to make it better.

Apart from that, you should accept that as you see this chemistry more often, it will 'stick'. To speed up the process, every time you see an acronym used for a compound, look it up and draw out the structure. It might not be elegant learning, but it is effective.

> Oh, and while we are at it, how would you make compound **2**? Maybe you should have a look at **Reaction Detail 15**. Draw a mechanism for the reaction.

---

[64] We are currently talking about this as an enolate alkylation, but as far as the alkyl halide is concerned it is a substitution.

SECTION 4 CARBONYL COMPOUNDS AS NUCLEOPHILES

ONE MORE EXAMPLE

Here is the next example, and I have deliberately chosen cyclohexanone again. I want to avoid any complexities with stereoselectivity and regioselectivity for now.

cyclohexanone → i) LDA, THF, HMPA; ii) CH2=C(CH3)CH2I → 2-(2-methylallyl)cyclohexanone, 67%

In essence, this is the same reaction we have already seen. I am taking this similarity as an opportunity to show you how you might see this in a research paper. We have enough information over (and under) the reaction arrow to tell you what is added, and in which order.

> We have acronyms rather than structures, but you know what to do about those!

This allows us to focus on one key difference—the electrophile. Instead of a *primary* alkyl halide, we have an allylic halide. This is considerably better in an S_N2 substitution reaction (**HTSIOC Reaction Detail 1**).

> In addition, the allylic iodide in this case cannot undergo a competing elimination reaction.

Of course, the yield is better, but not that much better.

WHAT NEXT?

Alkyl halides are decent electrophiles, and this does allow us to form a C–C bond.

> If this was all we could ever do with an enolate, it would still be useful.

But we know plenty of other electrophiles. **Section 3** was all about carbonyl groups acting as electrophiles. In **Reaction Detail 23**, we will see the aldol reaction.

> There, we have a carbonyl compound acting as a nucleophile reacting with a carbonyl compound as electrophile.

In **Applications 5** we will be back to 'simple' alkylation reactions, but we will find ways to exert regiochemical control.

Along the way, we will see some more facets of enol/enolate reactivity. These will include stereoselective reactions (**Stereochemistry 13**).

> Both the reactions in this chapter lead to the formation of a stereogenic centre. We must consider how this can be formed stereoselectively.

SECTION 4 CARBONYL COMPOUNDS AS NUCLEOPHILES

THERE'S ONE MORE THING

Here is the product of the first reaction again. I've drawn a reaction arrow and some reagents. You know what is coming next!

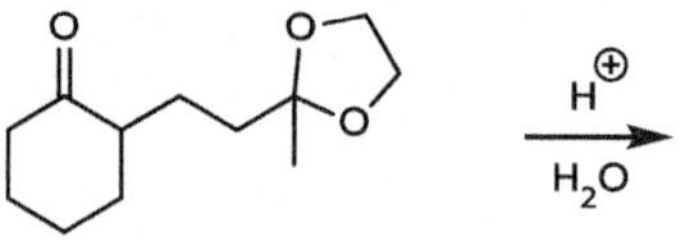

> Have another look at **Reaction Detail 15**, and draw a mechanism for the hydrolysis of the ketal.

Now, carry on working through this section. When you get to **Applications 7**, come back to this reaction, and consider what would happen if you were to add a bit more acid.

> Just keep drawing protonation steps and the curly arrows.

Then have another look at **Basics 6**, and see if there is anything in there that might help you.

> Wait, could this be a reaction that forms a ring?

Once again, I am trying to show you the 'connectedness' of organic chemistry. I should probably point out that when I chose the first example, I had no idea that I was going to finish the chapter like this. But organic chemistry is often about the interplay of functional groups within a molecule. It is about there being multiple possible products that *could* be formed, but some of the pathways are more favourable than others.

> Let's manage your expectations. You probably won't get the answer to this problem straight away. You might need to discuss it with your friends, or with your tutors. That's okay. By making you work for this, you will get more out of it. More frustration, certainly. But ultimately more understanding, and hopefully some satisfaction and confidence.

STEREOCHEMISTRY 13

Diastereoselective Alkylation of Enolates

TAKING STOCK

In **Reaction Detail 21** we looked at the alkylation of enolates. This is a reaction that, for suitable enolates, can lead to the formation of a stereogenic centre. Here is a very general reaction scheme.

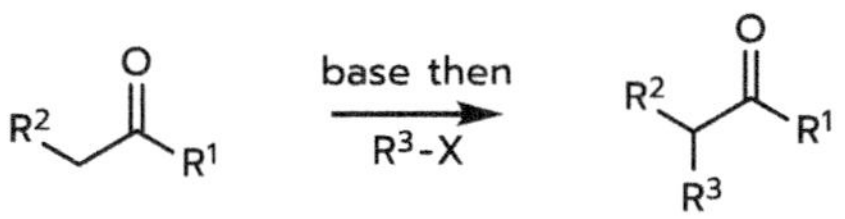

$$R^2\text{-CH}_2\text{-CO-}R^1 \xrightarrow{\text{base then } R^3\text{-X}} R^2\text{-CHR}^3\text{-CO-}R^1$$

> A stereogenic centre is a carbon atom with four different substituents. In time, and with practice, you will spot them at a glance.

In **Stereochemistry 2**, we looked at the criteria for stereoselective reactions. If we were to take the simple alkylation reaction shown below, we would form the product as a 1:1 mixture of enantiomers.

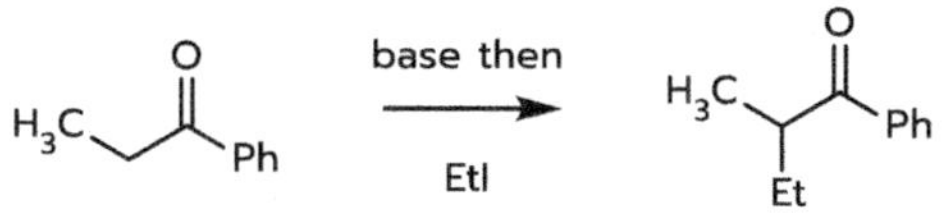

$$H_3C\text{-CH}_2\text{-CO-Ph} \xrightarrow{\text{base then EtI}} H_3C\text{-CHEt-CO-Ph}$$

This would happen because the starting material is achiral and the reagent is achiral.[65] The products are enantiomers, with exactly equal energy, and the transition states leading to the products are enantiomeric, with equal energy.

> To put it more simply, the entire reaction coordinate from starting material to product is exactly the same for the two possible stereoisomers of product.

You might have noticed that I started talking about enantiomer products, and then I changed to talk about stereoisomer products. Of course, enantiomers are stereoisomers, but by generalizing the term, we give ourselves the opportunity to think about what is needed for the reaction to not give a 1:1 mixture of stereoisomers.

ALKYLATION OF EVANS OXAZOLIDINONES

In the first instance, we will work at the carboxylic acid oxidation level. This doesn't matter.

> An enolate is an enolate.

---

[65] For completeness, we should probably state that the base is achiral.

293

SECTION 4 CARBONYL COMPOUNDS AS NUCLEOPHILES

We want to alkylate α- to the carbonyl group in compound **1**, and to do it in a way that leads to the production of only a single stereochemical isomer. Let's fix some parameters. We want **exactly** compound **2**, so both the methyl group and the electrophile (E) are fixed.

Okay, if we add a base to compound **1**, the first thing we will deprotonate is the carboxylic acid OH group.

> It is possible to deprotonate compounds more than once, and we will see this in Reaction Detail 24. For now, though, we will solve this problem in a different way.

Let's concentrate on stereochemistry. Compound **1** is achiral. Structure **2** has one stereogenic centre. The only stereoisomer of compound **2** is its enantiomer. We are going to solve 'the OH problem' and 'the chirality problem' at the same time. Here is a reaction scheme.

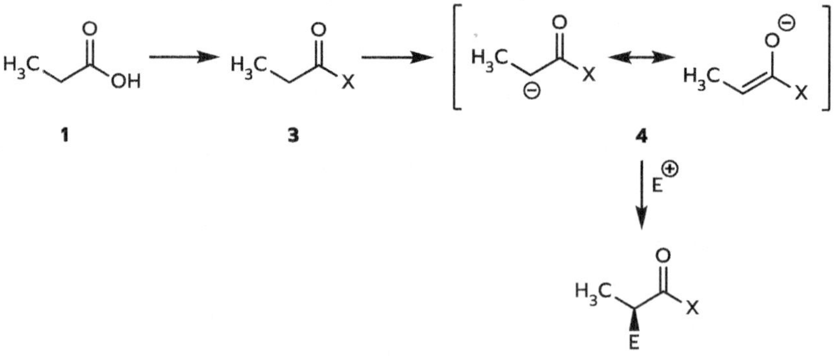

What we have done here is convert compound **1** into a derivative **3** which has an 'X' group that is capable of exerting stereochemical control.

> It can only do this if 'X' contains chirality.

We will then deprotonate α- to the carbonyl in **3** using a strong base. This will be followed by reaction with an electrophile, for example an alkyl halide, to give the product with a new bond and new stereochemistry. We won't worry about the steps that take us from **1** to **3** right now.

Right, here is the example. We are going to use a **chiral auxiliary** called an Evans oxazolidinone. These were developed by the group of Professor David Evans at Harvard. We are going to use compound **6**.

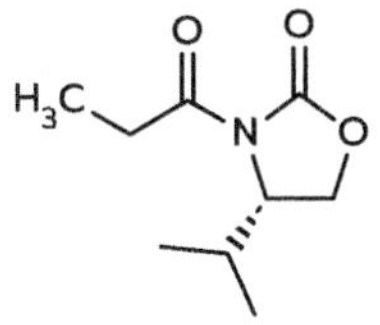

SECTION 4 CARBONYL COMPOUNDS AS NUCLEOPHILES

6

Let's take a moment to think about this compound. First of all, we could rotate around the C–N bond to give different conformers.

> We know that amides have some double bond character in the C–N bond. This one isn't quite an amide, but it is very similar. We might expect two conformers to be favoured, **6** and **7**.

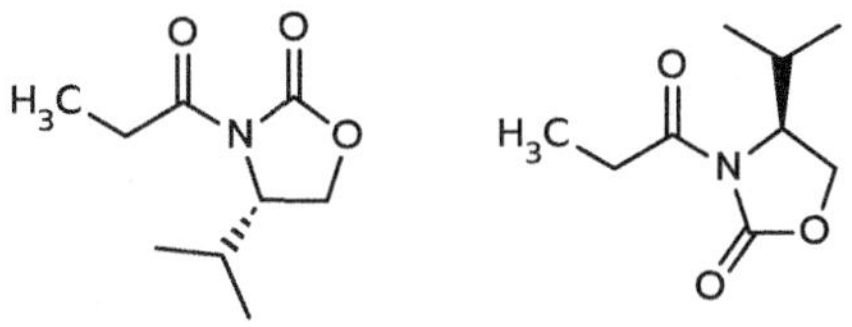

6   **7**

In conformer **6**, the isopropyl group is blocking the lower face. In conformer **7**, the isopropyl group is blocking the upper face.

> We really want only one face blocked, so that the electrophile approaches the other face.

Let's see what happens when we deprotonate compound **6** with lithium diisopropylamide (LDA), a strong base. The lithium is important!

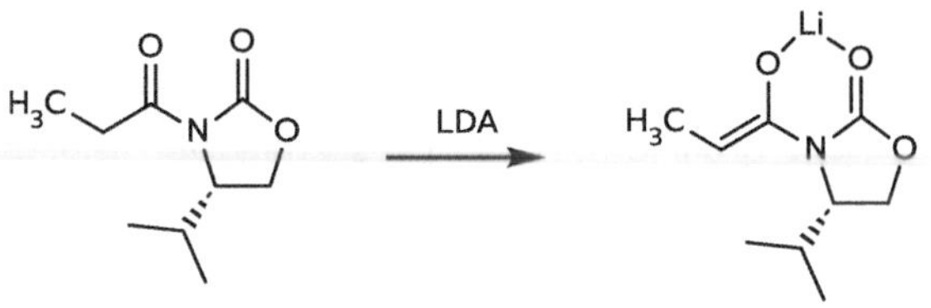

6 → LDA → **8**

Told you the lithium was important! It chelates the two oxygen atoms. Whatever the energy difference was between conformer **6** and conformer **7**, we have tipped the balance very much in favour of conformer **6**.

> The lower face of the enolate is blocked.

Let's take a deeper dive. It's really important to consider **everything**! Structure **8** is not the only enolate we could form from compound **6**. Structure **8** is the (Z)-enolate. Structure **9** is the (E)-enolate.

295

SECTION 4 CARBONYL COMPOUNDS AS NUCLEOPHILES

8 **9**

> Assign the stereochemistry of both of them. You may need to look up how to do this. It's in **HTSIOC Habit 6**.

Double bond isomers are diastereoisomers. They have different energies. They will be formed in different amounts or at different rates (which leads to the same thing). Enolate **9** is considerably less stable, because the highlighted CH₃ group is clashing with the isopropyl group. So this double bond isomer isn't formed!

The group that controls which face of the enolate is attacked by the electrophile also controls the geometry of that enolate.

> As an aside, I wonder how much of this was planned at the start of the project. When we write the papers that describe our research, we explain very carefully how we designed the experiment to give the desired outcome. Sometimes it's true. Other times, we just did the experiment, and once we've work out what happened, we pretend that's what we wanted and write it in a way that makes us appear clever!

Coming back to the point, why does enolate geometry matter?

> Over to you. Get your molecular models (or drawing skills) out and work out what happens if you have the electrophile reacting with the top face of enolate **8** or enolate **9**.

"What happens?" is quite a vague question, but this is deliberate. The point is that a competent organic chemist will know what to look for in this question, and you want to become a competent organic chemist. Let's show you the question, and then explain what to look for.

> In this case, "what happens?" means "what will the stereochemistry be?"

You should have worked out that each enolate will lead to a different stereoisomer of product. That wouldn't be much use! It's important that we can control the enolate geometry **and** the face attacked.

Let's finish the job. For this, we will use a specific electrophile, benzyl chloride.

8 → **10**

296

SECTION 4 CARBONYL COMPOUNDS AS NUCLEOPHILES

As we have discussed, we get attack from the top face because the isopropyl group is blocking the lower face.

> Make sure you can visualize why this stereoisomer is produced preferentially, rather than just drawing a wedge!

There are a couple of points to add. These reactions are often carried out in a solvent such as THF (tetrahydrofuran). We saw the structure of this in the last chapter, but here it is again.

The lithium in the enolate **8** can coordinate to more atoms. In a solvent such as THF, the solvent will coordinate. This solvation network increases the effective size of this lithium atom.

GETTING TO THE REAL POINT

Let's look at a reaction scheme that shows the potential of this methodology. We can start with carboxylic acid **1**. We can form acid chloride **11** from it. We saw how this works in **Reaction Detail 17**.

> Go back, draw it out again for this particular compound. Reinforce the learning!

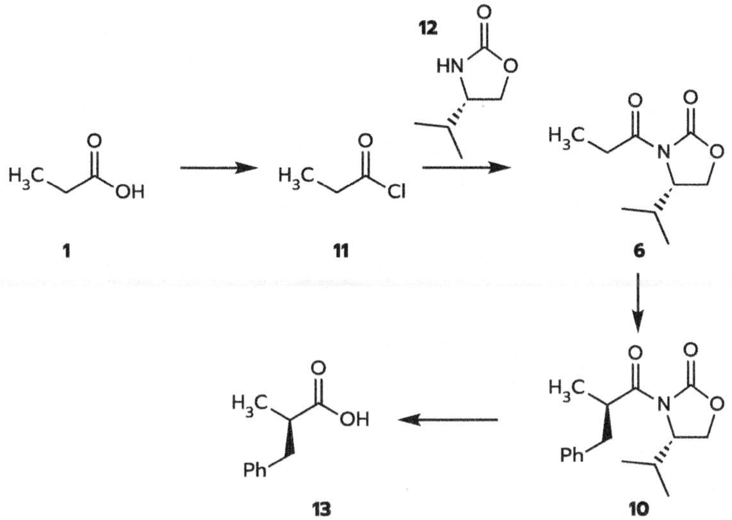

You can react this acid chloride with compound **12** to form compound **6**.

> Is compound **12** as nucleophilic as an amine?

Once we follow the chemistry above, we can form compound **10** with high levels of stereoselectivity. We can then hydrolyse compound **10** to give compound **13**. Let's focus on the overall transformation.

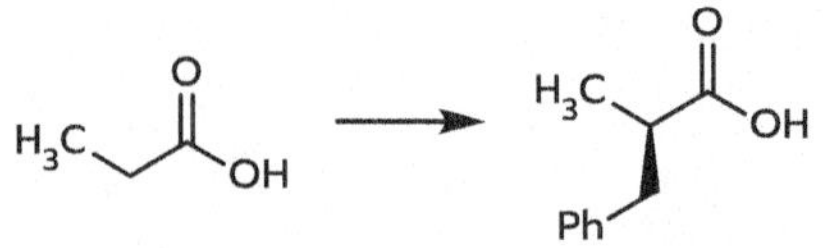

SECTION 4 CARBONYL COMPOUNDS AS NUCLEOPHILES

We have taken the achiral compound **1** and converted it into a single enantiomer (if the stereoselectivity is high enough!) of compound **13**. That is, we have accomplished, indirectly, an enantioselective synthesis.

> *In Stereochemistry 2 we saw that a reaction producing enantiomers will have the same energy profile for each enantiomer. It cannot be stereoselective.*

Here, we have found that by going *via* a chiral compound **6**, we can get around this limitation.

Let's have a slightly deeper look at the hydrolysis reaction. I want to use this to build connections. Compound **10** has two carbonyl groups, 'a' and 'b'.

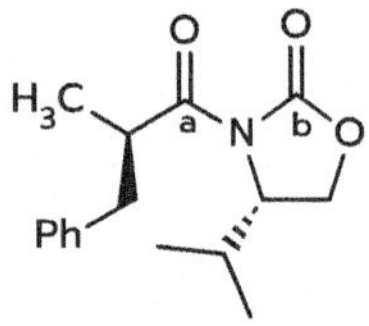

For hydrolysis to occur, we would need attack at carbonyl group 'a'. Fortunately, this is favoured. Carbonyl 'b' is quite a bit less reactive. We want to use a good nucleophile, but we don't want it to be too basic, as you could potentially deprotonate the α- position.

> *This would mess up the stereochemistry we have just put all the effort into forming.*

Sometimes in organic chemistry, it is quite easy to see what you need to do, but less easy to find the best reagents and conditions to do it. This is one of those times. You can use lithium hydroxide as a nucleophile for the hydrolysis of compound **10**, but it's actually better to use lithium hydroperoxide. This gives the 'peracid' **14** (we saw peracids in Reaction Detail 6). We then need to reduce this using sodium sulfite.

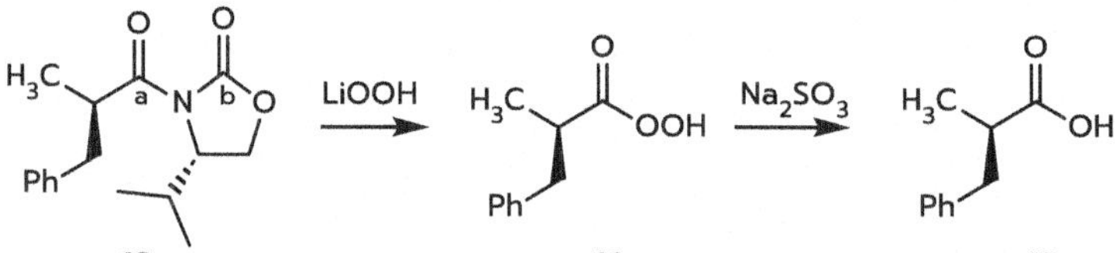

When you draw the mechanism for the formation of compound **14**, you just use the hydroperoxide anion (shown below).

SECTION 4 CARBONYL COMPOUNDS AS NUCLEOPHILES

Everything you need to draw this is in Reaction Detail 16.

> Have a go!

IN SUMMARY

This has been quite a deep dive into one particular reaction.

> *We looked at this reaction because it works well. It gives high levels of stereocontrol for a broad range of reactions.*

Of course, alkyl halides are the simplest carbon electrophiles we will encounter. What else could we add a nucleophile to?

> *A carbonyl group!*

We will find that this has chemical complexities, and then it has stereochemical complexities. We are going to deal with the bond-forming processes over the next few chapters, then we will look at the stereochemistry.

> *And **then**, we will come back to diastereoselective enolate alkylation in Stereochemistry 15, and we will add a little more depth.*

Of course, at this level, we are adding depth to illustrate key principles. It's about learning and understanding the patterns. It's not about 'learning more reactions'.

> *You need the framework to put those reactions into.*

SECTION 4 CARBONYL COMPOUNDS AS NUCLEOPHILES

EXERCISE 3

C- or O-Alkylation of Enolates?

THE POINT OF THE EXERCISE

We are going to calculate some enthalpy changes using bond dissociation energies. These calculations don't give perfect results in this case, but it is worth practising this skill again.

> We did this already in Exercise 1 and in Exercise 2.

If we alkylate an enolate selectively on either C- or on O-, this is a chemoselective process. We defined chemoselectivity in Recap 4.

THE EXERCISE

Let's look at four possible reactions of acetone with two different electrophiles.

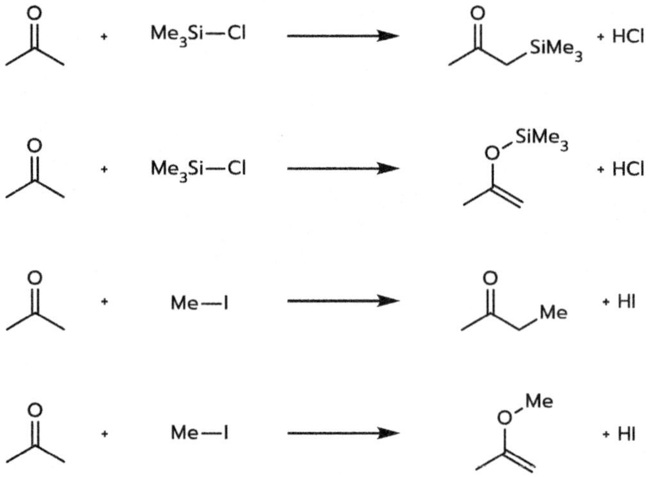

[Reaction 1: acetone + Me₃Si–Cl → α-SiMe₃ ketone + HCl]

[Reaction 2: acetone + Me₃Si–Cl → silyl enol ether (O–SiMe₃) + HCl]

[Reaction 3: acetone + Me–I → methylated ketone (α-Me) + HI]

[Reaction 4: acetone + Me–I → methyl enol ether (O–Me) + HI]

These correspond to C– and O–alkylation reactions. Sure, reaction with silicon isn't strictly alkylation, but it's close enough to use the same terminology.

> Again, we are building patterns and habits. If two things look similar, they will probably react in similar ways.

Use the following table of bond dissociation energies to determine the enthalpy of reaction in each case.

SECTION 4 CARBONYL COMPOUNDS AS NUCLEOPHILES

Bond Dissociation Energy / kJ mol^{-1}					
C–H	410	Si–Cl	360	H–I	298
C–C	350	C–Si	350	C=C	611
C–O	350	Si–O	460	C=O	732
C–I	240	H–Cl	432		

I don't find these calculations easy. If you replace a C=O bond with a C–O bond, you must consider it as if you have broken a C=O bond and formed a C–O bond. You cannot simply treat it as breaking 'part of the bond'.

The answers are at the bottom of the page.[66] With the numbers in hand, can you 'guess' the preferred outcome (*C*- or *O*-alkylation) in each case?

> *It turns out that you would be right in one case and wrong in the other. In fact, one of these reactions is too close to call.*

The problem is very simple, but not so simple to deal with. A C–O bond in an enol ether is very different to a C–O bond in an alcohol. The same applies to the other bonds. 'Average' bond-dissociation energies don't work very well! We have seen this before.

> *By getting you to do this calculation, I am really trying to get you to think more closely about the bonding.*

CALCULATED OUTCOMES

To provide a bit more perspective, we can calculate (or measure) the heats of formation of the two possible products in each case. Calculated data tend to relate to the gas phase, which isn't ideal, but it does still give us an idea.

> *The reaction with chlorotrimethylsilane is calculated to be favoured on oxygen by almost 30 kJ mol^{-1} compared to reaction on carbon. This didn't work so well when we used average bond-dissociation energies.*
>
> *The reaction with iodomethane is calculated to be over 100 kJ mol^{-1} more favourable on carbon than on oxygen. The average bond dissociation energies also gave a similar figure.*

---

[66] The numbers, in increasing numerical order, are −12 kJ mol^{-1}, −1 kJ mol^{-1}, +2 kJ mol^{-1} and +123 kJ mol^{-1}. I'm not going to tell you which one is which. It doesn't matter if it takes you a couple of attempts to get these numbers.

SECTION 4 CARBONYL COMPOUNDS AS NUCLEOPHILES

HARD AND SOFT ELECTROPHILES

We can define nucleophiles and electrophiles as being **hard** or **soft**. The definitions relate to the energies of the molecular orbitals. In this case, the difference is the electrophile, and we need the LUMO energy (Recap 1). A **softer** electrophile has a lower energy LUMO. The other point is the charge distribution in the electrophile. The Si atom in Me$_3$SiCl has a larger δ+ charge than does the carbon atom in CH$_3$I.

> *These definitions tend to be quite confusing at first. However, with practice you will find that you categorize nucleophiles and electrophiles as hard or soft simply by inspection of their structures. You'll be doing it by pattern recognition rather than worrying about the details. You know how I feel about that by now!*

A softer electrophile such as iodomethane tends to react on an enolate carbon atom, while a harder electrophile such as chlorotrimethylsilane will react on oxygen. This is useful!

APPLICATIONS 5

Kinetic and Thermodynamic Enolates

INTRODUCTION

The title of this short chapter is a little bit misleading. In a sense, that's the point. We will get to that in due course.

We looked at enols and enolates in Fundamental Reaction Type 4. We looked at whether an enolate reacts on carbon or on oxygen in Exercise 3. Because silicon forms strong bonds to oxygen, we can 'alkylate' an enolate on oxygen using chlorotrimethylsilane.

We know that more substituted alkenes are generally more stable (Recap 3).

Here are two reactions that are carried out under slightly different conditions. They give different regiochemical outcomes. The first one is carried out at low temperature, and it can be classed as kinetic control—the product that is observed is the one that is formed fastest. The second reaction is at higher temperature, so that equilibration can occur. The most stable product predominates.

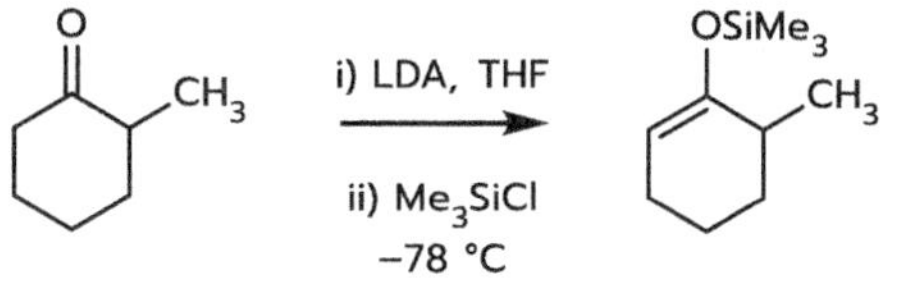

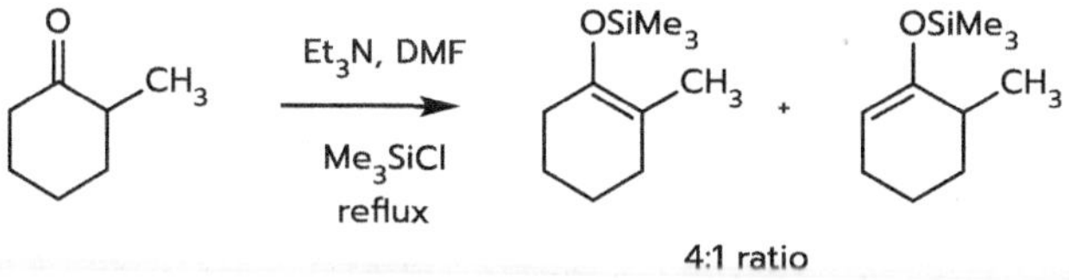

There is a key difference here. In the first reaction, you add one reagent, and then the other. In the second case, you just mix everything together and cook it!

LDA is lithium diisopropylamide, a strong (pK_a 35) base. THF is tetrahydrofuran, a solvent with an ether functional group. DMF is dimethylformamide, an amide solvent which is polar and has a relatively high boiling point.

I made the point, in Reaction Detail 21, that you need to look up any abbreviations that you are not familiar with. Are you doing it?

We need to be able to understand and explain the outcome of these reactions. Fundamentally, in both cases the overall equation for the reaction is the same. It is just the regioselectivity that changes.

303

SECTION 4 CARBONYL COMPOUNDS AS NUCLEOPHILES

> *If we cannot understand the outcome of these two reactions, how do we know what will happen if we try similar reactions on a different compound?*

We will break these reactions down into the traditional categories of 'kinetic enolate' and 'thermodynamic enolate'. It is certainly true that one gives the product that is formed fastest, and the other gives the product that is most stable.

KINETIC ENOLATE

This reaction is run at low temperature. We use a strong base.

At the level you are learning, we don't often consider detailed reaction conditions. Sometimes it helps. In this case, we take a flask of LDA and add 2-methylcyclohexanone.

> *The ketone is deprotonated quickly, so there is never any of the enolate present at the same time as the ketone. If we added the base slowly to the ketone, this would be a potential problem.*

> *What would happen if we had 50% of the enolate and 50% of the ketone in the flask at the same time? You can check your answer when you get to* **Reaction Detail 23**.

We have two different carbon atoms with acidic hydrogen atoms. We need to consider which one will be deprotonated.

> *The CH₂ group on the left is less hindered. We are using a bulky base. We will deprotonate the least hindered position.*

We can also consider a subtle electronic effect. First of all, let's reinforce what we understand by resonance. Here are the two possible carbanions that we would form by deprotonation at the two α-positions.

1a   **2a**

On the face of it, we would say that since a less-substituted carbanion is more stable, then **1a** would be more stable than **2a**.

But of course, if we consider the alternative resonance forms, we would say that **2b** is more stable than **1b** (**Recap 3** again).

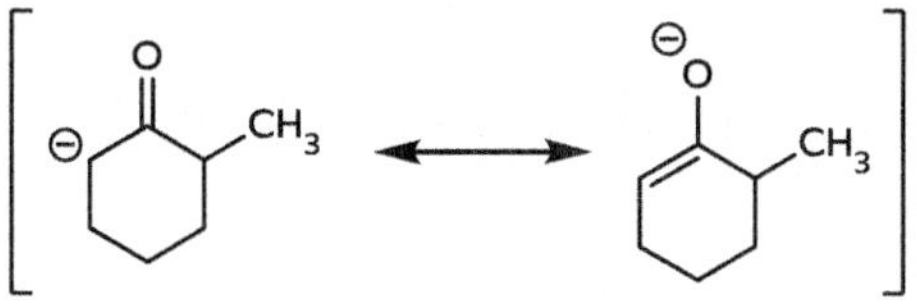

SECTION 4 CARBONYL COMPOUNDS AS NUCLEOPHILES

[Structures 1a and 1b shown as resonance forms]

1a   1b

[Structures 2a and 2b shown as resonance forms]

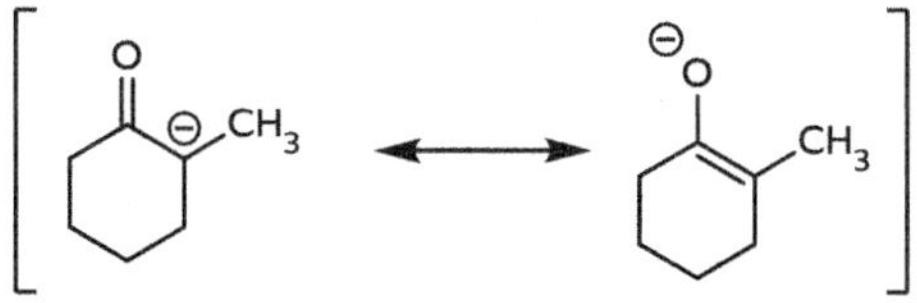

2a   2b

The reality is that 2 is more stable than 1. The reason that the hydrogen atom is acidic is that we can delocalize the electron-density onto the oxygen atom.

BUT.....there's something else that we might consider. As we start to deprotonate at the more-hindered side, to form **2a/2b**, we start to break the C–H bond. In the early stages of C–H bond breaking, we can consider the electronic effect of the methyl group destabilizing the build-up of electron-density at the α-position.

This is quite challenging to get used to. You won't get this the first time.

Of course, it is almost impossible to determine how much of the reason we get deprotonation of the less-substituted C–H is steric and how much is electronic. We should consider the electronic effect anyway, even if you do want to assume that steric factors dominate.

THERMODYNAMIC ENOLATE

There's a bigger problem here. We use the same substrate, a ketone, and we use triethylamine as the base. Triethylamine has a pK_a of 11.[67] The ketone has a pK_a of about 19.

What does this mean in terms of the following equilibrium?

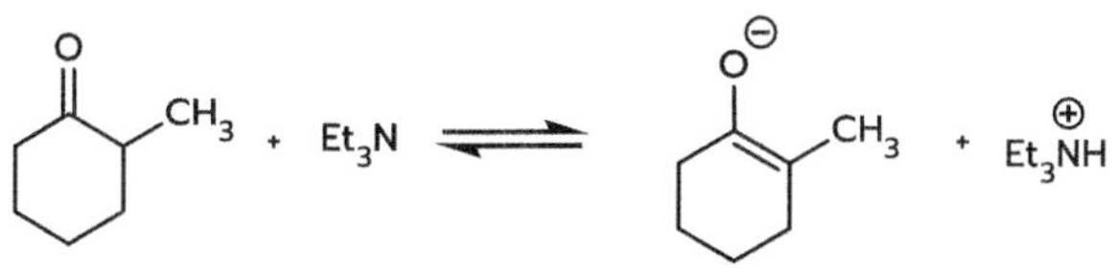

[Equilibrium: methylcyclohexanone + Et₃N ⇌ enolate + Et₃NH⁺]

The difference in pK_a is 8 units. This is a base 10 logarithmic scale. To put this another way, for the above equilibrium, $K_a = 10^{-8}$.

---

[67] Recall that when we talk about the pK_a of a base, we are really talking about the pK_a of its conjugate acid. Comparing the numbers is what is important.

SECTION 4 CARBONYL COMPOUNDS AS NUCLEOPHILES

> You cannot rule out a reaction in which very small concentrations of the enolate react very fast.

But perhaps there is an alternative mechanism which is better? Let's look at how we might analyse the problem. We have three things in the reaction—the ketone, triethylamine and chlorotrimethylsilane. The overall process is as follows.

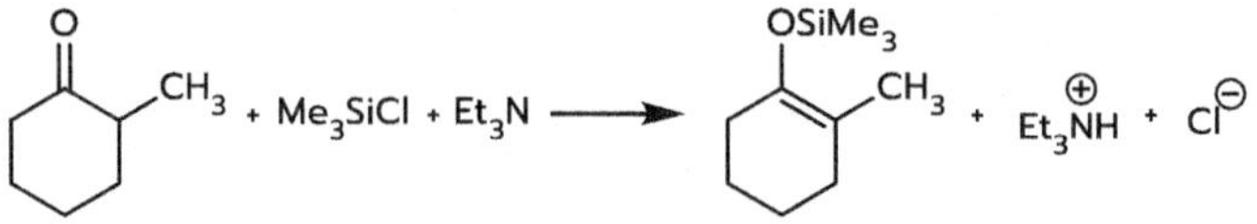

There is absolutely no doubt that the triethylamine is removing a proton from the ketone. We are breaking a C–H bond and the H is being picked up by the basic nitrogen.

> But this might not happen first!

We are definitely forming a bond between O and Si.

> If we don't form the enolate, then perhaps this doesn't happen last!

If we assume we only have two steps (not necessarily a safe assumption, but try it and see how far we get) then if Si–O bond formation doesn't happen last, it must happen first.

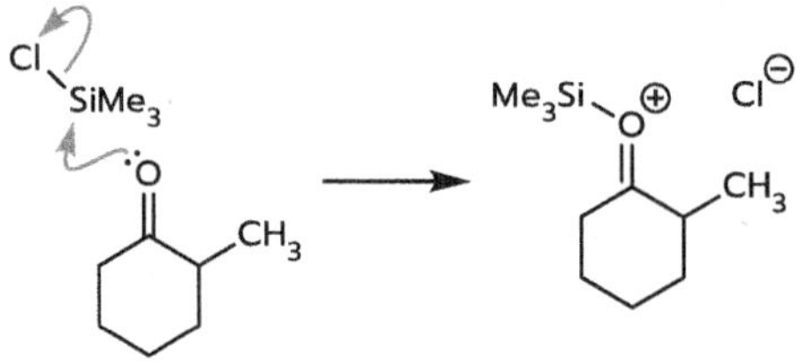

Let's see where this gets us. Here is the 'product' structure again, with three hydrogen atoms highlighted.

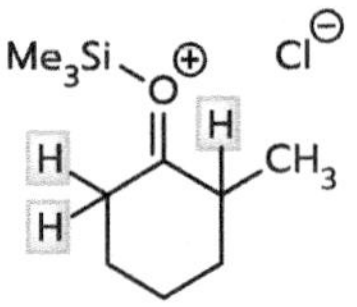

We have two hydrogen atoms on the left and one on the right. These are acidic. More importantly, they are a lot more acidic than they were before we silylated the oxygen atom.

> The reason these hydrogen atoms are much more acidic is because removal of any one of them takes us from a positively charged structure to a neutral structure. This is much more favourable. Now, a weak base such as triethylamine is absolutely fine.

In fact, it then doesn't matter which hydrogen atom we remove. We are heating this reaction, so that the products can equilibrate to favour the more stable product on the right.

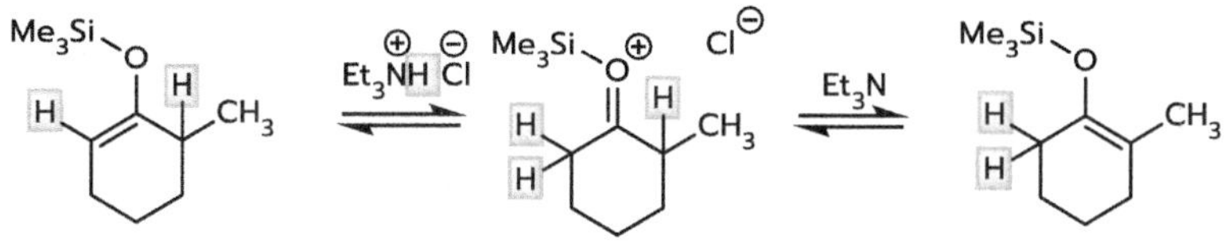

> Draw a curly arrow mechanism for the protonation of the compound on the left, using triethylamine hydrochloride[68] as the acid, followed by subsequent deprotonation to give the compound on the right.

MAKING THE LINK WITH REACTION PROFILES

There is no point to this exercise if we do not apply and link with basic principles. How general is this?

> *The short answer is that there are lots of reactions where products can interconvert to give an equilibrium mixture (which will favour the most stable), and there are lots of reactions that give mixtures of products based only on the rate of reaction.*

Which scenario a reaction falls into depends on whether there is a viable mechanism for interconversion of the products with a sufficiently low activation barrier that is accessible under the conditions for the reaction.

Let's have a look at some reaction profiles for the processes in this chapter. First of all, the 'kinetic' reaction. Product formation is *exothermic*. We have two possible intermediates with different energies. Each intermediate can only give one product. The profile is shown on the next page.

The important point here is that although the more substituted enolate is more stable, the transition state that precedes it is higher in energy. Therefore, it will be formed more slowly.

> This cross-over on reaction profiles is something we have not seen before.

---

[68] I haven't explicitly told you what this is, but the structure is right there on this page. Find it!

307

SECTION 4 CARBONYL COMPOUNDS AS NUCLEOPHILES

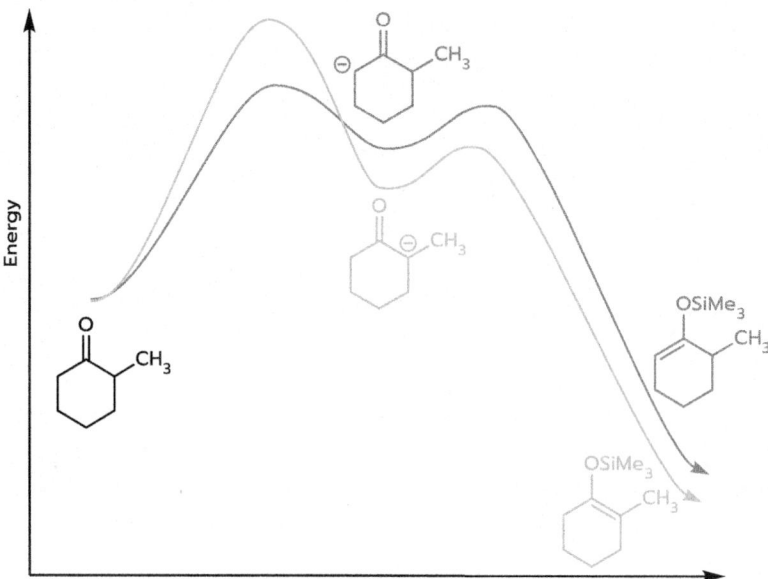

For the thermodynamic process, we have a different shape. We have a single intermediate that can give rise to the formation of either product.

We could debate the relative heights of the various barriers (the activation energies). But we are running this reaction at a much higher temperature than the kinetic reaction. We know, as an experimental fact, that the reaction reaches equilibrium. We want to ensure that we have enough depth to our analysis.

> But it's important to recognize when we have done enough!

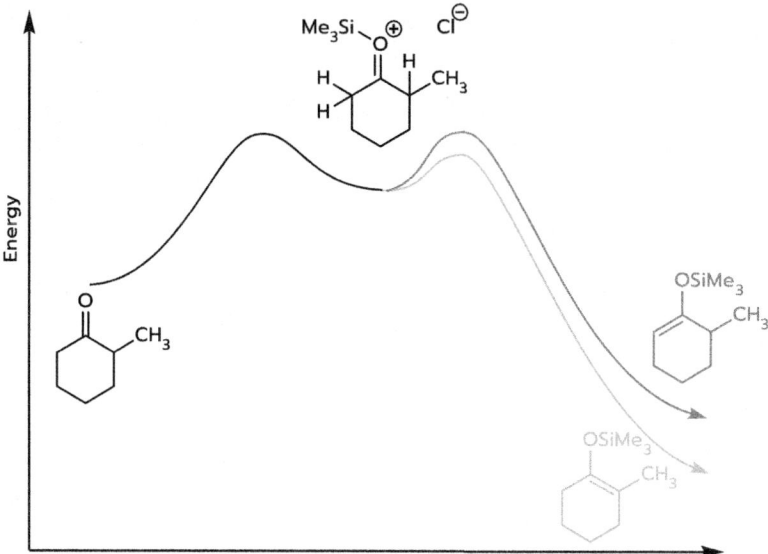

308

SECTION 4 CARBONYL COMPOUNDS AS NUCLEOPHILES

IN CONCLUSION

This is an important chapter. Sure, you will see reactions that are more widely used. You will encounter products that are much more important.

But this is a simple reaction that can be controlled to give different outcomes. If you understand, in detail, how this can be accomplished, you will be able to consider many different reaction mechanisms in great depth, and you will be able to apply the same principles to those reactions.

> *Don't lose sight of why we are doing this. You want to make organic molecules. You want your reactions to give high yields of the products you need, uncontaminated with impurities.*

SECTION 4 CARBONYL COMPOUNDS AS NUCLEOPHILES

REACTION DETAIL 22

Halogenation of Enols and Enolates

WHY?

This chapter covers a challenging reaction. All we are doing is adding an electrophilic bromine atom to a double bond, and we did that all the way back in **Reaction Detail 1**. But when the double bond is an enol/enolate, there are some rather subtle effects.

Don't expect to understand everything in this chapter on the first (or even the second or third) reading. This will take time to assimilate.

These reactions *do* get used in synthesis, but the main reason for covering these reactions here is that when you get different outcomes from a reaction under different conditions, you should understand why this happens.

INTRODUCTION

Let's start with the reactions, and then we should look at how we analyse them.

$$R-CO-CH_3 \xrightarrow{Br_2, \text{ acetic acid}} R-CO-CH_2Br$$

$$R-CO-CH_3 \xrightarrow{Br_2, \text{ NaOH}} R-CO-CBr_3$$

What we have is the same substrate and reagent. All we change is the pH, and we get a different outcome.

> We need to understand why this is.

These are very easy reactions to learn the mechanisms for without understanding. If I set an exam question featuring one of these reactions, I want to see that my students *really* do understand what is going on.

FIRST THINGS TO THINK ABOUT

We have a bond formed to the carbon atom α- to a carbonyl. This carbon atom can only be a nucleophile. The intermediate is either an enol, or an enolate (**Fundamental Reaction Type 4**). Which one will depend quite simply on whether we have a proton attached to oxygen.

> Right, very quickly we have established a possible reason for the outcome differing according to pH. But we still don't understand *why!*

310

SECTION 4 CARBONYL COMPOUNDS AS NUCLEOPHILES

THE FIRST STEP

At this stage, it is better to establish what the first step **isn't** rather than what it is. The first step, in either case, is **not** reaction of the organic substrate with bromine. The reason for this is deceptively simple. In order for this to happen, you have to remove a hydrogen atom from the α-carbon atom.[69]

> You cannot form a C–Br bond first and then break a C–H bond.

So, we **must** be talking about the removal of an α-hydrogen atom, under either acidic or basic conditions.

Under basic conditions, this is easy. Here is the reaction, with curly arrows.

We form an enolate. It doesn't matter which resonance form you draw for this. By now you should be confident that this is a stabilized anion, whichever way you draw it.

Note that we are not talking about complete removal of the α-hydrogen under such conditions. Hydroxide is the conjugate base of water, pK_a 15.7. The pK_a of a ketone is approximately 19. Nevertheless, we get enough enolate formed to permit further reaction.

From this point, the enolate can react with bromine as follows.

You will notice that we haven't drawn an intermediate bromonium ion here.

> Have a think about why we would not do that.

Now let's look at what happens under acidic conditions. In **Basics 2** we looked at acid catalysis in carbonyl chemistry. We shouldn't expect to change the rules here.

The first step, under acidic conditions, is protonation of the carbonyl oxygen atom. This is followed by loss of a proton from the α-carbon. This is keto-enol tautomerization yet again (**Fundamental Reaction Type 4**).

---

[69] Check that you understand what we mean by the 'α-carbon'. Look back at **Reaction Detail 21** if necessary.

SECTION 4 CARBONYL COMPOUNDS AS NUCLEOPHILES

We can now look at a similar bromination reaction to that shown above, but using the enol lone pair rather than the enolate negative charge. They are the same thing (a pair of electrons) but differ only in the extent of their reactivity.

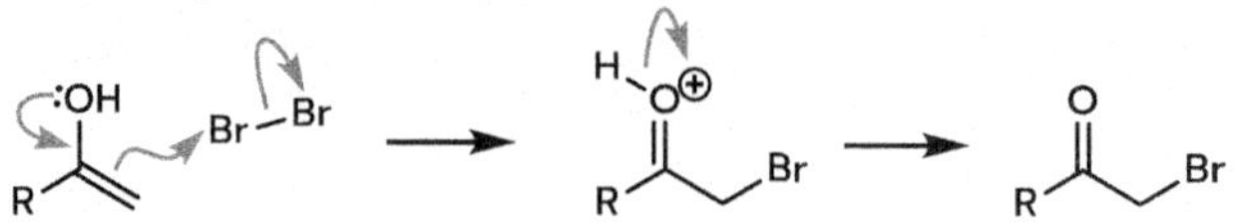

Once we lose the proton from oxygen, we will be at the same point as in the mechanism under basic conditions.

TAKING STOCK

It's good to take a step back at this point. We can now see how we can get bromination α- to a carbonyl group under acidic or basic conditions. If we were to carry on drawing curly arrows, we would be able to draw a mechanism for the triple bromination that is observed under basic conditions. What we have not done is address the question of **why** we only see a single bromination under acidic conditions, or **why** we get triple bromination under basic conditions.

> *The reasons must be simple.*

If bromination under acidic conditions only happens once, we can logically conclude that the second bromination reaction **must** be slower than the first. Similarly, if bromination happens three times under basic conditions, and cannot easily be stopped after the first bromination, we would assume that the second and third bromination reactions are faster than the first.

> *To understand the process, we should look very carefully at the relevant intermediates (and possibly the transition states) and make sure we draw appropriate comparisons. If we do this correctly, we would expect the answer to reveal itself.*

There! We have defined a strategy. Draw something, look at it and consider all aspects of stability (steric, electronic). This tends to be the default strategy for working out what happens in almost any organic reaction.

> *And that's why we do it!*

The main focus of this chapter is learning to solve problems in general, rather than just solving this specific problem.

ACIDIC CONDITIONS

We'll do the comparisons for the acid-catalysed reaction first. We have three distinct steps in the process.

1. Protonation of the carbonyl group
2. α-Deprotonation to form the enol
3. Reaction of the enol with bromine

SECTION 4 CARBONYL COMPOUNDS AS NUCLEOPHILES

We should compare each of these steps for the first and second bromination reactions, and see where this leaves us. Perhaps we can identify one (or more) steps that will be slower for the second bromination.

Here is the protonation for the two compounds, side-by-side. We might argue that because the bromine atom is electron-withdrawing, it will reduce electron density in the carbonyl group, and make **3** harder to protonate than **1**. However, this effect is likely to be marginal at best.

Now the α-deprotonation. In structure **4** you could argue that the electronegative Br would polarize the C–H bond and make deprotonation a little easier. This certainly wouldn't explain why the second bromination is slower.

> You might think we are not making progress, but by identifying the steps that are not likely to be affected, we are homing in on the one that is.

Now we should look at the actual bromination step. We would expect the bromine atom in **6** to withdraw electron-density from the enol, making this double bond less nucleophilic. It turns out that this is the step that makes the difference. Bromination of **6** is slower than bromination of **5**. Note that we have not worried about which double bond geometry we form of enol **6**.

Not that it really matters (since it never gets that far) but the third bromination reaction would be even slower.

> Before we proceed, let's express that key point another way. The enol **6** reacts more slowly because it is more stable.

In **Recap 3** we saw that it doesn't always follow that a more stable alkene is less reactive. It *is* the case here, because the intermediate formed in each case, a protonated carbonyl group, is comparable in stability.

SECTION 4 CARBONYL COMPOUNDS AS NUCLEOPHILES

BASIC CONDITIONS

Now we need to do the same with basic conditions. The only difference is, we need to see why each successive bromination reaction is easier (faster).

The C–H bond in compound **3** will be polarized as a result of the presence of the electronegative bromine atom. This means that the proton will be easier to remove (more acidic). We expect formation of **10** from **3** to be easier (faster) than formation of **9** from **1**.

> Note that this is the same argument that we made above. The enolate **10** is easier to form because it is more stable.

Enolates are more reactive than enols. They have a full negative charge, so they will attack nucleophiles, in this case bromine, much more readily.

While the enolate **10** will react slower than the enolate **9**, it turns out that they both react so quickly that it doesn't matter.

The rate-determining step for enolate bromination is formation of the enolate, and the more bromine atoms you have, the easier it is to form the enolate.

OVER TO YOU!

I have deliberately not shown you the full mechanism. The reaction schemes at the start of this chapter show you the overall outcome.

> Draw a full curly arrow mechanism under acidic and basic conditions.
>
> Sketch a reaction profile for each process (it isn't easy!) identifying all intermediates and transition states. For the reaction under basic conditions, try to put the relative energies of corresponding intermediates in the correct energetic order.

WHAT HAPPENS NEXT?

It turns out that this isn't quite the whole story. Under basic conditions, something else happens once the ketone has been exhaustively brominated.

> *Sodium hydroxide is a base. It is also a nucleophile.*

Once we have fully brominated the α-position of the ketone, assuming there are no acidic protons on the 'R' group, the hydroxide can no-longer act as a base. However, the CBr_3 group is electron-withdrawing, and makes the ketone **11** more susceptible to nucleophilic attack. Refer back to **Applications 1** for the reasoning.

SECTION 4 CARBONYL COMPOUNDS AS NUCLEOPHILES

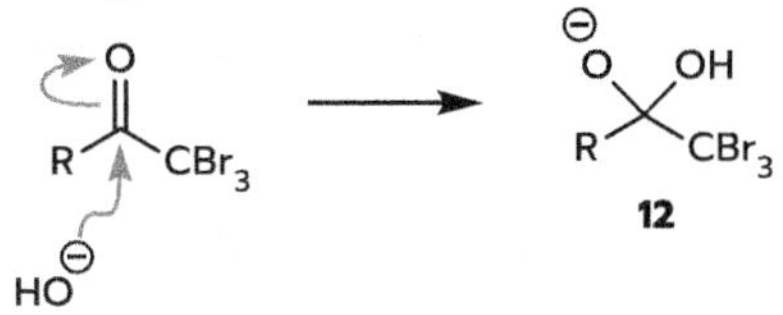

[Structure 11: R-C(=O)-CBr₃ with hydroxide attacking] → [Structure 12: R-C(O⁻)(OH)-CBr₃]

It turns out that compound **12** can do something interesting and useful. The tribromomethyl anion is not a bad leaving group. Each bromine atom stabilizes the negative charge a little as we have seen above. When we have three of them, this is a significant effect.

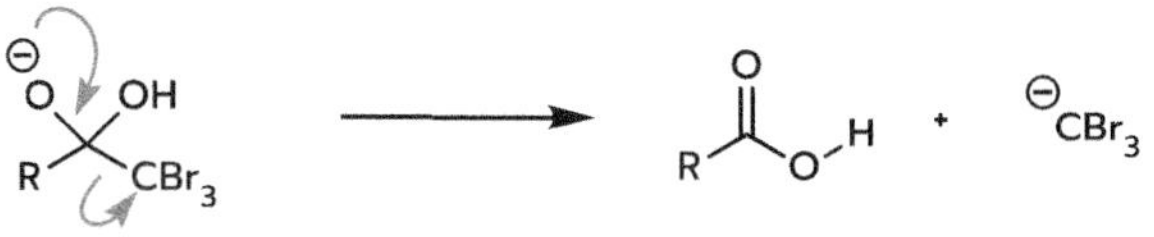

[Structure 12: R-C(O⁻)(OH)-CBr₃] → [Structure 13: R-C(=O)-O-H] + ⁻CBr₃

The formation of **13** from **12** is essentially **irreversible**.

| What is driving this to completion? |

Have a look at ester hydrolysis under basic conditions in **Reaction Detail 16**.

| What could the bromomethyl anion *do* to compound **13**? |

WHERE IS THIS GOING?

Fundamentally, what we have here is a range of enols and enolates reacting with nucleophiles. The relative reactivities that we have seen will apply to all electrophiles. Everything we have seen here is concerned with the fundamental reactivity of enols and enolates. We are going to see lots more reactions of these important intermediates.

SECTION 4 CARBONYL COMPOUNDS AS NUCLEOPHILES

APPLICATIONS 6

Predicting the Acidity of Enols

COMBINING BOND DISSOCIATION ENERGY AND pK_a

Here is a simple question, which we will work through. How acidic is an enol?

> If you've been paying attention up to this point, you'll probably realize that the answer is largely irrelevant, but how we might get to it is very important.

We need to start with what we already know. Here is the equation for keto-enol tautomerization.

$$H_3C-CO-CH_3 \; (1) \rightleftharpoons H_3C-C(OH)=CH_2 \; (2)$$

We know how acidic a ketone is. The pK_a for the α-hydrogen atom is about 19.

We know (HTSIOC Basics 35) how to work out enthalpy of reaction for reactions involving anionic species. We just don't (yet) have an equation involving an anionic species.

> Of course, we do. It's the acid-base dissociation reaction. We just haven't drawn it yet.

Here are the two key equations.

$$H_3C-CO-CH_3 \; (1) \rightleftharpoons H_3C-CO-CH_2^{\ominus} \; (3) + H^{\oplus}$$

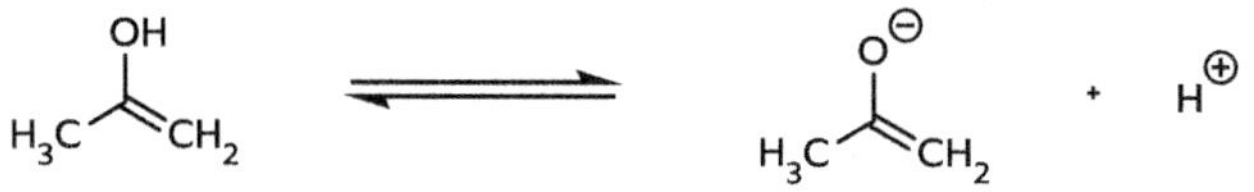

$$H_3C-C(OH)=CH_2 \; (2) \rightleftharpoons H_3C-C(O^{\ominus})=CH_2 \; (4) + H^{\oplus}$$

We know the pK_a of **1**, which means we know the K_a for the first equation. We don't know the pK_a/K_a for the second equation. This is what we want to find out.

> How do we compare the stability of structure 3 and structure 4?

Hopefully you realize that this is a trick question. They are the same thing, although I have drawn the two different resonance forms.

SECTION 4 CARBONYL COMPOUNDS AS NUCLEOPHILES

> I like to look at things graphically. It's not the only way to solve a problem. It's just the one I prefer. You may find another way that works better for you.

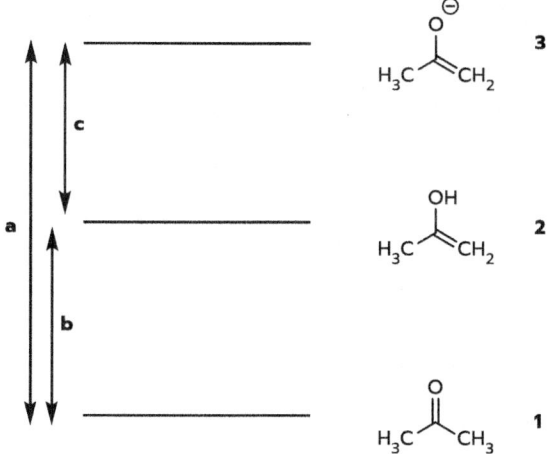

In the diagram above, we want to work out **c**, and we have enough information to work out **a** and **b**. It is hopefully starting to look a bit more manageable.

We know the K_a that corresponds to **a**. It is 10^{-19}, since pK_a = 19. We recall from our basic physical chemistry that

$$\Delta G = -RT \ln K$$

At 298 K, ΔG for this reaction is +108 kJ mol^{-1}. This looks okay. We expect a positive number, and this is significant but not too big.

> Have a go at the calculation. Make sure you get the same number. Check all the units. Remember, the logarithm in this equation is a natural logarithm.

How do we work out **b**? We have two stable (relatively) neutral species, so we can use bond dissociation energies to compare.

Bond Dissociation Energy / kJ mol^{-1}					
H–H	436	C–C	350	C=C	611
H–Cl	432	C–H	410	C≡C	835
H–Br	366	C–O	350	C=O	732
H–I	298	C–Cl	330		
H–O	460	C–Br	270	C≡N	898
H–S	340	C–I	240		

317

SECTION 4 CARBONYL COMPOUNDS AS NUCLEOPHILES

In going from **1** to **2**, we are breaking a C–C single bond, a C–H single bond and a C=O double bond. We are forming a C=C double bond, a C–O single bond and an O–H single bond.

$$\Delta H = (350 + 410 + 732) - (611 + 350 + 460) = +71 \text{ kJ mol}^{-1}$$

Again, this is a reasonable number.[70] Enols are generally less stable than the corresponding ketone, but the energy is not 'too much'.

> At this point, we have a ΔG and a ΔH. What do we do about these?

Strictly speaking, we need to consider the entropy change for each process. But as keto-enol tautomerization does not involve a change in the number of molecules, we can assume that the entropy change, ΔS, for this process will be close to zero. Therefore, as an approximation, $\Delta H = \Delta G$. We are almost there now.

$$\mathbf{a} = +108 \text{ kJ mol}^{-1}$$

$$\mathbf{b} = +71 \text{ kJ mol}^{-1}$$

$$\text{Therefore } \mathbf{c} = +37 \text{ kJ mol}^{-1}$$

We can now convert this back to a K_a value.[71]

$$+37{,}000 = -RT \ln K_a$$

$$K_a = e^{\left(\frac{37{,}000}{-8.314 \times 298}\right)} = 3.3 \times 10^{-7}$$

From this, we can deduce that $pK_a = 6.5$.

This looks like a reasonable number. We would expect an enol to be more acidic than a ketone. After all, it is more than half way (in energetic terms) there!

Structurally, we expect the enolate double bond to stabilize the negative charge.

> What do you need to get from this? Reinforcement of the relative stability of keto and enol forms of carbonyl compounds. Stopping to think about the relative acidity of keto and enol forms. Most of all, you get another worked example of a useful general method.

I don't recall seeing this calculation done in other textbooks. That doesn't necessarily mean that it is an omission on the part of the authors, or that it is an error to include it here. This isn't a trivial calculation, but it can become trivial once you take a few facts and a methodical approach.

> That's the point!

---

[70] Which doesn't automatically make it correct, so it's still important to check that you haven't done anything too silly.

[71] Perhaps you weren't expecting so much maths in this book. Neither was I!

SECTION 4 CARBONYL COMPOUNDS AS NUCLEOPHILES

REACTION DETAIL 23

Aldol Reactions Under Basic Conditions

INTRODUCTION

Okay, this is the big one! The aldol reaction is awesome. You form a carbon-carbon bond (sometimes a single bond, sometimes a double bond). You can form two stereogenic centres, often with high levels of stereochemical control. Nature uses aldol reactions to make many important compounds. 'Biosynthesis' is the term we apply to the synthesis of chemicals in biological systems such as plants, or in the human body. All the fats in your body are biosynthesized using aldol-type reactions to form the carbon-carbon bonds to build a long chain from several smaller precursors. Many important antibiotics and immunosuppressants are biosynthesized using stereoselective aldol reactions.

> *If I had to single out one reaction as being **the most important** organic reaction, the aldol reaction would probably be it! Whatever else you choose to push to one side, don't let it be this. It might not be the easiest reaction to learn, but if you nail this one, the rest will get easier.*

Let's start with a very simple statement of what an aldol reaction is, and then add some detail. Fundamentally, you take an enolate anion **2** and add it to another carbonyl compound **1**.

> *It doesn't matter which resonance form of the enolate you draw. I drew this one, because it saves me one curly arrow!*

This gives you intermediate **3**, which will be protonated to give the product **4**. It's an aldehyde, it's an alcohol. It's an aldol![72]

> *We can consider this to be a combination of Fundamental Reaction Type 4 and Reaction Detail 21. We are adding an enolate nucleophile to a carbonyl. We are 'alkylating' an enolate with a carbonyl electrophile. It is important to put this reaction into context. There is no new reactivity here!*

In the example above, we used an enolate. This would be formed under basic conditions. Let's explore this in a little more detail.

---

[72] We still use the term for the reaction, even when the product is not an aldehyde!

SECTION 4 CARBONYL COMPOUNDS AS NUCLEOPHILES

> It turns out that you can carry out the aldol reaction under acidic conditions as well. We will see this in Applications 7.

ALDOL REACTIONS UNDER BASIC CONDITIONS

The enolate **2** is an anion α- to a carbonyl group. It is formed by removal of a proton, and it is formed because the proton is acidic. If it wasn't, we wouldn't be able to remove it. We saw this in **Recap 7**. Therefore, the aldol reaction starts with the deprotonation of a neutral carbonyl compound as follows. We will use the hydroxide anion as base.

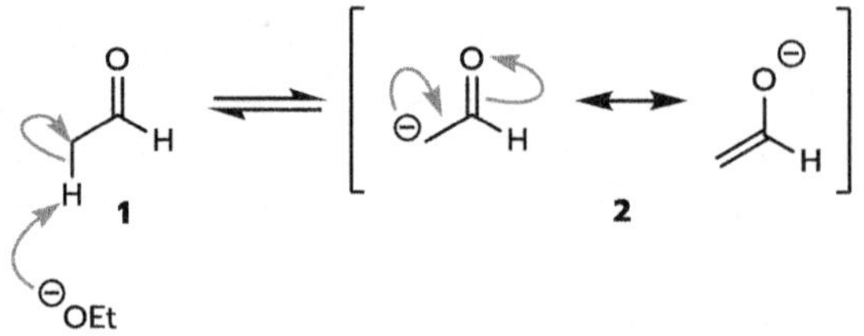

THE MOST COMMON MISTAKE

It is very common to claim that the hydrogen atom directly attached to the carbonyl carbon atom is acidic. It isn't! If you make this mistake, you will form the wrong bonds.

> What's the best way to check this? Draw it out!

The negative charge in the structure on the right is not stabilized.

> How do we prove that?

That's a very valid question. I could say "try to draw some resonance forms". The problem is, either you won't be able to draw resonance forms, or you will draw incorrect resonance forms. All I can suggest is keep looking at the resonance forms for structure **2**, above, and look at the 'wrong' anion and try to see why any curly arrow you push from the negative charge will be wrong.

> There is nowhere for the electrons to go!

Once again, this proves the value of a combination of learning, understanding and practice. You need to get used to seeing negative charges α- to carbonyl groups, and in order to fully do this, you need to be 100% confident that you know which is the α- carbon atom.

> It's not the carbonyl carbon atom—it's the one next to it.

SECTION 4 CARBONYL COMPOUNDS AS NUCLEOPHILES

BACK TO THE STORY

Now let's finish the job. If we treat an aldehyde with a base, some deprotonation will take place at the α- carbon atom.[73] This will form the key enolate intermediate that can then attack a second equivalent of the carbonyl compound. After addition of water as a source of protons, this will then give the aldol product. Here it is for the aldol reaction of acetaldehyde with itself.

ca. 50% yield

> Let's think about the first equilibrium for a moment. We need to know how acidic the α-hydrogen atom is. It has a pK_a of approximately 19. The "pK_a of hydroxide" is about 16, so there is a difference of 3 pK_a units, which is a factor of 1,000 (remember that pK_a is a logarithmic scale, with base 10). You will note that I put "pK_a of hydroxide" in inverted commas. This is because we should really refer to the pK_a of the conjugate acid, water. However, it is acceptable to talk about the pK_a of acids and their conjugate bases interchangeably.

The main point is that hydroxide is not a good enough base to completely deprotonate the aldehyde. However, it is good enough to generate a small equilibrium concentration of the enolate, and it is this that reacts (as above).

CROSSED ALDOL CONDENSATIONS

In the reaction above, we take acetaldehyde (ethanal) and add a base. One molecule of acetaldehyde is deprotonated, and the resulting enolate reacts with a second molecule of acetaldehyde.

> If this was all we could do with an aldol reaction, it wouldn't be much use.

How can we react two different aldehydes or ketones (or one of each) together, and control which is the nucleophilic partner (the enolate) and which is the electrophilic

---

[73] Assuming, of course, that there is a hydrogen atom here.

SECTION 4 CARBONYL COMPOUNDS AS NUCLEOPHILES

partner? Before we answer this question, here is a secondary question which illustrates the problem.

> What would happen if we took a 1:1 mixture of two different aldehydes, and add a base? We asked a very similar question in **Applications 5**.

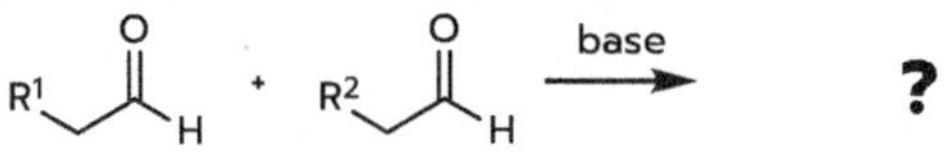

$$R^1-CHO \quad + \quad R^2-CHO \quad \xrightarrow{base} \quad ?$$

Hopefully now, you are starting to ask the right questions. Starting with "which aldehyde will be deprotonated?"

> Of course, we could make the mistake of looking at the aldehyde hydrogen atom that is explicitly drawn, and think we are deprotonating there—but we know that isn't what we mean!

If we simply mix two aldehydes, and add hydroxide, neither of them will be deprotonated much, and it would be hard to justify which will be deprotonated the most, unless R^1 and R^2 are very different.

> This places serious limitations on the reactions we could do successfully. And it gets worse, because the enolate will react more or less equally with either aldehyde. We would get a horrible mixture of four possible products.[74]

There's a very simple way to ensure that we only deprotonate one of the reacting partners. Look at the following scheme.

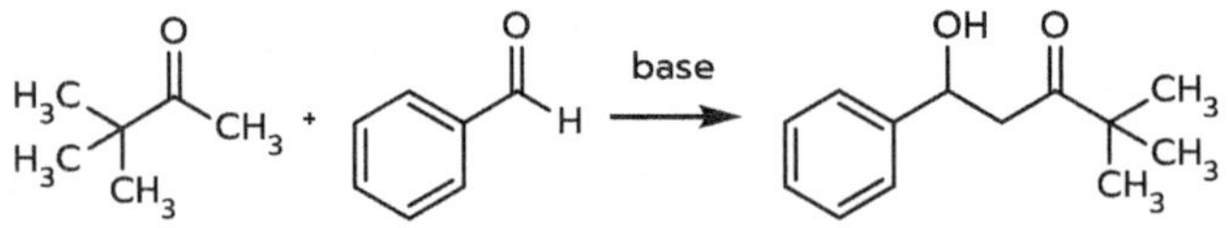

5   **6**   **7**

We can deprotonate compound **5**. We cannot deprotonate compound **6**.

> Make sure you look carefully at the structures, draw out the mechanism, and that you can see this is true—don't just take my word for it.

Aldehydes are more reactive than ketones. Even if we only have a small amount of the enolate, it will prefer to react with aldehyde **6** rather than with ketone **5**.

> Sure, aldehyde **6** is conjugated, which makes it less reactive. But ketone **5** is hindered, which also makes it less reactive.

What we are doing here is defining differences is reactivity types (which has an α-hydrogen atom) and in reactivity itself (which carbonyl group is more electrophilic).

---

[74] You did draw them all out, right? How many stereoisomers could you draw for those four different constitutional isomers? That reaction won't be pretty!

SECTION 4 CARBONYL COMPOUNDS AS NUCLEOPHILES

In doing this, we have defined a scenario in which we can mix two carbonyl compounds, add a base, and get a single aldol product. But it is still limited to certain reacting partners. We would really like to be able to react any **enolizable**[75] carbonyl compound as the nucleophile with any other carbonyl compound as the electrophile.

What we need is a flask of a pure enolate derived from one carbonyl compound, and to then add the second carbonyl compound. We saw that using hydroxide as base only gave a small equilibrium concentration of the enolate. This won't do! We need a stronger base, with $pK_a > 20$.[76]

> Even so, what happens if we take this first carbonyl compound, and add the base slowly?

Now we get to the answer to the question in Applications 5, and also on the previous page. Break it down. What will happen when you have added 10% of the base? You will have deprotonated 10% of the carbonyl compound, forming 10% of the enolate in the presence of 90% of the carbonyl compound. The 10% of the enolate will react with the carbonyl compound. Messy!

So, we need to make sure that there isn't carbonyl compound and enolate present at the same time. We start with the base in our flask, and add the carbonyl compound we want to deprotonate.

> There is always more base than carbonyl compound, and the base is more reactive, so we will form the enolate.

I actually added another qualifier here. We could use LDA as base. We saw this base before, in Applications 5. It has a pK_a of 35. The enolate has pK_a approximately 19. We can use the pK_a value as a proxy for reactivity.

Once we have formed the enolate, we can add the second (electrophilic) carbonyl compound, and get a single aldol product with high chemoselectivity. Here is the overall process shown schematically.

> To summarize, once you get the reaction conditions under control, you can deprotonate pretty much any carbonyl compound with an α- hydrogen atom, and you can then add the resulting enolate to more or less any other carbonyl compound.

---

[75] Something we could form an enol from (which means it has hydrogen atoms on the carbon α- to the carbonyl group).

[76] Remember, we are still talking about the pK_a of the conjugate acid.

SECTION 4 CARBONYL COMPOUNDS AS NUCLEOPHILES

Naturally, this works better if the electrophilic component is reasonably reactive (**Applications 1**). Aldehydes are good!

There's something else we need to mention. In the product in the above scheme, we have formed two new stereogenic centres. It turns out that we can get excellent stereoselectivity in aldol reactions. We will look at this in **Stereochemistry 14**.

THE CLAISEN CONDENSATION

There are many variations on the aldol reaction. These are relatively old reactions, and they have been associated with the names of their inventors.[77] The Claisen condensation is basically an aldol reaction but with an ester instead of an aldehyde. There are similarities and differences, and we should understand why there are differences.

> *Whenever you encounter reactions that give different outcomes despite obvious similarities, you should be able to work out the reasons by drawing plausible mechanistic steps. Don't try to reason out the entire problem without drawing. Just make a start.*

Here is a simple Claisen condensation.

[Structure: methyl acetate (CH₃C(O)OEt) → with NaOEt, EtOH → ethyl acetoacetate (CH₃C(O)CH₂C(O)OEt), 75%]

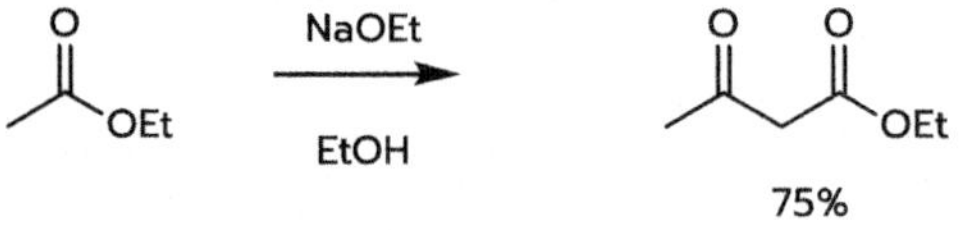

Compared to the aldol reaction, it does some of the same things. Clearly two molecules of the starting material have reacted together to give a single product. Since the starting materials are at different oxidation states (**Basics 3**), we would expect the products to be at different oxidation states. After all, we are not using an oxidizing agent or a reducing agent.

> *We didn't use hydroxide as base this time. We used ethoxide instead. If we had used hydroxide, we would have hydrolysed at least some of the ester (we saw this in **Reaction Detail 16**) and this would lead to complications. In this case, if ethoxide acts as a nucleophile, there is no overall change. We no longer need to wonder why the observed product requires ethoxide to be a base rather than a nucleophile.*

An ester is a carbonyl group, so you shouldn't be surprised that you can remove a hydrogen atom α- to the ester. The pK_a is a bit higher (25 rather than 19—**Recap 7**) which means that the hydrogen atom is quite a bit less acidic. We are using a base with a similar strength, so we will have an even smaller concentration of the enolate. However, we can draw the same mechanism with some relatively minor differences.

---

[77] Except aldol, which as we have seen is named after the product—even when it isn't an aldehyde or has an alcohol in it!

SECTION 4 CARBONYL COMPOUNDS AS NUCLEOPHILES

> Have a look back at ester formation/hydrolysis in **Reaction Detail 16**. Make sure you can see the similarities here. A nucleophile adds to form a tetrahedral intermediate, then the best leaving group leaves.

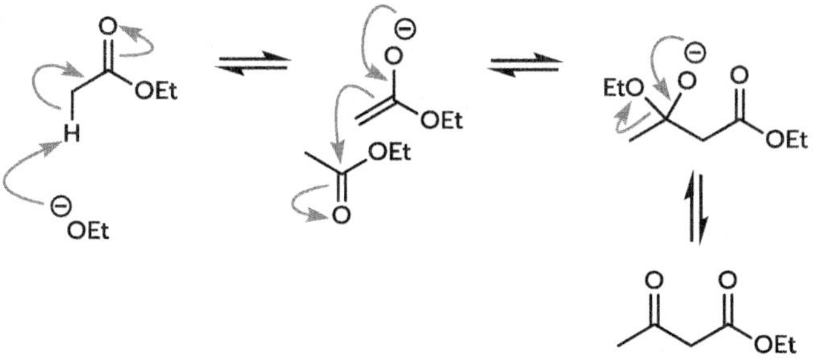

THERE'S JUST ONE COMPLICATION

It isn't even a big complication. But we will see in **Reaction Detail 25** how it can affect the position of equilibria.

> Sometimes, things happen, and we need to consider them, even if there aren't any real consequences. Because just one time, there will be!

After the enolate attacks the ester, we get an intermediate (top-right in the scheme above) which can lose ethoxide to give a keto-ester **8** with a CH_2 between the two carbonyl groups. The hydrogen atoms of this CH_2 group are much more acidic than if they were only α- to one carbonyl. They have a pK_a of about 12 (**Recap 7** again), so this happens.

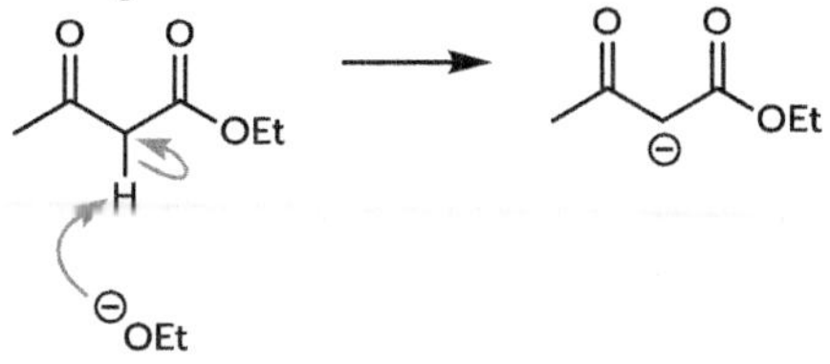

I've drawn the structure on the right as a "simple" carbanion but remember that it is an enolate and you can draw resonance forms. The most important point is that this deprotonation is complete and this "drives" the Claisen condensation to completion. In order to finish the reaction, you neutralize the base, and this protonates the enolate so that **8** is the final isolated product.

THE DIECKMANN CYCLIZATION

Another new reaction, except that it isn't really new. It's just a Claisen condensation in which both esters happen to be in the same molecule. Therefore, it forms a ring.

SECTION 4 CARBONYL COMPOUNDS AS NUCLEOPHILES

Here's an observation based on experience and lots of conversations with students. At first, many students struggle with reactions that form rings. As soon as we add ring formation to an existing process, it seems to become more difficult.

> You know what to do to make this easier!

At first, you won't be able to tell, at a glance, what size ring you are forming in the reaction below.

> Put numbers on the atoms. Draw a dotted line between the enolate carbon atom and the carbonyl carbon atom, so you are absolutely clear where the bond formation is. Work out the molecular formula of the starting material and product (you are losing two carbon atoms—ethanol) to check you haven't made a mistake.

All these steps are essential in order to develop your habits.

Ultimately, the mechanism is the same as we saw for the Claisen condensation, and the product **9** can be deprotonated between the carbonyl groups (remember, in this case there is only one hydrogen atom there—the carbon atom has three bonds to other carbon atoms, so there must be a hydrogen atom there as well!), so we have the same driving force.

> In this example, a six-membered ring is formed. This happens quite readily because six-membered rings are nice and stable.

Five-membered rings are pretty good too. Learn to look for reactions that form these ring sizes!

> What would happen if we tried to form a four-membered ring using this reaction? Would it work better/just as well/less well?

Hopefully you said it would work less well. The corresponding product with a four-membered ring would be less stable. The transition state for forming the ring would be less stable. We will quantify some of these aspects next, in **Basics 6**.

SECTION 4 CARBONYL COMPOUNDS AS NUCLEOPHILES

BASICS 6

Ring Strain and Ring-Forming Reactions

INTRODUCTION

We often find organic compounds with rings (cyclic compounds). These range from the very obvious and very stable benzene rings, to saturated rings. In **Reaction Detail 23** we have just seen a reaction that forms a six-membered ring.

Very early in your chemistry career, you will have encountered benzene rings, and you will have learned that they are very stable.

> *This is a good thing.*

Later, you will have encountered cyclohexanes, and also learned that they are pretty stable.

> *This is also good, as long as you don't confuse the aromaticity of benzene compounds with the 'not at all special' stability of cyclohexanes.*

STABILITY OF CYCLOALKANES

We saw in **Stereochemistry 1** that the bonds in cyclohexanes are as staggered as they could possibly be, and that each carbon atom has the perfect tetrahedral bond angle of 109.5°. In cyclopentane, we find that the C–C–C bond angles are around 106°, so that there is **bond strain** (deviation from the ideal bond angles), and there is some eclipsing of hydrogen atoms on adjacent carbon atoms (**torsional strain**).

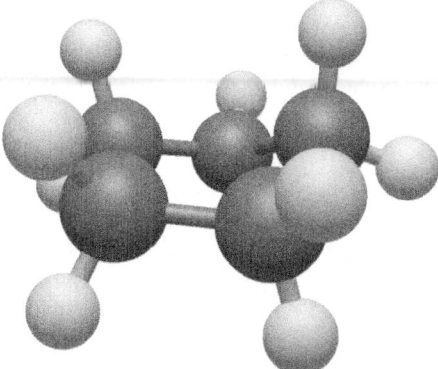

The net result of this is that cyclopentane is less stable than cyclohexane. We can quantify this as a strain per CH$_2$ group, which allows us to compare different ring sizes in a meaningful way.

Here is a summary of the strain and bond angles for a range of cycloalkanes.

327

SECTION 4 CARBONYL COMPOUNDS AS NUCLEOPHILES

Ring Size	Strain per CH$_2$ group/kJ mol^{-1}	C–C–C Bond Angle
3	38.8	60°
4	27.8	87°
5	5.5	106.3°
6	0	109.5°
7	3.8	115°
8	5.1	116°

Cyclopropane is a very extreme case. With only three carbon atoms, they must be in the same plane. The C–C–C bond angles must be 60°, which is quite a deviation from the ideal tetrahedral angle of 109.5°.

In fact, like so much of organic chemistry, it's not quite as simple as this. Have a look at the following diagram. What I have tried to show is the sp^3 hybrid orbitals of the C–C bonds. You can still get good orbital overlap without the orbital lobes being directly along the line between the atoms.

> In effect, the angles between the bonds are considerably larger than 60° even though the angles between the atoms are 60°.

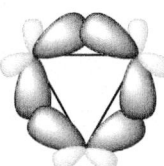

It is worth pointing out that hybridization is just one way to show the bonding. In **HTSIOC Perspective 3** we looked at how to 'make' molecular orbitals directly from the atomic orbitals without hybridization. We could do this with cyclopropane, although it is quite complicated.

> There is an intermediate model that is commonly used. In Reaction Detail 32 we will encounter carbenes. A carbene is a carbon atom with only two bonds. It is electron-deficient. You can think of cyclopropane as a 'trimer of carbene'. When you are ready, look up 'Walsh orbitals'.

When we get to the larger ring sizes, we find that there are unfavourable steric interactions. The following representation of cyclooctane shows this.

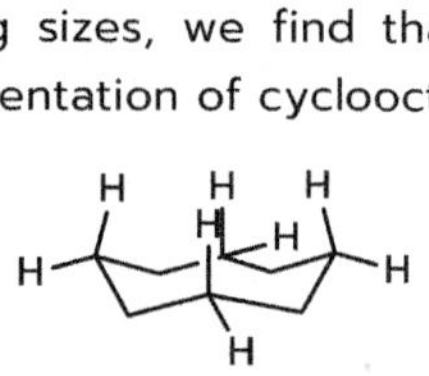

SECTION 4 CARBONYL COMPOUNDS AS NUCLEOPHILES

We refer to the cyclooctane structure as a crown. Don't over-think the terminology.

> Make a model of the various different cycloalkanes and 'explore' them.

You will find that depending on what type of molecular model kit you have, it probably won't be too easy to make a cyclopropane. The bonds in the model will bend. In cyclohexane, you will be able to look down the C–C bonds and check that everything is nicely staggered.

> *As always, actively learning this for yourself will have much more impact than reading it as a series of facts.*

If you know the ideal bond angles, you can see for any given cycloalkane just how much deviation there is. This is much easier to do with the smaller rings. If you had a sixteen-membered ring, I'm pretty sure you could get all of the C–C–C angles to be 109.5°. That alone won't make it stable. You then need to look for any eclipsing of bonds (**torsional strain**) and close interactions of atoms (**steric strain**). We defined these terms in **Stereochemistry 5**.

RING-FORMING REACTIONS

Cyclopropane is strained, but there are plenty of cyclopropanes known, and they are not actually that unstable.

> *They are thermodynamically unstable compared to cyclohexanes, but that doesn't mean they cannot be formed and isolated.*

What we need for a ring-forming reaction is the same as we need for any other reaction—overlap of orbitals.

In **HTSIOC Applications 2** we looked at cyclization reactions that form three-membered rings, specifically epoxides. In the following structure, we find that the conformer in which the O and the Br are on opposite sides of the C–C bond is favoured since this minimizes the dipole in the molecule. The cyclization reaction is an S$_N$2 substitution, so the nucleophile must 'attack' the σ* orbital of the C–Br bond.

> *It is perfectly lined up in this conformer to do this. This is a fast reaction!*

Of course, if a cyclopropane were formed in a similar way, the same factors would not apply.

If we want to form a five-membered ring, we have two conformers that we need to consider. The conformer shown below on the left is likely to be the most stable—all bonds are staggered, and the electronegative elements are as far apart as possible.

329

SECTION 4 CARBONYL COMPOUNDS AS NUCLEOPHILES

But in order to form the ring, we need the conformer on the right. This conformer is much more 'organized'.

As soon as we need a level of organization, this indicates that we are dealing with something of lower entropy. And, of course, the right-hand conformer has more bonds that are eclipsed or close to it.

> Make a model of the structure. Rotate the bonds, identify the staggered and eclipsed conformers.

So, does this mean that we form a three-membered ring more readily that we would form a five-membered ring?

> *Yes, for this particular reaction, we do!*

But we form a five-membered ring and a six-membered ring faster than we form the corresponding four-membered ring.

> *In the case of the four-membered ring, we have the strain energy, but we don't have the favourable low-energy conformation.*

Once again, this problem demonstrates the need for you to take a systematic approach to studying organic chemistry. There are too many reactions and too many compounds for anyone to be able to remember all of them. Instead, you need to be able to identify the factors that are relevant in any given reaction, and to apply them correctly.

Let's see how we can do this by comparing two reactions.

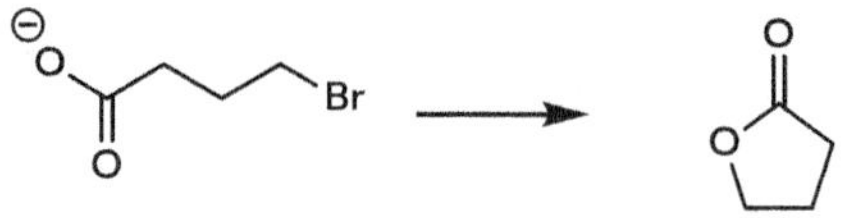

This is an ester-forming mechanism that we looked at in Reaction Detail 16. We found that it wasn't very good.

> *It's a lot better when it is **intramolecular**—when the nucleophile and the leaving group are in the same compound.*

SECTION 4 CARBONYL COMPOUNDS AS NUCLEOPHILES

Compound **1** is particularly strained. It has an sp² hybridized carbon atom in the ring, with ideal bond angles of 120°. Compound **2** isn't as bad as in the example we just looked at. We do still need to rotate a couple of bonds to get the oxygen atom close to the back of the C–Br bond, but now we have two oxygen atoms to choose from.

> The odds have just improved dramatically, although a carboxylate oxygen atom is inherently less nucleophilic (more stable, lower pK_a) than an alkoxide oxygen atom.

So now we form the five-membered ring about 10,000 times more rapidly than we form the three-membered ring.

FORMATION OF LARGER RINGS

Seven, eight and nine-membered rings are the hardest to form. We have considerable steric strain within the ring, and we have quite a lot of preorganization required for ring-formation to take place.

When we get to even larger rings, they tend to be formed more rapidly. We can usually find conformers that don't have so much steric strain. Of course, there is another fundamental problem!

> There usually is!

If the only thing we had to contend with was a slow reaction, we could simply wait, or we could heat the reaction to make it go faster. Look at the following reaction, which is the one we saw above, but in a more general sense.

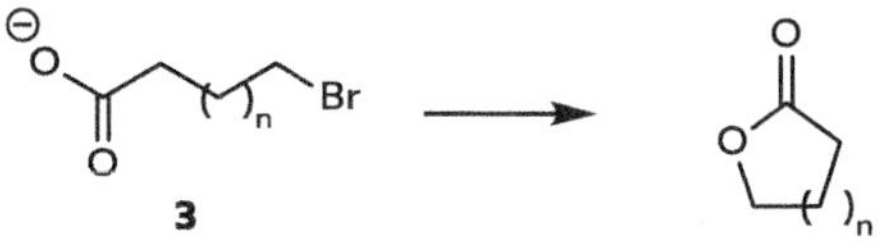

3

It's all about the questions you ask.

> What else could compound **3** do?

A molecule of compound **3** could react with **another** molecule of compound **3** in an intermolecular reaction. This would be a second order process. The cyclization reaction is first order.

> Write out the rate equations for both processes. Plug in some numbers. Draw a graph.

If we decrease the concentration of compound **3**, both reactions will be slower, but the intermolecular reaction will slow down **more**. If we want to form 'difficult' ring sizes, we tend to use high dilution conditions.

SECTION 4 CARBONYL COMPOUNDS AS NUCLEOPHILES

BALDWIN'S RULES FOR RING CLOSURE

In the 1970s, Baldwin formulated a series of 'rules' for cyclization reactions based on the size of the ring being formed and whether the centre being attacked is sp^3, sp^2 or sp hybridized. In the formulation of the rules, we refer to these as *tet*, *trig* and *dig* respectively.

> *All the cyclization reactions in this chapter are 'tet'. If we formed a ring by attacking a carbonyl group, this would be 'trig'.*

During the ring-closure, we must break a bond. If the bond is inside the ring that is being formed, the cyclization is classified as *endo*. If the bond is outside the ring, it is classified as *exo*.

> *All the cyclization reactions in this chapter are 'exo'. The following reaction is '6-endo-trig'. A carbon-carbon π bond is broken, with both atoms being within the ring that is formed.*

> *Draw an alternative 4-exo-trig reaction that the compound above could undergo.*

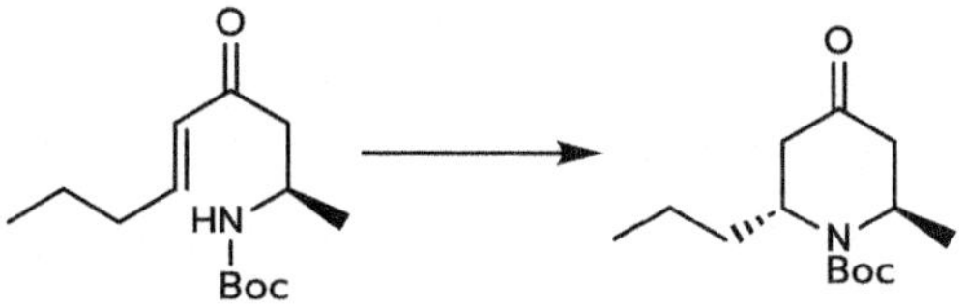

This takes us to the point of Baldwin's rules. They are a convenient summary of which reactions are favoured and which reactions are disfavoured.

We are not going to summarize Baldwin's rules here. I don't want to fill this book with data, even if it is quite useful data. All you need is to know the preferred angle of attack on a given functional group, and you can use your molecular models to predict which cyclization reactions will work well. If you need to check, you can easily find a table of Baldwin's rules online.

Don't lose sight of the fundamental principles. A 6-*exo-tet* cyclization forming a six-membered ring is favourable because it is possible to line up the nucleophile and the leaving group on opposite sides of the carbon atom.

> *For me, this is what is wonderful about organic chemistry. Baldwin's rules were formulated in the 1970s, and they were absolutely ground-breaking. And yet, if you understand bonding, you would be able to make all the rationalizations and predictions that Baldwin made.*

SECTION 4 CARBONYL COMPOUNDS AS NUCLEOPHILES

APPLICATIONS 7

Aldol Reactions Under Acidic Conditions

Now we have looked at aldol-type reactions under basic conditions (**Reaction Detail 23**), we can apply what we know to the same processes under acidic conditions. We just need to make a few adjustments.

ALDOL REACTIONS UNDER ACIDIC CONDITIONS

It turns out that we don't actually need an enolate to have an aldol reaction. We can use the enol instead (**Fundamental Reaction Type 4**). Enols are also carbon nucleophiles. Of course, they are not as reactive as enolates, but this turns out not to be a problem.

Let's look at the aldol reaction of acetone with itself (a self-condensation).

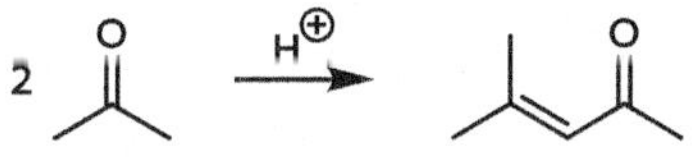

ca. 80%

Since we are working under acidic conditions (to catalyse formation of the enol), the first thing we need to do is protonate the oxygen atom of the carbonyl group. Here is the keto-enol tautomerization.

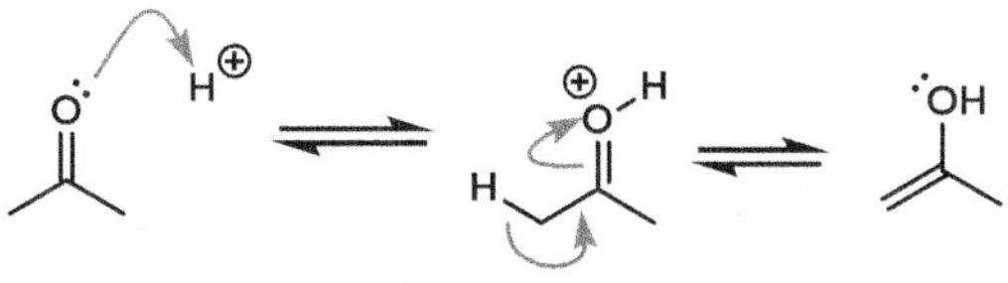

Now, we will use the enol in the same way that we used the enolate in the reaction under basic conditions. There is just one small difference.

> Let's see why this is by first of all drawing the reaction incorrectly.

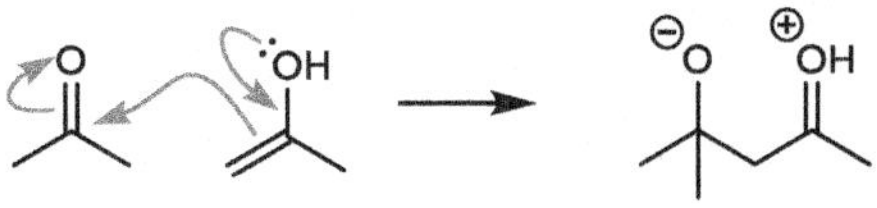

We are forming an intermediate with charge separation—a positive and a negative charge in the same molecule.

> It is easy to suggest that the negatively charged oxygen would be protonated quickly after formation. However, this doesn't pay us back for the energy needed to form it. We don't want to form an anion under acidic conditions.

SECTION 4 CARBONYL COMPOUNDS AS NUCLEOPHILES

If, instead, you have the enol attack a molecule of the protonated ketone, you do not have this problem. You start with a protonated compound. You finish with a protonated compound. Then it can lose the proton.

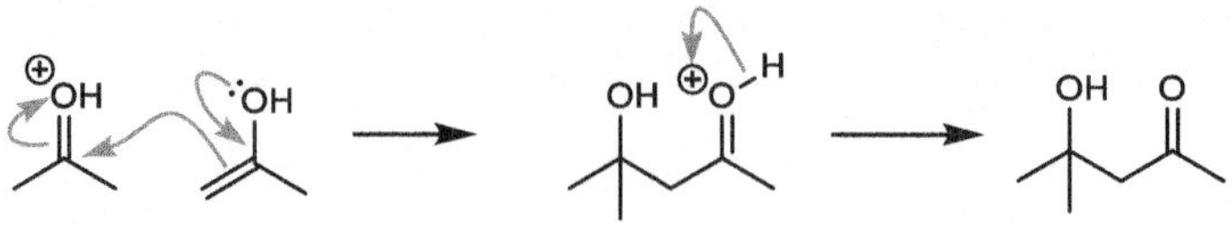

THERE IS ONE MORE COMPLICATION!

With the acid-catalysed reaction, it does not stop at the aldol. We have an acid present, and it could also protonate the alcohol in the product.

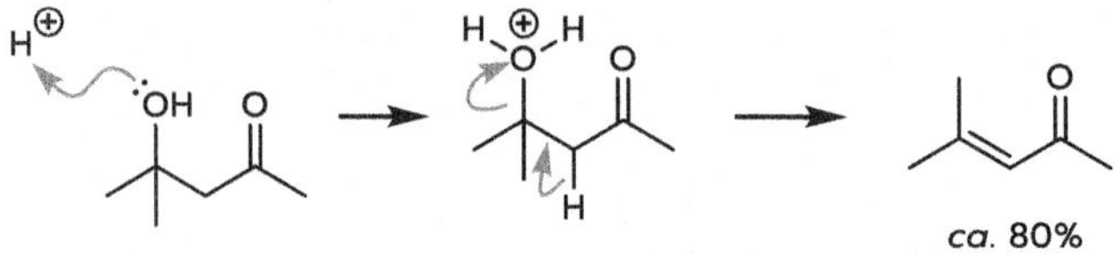

ca. 80%

The 80% yield is for the whole reaction all the way from acetone.

> *This version of the reaction is a condensation—water is lost. The aldol reaction under basic conditions is not a condensation reaction strictly speaking, although it is often described as such. The Claisen condensation, from the previous chapter, is also not strictly a condensation reaction.*[78]

We can add a little refinement to this. I have redrawn one of the structures from the aldol reaction scheme to show potential hydrogen bonding.

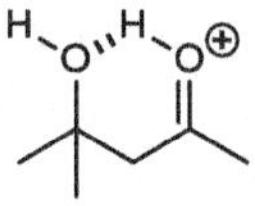

In effect, the oxygen atom was already protonated!

---

[78] In this case, there are two reactions associated with Claisen—the one we saw in the previous chapter, and a rearrangement that we will encounter in **Reaction Detail 37**. It wouldn't be any good to talk about the Claisen Reaction!

SECTION 4 CARBONYL COMPOUNDS AS NUCLEOPHILES

STEREOCHEMISTRY 14

Diastereoselective Aldol Reactions

Remember, the goal of this book is **not** to teach you everything there is to know about stereoselective reactions. Instead, it is to show you how the principles apply to stereoselective reactions.

In this chapter, we will look the factors that determine the stereochemical outcome in a diastereoselective aldol reaction.

> *This is definitely one to skip over once it gets too difficult (struggle with it a bit first though—that's how you will get it in the end) and come back to it later.*

GETTING STARTED

Here is a general aldol reaction, which we first encountered in **Reaction Detail 23**. It is the reaction of an enolate with a carbonyl compound, in this case an aldehyde. We aren't going to recap enolate formation here. We have seen plenty of enolates already.

Let's build this in stages. This time I have added some 'R' groups so that we form two new stereogenic centres.

> Draw these out with wedges and dashes, and establish how many stereoisomers you could form. Which are enantiomers and which are diastereoisomers of each other?

> *This is important. If you have two enantiomers, and try to explain why one is formed preferentially, you will always be wrong!*

It's definitely worth having another look at **Stereochemistry 2** to consolidate your understanding of **when** a reaction can be stereoselective based on the fundamental definitions.

In order to understand the stereochemistry of the reaction, we need to add a bit more detail. We often draw an enolate as a 'naked' $O^\ominus$. This is not a complete representation.

> *Whenever we have a charged species, positive or negative, there is always something with the opposite charge balancing it. What it is will depend on how we 'form the charge'. How closely associated it is will depend on how stable the charge is.*

335

SECTION 4 CARBONYL COMPOUNDS AS NUCLEOPHILES

In the case of an enolate, we form it by removal of an 'acidic' hydrogen atom. Remember, this is all relative. Compared to the simple carbanions in **Recap 5**, enolates are very stable. Compared to something like a chloride anion, they are very unstable!

Assuming we formed the enolate by deprotonation using a lithium amide base (e.g. LDA) then the counter-ion will be lithium. This will do for now. Here is the next level of representation.

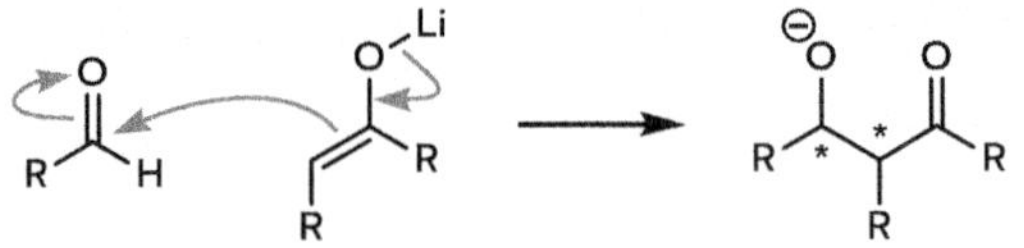

But surely the O⊖ in the product will also be associated with the lithium cation. The best way to resolve this is to have the lithium atom coordinating both oxygen atoms in the transition state.

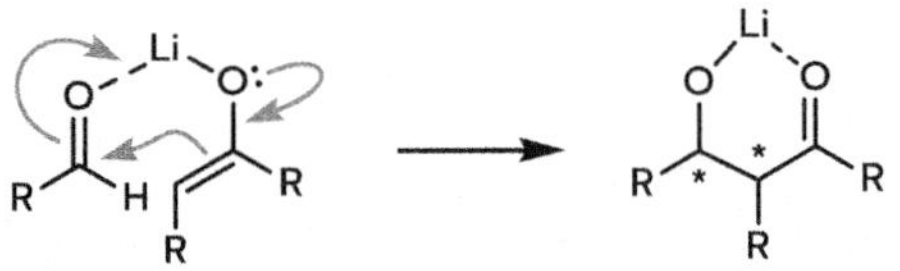

Now, we have a six-membered ring transition state that includes the enolate, the aldehyde and the lithium. We know (**Stereochemistry 1**) about the shape of six-membered rings.

ADDING THE STEREOCHEMISTRY

First of all, let's state the outcomes of diastereoselective aldol reactions in the simple cases. We have two possible enolate geometries, (E) and (Z).[79] The (Z) enolate gives the product shown below, which is described as *syn*. The OH and R² groups are on the same side of the chain in the orientation shown. I've taken another shortcut here. I have protonated the alkoxide at the end. The alcohol is the stable product.

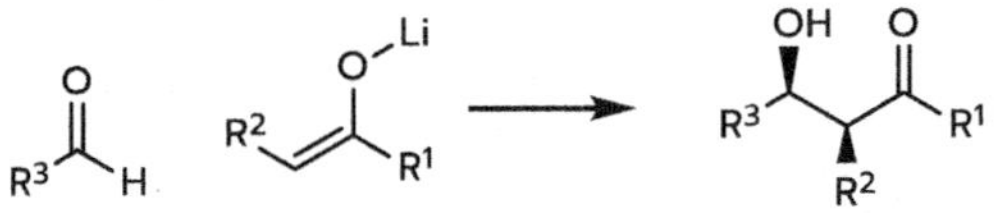

(Z)-enolate → syn product

> Of course, 'same side' is not an absolute. All we have to do is rotate one bond to put OH and R² on opposite sides of the molecule. The *syn* terminology specifically relates to the orientation shown.

---

[79] In this assignment, we are assuming that the oxygen takes priority over R¹. This would not be the case for an ester enolate. It is far more important to be able to understand the principles and to be able to apply them by correctly drawing the outcome of any given reaction.

SECTION 4 CARBONYL COMPOUNDS AS NUCLEOPHILES

The (*E*) enolate gives the *anti* product.

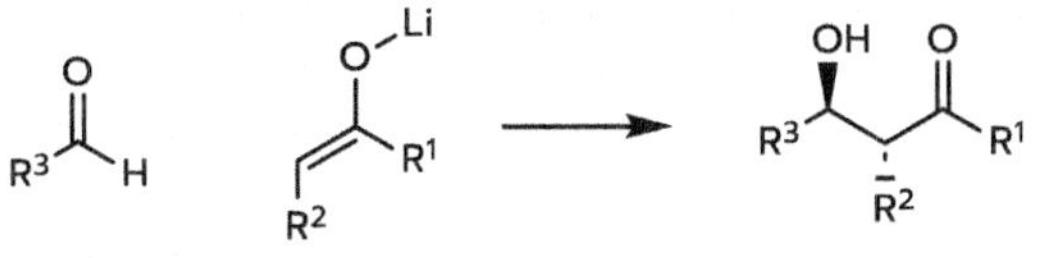

(E)-enolate → *anti product*

> Of course, since there is no chirality in either starting material, both of these products will be racemic—there will be an equal amount of the other enantiomer formed.

Draw it!

Here is a representation of the transition state for reaction of the (*Z*) enolate. The dotted line indicates the C–C bond we are forming. I decided (after some experimentation) to put the enolate at the back. In these representations, some orientations are easier to see than others. You tend to have to draw a few possibilities and get used to crossing out your work and starting again.

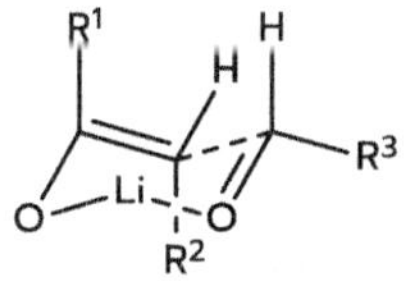

> You will note that R¹ and R² look as though they are axial. This isn't quite accurate, but they are where they are as a result of the enolate forming the ring, and having a specific double bond geometry.

The "R³" group of the aldehyde is equatorial. Here, we do have a choice. As this is effectively a cyclohexane we will, of course, put this group equatorial!

> The key to understanding this chemistry is knowing when you do and do not have a choice!

Let's add the outcome onto this diagram.

Hopefully you will be able to see that the product shown is the same as the *syn* aldol product above.

> Okay, what I really mean to say is that I hope you will be able to find a way to convince yourself that it is the same product. This isn't easy. I could redraw it in various stages, to try to show you, but even then, you might not see it.

SECTION 4 CARBONYL COMPOUNDS AS NUCLEOPHILES

> Make a model of the product. Convince yourself that the transition state shown above is okay. Unfold the model and convince yourself that the *syn* product is formed. It isn't easy learning, but it will be more effective.

NOW THE *ANTI* ALDOL PRODUCT

> Once you've finished sorting the *syn* aldol product, do the same for the *E* enolate giving the *anti* aldol product.

> Hint: Swap H and R^2.

A COMMENT

We've done quite a bit with cyclohexanes. In **HTSIOC Applications 3** and **HTSIOC Applications 5**, we saw substitution and elimination reactions of cyclohexanes. The defined bond angles make them good examples.

In **Stereochemistry 4** and **Stereochemistry 10**, we saw some more reactions of cyclohexanes (and cyclohexenes).

Now, we are applying the principles we learned from cyclohexane chemistry to things that are not cyclohexanes. In **Basics 6**, we saw that six-membered rings are particularly stable. Because they are so stable, reactions often proceed *via* six-membered ring transition states.

It is so important that you get used to the shape of these molecules, and that you become confident drawing them.

> *The effort will pay dividends, since these are skills that you can apply to a whole range of different reactions.*

We will come back to chair transition states in **Reaction Detail 37**, but we are now going to spend a couple of chapters building on the skills in this chapter, in combination with **Stereochemistry 10** and **Stereochemistry 13**. One is an 'Exercise' because you already know everything you need. One is 'Applications' because we need to make one minor change to our procedure.

EXERCISE 4

Combining Aldol Transition States and Felkin-Anh Stereochemistry

WHY?

We are adding complexity by combining two aspects (Stereochemistry 10 and Stereochemistry 14). We would only do this if there were a good reason. In this case, it's actually quite simple. The diastereoselective aldol reactions we saw in Stereochemistry 14 give racemic products. We saw the reasons for this, way back in Stereochemistry 2.

If we were to already have chirality present in the aldehyde, and if this were a single enantiomer, then we would have the potential to form products as single enantiomers. It's easier to see how this can happen if you work through an example.

> It's not going to be easy. It probably isn't going to be fun either, depending on your idea of fun!

Feel free to work through this chapter, but if you find it is too much at this point, skip it and come back to it when you are more confident.

> Not 'getting' this right now won't cause you any major problems.

INTRODUCING MORE COMPLEXITY—ALDOL REACTIONS WITH CHIRAL ALDEHYDES

Let's assume we have an aldehyde with a large α-alkyl group, a methyl group and a hydrogen atom. Here is the aldol reaction, with no stereochemistry shown for the outcome.[80]

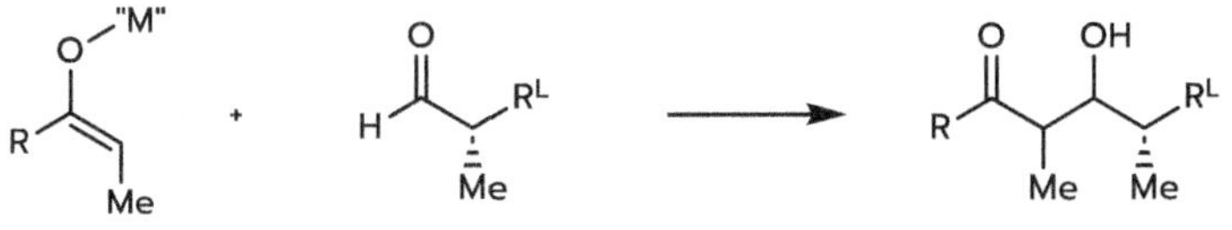

Let's define the problem, and the approach we will use. There are two distinct aspects to this, and we have seen both. In Stereochemistry 10 we saw the factors

---

[80] You will notice that in the previous chapter I had the aldehyde on the left and the enolate on the right. Now I have swapped them. This is deliberate. I don't want you to just copy structures from the previous chapter. I want you to take your time and get this right!

SECTION 4 CARBONYL COMPOUNDS AS NUCLEOPHILES

that determine the outcome in the reaction of nucleophiles with aldehydes containing a stereogenic centre at the α-position.

> An enolate is a nucleophile! You would expect the same principles to apply.

In addition, we now know (**Stereochemistry 14**) that aldol reactions proceed *via* chair transition states. The fact that R³ (from the example in the previous chapter) now has a stereogenic centre will not change that.

> So, we need to combine these two aspects, and we need to do it in a systematic way so that we don't make any mistakes.

The enolate as drawn (with an unspecified metal, "M") has *E* geometry. We expect, based on the discussion in the previous chapter, to get an *anti* product as the major diastereoisomer. The question is "which *anti* product?" The two possibilities are shown below.

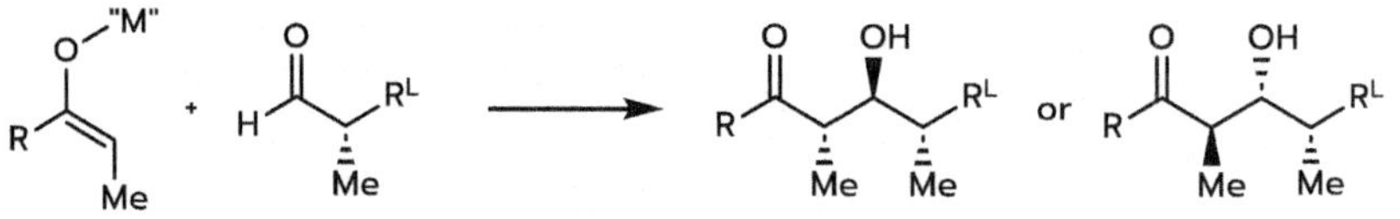

I have two comments to make at this point. First of all, I would always advise you not to focus too heavily on remembering which enolate gives which outcome. It's much better if you can derive this quickly by drawing the chair transition states.

> I've said this before, but it's worth repeating. If you need to remember an outcome because you find it difficult to work it out from the underlying principles, spending more time practising will be more effective than spending the time memorizing the outcome.

Speaking of which...

> Draw the aldol reaction, ignoring the α-stereochemistry, to convince yourself that you get the *anti* product in the transition state where the aldehyde "R" group is equatorial.

Right, that's one part of the problem solved. We know that the favoured aldol product will have *anti* stereochemistry between the two stereogenic centres we form, but we don't know which of the two possible *anti* products it will be.

The stereochemistry of the alcohol group will be determined by which face of the aldehyde is preferentially attacked. This is the Felkin-Anh rule! See if you can work it out before going to the next page.

Here is the aldehyde, drawn in a Newman projection. More importantly, it is the preferred Felkin-Anh conformer. Below, I have drawn the aldehyde in a Newman projection. I made sure I drew the same enantiomer as that shown above.

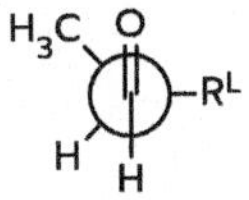

SECTION 4 CARBONYL COMPOUNDS AS NUCLEOPHILES

> Remember, we don't need to draw a chair transition state for this part of the problem.

From this, we expect attack of the enolate from the left. In the original orientation, that means attack from the back. This will give the following outcome, which isn't easy to see.

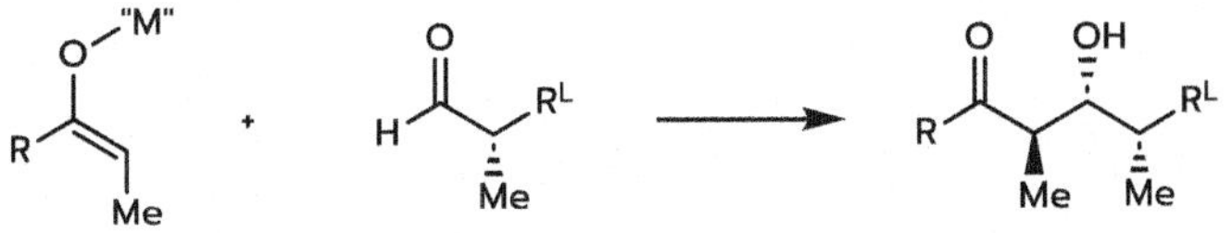

> Yes, the enolate attacks the aldehyde from the back, but the OH group ends up 'down'. Make a model of this. You don't need to use an enolate as nucleophile. It looks wrong, but it is right.

Let's draw the transition state. It isn't easy to show this with all relevant stereochemistry.

> Here is my best attempt—and it isn't very good!

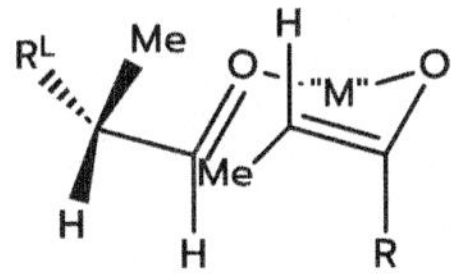

And that's why this is easier to solve using two separate steps.

IN CLOSING

Here we have a problem, and there are several ways to solve it. I would only ever do it this way, but you might not unless you see a 'guided' approach like this at least once.

> That's why it's here. Sure, making single stereoisomers of aldol products is important, but it's more important to learn how to solve problems.

341

SECTION 4 CARBONYL COMPOUNDS AS NUCLEOPHILES

APPLICATIONS 8

Aldol Reactions with Evans Oxazolidinones

WHY?

As with **Exercise 4**, making single enantiomers of chiral compounds is important. In **Stereochemistry 13**, we saw the Evans oxazolidinone, and learned that it is a chiral auxiliary.

> *Once we have done the stereoselective transformation, we can remove it.*

This is a bit different to **Exercise 4**, where the chirality in the aldehyde ended up in the product. Both approaches are useful. Which one you use depends very much on what you need to make.

But mostly, we are doing this because it forces you to look at molecular structures and to do some quite difficult 'stuff' with them that will help you build your skills.

EVANS OXAZOLIDINONES AGAIN—ALDOL REACTIONS WITH CHIRAL ENOLATES

Okay, if you've got this far, here's another warning. We are going to add one more factor. In itself, it isn't too complicated. The problem is, by end of this section, we will have looked at three different aspects of aldol reaction stereoselectivity. If you try to learn it all at once, there is a good chance of you getting confused.

> *This is another chapter that is okay to skip and come back to it when you are ready.*

Still here? Right then, in **Exercise 4** we looked at which face of the aldehyde is attacked by the enolate. Now let's look at which face of the enolate is attacked by the aldehyde.

We have already seen chiral enolates in **Stereochemistry 13**. We will build on that by using the same enolates, but instead of adding an alkyl halide as electrophile, we will add an aldehyde. Here is the enolate formation. We will use a boron enolate this time, but there is no difference really. We still get the (Z) enolate for the same reasons.

SECTION 4 CARBONYL COMPOUNDS AS NUCLEOPHILES

Whenever we vary the problem, we need to identify which bits stay the same, and which bits change. When we alkylated the above enolate, this was the shape, and we have alkylation of the front face.

> The key reason for this is that the boron chelates the oxazolidinone oxygen and the enolate oxygen to hold the enolate in place.

But we have seen that in an aldol reaction, we want chelation of the enolate oxygen atom with the aldehyde oxygen atom. We need to lose chelation of the oxazolidinone.

Here is my best drawing of the outcome, along with the transition state.

Once we lose chelation with the oxazolidinone oxygen atom, we have a C–N bond that we can rotate.

> We need to know which conformation is preferred.

This is good. It's the same thing that we did for epoxidation (**Stereochemistry 6**) and for nucleophilic addition to carbonyl groups (**Stereochemistry 10**).

In this case, the bond will rotate to move the oxazolidinone oxygen atom as far from the enolate oxygen atom as possible, to minimize dipole-dipole repulsion. Only when we have the reacting conformation can we place the aldehyde on the face of the enolate opposite the isopropyl group.

> Now 'unfold' the transition state to give the product shown.

This isn't easy, but we can take some shortcuts (as we did in **Exercise 4**). First of all, since the enolate geometry is (Z), we should know we are going to get a *syn* aldol product—we just don't know which one!

> If you remember which enolate gives *syn* and which enolate gives *anti*, you have a way of checking that your answer is sensible. Remember that *syn* and Z are both towards the end of the alphabet, while *anti* and E are towards the beginning.

SECTION 4 CARBONYL COMPOUNDS AS NUCLEOPHILES

What other changes might we make to our process? Here is the enolate again, but it's now in the correct conformation for an aldol reaction. I think the dipole repulsion is easier to see here.

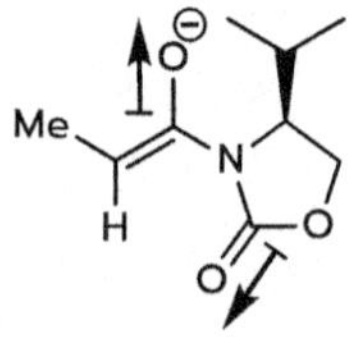

> The back face will be attacked.

Here it is drawn with a simple alkylation reaction.

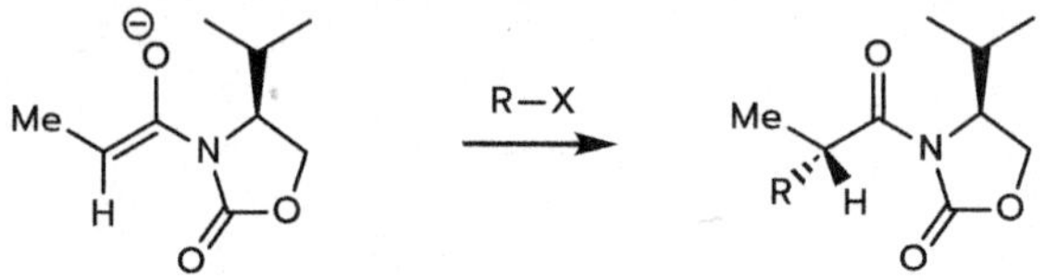

> Can you convince yourself that the stereochemistry is the same as I drew for the aldol reaction above?

Perhaps it is easier to see if I rotate a bond as shown.

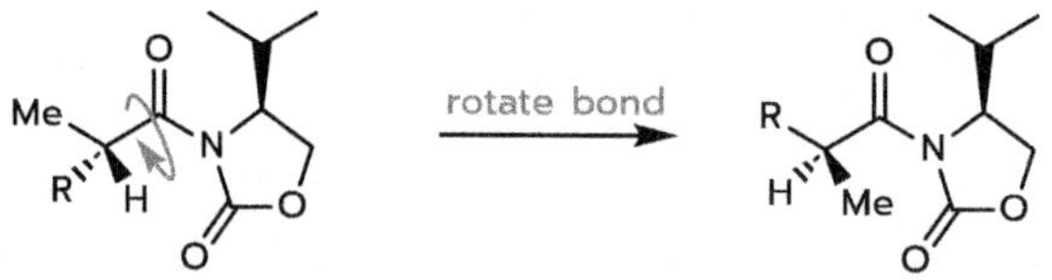

Once again, this is the key skill. If you see a structure in a textbook or in a research paper, you cannot control how the author drew it. You need to be able to interconvert structures.

From this point, knowing the enolate geometry, we know that the aldol product will have *syn* stereochemistry, so we can combine that with the diagram above to give the full stereochemical outcome?

> Which method is easier?

I honestly don't know. The **easiest** way is to make a model of the transition state and to unfold it to show the stereochemistry.

REACTION DETAIL 24

Dianion Chemistry and Decarboxylation

WHY?

This chapter describes some chemistry that isn't used very often these days. But there are important lessons to be learned from these reactions, so it is important that you understand them. We are going to start this chapter with a problem.

DIANION CHEMISTRY—BASIC PRINCIPLES

We have seen the formation of anions next to carbonyl groups, and we have seen the formation of anions between two carbonyl groups. What happens when we combine the two?

Look at the structure of ethyl acetoacetate below. I have drawn on approximate pK_a values for the hydrogen atoms attached to the key carbon atoms.

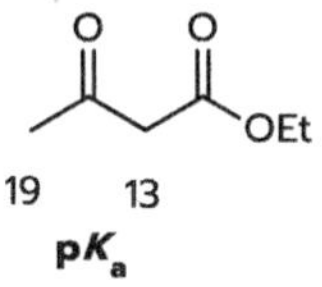

> Now let's ask a question. What will happen when we treat this compound with a base?

Don't over analyse this. The base will deprotonate the ethyl acetoacetate, and it will remove the most acidic proton.

Sodium hydride (NaH) is commonly used as a base in these reactions. This is what happens.

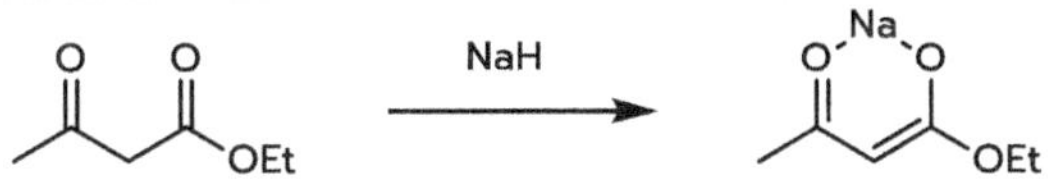

Don't get confused by this. We could just draw structure **3** to represent the anion. It's just that there will always be an associated cation.

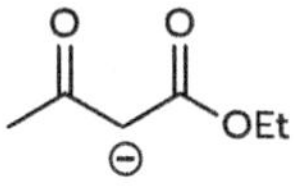

SECTION 4 CARBONYL COMPOUNDS AS NUCLEOPHILES

Now, here's the smart bit. If we treated compound **1** with *n*-BuLi (butyllithium) then we would probably expect the butyllithium to act as a nucleophile. But compound **3** is not readily attacked by a nucleophile.

> *It has a negative charge.*

The preferred reaction is a second deprotonation as follows. Butyllithium is a very strong base, as well as being a good nucleophile (Basics 5).

Now I am going to simplify the structures. Switching representations in this way will probably confuse you at first.[81] The butyllithium removes an acidic proton from the 'other' carbon atom. This hydrogen will now have a pK_a considerably higher than 19. After all, we are removing the proton from something that already has a negative charge.

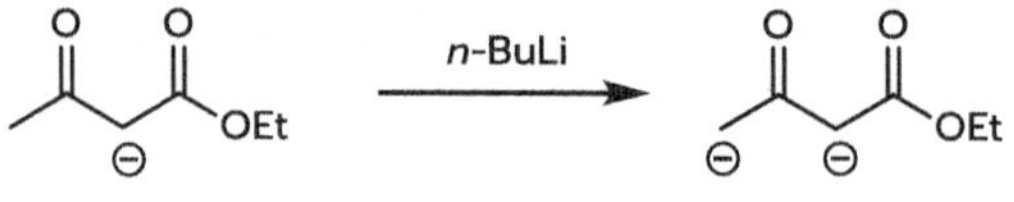

> *Structure 4 has two negative charges (if we ignore the counter-ions) so it is referred to as a dianion.*

SELECTIVE ALKYLATION OF β-KETOESTERS

Now we will look at why you might want to do this. If we react dianion **4** with an electrophile, such as an alkyl halide, we can form a new C–C bond. This is useful! Because the anion only stabilized by one carbonyl group is less stable, it is more reactive, so we would expect (and get) alkylation at this point. We would then, normally, add an acid to protonate the other anion.

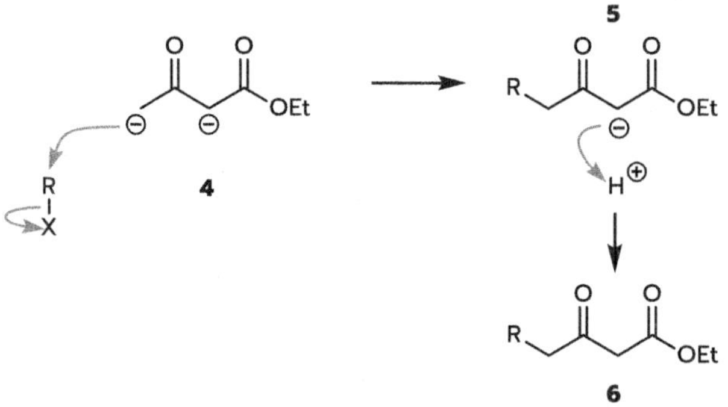

On the other hand, if we only form the mono-anion **3**, it too can undergo alkylation, and this time it will occur between the carbonyl groups.

---

[81] But you **will** encounter both representations. You're going to get confused at some point. We may as well get it out of the way now, so you can work through it and move on!

SECTION 4 CARBONYL COMPOUNDS AS NUCLEOPHILES

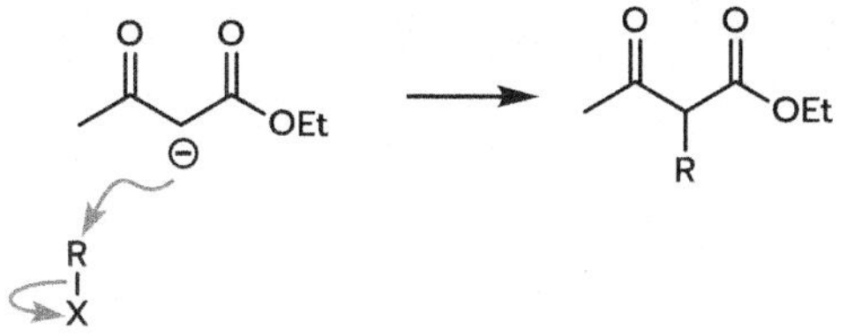

So, we can alkylate a β-ketoester at either position selectively.

> Being able to choose which site of a compound reacts is useful. This is chemoselectivity (Recap 4).

DECARBOXYLATION

There is another reaction that is commonly associated with dianion chemistry, although it is not directly related to it. First of all, let's take compound **6** or compound **7** and hydrolyse the ester. We saw how to do this in **Reaction Detail 16**.

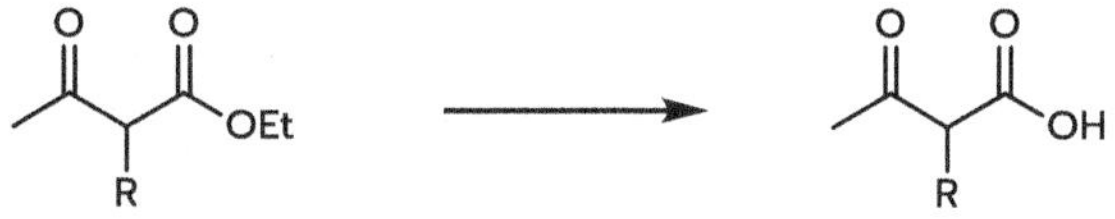

> Go back, revise the mechanisms and then draw the mechanisms as applied to these two specific compounds.

You will need to use acidic conditions. There are two reasons for this. The first is that a basic nucleophile such as hydroxide will simply deprotonate compound **6** or compound **7**. The second reason, we are coming to.

Once we get to the β-ketoacid, this is beautifully set up to lose carbon dioxide in a reaction known as decarboxylation. This forms the product as shown as the enol, and it then tautomerizes (**Fundamental Reaction Type 4**) back to the ketone.

I have drawn the mechanism for the 'new' step. You have already drawn the mechanism for the ester hydrolysis.

> Draw the keto-enol tautomerization mechanism. I know you've drawn it before. That doesn't matter!

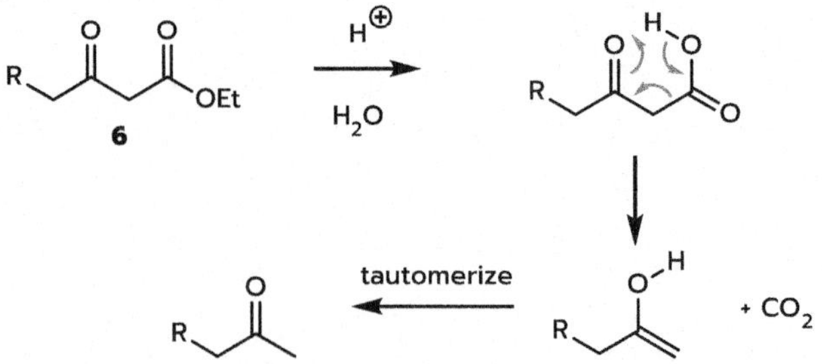

SECTION 4 CARBONYL COMPOUNDS AS NUCLEOPHILES

$$R\text{-COCH}_2\text{-COOEt} \quad \xrightarrow[H_2O]{H^+} \quad R\text{-CO-CH}_2\text{-COOH} \quad \rightarrow \quad R\text{-C(OH)=CH}_2 + CO_2 \quad \xrightarrow{tautomerize} \quad R\text{-CO-CH}_3$$

> Make sure you see the big picture here. You can form β-ketoesters using Claisen/Dieckmann reactions. You can alkylate them selectively (directed by the ester) and then hydrolyse and decarboxylate. It all fits together quite nicely.

DISPELLING MYTHS!

Dianion chemistry does cause some confusion. I want to highlight a common conceptual error. Let's look at structure **1** again.

1

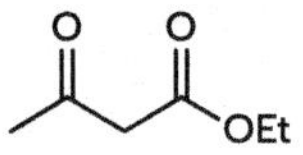

structure with pK$_a$ values 19 and 13

Here is the flawed logic.

> If we want to remove the proton with pK$_a$ 13, we use a weak base, but if we want to remove the proton with pK$_a$ 19, we use a strong base.

I see where this confusion comes from. We do indeed use weak (not so weak actually) and strong bases. To make sure you don't fall into this trap, keep asking fundamental questions. What happens if you treat compound **1** with NaH? You will remove the most acidic proton. What happens if you treat compound **1** with n-BuLi? You will still remove the most acidic proton![82]

> There is absolutely no scenario in which you could leave the most acidic proton in place and remove a less acidic hydrogen atom instead!

---

[82] Assuming no carbonyl addition reactions to complicate the outcome.

SECTION 4 CARBONYL COMPOUNDS AS NUCLEOPHILES

REACTION DETAIL 25

More Complicated Aldol Reactions

I have two confessions. First of all, the reactions in this chapter had a real impact on me when I was learning. The basically look the same, but have different outcomes.

> This makes them hard to learn.

With the benefit of hindsight, I can see exactly why they do different things. Hopefully I can explain this in a way that you will find accessible.

The second confession is that I didn't know what to call this chapter. Are they more complicated? Well, they are if you don't ask the right questions. The three reactions (plus one modification) in this chapter are named reactions, and they are all fundamentally aldol/Claisen condensations.

> *I very much doubt that anyone is going to ask you if you know (for example) the Hantzsch pyridine synthesis (coming up in Applications 9). They might ask you to work out what the reagents will do under the conditions, or how you would make a particular pyridine target. What you will find, as you build your knowledge framework, is that you will remember the names and associate them with the reactions. It just (in my experience) takes a little longer.*

THE KNOEVENAGEL CONDENSATION

Now there's a Scrabble word! Don't worry about how to pronounce it. I have heard so many variations. The most common one is 'no-ven-ar-gel', but I have also heard it pronounced 'ker-nuff-nar-gel'. Oh well, as long as you know the reaction it doesn't matter.

This one follows on nicely from the aldol and Claisen condensations. In the simple example we take a 1,3-diester and an aromatic aldehyde and we treat them with a base, in this case piperidine. Here's what happens overall:

> *We have a base. The reaction doesn't work without one. The base must be doing something. It isn't a foregone conclusion that the first step is deprotonation, but it is a good starting point. Don't look for complications!*

The aromatic aldehyde doesn't have any acidic hydrogen atoms.

349

SECTION 4 CARBONYL COMPOUNDS AS NUCLEOPHILES

> By this point, two things should already have happened. You should see the abbreviation 'Ph' and visualize exactly what it is and where all the hydrogen atoms are. You should know that the hydrogen attached to the aldehyde carbon is not acidic (and you should understand why!).

Now we have established where there are no acidic hydrogen atoms, let's find one. There is an acidic hydrogen (actually there are two of them, but one at a time!) between the two esters. It is α- to both of them, and it has a pK_a of around 12. This is pretty acidic—more acidic than water!

We use the piperidine to deprotonate the diester **1**. The resulting anion (**2**) reacts with the aldehyde (benzaldehyde) to give intermediate **3**. This can then be protonated, as shown, to give **4**. The ammonium species that is being used to protonate **3** was formed in the first step.

> It is better to use this rather than simply drawing H$^{\oplus}$, since we are working under basic conditions.

> Propane-1,3-dioic acid is commonly known as malonic acid. The corresponding ester **1** is usually described as a malonate ester. It helps to be familiar with these trivial names. We will see the parent acid, malonic acid, shortly.

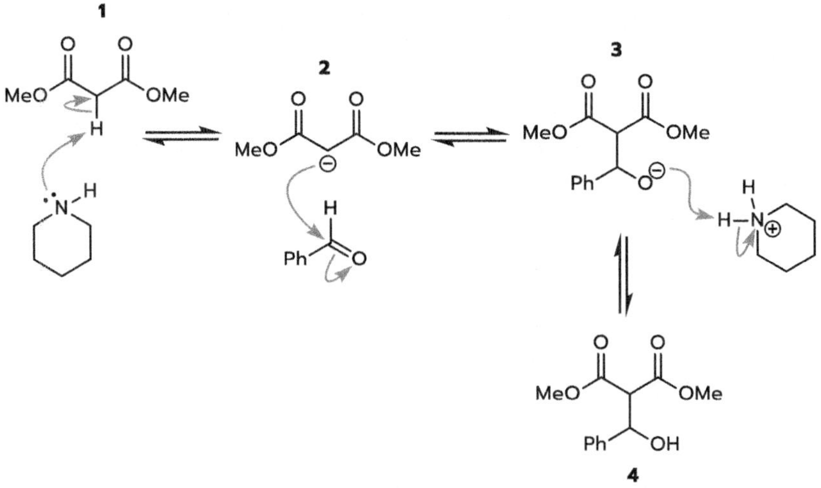

Now we can carry on with the mechanism. Compound **4** has a hydrogen atom α- to two esters (again, pK_a about 12). We removed one like that in the first step. Why should we not do the same again? Removal of this hydrogen atom will give enolate **5**. Now we can lose hydroxide to give the product **6**.

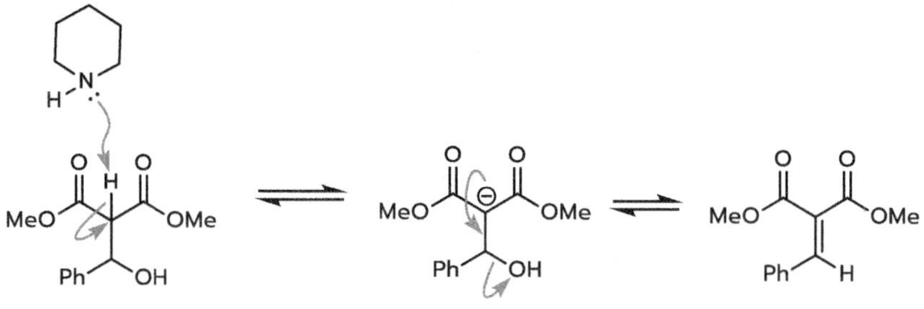

SECTION 4 CARBONYL COMPOUNDS AS NUCLEOPHILES

[Structures 4, 5, 6 showing equilibrium between hemiaminal-like intermediate, enolate, and α,β-unsaturated diester]

I have drawn this elimination reaction as an E1cb mechanism (**HTSIOC Fundamental Reaction Type 2**). Is this correct? Well, there is little doubt that enolate **5** is more stable than alkoxide **3** (the corresponding alcohol has a pK_a of about 16).

[Structures 3 and 5]

It is certainly plausible for this to be an E1cb reaction. However, let's not forget that we should probably think of mechanisms as a continuum (**HTSIOC Perspective 4**) rather than as absolutes. Perhaps this is an E2 elimination with significant E1cb character?

> We are about to see that it probably doesn't matter anyway.

BASE OR NUCLEOPHILE, OR BOTH?

The mechanism shown above is not complete. Here is the problem step.

[Structure 2 with curly arrow mechanism showing enolate attacking benzaldehyde]

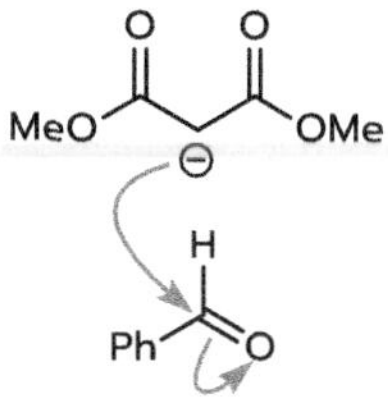

"What's wrong with this?", you might ask. Well, it's a perfectly good curly arrow mechanism, but we are taking a relatively stable C$^{\ominus}$ and replacing it with a less stable O$^{\ominus}$. The difference is not dramatic, and there would be a reasonable concentration of the resulting anion **3** which would rapidly be converted into anion **5**.

But it turns out that there is an alternative mechanism that has a lower overall barrier.

> Now it's time for you to do some work!

351

SECTION 4 CARBONYL COMPOUNDS AS NUCLEOPHILES

It turns out that the electrophile is not benzaldehyde, it is the iminium ion **7**.[83]

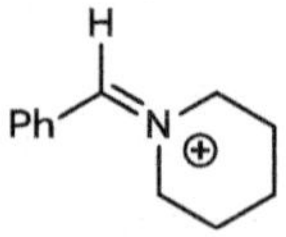

[structure of iminium ion 7: Ph-CH=N⁺(piperidine ring) with H on carbon]

7

> Draw a mechanism for the formation of iminium ion **7** from benzaldehyde and piperidine. You might need to have another look at **Applications 2**.
>
> Draw a mechanism for the reaction of iminium ion **7** with anion **2**. You should get the structure shown below.

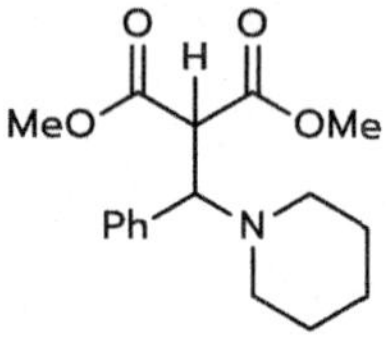

[structure: MeO-C(=O)-CH(-C(=O)-OMe)... with central CH bearing H, attached to CH(Ph)-N(piperidine)]

> Now we need the elimination reaction to give the final product **6**. Make sure you don't have a negative charge on the nitrogen atom at any point in your mechanism.

Hopefully that wasn't too difficult. Your mechanism should look very much like the one above. If your mechanism looks dramatically different, it is wrong! There's a bit more 'inspiration' coming up in the next few pages, if you are still having trouble.

THE VERLEY-DOEBNER MODIFICATION OF THE KNOEVENAGEL CONDENSATION

What happens if we carry out the Knoevenagel condensation using malonic acid instead of a malonic ester? In **Reaction Detail 24** we saw that a carboxylic acid group β- to a carbonyl group can be induced to decarboxylate. It turns out that we have to use pyridine as base, and heat the reaction, but here is the overall process.

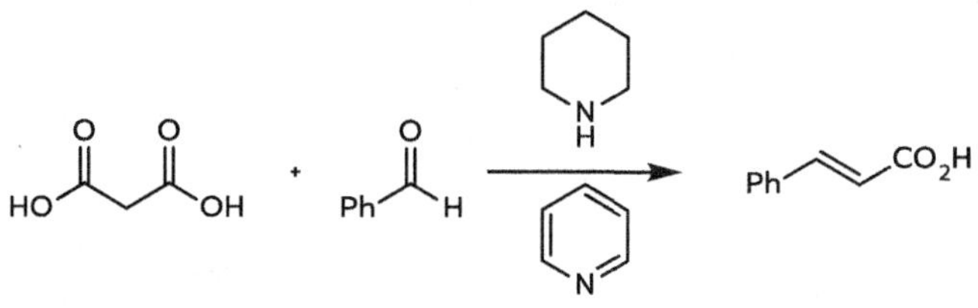

[reaction scheme: HO-C(=O)-CH₂-C(=O)-OH + Ph-CHO, with piperidine and pyridine as reagents, gives Ph-CH=CH-CO₂H]

---

[83] A 50% share in the 2021 Nobel Prize in Chemistry was awarded to David MacMillan for his seminal work on stereoselective iminium ion catalysis. The reaction we are looking at here is a simple example of using an iminium ion to find a more favourable reaction pathway.

352

SECTION 4 CARBONYL COMPOUNDS AS NUCLEOPHILES

> Draw the mechanism up to the following point. You should do this by comparison with the Knoevenagel condensation that we have just seen.

> Leaving this for you to do is important. When you encounter two reactions that have similarities and differences, drawing the mechanisms is the best way to make sure you see the similarities. This allows you to focus on the reasons for the different outcome.

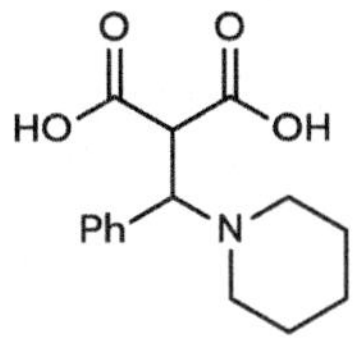

Now let's think about what needs to happen in order for the product to be formed. We need to lose CO_2 and piperidine. We will definitely need to protonate the nitrogen atom in order for it to leave. Overall, the reaction conditions are basic, but when we use an amine (including pyridine or piperidine) to deprotonate something, we will generate an ammonium ion which can deliver the proton back.

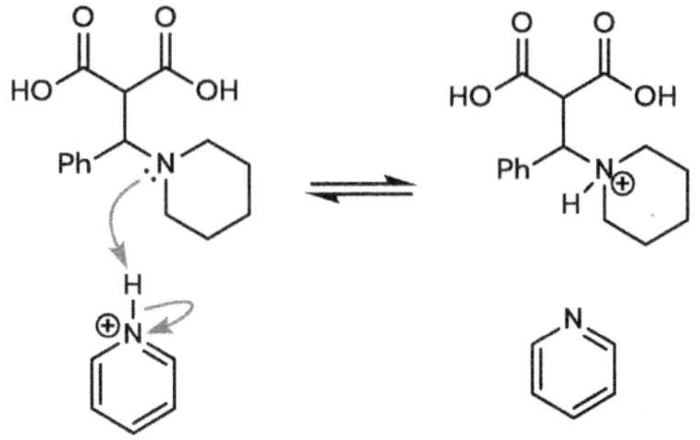

Of course, we also have two carboxylic acid groups in the structures above. Will they still be protonated or will they be deprotonated?

> If we focus on every detail, the mechanism will get complicated. Sometimes it is better to simply draw the curly arrows.

Now, we can deprotonate a carboxylic acid to initiate the decarboxylation and the elimination. Here is a curly arrow representation of this process.

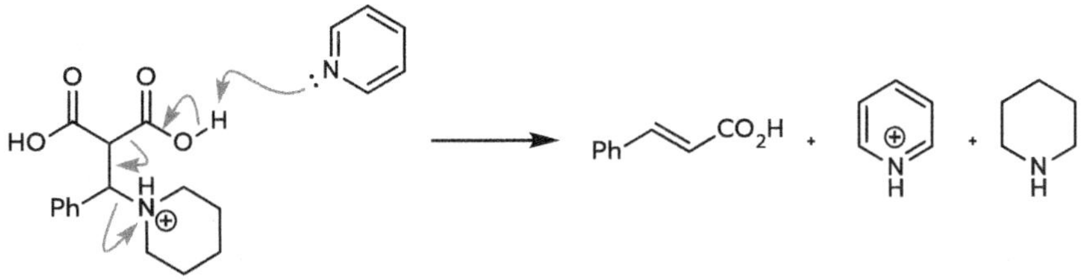

This is an E2 elimination (**HTSIOC Reaction Detail 4**). But instead of simply losing $H^⊕$ as one component, we are also losing CO_2. It is an E2 elimination combined with the decarboxylation reaction we saw in **Reaction Detail 24**.

SECTION 4 CARBONYL COMPOUNDS AS NUCLEOPHILES

> Now, dig out your Newman projections (**Stereochemistry 5**) and work out why we get the (*E*)-alkene geometry. It isn't enough that this is the more stable product. We should be able to work out why it is formed by considering the reaction, not just the product.

Hint—you can decide which of the two carboxylic acid groups to lose. Go for the one that will give a pathway with a lower overall barrier.

THE STOBBE CONDENSATION

Remembering back, I think this is the one that really confused me. Let's see the reaction.

$$\text{8} + \text{Ph-CO-Ph} \xrightarrow[\text{then } H^\oplus]{\text{KO}t\text{-Bu}} \text{9}$$

8: EtO₂C-CH₂-CH₂-CO₂Et (diethyl succinate)
9: EtO₂C-C(=CPh₂)-CO₂H

> As before, we should look at the reaction and identify the issues. The starting material, diethyl succinate, has two ester groups. The product has only one ester group, and one carboxylic acid group. It would appear that one of the ester groups has been hydrolysed. If you draw a mechanism, you would need to be able to explain how and why this happens.

Before we see what the mechanism *is*, we should see what it is *not*.

WHAT ARE THE PITFALLS?

On the face of it, this reaction looks very much like the Knoevenagel condensation. This is the cause of the problem! It is quite common, and very understandable, to draw a Knoevenagel mechanism to give the following outcome.

> Try drawing this! The curly arrows are fine.

8 + Ph-CO-Ph → **10** (EtO₂C-C(=CPh₂)-CO₂Et)

When we looked at the Knoevenagel condensation, we saw that the reason we were able to get the elimination under basic conditions is because anion **5** is considerably more stable than anion **3**. This is because anion **5** is α- to two carbonyl groups.

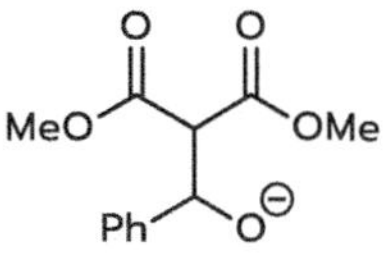

SECTION 4 CARBONYL COMPOUNDS AS NUCLEOPHILES

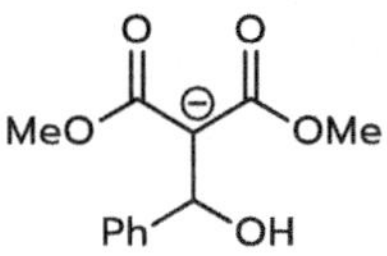

> You will notice that I am sticking to the mechanism that doesn't involve an iminium ion.

This is partly because the Stobbe condensation doesn't involve an amine base, and partly because I want to keep things simple. Using the iminium ion mechanism won't change the conclusions.

When you drew the corresponding mechanism for the Stobbe condensation, you should have drawn the following two structures.

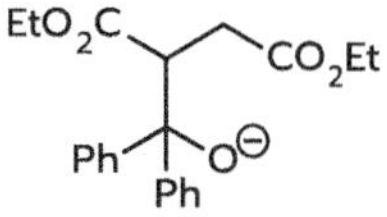

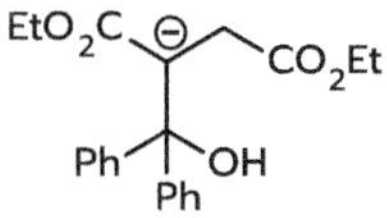

The problem is that structure **12** is much *less* stable than structure **11**. The anion in structure **12** is α- to only one carbonyl group, and it would have a pK_a of about 25.[84]

> This example really does highlight the challenges you face in learning organic chemistry. Two structures look very similar, and you might expect them to do the same thing. The placement of the carbonyl groups (1,3- versus 1,4-) makes a huge difference.

Once you get to compound **10**, it is likely that you will then draw the following reaction.

> This is a hydrolysis reaction of a carboxylic ester (Reaction Detail 16).

The problem is, we are not working under conditions that would hydrolyse an ester. We have *t*-butoxide present as a base. The only hydroxide present would be that formed as a result of the elimination reaction from intermediate **12**.

It is true that if we had to choose, we would expect the non-conjugated ester in compound **10** to be hydrolysed more readily than the conjugated ester. However, this isn't enough to offset all the problems we would face getting there.

---

[84] Strictly speaking, the H that has been removed had a pK_a of 25.

SECTION 4 CARBONYL COMPOUNDS AS NUCLEOPHILES

BACK TO THE MECHANISM

Now we have seen what *doesn't* happen, let's look at what does happen. It is perfectly reasonable to draw the deprotonation of compound **8**. We are using potassium *t*-butoxide, which is a pretty strong base. From this point, the ester enolate can attack the ketone carbonyl. This gives us anion **11** which we saw before.

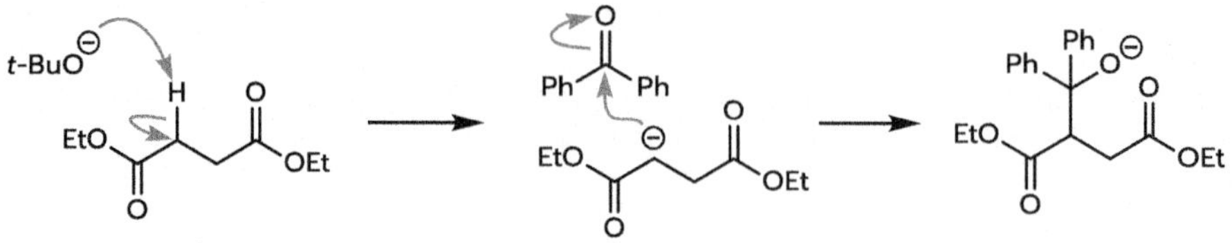

> We established what doesn't happen, so let's now ask the key question. What can anion **11** do?

Well, we could protonate it, but this doesn't do much, and we are working under basic conditions anyway. If we did protonate it, it would be acting as a base. If it isn't a base, could it act as a nucleophile?

> It can, and it does.

The negatively charged oxygen atom can attack a carbonyl group intramolecularly.

More importantly, it will attack the ester carbonyl shown, because this leads to the formation of a five-membered ring. Attack on the other carbonyl group would form a much less stable four-membered ring (**Basics 6**).

> Recall that one of the ester groups is 'hydrolysed' during this transformation. We have just identified a good reason why one of the ester carbonyl groups can react while the other does not. You've got to be feeling optimistic that we are on the right lines!

We can now kick out ethoxide, and then consider what it might do!

SECTION 4 CARBONYL COMPOUNDS AS NUCLEOPHILES

[Structures 13 → 14 → 15, with 15 → 9]

Yes, the ethoxide can facilitate an elimination reaction to give carboxylate anion **15**. This is a thermodynamically favoured process, as the carboxylate is a great leaving group ($pK_a \approx 4$). However, we might question whether this is an E2 elimination (as drawn) or an E1cb elimination. For now, we've done enough hard work. To finish the process, we simply protonate to give product **9**.

> Okay, there is quite a lot to this mechanism, but each step makes sense. In order to get these mechanisms right, you have to be very disciplined, and stop drawing as soon as you draw something that doesn't look sensible.

Now let's work through one more, which looks different but applies the same principles.

THE PERKIN CONDENSATION

This time, I'm going to get you to do more of the work, with a bit of guidance of course. Here is the overall reaction.

[PhCHO + (CH₃CO)₂O → (KOAc) → Ph-CH=CH-CO₂H]

Potassium acetate is a base. It's not a very strong base, but it will remove an acidic hydrogen. It might only generate a **very small** equilibrium concentration of the corresponding anion, but it will be enough to react further.

> Why would you use such a weak base in this reaction?

I would say that is as much a philosophical question as a practical one. It may simply be that this is what was used the first time, and it worked well so why change it? However, we can analyse a bit deeper. We have an acid anhydride in this reaction. These are very reactive carbonyl compounds, comparable in reactivity to acid chlorides. Trying to find something that would act as a base, and not as a nucleophile, would be difficult.

> You couldn't use an amine base—you would form an amide. Draw a mechanism!

But if acetate attacks acetic anhydride, you will form acetic anhydride! So, while this may be the favoured process, it doesn't do anything.

357

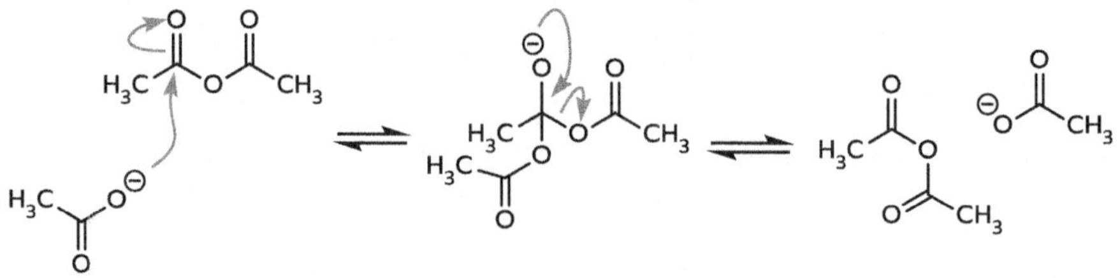

SECTION 4 CARBONYL COMPOUNDS AS NUCLEOPHILES

When acetate acts as a base, it does do something!

> Look at the reaction scheme above. Identify the most acidic hydrogen atom and draw a mechanism for the deprotonation step. Then use the anion to form a C–C bond. The product structure should guide you towards the correct bond to form.

I am not going to give you the full answer, but you should have drawn the following structure.

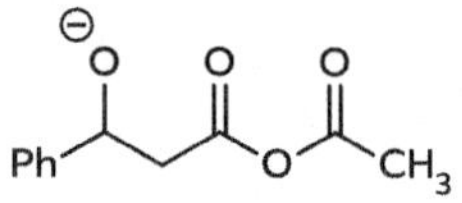

We have the same problem that we encountered with the Stobbe condensation. Clearly, there is an elimination reaction involved, but we do not have a good leaving group. Rather than drawing random protonation reactions, we should ask the more fundamental question "what will the O$^\ominus$ do?"

> *Perhaps the oxygen can act as a nucleophile? Can we use this to make it a better leaving group?*

I am not going to give you the answer to this problem. You know what it is called, so you can look it up.

> First, though, try to come up with a plausible solution.

You'll feel so much more confident if you work this out for yourself.

SECTION 4 CARBONYL COMPOUNDS AS NUCLEOPHILES

REACTION DETAIL 26

Alkene formation—the Wittig Reaction

WHY?

Throughout this book, we have seen quite a few reactions of alkenes, giving useful products. You haven't **really** seen the value of the products yet, but you are starting to get some idea.

> We need to know how to make alkenes.

In **HTSIOC**, we looked at elimination reactions. These give us a really good way to make alkenes. But they are not the only way. In this chapter, we will look at a reaction developed in the first instance by Georg Wittig, who shared the Nobel Prize in Chemistry in 1979 with Herbert Brown. We have already seen Brown's work on hydroboration in **Reaction Detail 5**.

INTRODUCTION

We must not view this as yet another reaction to add to our list. Instead, I would like to start with the reaction and establish what we already know.

Here is a typical reaction scheme for a Wittig reaction.

$$Ph-CHO + CH_3CH_2Br + Ph_3P \longrightarrow Ph-CH=CH-CH_3 + Ph_3PO + HBr$$

> We do not simply mix all these and hope for the best. We will get to the order of addition in a moment.

One of the reagents is an aldehyde. We **know** how aldehydes react. We are forming a C–C bond to the aldehyde carbon atom.

> Yes, it happens to be a double bond rather than a single bond. But that doesn't change aldehyde reactivity. Let's go with the flow.

We have triphenylphosphine as a reagent, and it is being converted into triphenylphosphine oxide. Phosphorus-oxygen bonds are strong. This is a (the!) thermodynamic driving force.

Since the triphenylphosphine is being oxidized, something else is being reduced.

> Go back to **Basics 3** and work out the oxidation states of the relevant carbon atoms in the reagents and products in this reaction.

Did you work out what I meant by 'relevant'?

359

SECTION 4 CARBONYL COMPOUNDS AS NUCLEOPHILES

Now let's look at the bromoethane. There is absolutely no doubt that we have broken a C–Br bond, giving bromide. We have also lost a hydrogen atom from the same carbon.

> It is probably being lost as $H^{\oplus}$, not as $H^{\ominus}$. That will give us HBr. Especially since we use a base in this reaction.

You don't have to do this with every new reaction you see, but it's good to be able to take things back to basics. Right, let's move things on a little, with a couple of questions.

> Will we break the C–H bond or the C–Br bond first?

If we 'just' break the C–H bond, we would get the following carbanion.

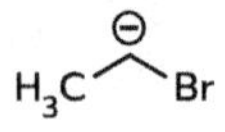

$$H_3C{-}^{\ominus}\!CH_2{-}Br$$

> That doesn't look very stable at all.

On the other hand, if we 'just' break the C–Br bond, we would get the following carbocation.

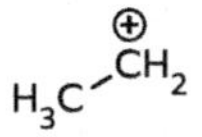

$$H_3C{-}\overset{\oplus}{C}H_2$$

> That doesn't look good either.

At least the first one gives us something that looks like a carbon nucleophile. But not one that we could easily form. We need to keep asking questions.

> What does Ph₃P do?

The phosphorus atom has a lone pair. It is a potential nucleophile. In fact, it is a very good nucleophile. We didn't quite see it in **HTSIOC Reaction Detail 1**, but everything you need is in there.

> Whenever you encounter an unfamiliar reagent, try to identify its fundamental reactivity.

We are now in a position to draw a plausible reaction.

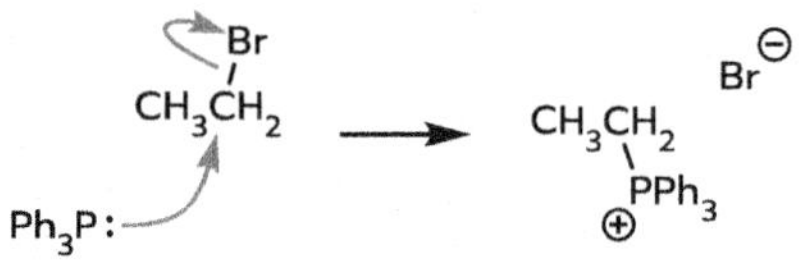

$$Ph_3P{:} \;+\; CH_3CH_2{-}Br \;\longrightarrow\; CH_3CH_2{-}\overset{\oplus}{P}Ph_3 \;\; Br^{\ominus}$$

1

The product of this is a phosphonium salt, **1**. We have seen ammonium salts at various points in this book, so a phosphonium salt looks reasonable.

SECTION 4 CARBONYL COMPOUNDS AS NUCLEOPHILES

> Typically, we would react the bromoethane with triphenylphosphine first, and isolate the phosphonium salt.

We can now deprotonate. In **Recap 7**, we saw that compound **1** is slightly acidic, with a pK_a of about 30. We need a very strong base for this transformation, with *n*-butyllithium (**Basics 5**) typically being used.

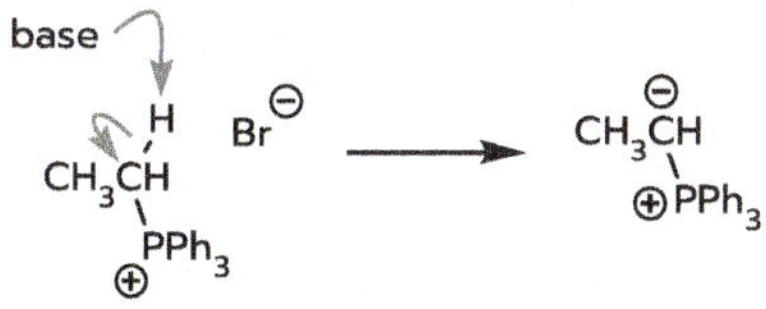

Compound **2** has a special name. It is called an ylide.[85] This term is used for a species that has a positive and a negative charge on adjacent atoms. As this is phosphorus, and it has d orbitals, we can draw a resonance form to show this stabilization. The resonance form on the right is described as a phosphorane.

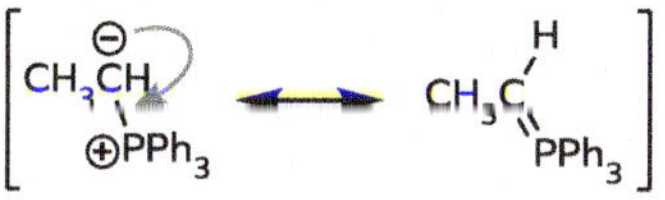

Really important point coming up!

> If you drew this resonance form in any of the structures up to this point, it would have been wrong.

But it's okay with phosphorus, because phosphorus can form five bonds.

Let's relate this back to a general principle. In **Recap 7**, we saw a diagram like this for some stabilized anions. We didn't draw the diagram for stabilization by phosphorus there, so here it is. I've changed the empty orbital to a d orbital on phosphorus.

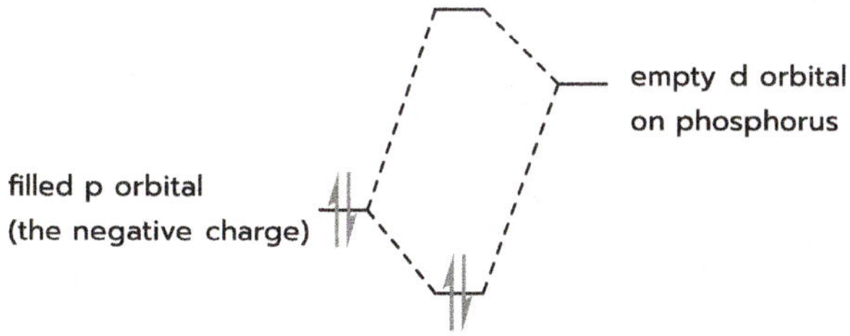

What we are seeing here is exactly the same as we have seen every other time we have stabilization of a negative charge. The electrons must have somewhere to go. This has to be an empty orbital of the right energy. The curly arrows show the sharing of electrons in exactly the same way as the molecular orbital diagram.

---

[85] Sometimes spelled 'ylid', and always pronounced ill-id, not why-lid.

SECTION 4 CARBONYL COMPOUNDS AS NUCLEOPHILES

Right! We are getting somewhere. We have a stabilized (but not too stable) carbanion. It does what stabilized carbanions do. I'm going to draw this as two steps. You'll see why in a moment.

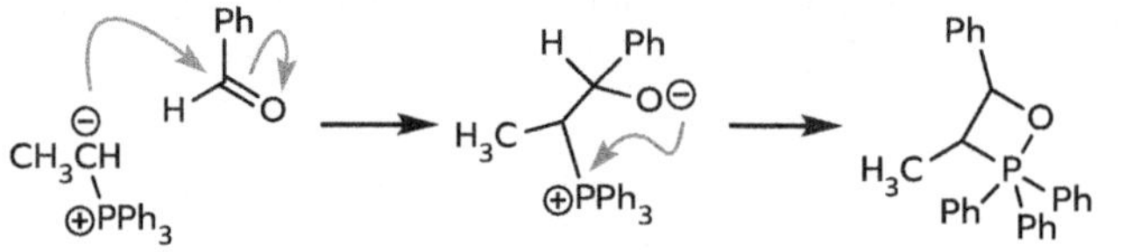

I have drawn ylide **1** acting as a nucleophile, attacking benzaldehyde to give intermediate **2**. I have then drawn this cyclizing to give structure **3**, which we refer to as an oxaphosphetane.

You probably don't like this. And you would be right to be concerned. We established in **Basics 6** that a four-membered ring is strained. This one isn't quite so bad. The C–P and P–O bonds are quite long.

There's another detail to cover. When we talk about the stereochemistry of this reaction, next, this will become important. The C–C and P–O bonds form at the same time. We would be better to draw it like this.

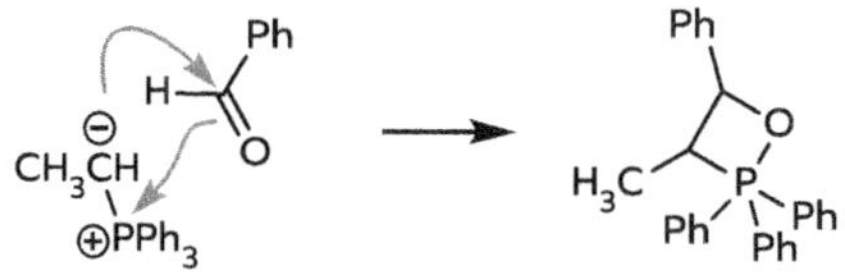

Finally, we lose triphenylphosphine oxide as follows.

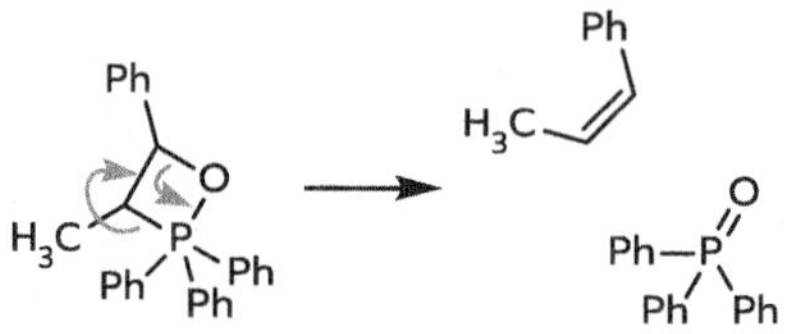

> Right, that's the full mechanism. Go back and draw the whole thing out a couple of times, just to get used to the 'shape' of it.

SOME TERMINOLOGY

The hydrogen atom we removed from compound **1** is not very acidic. It has a pK_a around 30. So that ylide **2** is not stabilized much.

> We refer to this as a non-stabilized ylide!

It's all relative. As phosphonium ylides go, this is unstable. We are about to see some much more stable ylides.

SECTION 4 CARBONYL COMPOUNDS AS NUCLEOPHILES

STEREOCHEMISTRY OF THE WITTIG REACTION

I drew the product of the above Wittig reaction as the *cis* alkene. The reason for that is simple. Despite being less stable than the corresponding *trans* alkene, this is the product that predominates.

> The reaction **must** be under kinetic control (Applications 5).

Let's have a closer look at oxephosphetane **4**. There are, in fact, two possible diastereoisomers, **4**-*cis* and **4**-*trans*, shown below.

4-*cis*   **4**-*trans*

We **must** be getting **4**-*cis*, since this is the one that will lead to the *cis* alkene product. So, we have changed the question. Instead of asking why we get the *cis* alkene, we ask why we get **4**-*cis*.

The answer to this problem is relatively complex. I'd like to focus on the principles rather than the detail. The first point is that it isn't just that the *trans* alkene is more stable than the *cis* alkene. In fact, the *trans* oxaphosphetane is more stable than the *cis* oxaphosphetane.

> But reactions under kinetic control give the product that is formed fastest, and rate of reaction is determined by the activation energy. This relates to the energy of the transition state.

What we know, with absolute certainty, is that the transition state leading to the *cis* oxaphosphetane **must** be lower in energy than that leading to the *trans* oxaphosphetane.

These are the principles that I believe you should understand. The question of what the transition states look like and **why** the transition state leading to the *cis* oxaphosphetane is lower in energy has been determined as a result of computational chemistry studies. I will simply present the outcome.

Here are the lowest energy calculated transition states in each case. I've used dashed lines to indicate the bonds that are being formed. The transition state leading to **4**-*cis* is very puckered. I have indicated a steric clash between the H of the aldehyde and one of the triphenylphosphine Ph groups.

> This isn't too bad.

363

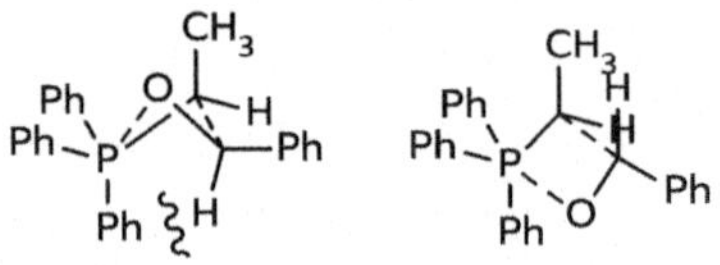

SECTION 4 CARBONYL COMPOUNDS AS NUCLEOPHILES

TS-4-cis   **TS-4-trans**

If we simply swap the H and Ph of the aldehyde, this would give **4-trans**, which is found to be much higher in energy.

> The steric clash would be worse. While you couldn't predict the detailed shape of this transition state, you could predict what would happen to the energy if you simply swap H and Ph.

The lowest energy transition state leading to **4-trans** isn't particularly puckered. It is shown above. There is a lot of eclipsing of bonds (**Stereochemistry 5**) in this transition state.

> Again, this isn't something you could have predicted. But if I'm going to tell you the outcome, I'm going to tell you why we get that outcome.

The lowest energy transition state leading to **4-cis** is lower in energy than that leading to **4-trans**.

INTERLUDE

This section is about carbonyl compounds acting as nucleophiles as a result of their enolate/enol-type reactivity. What we have here is something that isn't an enolate, but it does have **broadly** similar stability, and it reacts like an enolate.

We are also adding a nucleophile to a carbonyl, so this reaction could just as easily have ended up in **Section 3**.

> I felt that you needed to see a few simpler examples of this type of reactivity before we dealt with the subtleties and complexities of the Wittig reaction.

We will now consider what happens if the ylide (carbanion) is more stable.

WITTIG REACTIONS WITH STABILIZED YLIDES

How do we make the ylide more stable? We are back in **Recap 7** territory. We need to add something else to further stabilize the negative charge.

> In this case, it will be a carbonyl group.

Here is exactly the same reaction, but now it gives almost exclusively the *trans* alkene product.

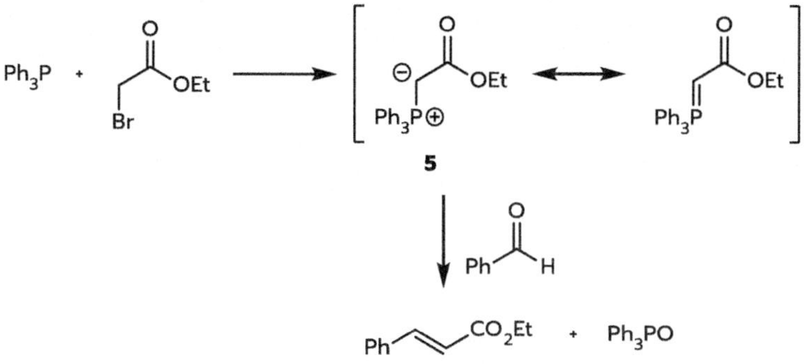

SECTION 4 CARBONYL COMPOUNDS AS NUCLEOPHILES

$$Ph_3P + \underset{Br}{\overset{O}{\underset{|}{\bigwedge}}}OEt \longrightarrow \left[\underset{Ph_3P^{\oplus}}{\overset{O}{\underset{\ominus}{\bigwedge}}}OEt \longleftrightarrow \underset{Ph_3P}{\overset{O}{\underset{||}{\bigwedge}}}OEt \right]$$

5

$$\downarrow Ph\overset{O}{\underset{}{\bigwedge}}H$$

$$Ph\diagup\!\!\!\diagdown CO_2Et \quad + \quad Ph_3PO$$

Once again, we have to identify the correct questions to ask ourselves.

> Does it give the *trans* alkene because it is the most stable product? If so, this would imply thermodynamic control and a mechanism by which the *cis* alkene product is converted into *trans* alkene product.
>
> Does it give the *trans* alkene product directly because the ylide is more stable?

The accepted wisdom is that dipole-dipole repulsion in the transition state, as shown below, is responsible for the transition state leading to the *trans* oxaphosphetane being more lower in energy (faster reaction!).

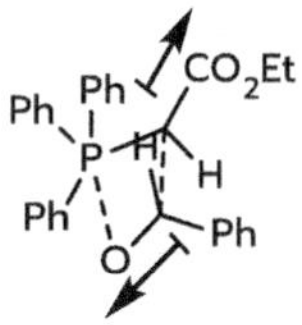

So, this reaction is also under kinetic control, but it happens to give the most stable product.

Incidentally, when I was taught this chemistry, about 35 years ago, I was given a different explanation for the stereochemical outcome[86] in each of these cases. It is rare (but not impossible) for the outcome of a reaction to be revised in light of new data. It is much more common for the underlying reasons for the outcome to be revised as a result of improved understanding.

The mechanisms given in this chapter follow from a great deal of experimental and theoretical work from a number of chemists over almost 50 years. I doubt the current 'conventional wisdom' will change, but you can never be sure!

---

[86] I keep using the term 'stereochemistry' even though we don't have any stereogenic centres in the products as drawn. Double bond geometry is **still** stereochemistry.

SECTION 4 CARBONYL COMPOUNDS AS NUCLEOPHILES

WHAT'S WRONG WITH THE WITTIG REACTION?

The Wittig reaction is a great way to make alkene bonds. These bonds can be formed with good levels of stereochemical control. There are a wide range of modified reaction conditions that improve on the original findings.

The problem is triphenylphosphine oxide. It has a molecular weight of 278. In many cases, that will be more than the molecular weight of your product.

> *If so, you will end up throwing away more than 50% of the mass of the chemicals that you use.*

This isn't very efficient. Also, triphenylphosphine oxide can be difficult to completely remove from your product.

We haven't actually considered all that many practical aspects during our discussion. Some practical aspects are really important, but you'll soon get to grips with those once you are spending a lot of time in a chemistry laboratory.

Of course, chemists have come up with 'better' ways to make alkenes, and we will look at one of these in Reaction Detail 27, next.

SECTION 4 CARBONYL COMPOUNDS AS NUCLEOPHILES

REACTION DETAIL 27

Alkene Formation–Metathesis

INTRODUCTION

Let's have a very brief recap of the mechanism from the last chapter, and remind ourselves of the problem.

I've drawn it here using the phosphorane resonance form rather than the more usual ylide. We will see why in a moment.

We can describe this process as **metathesis**. In general terms, the word means a "change of place and condition". In chemistry terms, we are moving the double bonds from one place to another.

A SHORT INTERLUDE

When we get to Reaction Detail 34, we would describe these reactions as [2 + 2] cycloadditions, and we would find that they do not happen thermally. This is the case when we are simply talking about carbon-carbon π bonds. When we deal with phosphoranes, and the other systems in this chapter, this does not apply.

> *It isn't a contravention of the rules on orbital symmetry. We are just dealing with different orbitals.*

THE PROBLEM

We dealt with this in the last chapter, but perhaps the most pressing problem is that the by-product of the reaction, triphenylphosphine oxide, has a molecular weight of 278. It is waste, and wouldn't be easy to recycle to give something useful. You **could** potentially reduce it to regenerate triphenylphosphine, but this would require time, energy and more reagents. It's still not a win.

There **are** modified Wittig reactions which generate considerably less waste. We will see one of these briefly in Total Synthesis 2.

THE SOLUTION

We have actually seen one solution. The aldol reaction, when carried out under acidic conditions (Applications 7) forms a double bond, with loss of a much smaller molecule, water.

SECTION 4 CARBONYL COMPOUNDS AS NUCLEOPHILES

> But this does place considerable constraints on the two reacting partners. We need something that can form an enol.

What we need is a compromise—something that gives us much more flexibility in terms of what we can make, while not generating significant waste.

The solution to this (or at least one solution) is transition-metal catalysed alkene metathesis. Here is a general scheme, with the metal as 'M'. We will look at a couple of specific examples in a moment.

We start with an alkene **1** and a metal alkylidene (don't worry about the name) **2**. These react as shown to give compound **3**, with a four-membered ring. We then get a reversal of this process, generating a different metal alkylidene **4**.

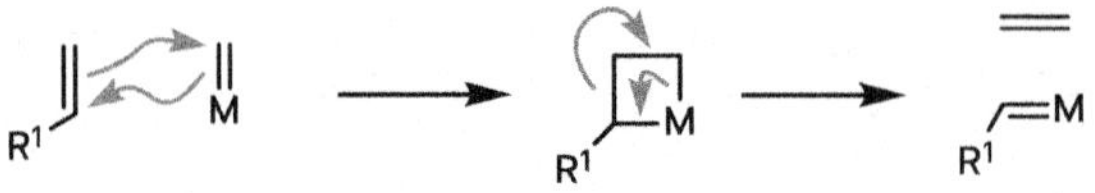

> The by-product is ethene, with a molecular weight of 28. It is a gas, so it won't contaminate the reaction product.

Metal alkylidene **4** then reacts with a second alkene **5** to give compound **6**, which does exactly the same thing in reverse (again!). This generates our target alkene **7**, and regenerates our original metal alkylidene **2**. We only need a catalytic amount of the metal alkylidene.

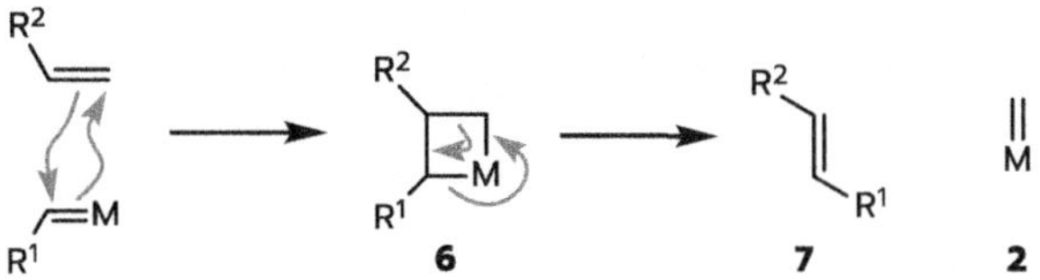

Ignoring the catalyst, the overall reaction is as follows.

This is a much more 'economical' process than the Wittig reaction. Chauvin, Grubbs and Schrock shared the Nobel Prize in Chemistry in 2005 for developing this process to the point where it was useful for the preparation of a wide range of alkene products.

Perhaps you would expect alkenes **1** and **5** to show different levels of reactivity with the catalyst **2**.

SECTION 4 CARBONYL COMPOUNDS AS NUCLEOPHILES

> Although we aren't going to look at this in detail, it is a reasonable expectation, depending on the nature of R¹ and R². This can be addressed to a large extent by a choice of catalyst.

There's another selectivity aspect to consider—regioselectivity (Recap 4). Here's an exercise for you.

> 'Play around' with the mechanism given above. The reaction of **1** with **2**, or that of **4** with **5** is shown with one regiochemical outcome. If we had the alternative regiochemical outcome, would it actually matter? Draw some structures and curly arrows, and see if you can work this out!

Hopefully you came to the conclusion that it doesn't matter! The only issue is if we get the metathesis product of **1** with **1** or **5** with **5**. If the alkenes are sufficiently different, this can be controlled. We see another way to control this in Total Synthesis 2.

> Again, I don't want to go into the details. The point here is to establish that there is a neat way to make an alkene bond, and why it is better than some of the methods that preceded it.

EXAMPLES

Many different catalysts have been used in metathesis reactions. It is probably fair to say that those featuring ruthenium as the metal originated with Grubbs, while those with molybdenum originated with Schrock.

In the example below, the ruthenium catalyst **11** has a benzene ring attached to the metal alkylidene double bond. This doesn't really change anything.

AcO~~~~~⧸ + ⧸~CO₂Me — **11** (5 mol%) → AcO~~~~~~CO₂Me

8   **9**   **10**, 94%

11

This is a really good catalyst for 'cross metathesis' reactions,[87] in which two different alkenes react together. I haven't given many reaction yields in this book, but it's

---

[87] For anyone who is interested, catalyst **11** is the second-generation Grubbs catalyst. There are lots of different Grubbs catalysts. Some work better for certain types of metathesis reaction. Don't worry about that for now.

SECTION 4 CARBONYL COMPOUNDS AS NUCLEOPHILES

worth highlighting how good this reaction is, even when the same number of moles of substrates **8** and **9** are used. Only 5 mol% (5% of the number of moles of **8** or **9**) of catalyst **11** was used.

The two double bonds don't need to be in different compounds. If they are in the same compound, it is possible to form a ring. The example below forms an eight-membered ring, which we have established (**Basics 6**) is not the 'easiest' of ring sizes.

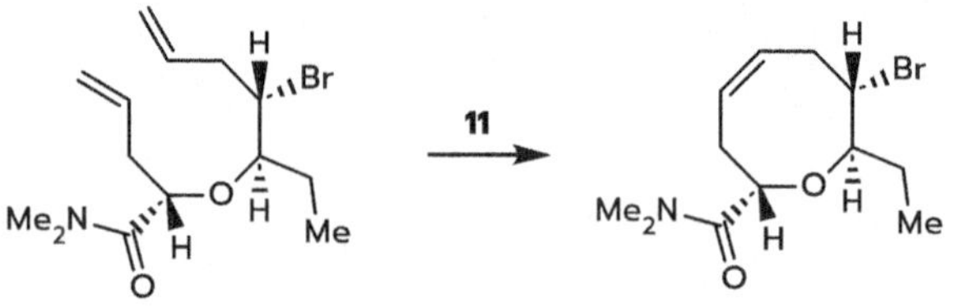

12 → (**11**) → **13**

This example uses the same catalyst as in the previous case.

IN CLOSING

This is another chapter that doesn't fit within the theme of this section, but in terms of a logical flow of the chemical story we are building, it is in the right place. That's good enough for me!

We are barely scratching the surface of what is possible with alkene metathesis. It is possible to form one ring, or even multiple rings, in one step. It is possible to form polymers with high levels of control.

And if we didn't actually want a double bond, we could always just hydrogenate it (**Reaction Detail 7**), which means alkene metathesis can be used indirectly to form carbon-carbon single bonds.

For now, don't worry about all that. Make sure you understand what a metathesis reaction is, and that you can draw a curly arrow mechanism.

> *Details, such as which catalyst you would choose to get a good yield in a specific metathesis reaction, can wait a while.*

REACTION DETAIL 28

Conjugate addition Reactions

INTRODUCTION AND ENERGETICS

In **Basics 1**, we looked at why an alkene reacts with an electrophile but a carbonyl group reacts with a nucleophile. We saw that the problem is a little more subtle, and we identified a situation in which an alkene *might* prefer to react with a nucleophile. It's now time to talk about this chemistry.

Here is a reaction profile that we saw in **Reaction Detail 8** for aromatic electrophilic substitution. It is exactly the same profile, but I have removed the structures and changed one word.

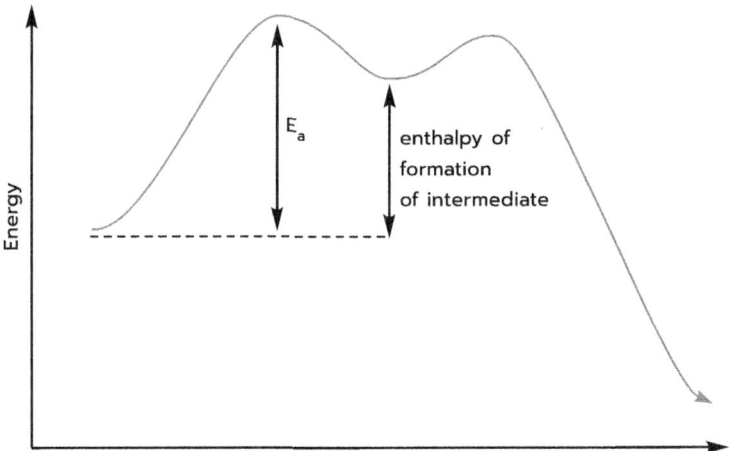

The key point here is that this could apply to any reaction that has an intermediate. It doesn't matter what the intermediate is.

> *We define an intermediate as something along the reaction coordinate that it is higher in energy than either the starting material or the product, but does have a discrete existence.*

The intermediates we have encountered so far are carbocations and carbanions.

In **Fundamental Reaction Type 2** we looked at the idea of adding electrons to an antibonding orbital in order to break a bond.

> *Nucleophiles add to carbonyl groups because this process leads to the formation of a stable intermediate.*

There is a corollary to this. If addition of a nucleophile to an alkene double bond gave a stable intermediate, it could also be a viable process.

SECTION 4 CARBONYL COMPOUNDS AS NUCLEOPHILES

ENOLATES IN CONJUGATE ADDITION

We will look at a reaction, and then we will add some detail.

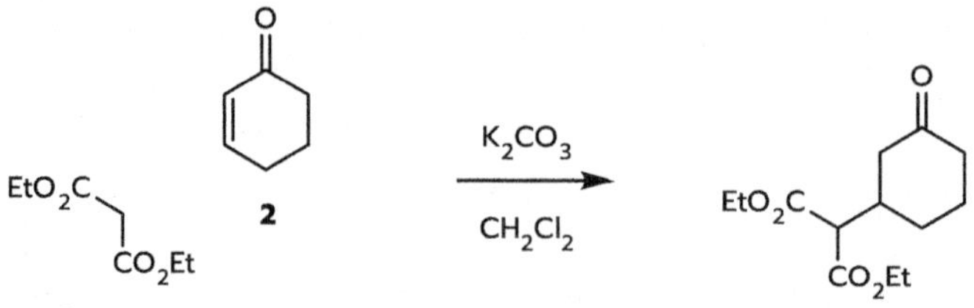

Under basic conditions (potassium carbonate is a weak base, dichloromethane is 'just' the solvent) compound **1** (diethyl malonate) forms a nucleophile, which then adds to compound **2** to form product **3**.

When you are learning this chemistry, you might have a number of questions. First of all, you want to know what the nucleophile is. We saw compound **1** (actually it was the corresponding methyl ester) in **Reaction Detail 25**.

> Don't be surprised if it does the same thing here.

In that chapter, we formed a nucleophilic anion, and it added to a carbonyl carbon atom.

> Why does it do something different here?

As always, we have a process—draw a curly arrow mechanism. We are starting with deprotonation of compound **1**. Note that I have not drawn the ester groups out fully, and as a result, the resonance forms that show the stabilization cannot be drawn. This is deliberate, as you need to get used to abbreviations such as CO₂Et (even when they are drawn 'backwards' (EtO₂C)—something else I have deliberately done).

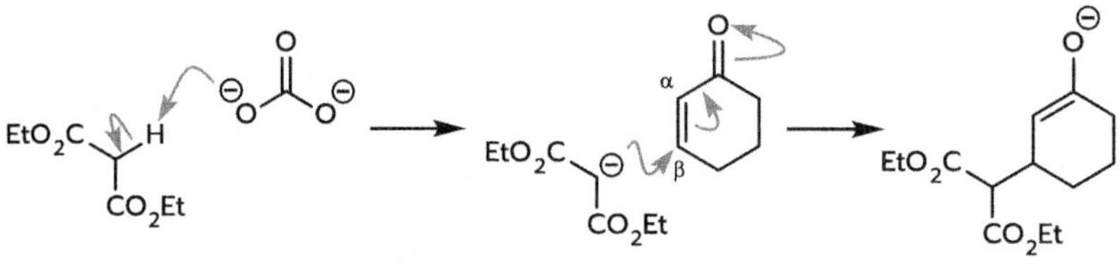

> Draw the diethyl malonate anion out fully, and draw the resonance forms.

The anion formed then adds to the alkene double bond of cyclohexenone at the β-carbon, to give intermediate **4**, an enolate. Of course this isn't finished. We could simply protonate the enolate by addition of water/acid at the end of the reaction, but I prefer to draw the following.

SECTION 4 CARBONYL COMPOUNDS AS NUCLEOPHILES

Anion **5** is much more stable than enolate anion **4**, so we would expect the proton transfer shown to take place long before we add the acid.

> If you look back at **Applications 2**, you will see that I have broken one of my rules. I have drawn this as a direct proton transfer, which would require a four-membered ring transition state.

This is something I have done deliberately. It is more likely that we have separate protonation/deprotonation steps, but sometimes it is easier to draw one step, even if we know it isn't quite right.

> All the way back in **Basics 1** we considered that an alkene could react with a nucleophile if this process led to a stable intermediate.

We have now seen what you have to 'put on an alkene' in order to make this process possible.

Let's take stock of where we are. We have a good curly arrow mechanism for the formation of product **3**. But we could equally well have drawn the nucleophile adding directly to the carbonyl carbon atom, and this would give a stable product via an intermediate of similar stability. In this case, it isn't an enolate, but we do still have a negatively charged oxygen atom.

> Draw a full curly arrow mechanism for this alternative process.

What we now need to do is explain **why** the nucleophile attacks the β- carbon atom rather than the carbonyl carbon atom. Let's have a closer look at cyclohexenone.

We have two resonance forms, **A** and **B**, with positive charges on carbon. These are the two carbon atoms that we have identified as being potentially attacked by a nucleophile.

> I didn't draw the curly arrows for these resonance forms. I left that for you to do!

But we also have an inductive effect. The carbonyl carbon atom has more positive charge as a result of being directly attached to oxygen.

373

SECTION 4 CARBONYL COMPOUNDS AS NUCLEOPHILES

> *Any reaction that is dominated by electrostatic interactions will favour attack at the carbonyl carbon atom. We will see one of these in a moment.*

In **Recap 1** we saw that a reaction of a nucleophile with an electrophile proceeds by attack of the nucleophile HOMO with the electrophile LUMO. In **Recap 6** we looked at the molecular orbitals of butadiene as a typical conjugated system. We saw that the LUMO of butadiene looks like this.

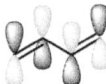

This only shows the symmetry. There is one detail that we missed—the sizes of the orbital coefficients. When we add those, and apply it to cyclohexenone, it looks more like this.

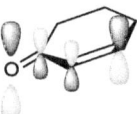

The orbital coefficient on the β-carbon atom is larger than that on the carbonyl carbon atom, and a larger orbital coefficient leads to better orbital overlap.

HARD AND SOFT NUCLEOPHILES

In **Exercise 3** we saw the concept of hard and soft electrophiles. We now do the same for nucleophiles. A hard nucleophile has a larger concentration of negative charge on the nucleophilic atom. All other things being equal, sulfur is a softer nucleophile than oxygen.

In the example above, the diethyl malonate anion is delocalized. The negative charge is distributed between the carbon atom and the two carbonyl oxygen atoms.

> Check this statement against the resonance forms I asked you to draw.

The diethyl malonate anion is a soft nucleophile, and reacts with cyclohexenone at the site with the largest LUMO coefficient.

REFLECTION

Most organic chemists do not routinely calculate molecular orbital coefficients. However, they will be able to draw resonance forms and consider the interplay of inductive and mesomeric effects to determine where there is more positive/negative charge within a structure.

> *Don't expect to be able to glance at a structure and see which atoms will have the larger orbital coefficients—yet.*

In the first instance, it is better to learn patterns of reactivity. With experience, you will glance at a nucleophile and you will know whether it is hard or soft.

SECTION 4 CARBONYL COMPOUNDS AS NUCLEOPHILES

Let's look at another reaction in order to consolidate the pattern.

CONJUGATE ADDITION OF ORGANOMETALLICS

In **Reaction Detail 12** we saw the addition of carbon nucleophiles such as Grignard reagents to carbonyl compounds. If we compare an organocopper reagent with a Grignard (organomagnesium) reagent, we find that copper is less electropositive. As a result, the δ− charge on carbon in the organocopper reagent will be less than that in the Grignard reagent.

> *Organocopper reagents are softer nucleophiles than organomagnesium reagents.*

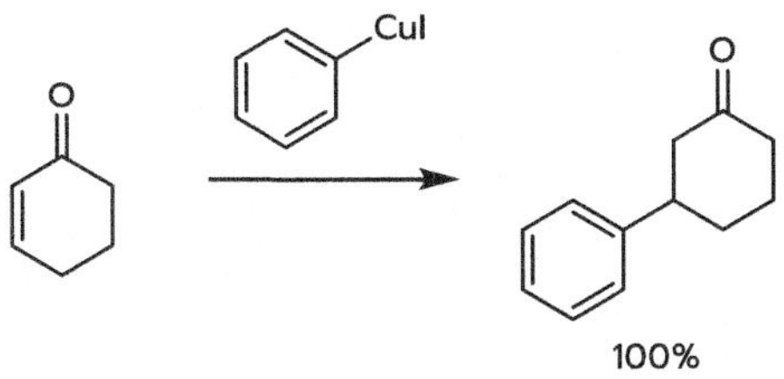

In the reaction above, there are some additional reagents. The key point, though, is that we use an organocopper reagent.

If we use an organolithium reagent, this happens.

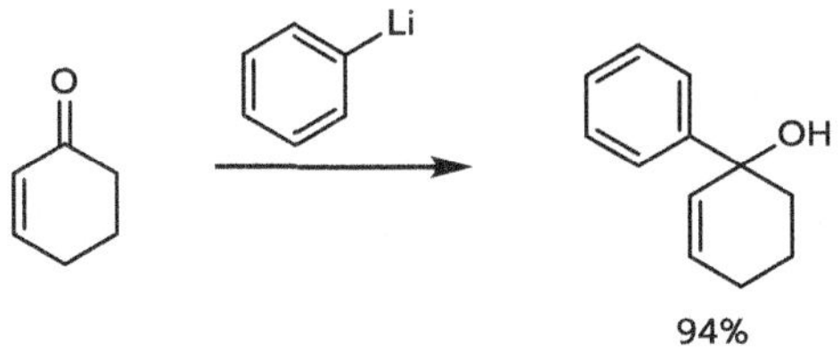

CONTROL IN SYNTHESIS

I have not routinely given yields for reactions in this book. Most of the reactions have been done many times, and work well.

> *They wouldn't be here if they didn't work well!*

In the example above, I wanted to make a very clear point that by making one small change (copper versus lithium) we can completely change the course of a reaction. This is **really** important.

> *How do you determine the best way to carry out a particular reaction?*

There are several answers to this question. If you wanted to carry out a particular transformation in the laboratory, there are databases in which you can search for reactions. I would generally search for something similar that is likely to have been done. The example above is a case in point. Perhaps the 'real' reaction uses a

SECTION 4 CARBONYL COMPOUNDS AS NUCLEOPHILES

cyclohexenone with different substituents, or it requires a benzene ring with additional substituents. If it hasn't been done with phenyl and cyclohexenone, it probably doesn't work!

With experience, you will find that you can make a pretty good guess which reactions are most likely to have been done. The databases are relatively easy to search, so that you can identify reactions that are similar to the one you are searching for.

It is much easier to find the right reaction conditions if you already know the 'type' of reactivity you are looking for. In this case, we have two sites in a molecule that can both be attacked by a nucleophile.

> *We know we need a nucleophile—we just need to find out which one is best to use.*

We might determine the 'best' nucleophile according to the yield, or simplicity of the experimental procedure. It may even come down to which chemicals we already have in our store, and which we need to purchase.

Either way, we need to know the general patterns of reactivity, and the types of change we can make which will achieve a particular outcome.

SECTION 4 CARBONYL COMPOUNDS AS NUCLEOPHILES

REACTION DETAIL 29
Aromatic Nucleophilic Substitution

INTRODUCTION

We saw the addition of nucleophiles to alkene double bonds in Reaction Detail 28.

> *Of course, it doesn't work for every alkene.*

When a nucleophile attacks 'anything', there is an interaction of the nucleophile HOMO with the LUMO of the thing it is reacting with. In Reaction Detail 28 we looked at orbital coefficients as a key factor in determining where the nucleophile attacks.

We also saw that we will only get nucleophilic attack onto an alkene if we can get a stabilized intermediate. The intermediate is preceded by a transition state, which is close in energy to the intermediate.

> *If the intermediate is stabilized, the transition state that precedes it will be stabilized—the Hammond Postulate (Recap 2).*

There is something else we could consider—the energy of the LUMO. However, I don't want to over-complicate things. It turns out that the structural factors that lead to the 'right' LUMO energy will also lead to stabilization of the anion intermediate.

> *It's easier to look at the intermediate.*

NUCLEOPHILIC ATTACK ON A BENZENE RING

Benzene rings have more electrons than alkenes. This might mean that when you think of a nucleophile attacking a benzene ring, you are less comfortable with this process. But a benzene ring still has a LUMO and if it has the 'right' energy, attack of a nucleophile will be favoured.

We saw cross-coupling reactions of aromatic compounds in Reaction Detail 10. These are reactions of benzene compounds in which aromaticity is **not** lost during the course of the reaction.

It isn't quite fair to call these 'unusual' reactions, but at this level, aromatic electrophilic substitution (Reaction Detail 8) is more typical, and we do lose aromaticity in forming an intermediate. In that case it is a carbocation. We then regain aromaticity by loss of a proton at the end of the reaction.

> *Addition of a reagent, giving a stabilized, but non-aromatic, intermediate is perfectly okay.*

SECTION 4 CARBONYL COMPOUNDS AS NUCLEOPHILES

There is no fundamental reason why the intermediate needs to have a positive charge. It could equally well have a negative charge. We know the factors that stabilize negative charges (Recap 7).

So, we have established the 'rules' for attack of nucleophiles onto benzene rings. There is one more thing to consider, and we will deal with this as we work through some examples.

EXAMPLES

Let's start with a classic. 2,4-Dinitrophenylhydrazine is a compound that has a long and rich history in organic chemistry. You can make it in the following reaction.

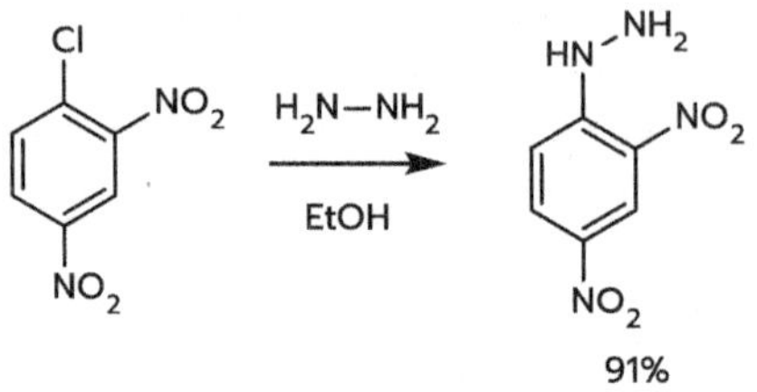

The chlorine and nitro substituents all do something really important—they lower the benzene ring LUMO energy so that overlap with the hydrazine HOMO is favourable.

> *I want to highlight this, but it won't be our focus.*

This substituted benzene ring has six different carbon atoms. Attack at only one of these—the one shown—leads to a productive reaction. If we are good at drawing curly arrow mechanisms and resonance forms, we should be able to work out why this is the case. Let's make a start.

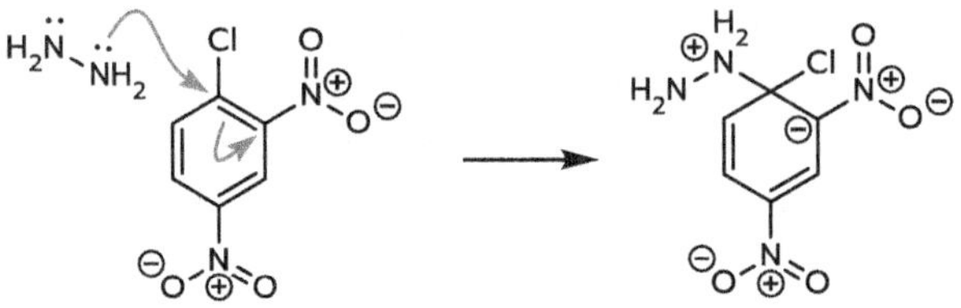

First things first—I drew out the nitro groups 'fully'. We need this in order to draw resonance forms.

> *Of course, what I am trying to do is help you get to the point when you know exactly when to draw functional groups out fully, and when an abbreviation is sufficient.*

The negative charge on carbon is resonance stabilized.

> I think there are four *other* resonance forms that you can draw. If you have drawn more than this, you may have made a mistake (or I have missed one!). If you haven't drawn four, keep looking.

SECTION 4 CARBONYL COMPOUNDS AS NUCLEOPHILES

You should have determined that both of the nitro groups can stabilize the negative charge by resonance. To put it another way, we can draw curly arrows for resonance that directly involve the nitro groups.

> There are two other positions on the benzene ring for which this is also the case. Work out which positions they are.[88]

Chlorine is electronegative. It makes the carbon atom that it is attached to more positive due to an inductive effect. The nucleophile is more likely to attack at this carbon atom.

More importantly, when we get attack at this position, we have a leaving-group. We can lose chloride (and then a proton) as shown below. We get our aromatic ring back.

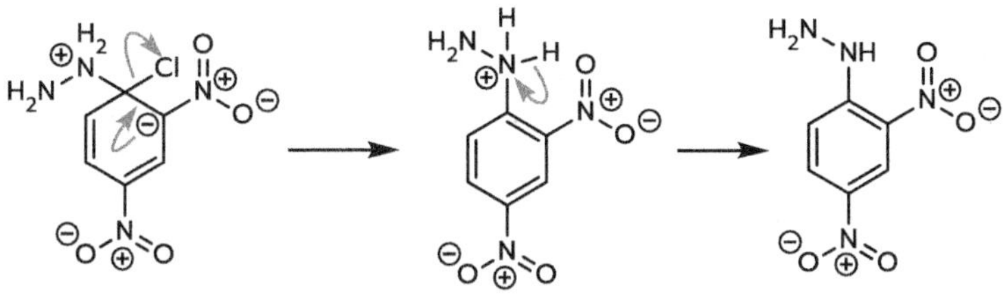

> For the other two possible sites of nucleophilic attack that you identified above, you would have to lose a $H^{\ominus}$. This is a *very bad thing*! We looked at this in more detail in **Reaction Detail 19**.

So, we get attack on the benzene ring in this compound because the LUMO has the 'right energy'. This gives us good orbital overlap, which means that the activation energy is not too high. Another way to see this is that the intermediate is stabilized, and the transition state that precedes the intermediate is also stabilized (the Hammond Postulate, yet again!).

> The resonance forms are much more 'accessible' than the orbital energies.

I mentioned that 2,4-dinitrophenylhydrazine is an important compound. This is because it reacts with pretty much any aldehyde or ketone to give a crystalline derivative. This derivative is an imine (**Applications 2**) with an extra nitrogen. We call it a hydrazone.

> Draw the mechanism of the reaction of 2,4-dinitrophenylhydrazine with cyclohexanone to give the corresponding 2,4-dinitrophenylhydrazone derivative. Which nitrogen atom of the hydrazine is more nucleophilic?

You might need to look up some of the compounds to complete the exercise above. That's okay. By having to do a little more for it, you will get more benefit.

---

[88] Draw some curly arrows for nucleophilic attack onto 1-chloro-2,4-dinitrobenzene, then draw resonance forms for any intermediates you draw.

SECTION 4 CARBONYL COMPOUNDS AS NUCLEOPHILES

Because 2,4-dinitrophenylhydrazone derivatives are crystalline, their melting points can be measured. It's a very old way of identifying an aldehyde or ketone by making a derivative. These days we just record an NMR spectrum.

When we looked at S_N1 and S_N2 substitution mechanisms in HTSIOC, we found that iodide was a great leaving group and fluoride was a terrible leaving group. Now, when we look at nucleophilic aromatic substitution, we find that fluoride appears to be a *much* better leaving group than iodide.

> There has to be a reason for this, but the 'fundamental nature' will not have changed.

Here is an example reaction. We have a few points to make—as usual!

All other things being equal (*i.e.* the nucleophile and solvent), this aromatic compound reacts about 1000 times faster than the previous one. All we have changed is the halogen.

Fluorine is more electronegative than chlorine. It makes the carbon it is attached to more electropositive. This results in faster attack of the nucleophile.

> I'm going to get you to draw a reaction profile in a minute. Before we get to that point, I've got a few questions.

The purpose of asking you these questions now, is that these are the sort of questions you can ask yourself (and then answer) in an exam.

> This reaction has a two-step mechanism. Which step do you think is rate-determining (slow)?

Is it the second step, in which the leaving group leaves, and we have established that the reaction with the poor leaving group is overall faster? Or is it the step where we start with a nice stable aromatic compound and get an intermediate in which aromaticity has been lost?

> Hopefully when I put it like that, you can make the decision relatively easily.

What we have in this chapter is a different type of reactivity of aromatic compounds to that in Reaction Detail 8, but there are some absolutes here. Aromaticity provides stability to a compound. In most reactions that affect the aromatic ring directly, we temporarily lose the aromaticity.

SECTION 4 CARBONYL COMPOUNDS AS NUCLEOPHILES

> *The reactions in Reaction Detail 10 are an exception, and we didn't see the mechanism of those reactions anyway.*

The step in which we lose the aromaticity will have a relatively high activation energy. We will get an intermediate which is **always** stabilized by resonance within the ring, and **usually** stabilized further by resonance involving substituents attached to the ring.

> Here is another question. For the reaction above, would we necessarily get a lower yield if we used Cl instead of F?

Very often, we ask one question, but the more helpful question is slightly different. In this case, what changes when we switch halogens?

> *It's the rate of reaction.*

If the reaction with F takes 10 seconds to reach completion, the reaction with Cl (1000 times slower) will take almost 3 hours to reach completion. And 10 seconds is already a pretty quick reaction.

To answer the question, a slow reaction has as much chance as a fast reaction of giving 100% yield, if you can leave it for long enough. But it might not be possible to leave it for long enough in practice.

> Now draw the reaction profile, paying attention to which step has the higher activation energy.

Now to come back to the key point, the compound with a fluorine atom reacts more rapidly than that with a chlorine atom because the rate-determining step is faster. Fluoride is still a poor leaving group, but that doesn't actually matter for these reactions.

MAKING MORE CONNECTIONS

The reactions in this chapter feature benzene rings that react well with nucleophiles. These compounds would not react well with electrophiles.

Conversely, the aromatic compounds in Reaction Detail 8 that react well with electrophiles will not react with nucleophiles.

> *This makes sense, as the structural features that lead to stabilization of positive charges will generally lead to destabilization of negative charges.*

SECTION 4 CARBONYL COMPOUNDS AS NUCLEOPHILES

BASICS 7

Heterocyclic Compounds are Important

Here are the structures of five commonly prescribed drug molecules.

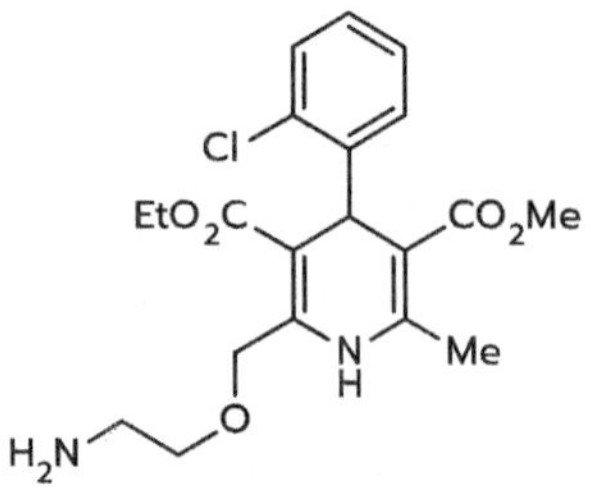

Amlodipine
(dihydropyridine)

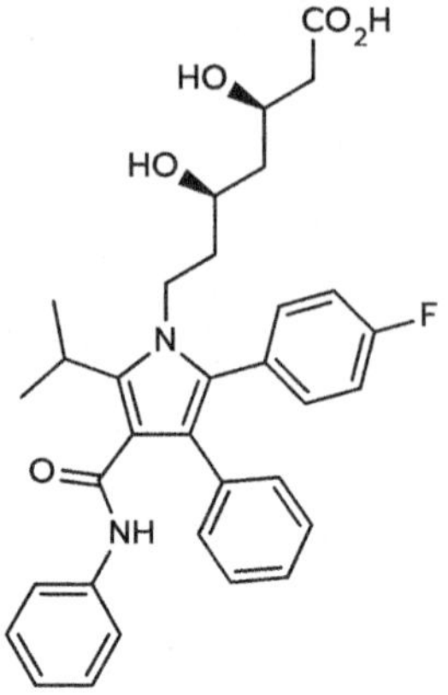

Atorvastatin
(pyrrole)

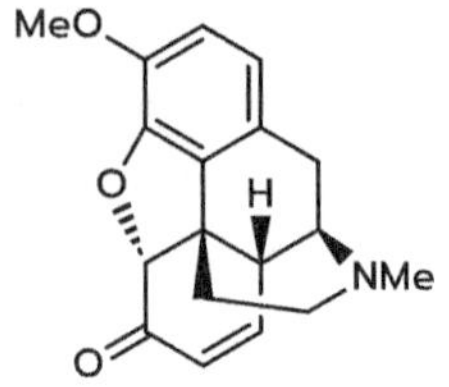

Hydrocodone
(dihydrofuran, piperidine)

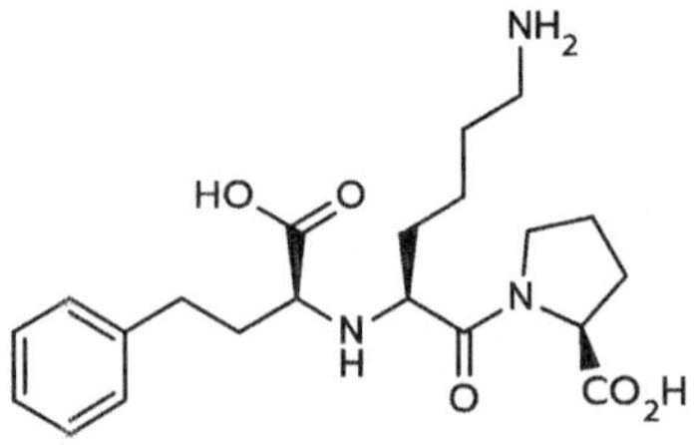

Lisinopril
(pyrrolidine)

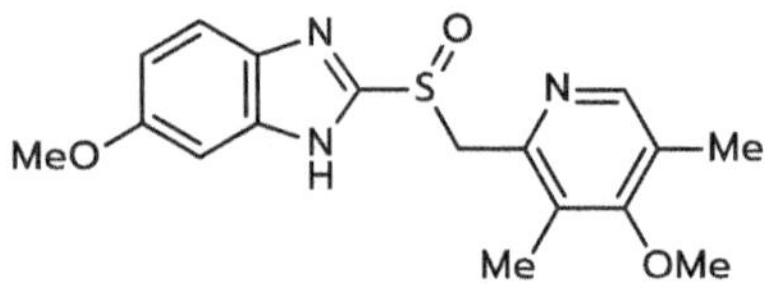

Omeprazole
(benzimidazole, pyridine)

When I say commonly prescribed, these were in fact five out of the six top selling drugs in 2016. Together, they accounted for 433 **million** prescriptions worldwide. Only one of the six top selling drugs in that year did not contain a heterocyclic ring.

> Heterocycles really are important.

It isn't too difficult to see why this would be the case. Heterocyclic compounds have heteroatoms that can participate in hydrogen bonding. Planar aromatic heterocycles have very defined shapes that can fit into enzyme active sites. Heterocycles are the perfect compounds for pharmaceutical research, and methods for their formation are equally important.

Every variation of heteroatoms in a ring has a name. In the diagram above, I've put the names of the heterocyclic rings in brackets.

SECTION 4 CARBONYL COMPOUNDS AS NUCLEOPHILES

A five-membered ring with one nitrogen atom can be a pyrrole,[89] pyrroline or pyrrolidine, depending on how many double bonds are present.

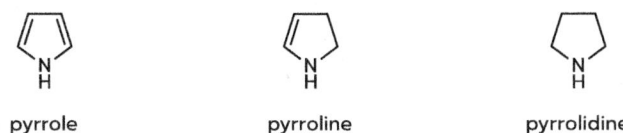

pyrrole   pyrroline   pyrrolidine

Pyrrole is aromatic, but that won't really concern us here.

> It is aromatic because it has six π electrons (The Huckel rule—Recap 6—we have 4n + 2 π electrons with n = 1) in a fully delocalized cyclic system. Two of these electrons are a lone pair on nitrogen.

The real point, at this stage, is that when you look at the heterocycles on the previous page, you will find functional groups that we have seen before—imines and enamines. We can also find enol ethers, and their sulfur equivalents.

Let's make a couple of connections. We know how to make imines and enamines (**Applications 2**). Aromatic compounds have added stability (**Recap 6**—this stability tends to be less for heterocycles than for benzene, but is still significant). If we are making an imine or enamine that happens to be aromatic, that will be a favourable process. In effect, we can add the energy for 'aromaticity' to the values we saw in **Exercise 2**. We did the same thing for aromatic electrophilic substitution in **Fundamental Reaction Type 3**.

> Remember that the steps for imine and enamine formation are reversible.

Formation of aromatic heterocycles constitutes a large 'thermodynamic hole' for a reaction to fall into. As long as you draw sensible curly arrows, you tend not to have to worry about which bond gets formed first.

> In many cases it isn't known for sure. In other cases, the detailed order of steps has been determined by very careful experimentation.

In the next chapter, we are going to look at the synthesis of a couple of heterocyclic ring systems. In **Section 7**, we will see some more complex examples.

> The point of all this is that if you are given a heterocyclic compound and its precursors, you will be able to identify which bonds are being formed, and to draw a plausible reaction mechanism.

---

[89] You **can** make the arsenic analogue of pyrrole. It **is** called arsole!

SECTION 4 CARBONYL COMPOUNDS AS NUCLEOPHILES

APPLICATIONS 9

Imines, Enamines and Heterocycles

This isn't a book about heterocyclic chemistry. In the previous chapter, we saw why heterocycles are important. For each heterocyclic ring system, there are numerous different syntheses.

> It isn't easy to learn them all.

What I want to do, in this short chapter, is to provide a framework for you to link your fundamental chemistry (imines, enamines, substitution reactions, oxidation states) with heterocyclic chemistry.

If you can predict how a particular heterocyclic compound *might* be formed, it will be a lot easier to learn the reactions.

THE PAAL-KNORR FURAN SYNTHESIS

I have deliberately used the above heading to highlight the cognitive problem. First of all, you need to know what a furan is. Secondly, if you know five reactions for making furans, you need to remember which one is the Paal-Knorr synthesis.

> With a little practice, you'll soon remember which heterocycle is which. And I don't think anyone is going to simply ask you (in an exam) to draw an example of the Paal-Knorr furan synthesis. You are much more likely to be given reagents and asked to predict what happens.

I'm going to use an example with an added complication. Let's treat compound **1** with an acid. Don't over-think this. We have two carbonyl groups. One of them will be protonated. This makes the carbonyl group more susceptible to nucleophilic attack, and we have a nucleophile within the compound—the other carbonyl group.

Within this scheme, we have a number of 'housekeeping' issues. Compound **1** has a methyl group attached to each carbonyl atom. I have explicitly drawn the one on the left, but not the one on the right.

> Consistency would have been better, but adding 'CH$_3$' on the right made the drawing a bit cluttered. And I wanted to make the point!

The two carbonyl groups are equivalent. It doesn't matter which one we protonate.

SECTION 4 CARBONYL COMPOUNDS AS NUCLEOPHILES

The cyclization step is 5-*exo-trig* according to Baldwin's rules, which we saw briefly in Basics 6. It turns out that this is one of the better cyclization reactions, but you could probably have worked this out by making a model of compound **2** and aligning it so that the oxygen lone pair approaches the carbonyl group at the Bürgi–Dunitz angle (Stereochemistry 10).

From this point, we can lose a proton from compound **3** (note that the bond to the ethyl group in structure **4** is no-longer a wedge—the stereochemistry has gone!), and then we have an elimination reaction as shown below.

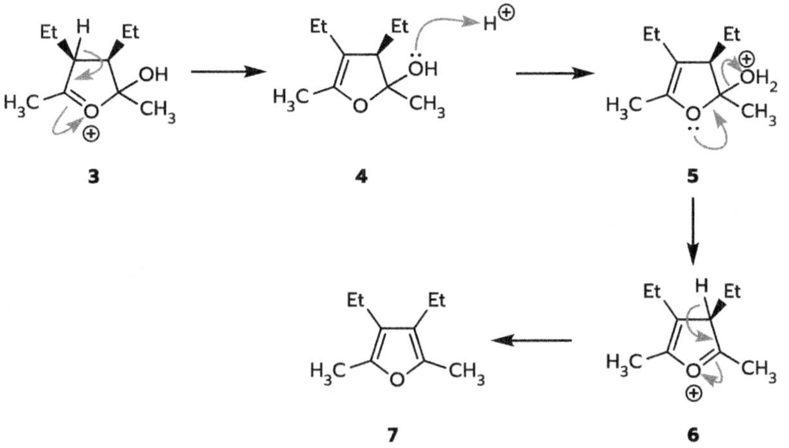

I have drawn this as an E1 elimination, pulling some aspects from Reaction Detail 15. After all, it's basically the same reaction, but with an added driving force—aromaticity—which over-rides some of the considerations we looked at in Exercise 2.

> Once you get past 'all these reactions' and you are able to see the underlying principles, organic chemistry is a beautifully 'connected' subject.

There's one more thing. You're getting used to that by now! Compound **8** is a diastereoisomer of compound **1**. They have different energies (Stereochemistry 2).

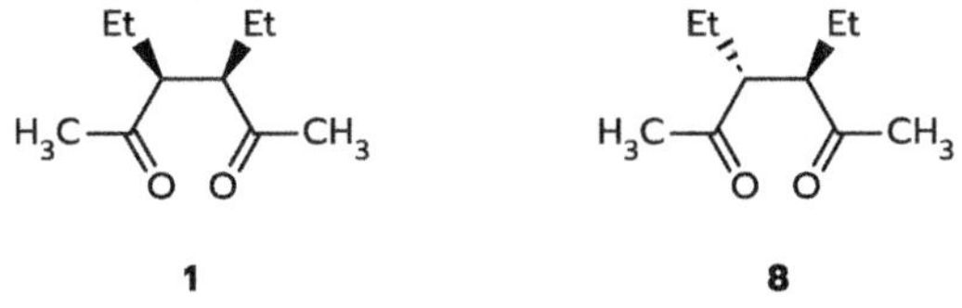

They form the same furan (**7**) but at different rates. Both will form compound **4** as an intermediate, and from that point, the intermediates are the same. The rate-determining step *must* be before the formation of structure **4**.

> It is instructive to look at this example, as it gives an insight into some of the experiments organic chemists carry out to work out the mechanisms of the reactions in this book.

SECTION 4 CARBONYL COMPOUNDS AS NUCLEOPHILES

THE HANTZSCH PYRIDINE SYNTHESIS

Okay, this one will take a little while. There are quite a few steps involved. Here is the first part of the reaction. We take two molecules of ethyl acetoacetate, one molecule of formaldehyde, one molecule of ammonia, and we heat it all together in a solvent. Very often, the solvent is water.

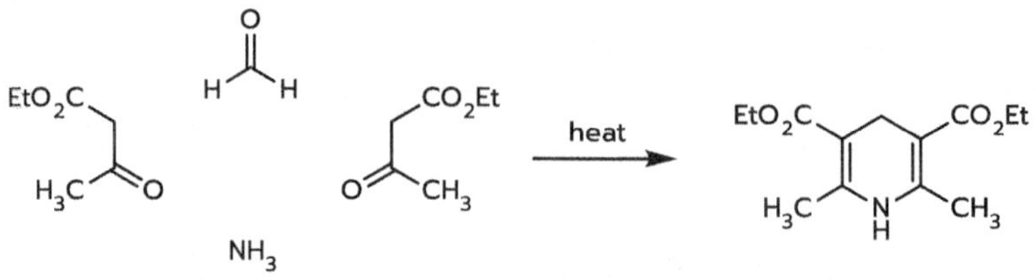

> Draw the product, showing every hydrogen atom, and work out which ones are lost from the starting materials.

Overall, we are losing three molecules of water, which recoups some of the entropy we would otherwise lose by having four molecules react to give one.

The product is not aromatic. This is relatively unusual for a 'classic' heterocycle synthesis. We saw one very important application of a dihydropyridine in **Reaction Detail 19**. It can be used as a hydride donor in biological systems. They are pretty easy to oxidize. Here, we are using iron(III) chloride.

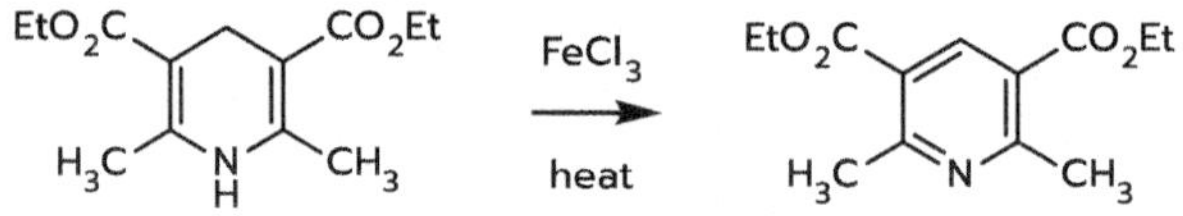

We need to think about the precise order of steps in the mechanism for heterocycle formation. We are forming two enamines (**Applications 2**). We are forming two carbon–carbon bonds.

> We could worry about this forever, or we could just start drawing.

Let's form a C–N bond. The ammonia is not becoming bonded to the formaldehyde, so we can safely start with ammonia reacting with ethyl acetoacetate.

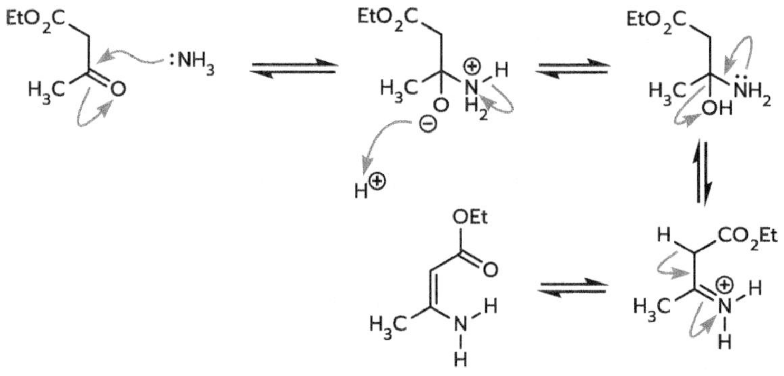

SECTION 4 CARBONYL COMPOUNDS AS NUCLEOPHILES

This is, very simply, an enamine formation. There is absolutely nothing that you have not seen before. It's actually a pretty good enamine formation. We form **A**, which is almost halfway to the dihydropyridine product.

Here is a slight digression. I drew intermediate **A** as the (Z) double bond isomer. Most mechanisms I looked at showed the (E) isomer.

> First of all, make sure you understand why **A** is the (Z) isomer. Draw the (E) isomer.

I drew the (Z) isomer because it looks more stable. There is going to be a hydrogen bond between the hydrogen atom on nitrogen and the carbonyl oxygen atom.

> *I have almost drawn it!*

Next up, we will react the other molecule of ethyl acetoacetate with the formaldehyde. This is a Knoevenagel condensation (Reaction Detail 25). We could use ammonia as the base in this mechanism.

> *It's a base as well as a nucleophile, and it's in the flask!*

We have seen before (Reaction Detail 25) why carbanion **C** is more stable than alkoxide anion **B**. From this point, we get intermediate **D**, which is an α,β-unsaturated ketone. We know these undergo conjugate addition reactions (Reaction Detail 28).

> *We are building a lot of connections here, but I suspect it's starting to get a bit intimidating. I wish there was a shortcut. There isn't. Look at the mechanisms in this chapter. Go back to the earlier examples and the fundamental explanations. Take the time to draw the mechanisms for yourself, and make sure you understand the curly arrows and the principles.*

If you do go back and look at Reaction Detail 25, you might wonder why I am not drawing an iminium ion mechanism for the Knoevenagel condensation. I am keeping it simple!

> *There is a very good chance that the ammonia does in fact react with the formaldehyde, but it is lost in a subsequent step.*

> Once we get to the end of this synthesis, go back to the start and draw the whole thing again starting with the reaction of ammonia with formaldehyde.
>
> When you've done that, draw another version in which intermediate **A** reacts with formaldehyde or with the derived iminium ion.

What I am trying to show you (or get you to prove to yourself) is that the individual steps can be carried out in various different orders, and eventually they all get to the same point.

> *If this was an exam question, you should get full marks for any sensible approach.*

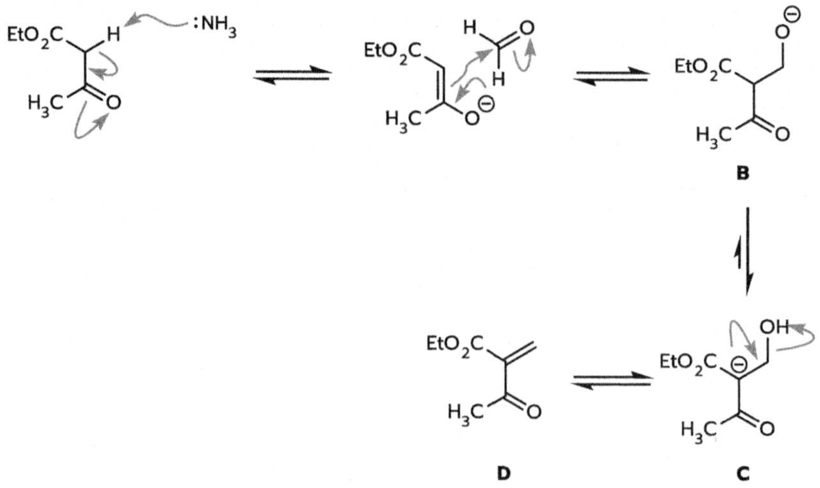

SECTION 4 CARBONYL COMPOUNDS AS NUCLEOPHILES

Almost there now. Here is the conjugate addition reaction.

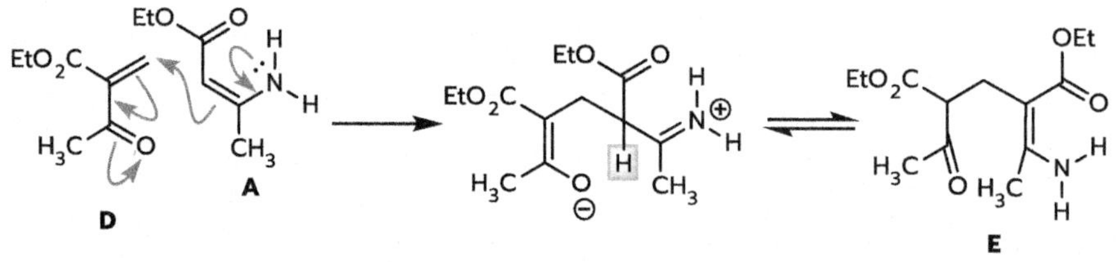

> Draw the curly arrows for the second part of this process. In this case, you might want to initially form the enol by using the O⁻ to deprotonate the indicated hydrogen atom. It would be a six-membered transition state.

You will see some strange bond angles drawn on structure **E**. I could have rotated some bonds to make it tidier, but I prefer to keep it in the same orientation as the product. Now we can see a problem with the enamine geometry. There is no way the nitrogen atom could reach the ketone carbonyl group with this geometry.

> This raises several questions. Is the mechanism correct? Is the double bond geometry correct?

Spectroscopic studies have identified structures **A** and **D** in the reaction mixture. They can be prepared independently and react under the same conditions to give the product. The double bond geometry **E** looks plausible, but that doesn't mean it cannot isomerize to give **F**, which would be able to cyclize.

> Draw a mechanism for the interconversion of **E** and **F**. Look at the mechanism for the formation of **A** for inspiration. You will need to protonate on carbon, rotate a single bond, and then deprotonate again.

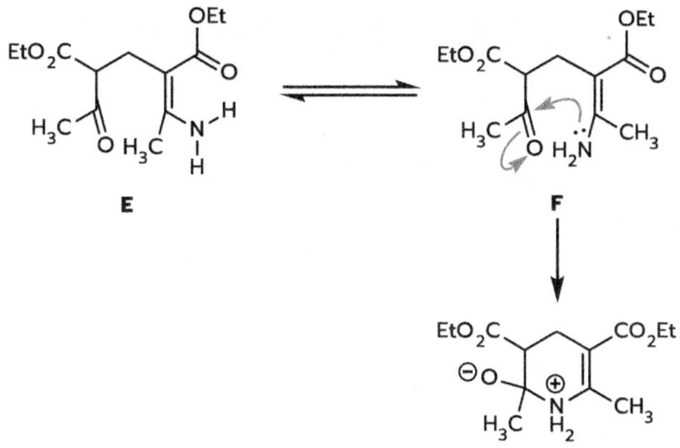

SECTION 4 CARBONYL COMPOUNDS AS NUCLEOPHILES

[Structures E and F shown in equilibrium, with F cyclizing to a six-membered ring intermediate containing N⁺H₂, O⁻, CH₃, EtO₂C, CO₂Et, and CH₃ substituents.]

We would then expect the cyclization step shown above to be rapid. I've drawn this as a 'reaction arrow' rather than an 'equilibrium arrow'.

> Any reaction is reversible, but this one looks pretty favourable, so I prefer to make a distinction (and a point!).

> Finish the job! We saw the product structure at the start of this section. You need to lose a proton from nitrogen, add a proton to oxygen, and then form another enamine. Look back at **Applications 2** for general guidance, although everything you need has already been covered in this section.

Make sure you didn't take too many shortcuts. You could have drawn the direct transfer of the proton from the nitrogen atom to the oxygen atom, but this would involve a four-membered ring transition state. We saw why this would be wrong in **Applications 2**. Make sure you 'use' the nitrogen lone pair to 'push' the oxygen atom out. It should be an E1 elimination mechanism (**HTSIOC Fundamental Reaction Type 2**) because the cationic intermediate (we don't really think of it as a carbocation) is stabilized.

> You would probably prefer me to give you the answer, so you can check if you've got it right. I prefer you to have to keep referring back to the earlier chapters to ensure you have built all the right connections. Sorry!

THE HANTZSCH THIAZOLE SYNTHESIS

And we further compound the challenge by recognizing that some of the pioneers of heterocyclic chemistry devised more than one synthesis, working on more than one heterocyclic ring. A thiazole is a five-membered ring with one nitrogen atom and one sulfur atom. Here is the overall reaction scheme.

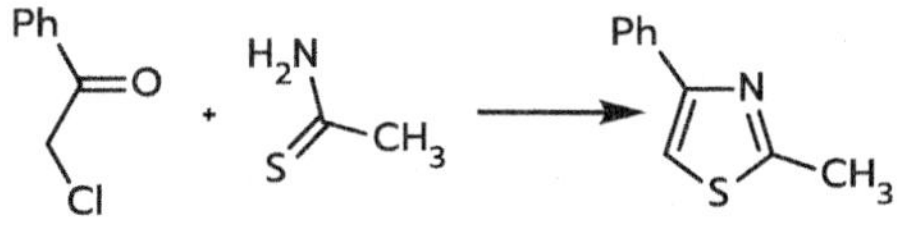

[Reaction scheme: PhC(O)CH₂Cl + H₂N-C(=S)-CH₃ → 2-methyl-4-phenylthiazole]

389

SECTION 4 CARBONYL COMPOUNDS AS NUCLEOPHILES

Having already looked at two heterocyclic synthesis mechanisms, perhaps we can see what is happening here. We are forming a C–S bond from a thioamide sulfur atom. This looks like an S$_N$2 substitution, and we saw in **HTSIOC Reaction Detail 1** that a chlorine atom α- to a carbonyl group reacts very rapidly in S$_N$2 substitution reactions. Sulfur is a very good nucleophile, so this is a good reaction. We are also making an enamine from an amide[90] and a ketone. We saw this process in **Applications 2**, although not with an amide. Amides are much less nucleophilic than amines.

> Draw the resonance forms to explain why this is the case.

Perhaps there is a **good reason** why it works better for this compound! As always, we can spend a lot of time analysing the problem, or we could start drawing and see what happens.

Let's start with the S$_N$2 substitution, because sulfur is such a good nucleophile, and the sulfur lone pairs are 'available', since they are not delocalized into the thioamide. This gives us intermediate **9**, with a positive charge.

If you want to get rid of a positive charge, the simplest way is to lose a proton. Make sure you can see why the next step is effectively a keto-enol tautomerization (**Fundamental Reaction Type 4**).

From structure **10**, we can protonate the oxygen atom ready for the nitrogen lone pair in **11** to cyclize.

> We were worried that the thioamide nitrogen atom might not be a good nucleophile. It turns out that we shouldn't have worried, as by the time we get here, it isn't a thioamide any more. This lone pair is much more 'available'.

---

[90] Okay, it's a thioamide, which does make a difference, but it's closer to an amide than to an amine.

SECTION 4 CARBONYL COMPOUNDS AS NUCLEOPHILES

We have got to compound **12**. We are not yet at the aromatic thiazole on page 389, but you've got enough experience now to finish the job. You need to lose a proton and eliminate water.

> Complete this mechanism. Take your time. Make sure the curly arrows and intermediates make sense.

Were you tempted to 'abbreviate' the loss of a proton? Check again and make sure the curly arrow starts at a N–H or C–H bond.

> Now try to draw an alternative mechanism in which the C–N bond is formed first.

Your mechanism will probably look okay, although it isn't the pathway with the lowest overall energy. At this stage, make sure that *any* mechanism you draw is sensible—good curly arrows and reactions that could work. Once you can do this with confidence, you will start to think about which of the possible pathways might be best.

IN CLOSING

The three heterocycle syntheses we have looked at are representative. They are all aromatic heterocycles, so we have a clear driving force for their formation. But most of all, we have functional groups within these heterocycles that are becoming familiar to us—imines, enamines, enols. When you encounter a 'new' heterocyclic synthesis, try to relate it back to the 'simpler' reactions you have seen. This will allow you to place it clearly in context.

SECTION 4 CARBONYL COMPOUNDS AS NUCLEOPHILES

BASICS 8

The Curtin-Hammett Principle

In **Applications 9**, we looked at the Hantzsch pyridine synthesis. In doing so, we saw a 'problem step'. Let's look at this in a little more detail. We expected the compound below to be formed as the double bond isomer **E**, which is more stable than isomer **F** due to the presence of a hydrogen bond.

> However, isomer **E** cannot undergo cyclization. Isomer **F** can.

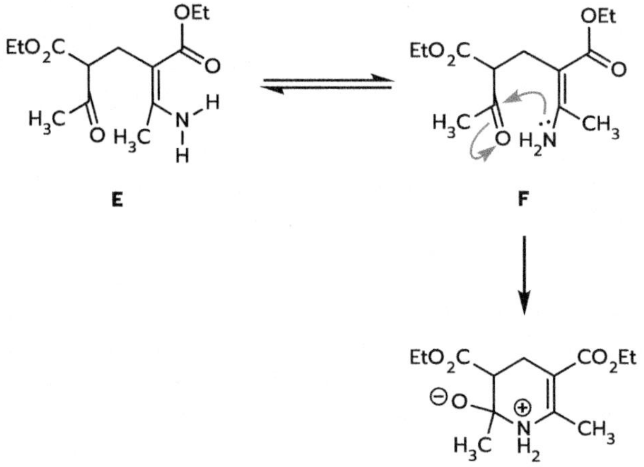

What we have here is a nice example that allows us to introduce a more general principle, the Curtin-Hammett principle. It may not actually fit all aspects of the definition fully, but bear with me.

When we say that isomer **F** can undergo cyclization but isomer **E** cannot, we are seeing an extreme case of something we can express in energetic terms.

> The activation energy for cyclization of isomer **E** is **much** higher than that for cyclization of compound **F**.

Here is the usual energy profile used to show the Curtin-Hammett principle. I have added two of the structures from the Hantzsch pyridine synthesis, and I will try to explain the idea behind the Curtin-Hammett principle.

> This example is not really a Curtin-Hammett situation, and I will also try to explain why this is the case, so you know when you can apply this idea.

SECTION 4 CARBONYL COMPOUNDS AS NUCLEOPHILES

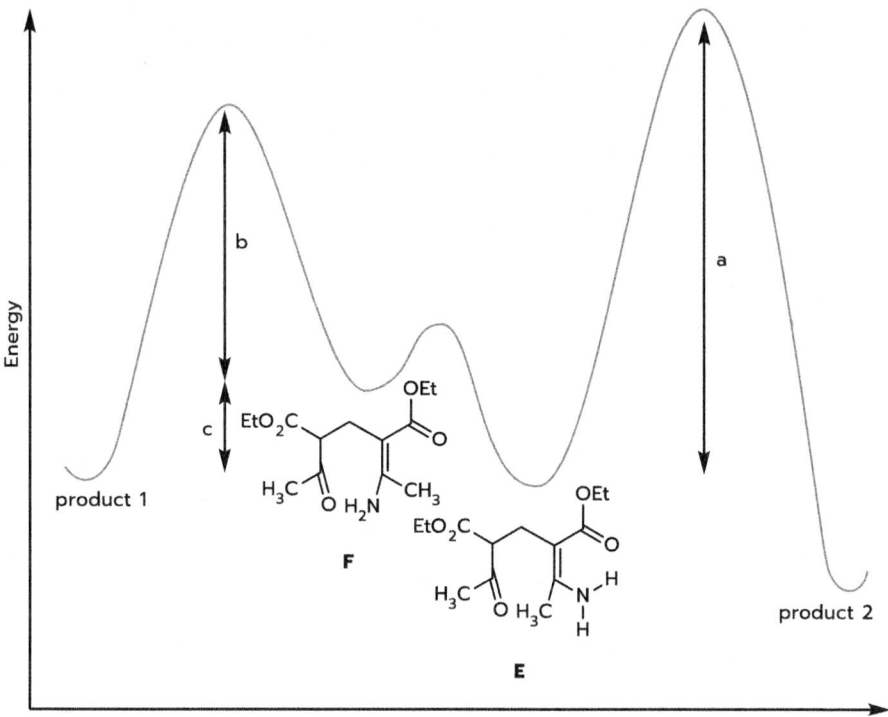

In a Curtin-Hammett situation, the intermediates are in rapid equilibrium. The energetic barrier to their interconversion is low compared to the barrier to further reaction. As a result of this, the product formed does not necessarily come from the most stable intermediate. Instead, it depends on the overall activation energy ('b + c' compared to 'a' on the diagram above).

> We can form more of the unstable product via a lower overall barrier.

In the above case, the barrier, 'b' will be relatively low. The barrier 'a' will be much higher. Let's see why this is.

> Try to draw the **direct** product of cyclization of intermediate **E**.

Before I give you the answer, there is one more thing to try.

> Make a model of intermediate **E**, and try to get the nitrogen atom close to the ketone carbonyl carbon atom.

In making a molecular model, and failing to get the nitrogen anywhere near the ketone carbon atom, you have determined that the energetic barrier, 'a', for this process is high—much higher than 'b + c'.

What did you draw for the structure of the product of cyclization of **E**? Was it exactly the same as product 1?

Product 2 would have a *trans* double bond in a six-membered ring. We encountered ring strain in **Basics 6**.

393

SECTION 4 CARBONYL COMPOUNDS AS NUCLEOPHILES

> You cannot have a *trans* double bond in a six-membered ring. Product 2, and the transition state that precedes it, would have **very** high energy.

If you drew the same structure for product 2 as for product 1, you made a change in the double bond geometry. There would have to be a reason for this to happen.

WHY IS THIS A BAD EXAMPLE?

Now let's see why this particular example isn't quite a Curtin-Hammett situation. In a true Curtin-Hammett situation, the intermediates (**E** and **F** in this case) are in rapid equilibrium (very low barrier) and each can react to give a different product.

> In this case, it really doesn't matter. Isomer F can cyclize. Isomer E cannot cyclize. It really doesn't matter whether the equilibration of E and F is slow, as long as it can happen.

In addition, there isn't really the possibility of forming two different products. The barriers, 'b + c' and 'a' are dramatically different.

> There is really only one possible outcome.

Of course, we are looking at this reaction because it ***does*** work. Isomers **E** and **F** can exist in equilibrium *via* a mechanism you drew in **Applications 9**. Isomer **F** can cyclize, and it gives the product. Isomer **E** cannot do anything, until it isomerizes to give isomer **F**.

A BETTER EXAMPLE

The Curtin-Hammett principle is generally applied to systems in which the barrier to interconversion of the intermediates is low. The intermediates themselves need to be of similar (but not the same) energy. In most cases, this means that we are looking at conformational isomers (**Stereochemistry 5**) which are interconverted only by rotation around one or more bonds.

There is a classic example of the Curtin-Hammett principle, and I like to use the classic examples. First of all, have a look at the following two 'conformers'. It isn't quite bond rotation. The inversion of the nitrogen lone-pair is known as *pseudorotation*. The point here is that the difference in energy of the two structures is relatively small, and the barrier for interconversion is also relatively small.

The conformer on the right is more stable.

> There is more room on the left for the methyl group. It is in an equatorial position if you consider the six-membered ring highlighted.

SECTION 4 CARBONYL COMPOUNDS AS NUCLEOPHILES

> Draw the relevant hydrogen atoms that 'clash' with the methyl group in each conformer. Can you see why the conformer on the right is preferred?

When we add ^{13}C labelled iodomethane, it reacts in the equatorial position.

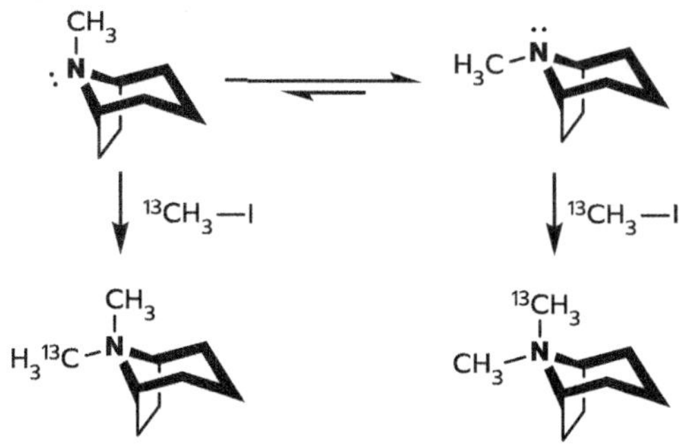

There is more room in the equatorial position for the attacking methyl group. During the approach, we might consider it to be bigger than the methyl group already present.

Have a look at the energy profile earlier in this chapter. It applies directly to this reaction, with some minor modifications. The key modification is that apart from the isotopic label, the two products are the same. They will have the same energy.

> Draw the Curtin-Hammett profile for this example, and add all structures with plausible relative energies.

As an additional exercise, find this example in another textbook, or on a website. You will see the ring system drawn as follows.

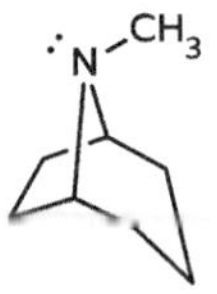

This is the core ring system of the tropane alkaloids, of which cocaine is a prominent example. It is usually drawn this way.

> Can you spot when the methyl group is axial or equatorial as easily on this structure?

I cannot, which is why I chose to draw it differently.

BUT WHAT ABOUT STEREOCHEMISTRY 6?

In **Stereochemistry 6**, and in **Stereochemistry 10**, we rationalized the stereochemical outcome of reactions by considering which conformer will be most stable. We then claimed that the product will be formed from the most stable conformer.

SECTION 4 CARBONYL COMPOUNDS AS NUCLEOPHILES

> And now we are saying this doesn't have to be true!

What we really need to consider is the energy of the transition states leading to each stereoisomer of product. Here is conformer **A** from **Stereochemistry 6**. Alongside it is a representation of the transition state for the epoxidation of this conformer, which we also saw in that chapter.

The point we made there, which we are reiterating here, is that because of the shape of the transition state for epoxidation of alkenes, we find the same $A^{1,3}$ interaction in the transition state.

> Sure, it isn't really allylic any more, but principles beat labels every time.

Now to the really big question. How do you know when it is safe to consider only the conformational bias of the substrate, and when do you explicitly need to fully consider the transition state?

I imagine you are hoping for something 'deep' here. I was too! There is a reason why we teach these reactions in terms of $A^{1,3}$ strain or Felkin-Anh. It helps you define a process to follow, and that process works. And then you encounter a situation where that process doesn't work. You find there is a more rigorous approach.

My experience is that it takes much longer to get the hang of the Curtin-Hammett principle.

> I'm still getting there!

I've given you some pointers. All I can do at this point is to ensure that you are aware of the Curtin-Hammett principle, and that you are also aware that you should not yet expect to be able to apply it rigorously in all situations.

Here is another (horrible) example to show you why you cannot yet be confident. This is a conjugate addition reaction (**Reaction Detail 28**) but this time using a methyl radical (don't worry about this distinction—we aren't covering any free-radical chemistry in this book).

> I am deliberately not showing the full mechanism here, but we are adding a methyl group to the double bond. I have deliberately indicated the methyl group in the starting material as 'CH_3' and the methyl group from the radical as 'Me' to make the distinction.

SECTION 4 CARBONYL COMPOUNDS AS NUCLEOPHILES

The lowest energy conformer of compound **1** is structure **2**. The next conformer, in energetic terms, is conformer **3**. You could have predicted this from **Stereochemistry 5**. We do actually get the major product (**4**) from conformer **2**.

> So, what is the problem?

The problem is that if the proportion of conformers **2** and **3** was reflected in the product distribution, we would expect to get almost exclusively product **4**.

This is a challenging example. We have cyano groups on the double bond. Cyanide is sp hybridized. It is small!

When the radical adds to conformer **2**, there is a significant interaction with the existing methyl group. This interaction is comparable (in energetic terms) with the interaction between the H and CN groups.

When the nucleophile adds to the higher energy conformer **3**, the incoming nucleophile has an interaction with a hydrogen atom. This is a much smaller interaction, although we have already 'paid the price' in energetic terms by going to conformer **3** in the first place.

> Could you have predicted this situation?

I would say definitely not! In fact, because the major conformer does give the major product, I don't think you would even know to look for a Curtin-Hammett situation given the experimental data.

> So, what do you do?

What you don't do is 'throw the baby out with the bathwater'! The application of $A^{1,3}$ strain (and other conformational bias such as Felkin-Anh) allows you to make some really good predictions about the outcome of reactions. You should learn it, understand it, and use it. What you should then do is be aware that there are limitations, and that there is another approach you can take as and when the lowest

SECTION 4 CARBONYL COMPOUNDS AS NUCLEOPHILES

energy conformer does not give the major product, or does not give enough of the major product.

Most of all, though, you should accept that this is complicated, and it will take a good while for you to be able to apply this with confidence.[91]

---

[91] By 'a good while', I mean after you graduate!

SECTION 4 CARBONYL COMPOUNDS AS NUCLEOPHILES

STEREOCHEMISTRY 15

Improving Stereoselective Enolate Reactions

We looked at diastereoselective enolate alkylation reactions in Stereochemistry 13. These reactions are pretty good. They give high yields of products, and with high levels of stereochemical control.

> *But there is still room for improvement.*

Don't lose sight of the bigger picture. Every reaction has been reported at some time, and has then been further developed by either the original researchers or by others. Reaction optimization is a standard part of organic chemistry research.

What I am going to do in this chapter is show you how Professor Steve Davies at Oxford University identified opportunities to improve on Evans' work on the chiral auxiliaries that bear his name.

WHY?

Let's be clear, this is nice chemistry, but at this level, this isn't a reaction you would be expected to learn.

We are doing this because it is important to get better at 'looking at molecular structures' and 'seeing everything', and this is a nice example that we can use to build your skills.

> *In showing you how this optimization was done, I am guiding you towards how you might think about the reaction optimization process more generally.*

To some extent, once you've gone through this chapter, you can probably forget about this example.

> *Then again, if you've understood everything in this chapter, you'll probably find you remember it anyway.*

GETTING STARTED

As always, the best place to start is with an example. Here is a stereoselective alkylation reaction.

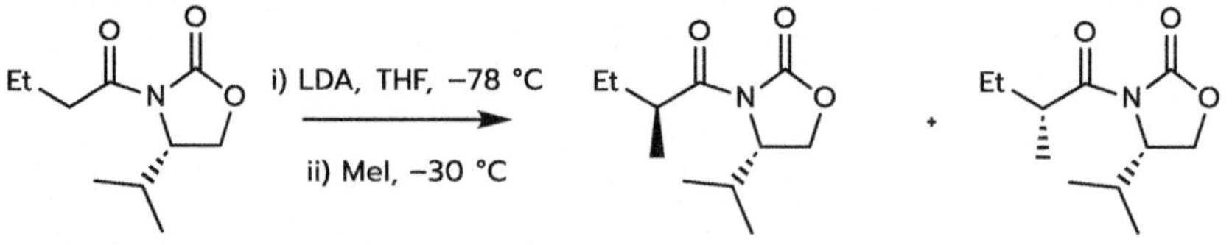

SECTION 4 CARBONYL COMPOUNDS AS NUCLEOPHILES

The ratio of stereoisomers is 91:9. This is pretty good, but there is still quite a lot of room for improvement. This chiral auxiliary is made from L-valine, a relatively inexpensive naturally occurring aminoacid.

Here is another example, in which we make the group on the chiral auxiliary bigger.

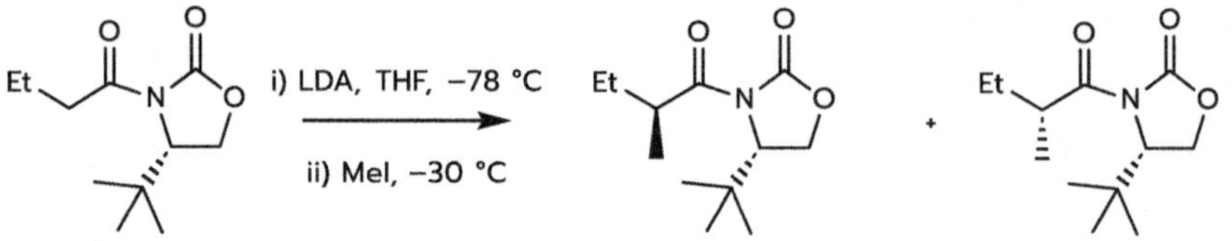

Now we get a 99:1 ratio of stereoisomers.

> Remember that when we remove the chiral auxiliary, we will get exactly the same product in each case.

The problem with this is that the chiral auxiliary used in the second example is derived from L-*tert*-leucine, which at the time of writing costs 40 times more than L-valine.

> We can improve the level of stereoselectivity, but (quite literally) at a cost.

In **HTSIOC Basics 32** we looked at the steric impact of some groups in cyclohexanes. We saw that an isopropyl group isn't all that much bigger than a methyl group. The reason for this is summarized in the following diagram.

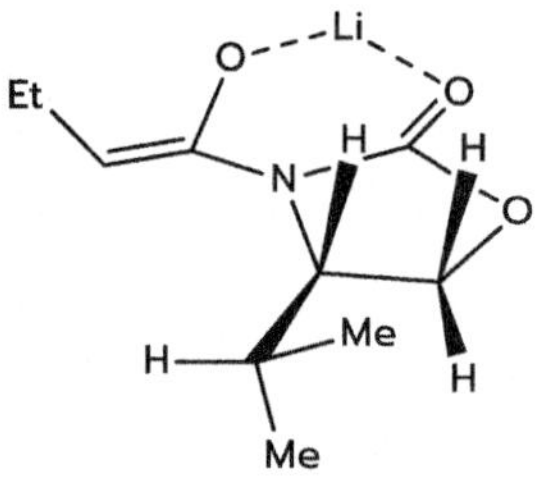

The isopropyl group can rotate (**Stereochemistry 5**). We have two methyl groups and one hydrogen atom on the isopropyl group, but it isn't the methyl group that is 'getting in the way' of the electrophile reacting with the lower face of the enolate double bond. Instead, it is a hydrogen atom.

SECTION 4 CARBONYL COMPOUNDS AS NUCLEOPHILES

> We can now see why *tert-butyl* works much better. Hydrogen versus methyl is a large energy difference.

Now look at the following example.

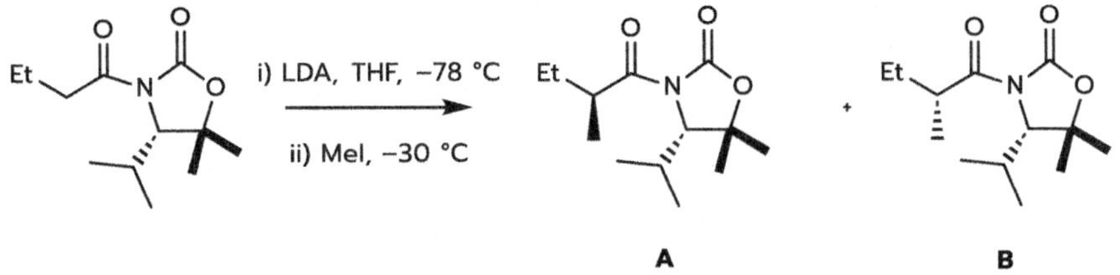

i) LDA, THF, −78 °C
ii) MeI, −30 °C

A   **B**

98:2 ratio

Now, we have added another two methyl groups (bold) to the chiral auxiliary, and we get a 98:2 ratio of stereoisomers.

> *Isopropyl is almost as good as tert-butyl now.*

Here it is in 3D. You should get your molecular models and convince yourself that this looks 'right'.

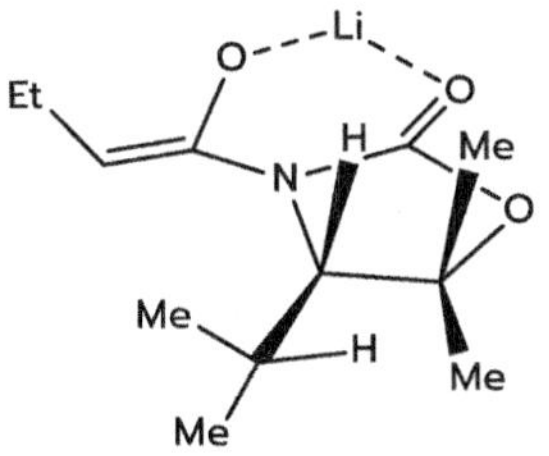

Addition of the methyl groups has changed the preferred conformer of the isopropyl group to increase the steric impact **where it matters**. I think this is a really nice illustration of how a good understanding of basic principles can allow you to improve the outcome of a reaction.

> I remember the first time I saw this chemistry. My reaction was 'why didn't I think of that?'. For me, that is a definition of good science. It doesn't have to be so complicated that no-one else could do it. It should be so elegant that everyone can understand it.

THERE'S MORE!

In **Stereochemistry 13** we also looked at cleavage of the chiral auxiliary. Here is a scheme from that chapter.

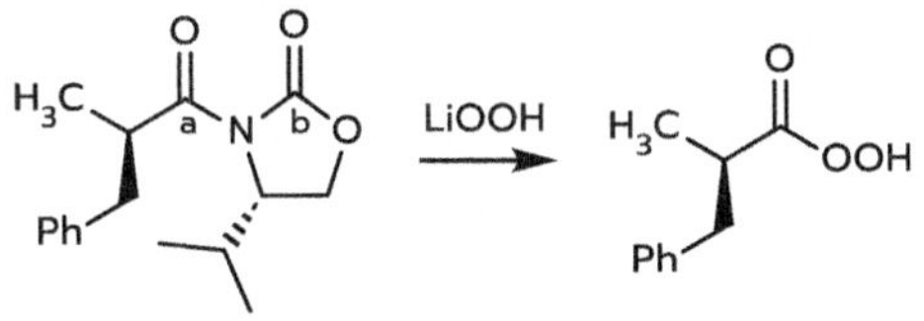

LiOOH

SECTION 4 CARBONYL COMPOUNDS AS NUCLEOPHILES

In order to cleave off the chiral auxiliary, we need the nucleophile to attack at carbonyl group 'a', rather than at carbonyl group 'b'.

Let's make some connections (again!). In **Stereochemistry 10** we saw the angle that a nucleophile approaches a carbonyl group. Have a look at the methyl groups we added.

> They will hinder the approach of a nucleophile to carbonyl group 'b'.

Not only do the methyl groups make the alkylation reaction more stereoselective. They also make the attack of a nucleophile more chemoselective, so that a higher yield of the desired product is formed in both steps.

> This is great!

In general, if you had a two-step synthesis and you were asked to make one change that would improve both steps, it would be difficult to do this. It is pretty unusual to be able to dramatically improve a synthesis in this way, but it works here, and we can understand why it works.

SECTION 4 CARBONYL COMPOUNDS AS NUCLEOPHILES

REACTION DETAIL 30

The Favorskii Rearrangement

This chapter focuses on a reaction that is quite esoteric, but there are important lessons to be learned from the mechanism.

> I have never personally carried out a Favorskii rearrangement. There are relatively few instances of the reaction being applied in synthesis. It is an interesting reaction but not a very useful reaction!

Here is the reaction. The first thing you will notice is that we start with a six-membered ring and end up with a five-membered ring.

> There must have been 'some sort' of rearrangement.

Perhaps the second thing you will notice is that we start with a ketone and end up with an ester. An ester is at a higher oxidation state (**Basics 3**) than a ketone, so does this mean we have an oxidation reaction?

> Work out the oxidation states of all carbon atoms in the starting material and the product. Is there a net change?

The other question you might ask is whether an oxidizing agent has been used.

The final change to recognize is that one molecule of methanol, used as solvent, has been incorporated into the product.

We are going to work through the possible reactivity of 2-chlorocyclohexanone with hydroxide/methanol, and I will then add the key piece of evidence that allowed the mechanism to be determined.

First, we will consider the reaction conditions. We have potassium hydroxide in methanol. These conditions are basic/nucleophilic, and we have hydroxide and methoxide present. The following equilibrium is inevitable.

$$\text{MeOH} + {}^{\ominus}\text{OH} \rightleftharpoons \text{MeO}^{\ominus} + \text{H}_2\text{O}$$

If a mechanism we draw works with hydroxide but not with methoxide, or *vice versa*, then we would have to question whether it is correct.

Right, let's have a closer look at 2-chlorocyclohexanone. There are two sites that could be attacked by a nucleophile. Here is the first reaction.

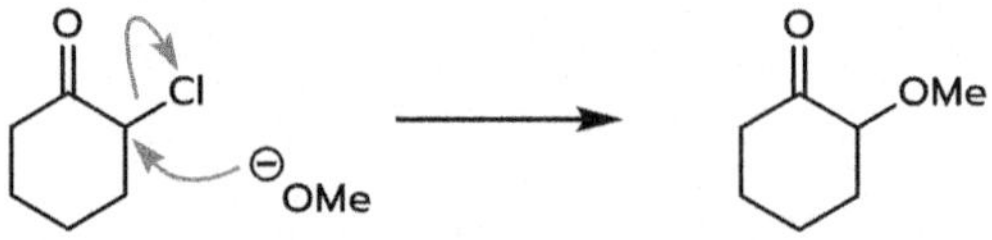

SECTION 4 CARBONYL COMPOUNDS AS NUCLEOPHILES

This is an S$_N$2 substitution reaction. It's actually a pretty good one, as the nucleophile attacks α- to a carbonyl group and the build-up of negative charge in the S$_N$2 transition state is stabilized by the carbonyl (**HTSIOC Reaction Detail 1**).

> The problem is, although this looks like a good first step, there isn't an obvious way to get from here to the observed product.

Here is a second possibility.

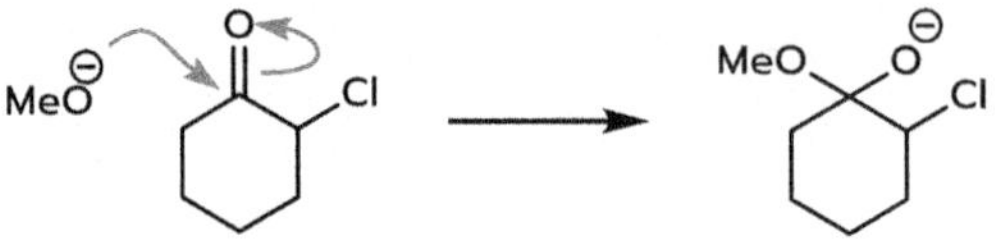

This is a carbonyl group doing what carbonyls do. Whatever else you do, don't draw incorrect curly arrows in the hope that you will get to the product.

It turns out that there is a rearrangement mechanism which will get us from here to the product.

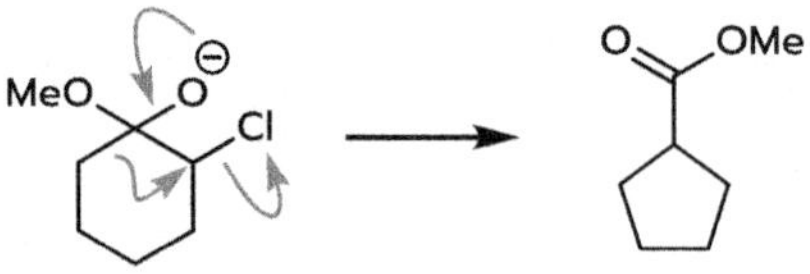

In fact, this looks a lot like a reaction we are going to see in **Reaction Detail 31**. The curly arrows look a bit 'different', and we will see what they mean. For now, bear with me and accept them.

> So, this process looks plausible, from the perspective that the curly arrows make sense, and it gets to the product.

There's a complication! If we label the carbon indicated with a '*' in the starting material with ^{14}C, then the label is equally distributed between the two '*' carbons in the product.

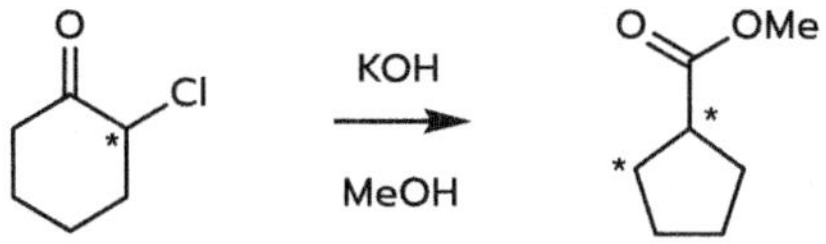

This isn't consistent with the mechanism above. All it takes is one piece of contradictory evidence, and we can disprove a mechanism.

SECTION 4 CARBONYL COMPOUNDS AS NUCLEOPHILES

> Draw out the mechanism above. Confirm for yourself where the ^{14}C label would end up.[92] It will only be on one carbon atom.

So, we are gradually working through the possible mechanisms that make chemical sense. We have two sites that can be attacked by a nucleophile, but neither of them gives the observed product. Perhaps we should consider deprotonation instead.

> *I think I've made enough of a point now, so I'm going to go to the 'right answer'. There is another site for deprotonation, but it doesn't get us anywhere.*

We can deprotonate α- to the carbonyl on the left. The resulting anion reacts in an S_N2 manner to kick out the chloride.

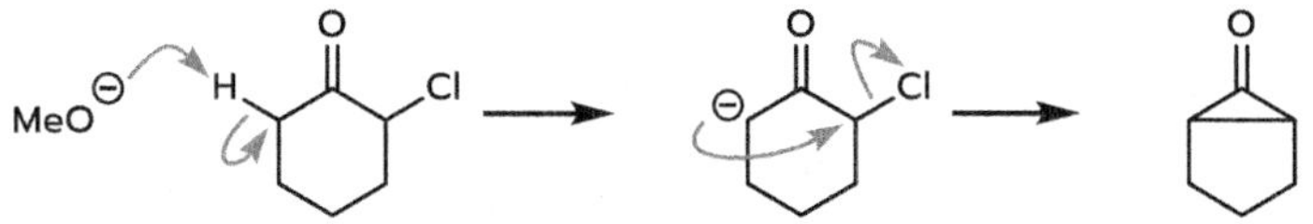

> We are alkylating an enolate (Reaction Detail 21).

This one looks a bit different, because it is intramolecular. It looks even stranger because we are forming a very strained three-membered ring (**Basics 6**). It is even worse, because the three-membered ring has an sp^2 carbon atom, with ideal bond angles of 120°, in it.

> *You can bet this is a reactive ketone!*

We will add a nucleophile, and methoxide is the logical choice. The carbonyl carbon atom goes from sp^2 to sp^3 hybridized. A cyclopropane has bond angles of 60°. The 'ideal' bond angle has been reduced from 120° to 109°, so that attack of the nucleophile has removed some of the strain.

However, there is still a lot of strain, which facilitates the ring-opening as shown.

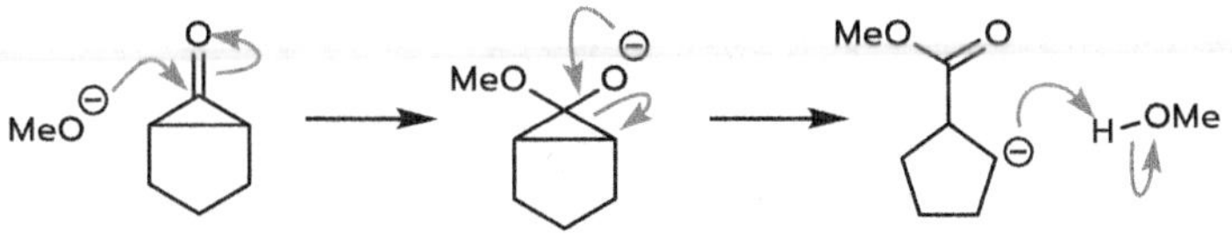

Now we can explain the labelling experiment. In the structure below, I have highlighted the labelled carbon atom, and two bonds we could break. Remember, the intermediate, shown again below, is symmetrical apart from the isotopic label!

---

[92] As a brief digression, we wouldn't use a ^{14}C label these days. We would use ^{13}C instead, as it is not radioactive, and we can work out where it ends up in the product using ^{13}C NMR spectroscopy. This study would be much easier now!

405

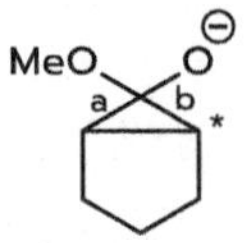

SECTION 4 CARBONYL COMPOUNDS AS NUCLEOPHILES

Draw a mechanism in which either of bond 'a' or bond 'b' is broken. Make sure you can see why this would give the distribution of label that I showed you earlier in the chapter.

ONE MORE THING!

There is usually one more thing! In this case, it is the following step.

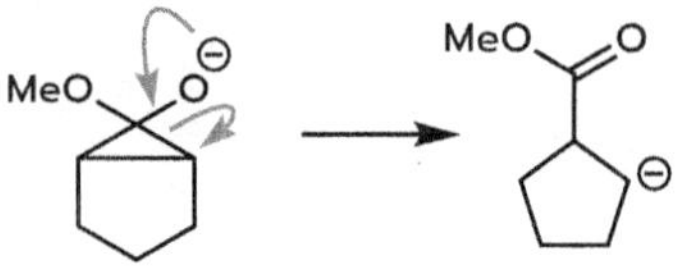

We are going from an alkoxide anion to a non-stabilized carbanion. The cyclopropane is strained, but not *that* strained. Energetically, I would say this is going uphill, by quite a way.

Go back and have another look at Applications 4. Towards the end of that chapter, we formed an unstable carbanion, and I suggested that the protonation may start before the carbanion is fully formed.

Perhaps that is the case here. We are working in a protic solvent (methanol). If there is a proton near the C–C bond as we are breaking it, perhaps we don't fully form the carbanion. We could draw it like this.

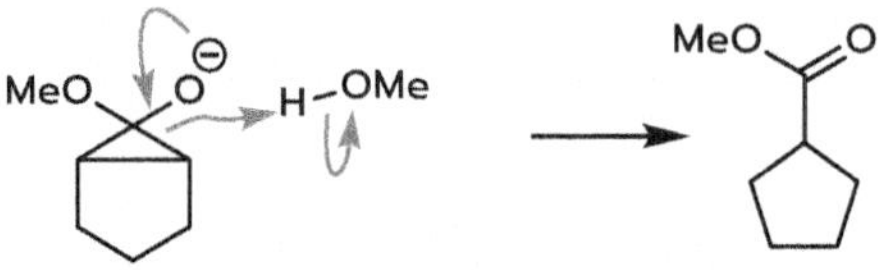

The point of this discussion is to ensure that you think about the energy of any structure you draw, rather than just drawing the curly arrows.

Of course, the curly arrows need to be correct, and the mechanism needs to fit all available experimental observations.

SECTION 5

REARRANGEMENT REACTIONS

SECTION 5 REARRANGEMENT REACTIONS

INTRODUCTION

The primary focus of this book is the 'typical' reactivity of alkenes, aromatics, alkynes and carbonyls.

There is another class of reactions that we have touched on briefly. During Friedel-Crafts alkylation reactions (**Reaction Detail 9**), we saw that the alkyl group can undergo rearrangement. In the last chapter (**Reaction Detail 30**) we saw a mechanism that also included a rearrangement (although this turned out not to be the correct mechanism).

Now it's time to look at a few rearrangement reactions in more detail.

SECTION 5 REARRANGEMENT REACTIONS

FUNDAMENTAL REACTION TYPE 5

Rearrangement Reactions of Carbocations

INTRODUCTION

In **Reaction Detail 9**, we saw that there are two particular problems with Friedel-Crafts alkylation reactions. We didn't draw the curly arrows at the time, but we established that carbocations can undergo rearrangement reactions.

Here is one of the examples we used.

Very soon, we will draw the curly arrows for this process.

> They are similar to the curly arrows we have drawn before. You would expect this, because curly arrows *always* represent the sharing of electrons.

But there's a subtle difference as well, which is why I've left this until now. First of all, we will look at the structure and molecular orbitals of a carbocation.

THE ETHYL CARBOCATION

In **Recap 2** we looked at the factors affecting carbocation stability. Everything we said there was correct, but it wasn't the whole truth. Here is the ethyl carbocation.

$$H_3C-\overset{\oplus}{C}H_2$$

The problem is, at least when we calculate the shape of this carbocation, it actually looks like this.

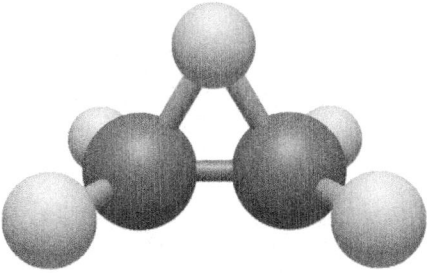

This structure is entirely symmetrical, with the hydrogen atom bridging the two carbon atoms. Don't worry about the fact that the hydrogen atom on top has two bonds. We have said before that the hybridization bonding models is a simplification.

In this case, we could draw resonance forms as follows.

409

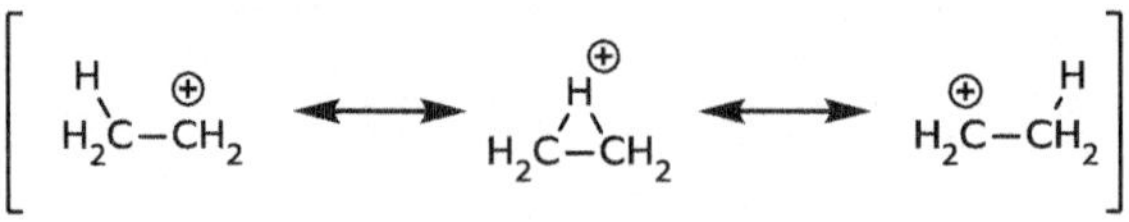

SECTION 5 REARRANGEMENT REACTIONS

$$\left[\begin{array}{ccc} \overset{H}{\underset{H_2C-CH_2}{\diagdown}}{}^{\oplus} & \longleftrightarrow & \overset{\overset{\oplus}{H}}{\underset{H_2C-CH_2}{\diagup\diagdown}} & \longleftrightarrow & \overset{\oplus}{\underset{H_2C-CH_2}{\diagup}}{}^{H} \end{array}\right]$$

The structure in the middle is what we would get initially if we protonated ethene.

Of course, if we had an unsymmetrical system such as the one below, then the resonance form on the right would be more significant than the one on the left.

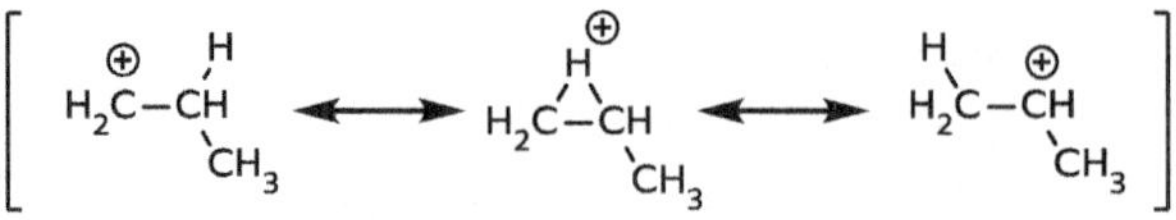

$$\left[\begin{array}{ccc} \overset{\oplus}{H_2C}-\underset{CH_3}{\overset{H}{\underset{|}{CH}}} & \longleftrightarrow & \overset{\overset{\oplus}{H}}{\underset{H_2C-CH}{\diagup\diagdown}}\underset{CH_3}{} & \longleftrightarrow & \overset{H}{\underset{H_2C-CH}{\diagdown}}{}^{\oplus}\underset{CH_3}{} \end{array}\right]$$

> That is, the resonance form on the right would be a more significant contributor to the 'real structure'.

We saw the same thing with bromonium ions in **Reaction Detail 1**. In that case, we invoked the bromine lone pair to explain the bonding. Here, we cannot do that—the only electrons we have are in the C–H bond.

> In fact, what we are now seeing is that the situation with bromonium ions is actually much more general, and does not actually require an atom with a lone pair. It just requires orbital overlap.

CARBOCATION REARRANGMENT

This leads us nicely on to the next type of reaction—rearrangement of a carbocation.

Let's define what we are talking about. We have looked at carbocation stability in **Recap 2**. We know that *primary* carbocations are not very stable. If we form the *primary* propyl carbocation, it will rearrange rapidly to give the *secondary* carbocation as follows. The middle resonance form drawn above would essentially be the transition state for this process.

$$\overset{\oplus}{H_2C}\cdot HC\overset{H}{\underset{CH_3}{\diagup}} \quad \underset{\longleftarrow}{\longrightarrow} \quad \overset{H}{\underset{H_2C-CH}{\diagdown}}{}^{\oplus}\underset{CH_3}{}$$

Now let's look at how this happens. After all, every reaction involves orbital overlap. Recall that carbocations are stabilized by hyperconjugation—overlap of an adjacent C–H bond with the empty p-orbital.

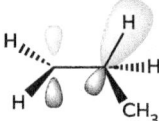

What we are seeing here is an extreme case of this orbital overlap. The hydrogen atom moves to the left, gets to the middle point (the transition state), and carries on.

> Draw an energy profile for this process. Draw all relevant structure on it.

SECTION 5 REARRANGEMENT REACTIONS

Now we are ready to draw the curly arrows. The orbital overlap involves the electrons in the C–H bond. This is where we need to start our curly arrow.

I have put a little curve in the curly arrow, so that it is clear that the hydrogen atom is migrating **with** its electrons.

> We call this a hydride shift. We will see a different type of hydride shift in **Reaction Detail 36**.

Compare this to the following.

This curly arrow looks like loss of a proton, so we might expect the product to be an alkene. In the migration reaction above, the product structure tells you what must be happening.

> Once again, this is a case where you internalize the concept, so that if you see it drawn in a less-than-perfect manner, you still understand what is going on. You will be able to dismiss any alternative reactions as being much less likely.

Now let's broaden the scope of these reactions slightly, and make the link with **Reaction Detail 9**. Consider the formation of the following carbocation. We aren't going to worry too much about how you would make it at this stage.

This carbocation is not very stable. This means that it will have a high barrier to its formation (the Hammond Postulate, **Recap 2**, yet again!).

This doesn't necessarily mean that it cannot be formed—just that it would be formed slowly (see **HTSIOC Basics 15**).

Because it is unstable, if it is formed, it will react quickly.

> The key question is 'what will it do?'

You could, in principle, add a nucleophile. This would make it an overall S_N1 substitution reaction (**HTSIOC Reaction Detail 1**). But the reality is, this will not happen with a *primary* alkyl halide. For this to happen, you would need the unstable *primary* carbocation to 'hang around' until it 'meets' a nucleophile. Instead, the carbocation will rearrange quickly.

SECTION 5 REARRANGEMENT REACTIONS

> Rearrangement is a unimolecular process with a low barrier. Attack by a nucleophile is bimolecular. The nucleophile must get to the carbocation before it can do anything else. This is unlikely.

Here is the rearrangement. The product is a *tertiary* carbocation, which is much more stable, so this is energetically favoured.

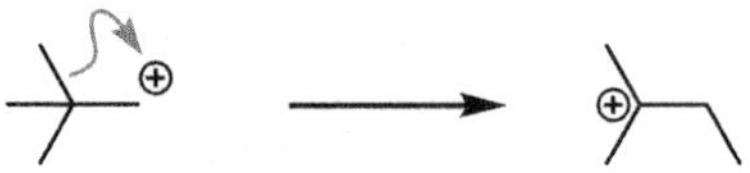

The curly arrow is the same as we used above. A methyl group can migrate in the same way as a hydrogen atom.

> We will eventually need to consider which group migrates when we have more than one possibility.

Now let's look again at the example from **Reaction Detail 9**. This starts with a hydride shift. There is no doubt that the *secondary* carbocation formed would be more stable than the *primary* carbocation drawn.

Remember, in **Reaction Detail 9**, we considered the possibility that we didn't actually form a *primary* carbocation, but the hydride could migrate at the same time as the chloride leaves.

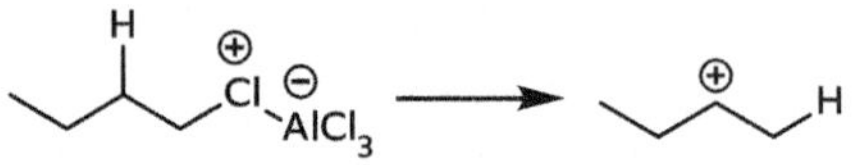

> Add curly arrows to the above two reactions. Can you see how these mechanisms parallel those for S_N1 and S_N2 substitution?

In **Reaction Detail 9**, we saw that when we react benzene with 1-chlorobutane, we get more of compound **2** than we get of compounds **1** or **3**.

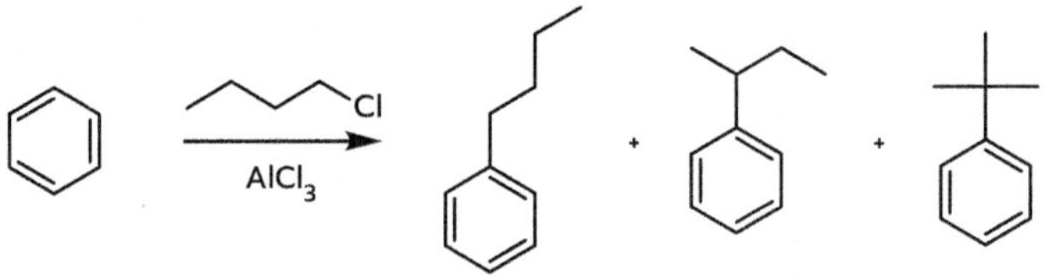

We saw that compound **2** is formed from the rearranged carbocation above. Compound **3** must be formed from the *tert*-butyl carbocation.

SECTION 5 REARRANGEMENT REACTIONS

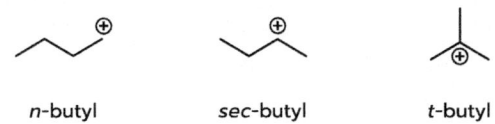

n-butyl   sec-butyl   t-butyl

> Draw curly arrows for the rearrangement of sec-butyl to t-butyl on the reaction scheme below. Draw an energy profile for this rearrangement.

> Can you now see why the major product is formed from the sec-butyl carbocation even though the t-butyl carbocation is more stable?

WHAT NEXT?

Any time we can form an unstable carbocation, it could rearrange to form a more stable carbocation. We saw this happen in **Reaction Detail 9**. Rearrangement reactions have been widely used in some very elegant total syntheses—we will see one example in **Total Synthesis 5**. They can also limit the potential of some useful reactions, as we have already seen.

> But then you know the problem, so you would take account of this when planning your synthesis.

SECTION 5 REARRANGEMENT REACTIONS

REACTION DETAIL 31

Carbocation Rearrangements

PINACOL REARRANGEMENTS

In **Fundamental Reaction Type 5** we defined carbocation rearrangement. Here is a very specific carbocation rearrangement. First of all, let's ask an important question.

> *What's the most stable carbocation you can think of?*

When we look at the examples in **Recap 2**, we see that a heteroatom with a lone pair provides excellent stabilization to a carbocation. This is important. If we want to use carbocation rearrangements, we need to identify the most stable carbocations.

Here are the resonance forms for a protonated carbonyl group.

If only we could carry out a rearrangement to give resonance form **B**, it would work really well!!!! We would have a thermodynamically-driven process which will ultimately give us a carbonyl group. Of course, it has been done, and it works very well. Let's explore this reaction, and see what we can learn.

The prototype pinacol rearrangement is shown below. The name of this reaction comes from the trivial name of the starting material, pinacol. The trivial name of the product is pinacolone.

I will break it down, so that you can focus on the actual rearrangement step. The first step is formation of a carbocation from the alcohol. As a result of protonation, the oxygen atom is a better leaving group.

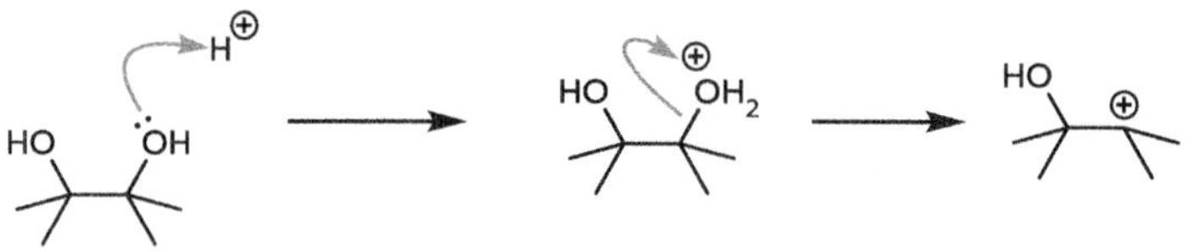

We have formed a *tertiary* alkyl carbocation, which is pretty stable. But it's not as stable as the carbocation formed in the rearrangement reaction shown below.

Of course, as we have seen above, the carbocation drawn above-right is not the best representation. Here are the resonance forms. It's a protonated carbonyl group.

SECTION 5 REARRANGEMENT REACTIONS

Here is the overall reaction. I've drawn the rearrangement to give the 'better' resonance form immediately.

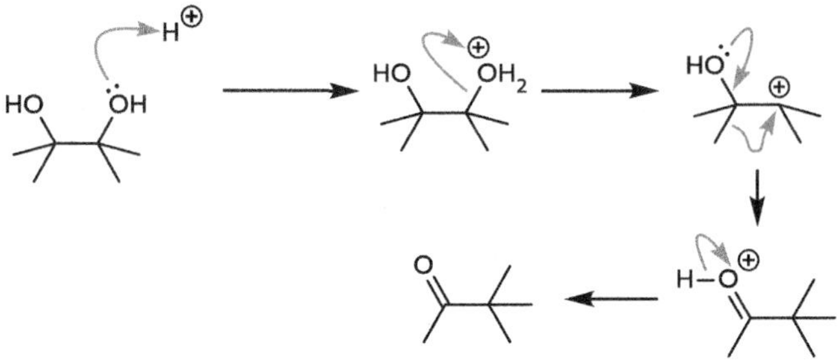

> Another way to express this is that the oxygen lone pair is 'pushing' the methyl group across to the other carbon atom.

Don't lose sight of the bigger picture. We have a stable carbocation formed (*tertiary*), but it can rearrange to give an even more stable carbocation.

> There is a very good reason why this happens!

TERMINOLOGY

The rearrangement reactions we are looking at here are migrations of alkyl groups from one carbon atom to an adjacent carbon atom. They are '1,2-shifts'.

Much of the pioneering work on 1,2-alkyl shifts was carried out by Wagner and Meerwein. They are therefore also known as Meerwein-Wagner shifts.

MOLECULAR ORBITAL INTERACTIONS

Recall (**Recap 2**) that a carbocation is stabilized by hyperconjugation. Normally this refers to the overlap of the C–H bond orbital with the empty p-orbital.

The C–C bonding orbital to the migrating group can overlap with the empty p orbital as shown below. This leads to migration of the methyl group.

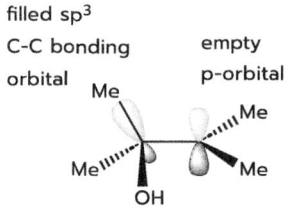

In this case, we have two methyl groups that could migrate, and a bond that can rotate freely. Either methyl group could migrate, although it would be impossible in this case to tell the difference.

415

SECTION 5 REARRANGEMENT REACTIONS

Of course, if the bonds cannot freely rotate, it is possible that you could have a compound in which only one group could migrate.[93] We will look at an example of this in Total Synthesis 5.

But what if we have a simpler case, in which we have two different groups that could migrate.

MIGRATORY APTITUDE

Once we form a carbocation, and a number of groups could migrate, we have to be able to decide which will actually migrate.

For simple alkyl groups, the trend is as follows:

$$Me_3C > Me_2CH > MeCH_2 > Me$$

You will probably have realized that this is exactly the trend for stability of the corresponding carbocation (Recap 2). This isn't surprising really. If the carbon atom is able to stabilize a positive charge, we will get better orbital overlap as shown above.

> *A more colloquial way to put this is that the group is better able to share those electrons.*

PHENYL

It turns out that a phenyl group migrates even more readily than a *t*-butyl group.

$$\text{Ph}^+ \text{ is } \textbf{not} \text{ more stable than } (CH_3)_3C^+$$

When we see a trend with an exception, we have to wonder what else might be happening. We saw that for the group to start migrating, we need overlap of a filled orbital with the carbocation p orbital. This is usually a C–C σ-bond orbital as we saw above.

> But it doesn't have to be!

To explain this apparent exception, we need to ask a fundamental question—which orbitals?

> *It's a benzene ring. Take a wild guess!*

Here it is, just with the curly arrows. The benzene ring uses its π electrons to bridge the two carbon atoms. This makes the curly arrows for the next step look a bit complicated. Take your time and make sure you can follow them.

---

[93] Are you thinking about cyclic compounds? Tell me you're thinking about cyclic compounds!

416

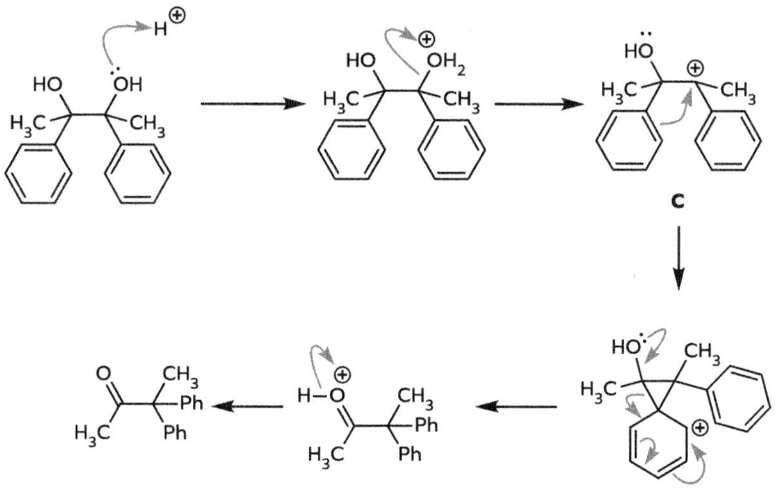

SECTION 5 REARRANGEMENT REACTIONS

Oh, and I switched to 'Ph' towards the end of the mechanism, partly because the structures were getting a bit crowded, and partly because I wanted to remind you what Ph was! Doesn't this mechanism look a little bit like aromatic electrophilic substitution (Reaction Detail 8)?

IN CONCLUSION

We have barely scratched the surface of carbocation rearrangements. They are possible whenever a carbocation is formed. All we would need is the potential to form a more stable carbocation by migration of one of the groups from an adjacent atom.

They happen for a good reason. A more stable carbocation will not rearrange to give a less stable carbocation.

As is often the case, I'm going to add a bit of added detail and to qualify the statement above. I want to ensure there are no contradictions, and we make all the necessary connections.

In **Fundamental Reaction Type 3**, we found that we would expect a carbocation such as **D** (in the scheme above) to be of broadly similar stability to a *tertiary* alkyl carbocation. This carbocation also has a strained three-membered ring (**Basics 6**). Carbocation **D** is going to be a bit less stable than carbocation **C**.

But once we get to carbocation **D**, it can rearrange rapidly to give a *much* more stable product. It isn't quite a Curtin-Hammett situation (**Basics 8**) but you can use some of the same reasoning.

417

SECTION 5 REARRANGEMENT REACTIONS

REACTION DETAIL 32

Rearrangement to Electron-Deficient Nitrogen and Oxygen

INTRODUCTION

In **Fundamental Reaction Type 5** we encountered rearrangement reactions to electron-deficient carbon atoms, specifically of carbocations. Rearrangement reactions are really useful, as they can (sometimes dramatically) change the connectivity within a molecule.

It turns out that we can also carry out rearrangements to other atoms, including oxygen and nitrogen. We are going to start with oxygen, because it's easier to draw parallels with carbocation rearrangements.

THE BAEYER-VILLIGER OXIDATION

Here is yet another named reaction. The overall process is an oxidation which inserts an oxygen atom into a C–C bond adjacent to a ketone.

> *When you put it like this, it isn't at all obvious that the mechanism is a rearrangement.*

> Work out the oxidation state (**Basics 3**) of carbon atoms a and b in the reaction above.

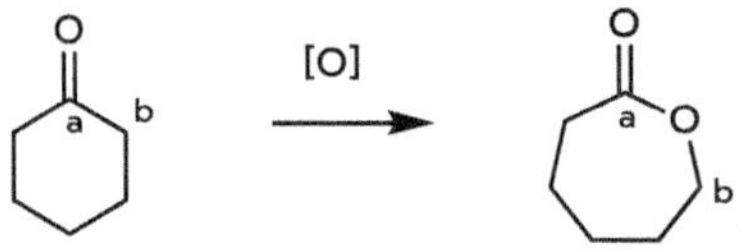

A peracid oxidant is often used. We are using *m*-CPBA, which we saw being used for the epoxidation of alkenes in **Reaction Detail 6**. Here is the structure of *m*-CPBA again. Remember that the driving force in the epoxidation reaction was breaking of the weak O–O bond.

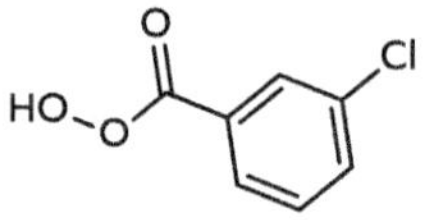

> *It's the same driving force in the Baeyer-Villiger oxidation, even though it is reacting with a different functional group.*

A typical mechanism is shown below. We protonate the carbonyl group (**Basics 2**) and *m*-CPBA reacts as a nucleophile.

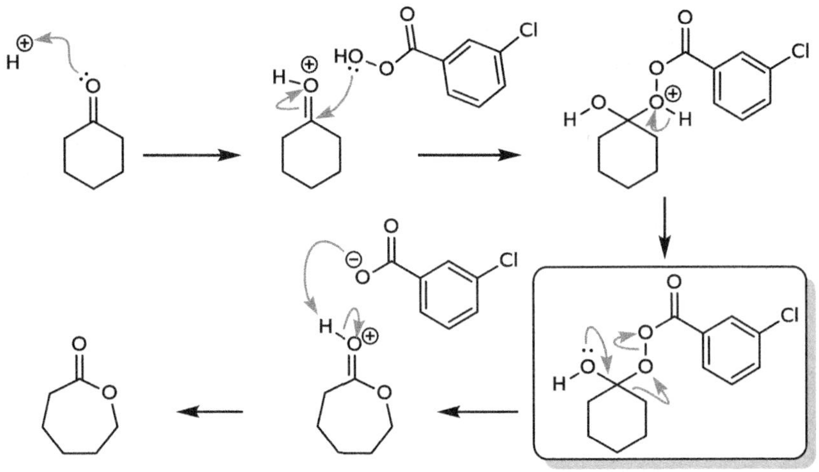

SECTION 5 REARRANGEMENT REACTIONS

The key migration step is shown in the box. Compare this to the Beckmann rearrangement later in this chapter, and with the pinacol rearrangement in **Reaction Detail 31**. I want you to see that they are basically the same.

> We have a leaving group on the migration terminus, and a group that can migrate to give a relatively stable carbocation.

Perhaps you can also see a similarity with the Friedel-Crafts reaction in **Reaction Detail 9/Fundamental Reaction Type 5**. In that case we argued that the Cl leaving and rearrangement probably happen at the same time, because we don't want to form a *primary* carbocation. Here, we **could** break the O–O bond to form an O$^{\oplus}$, but it wouldn't be very stable.

I think this is quite a challenging mechanism.

> Draw it out a couple of times. Draw the pinacol rearrangement (**Reaction Detail 31**) alongside it, and convince yourself that they are basically the same thing.

Of course, when you look at the starting materials and reagents, they look very different, which is part of the problem.

THE SCHMIDT REARRANGEMENT

In fact, there are two Schmidt rearrangements. The other one uses a carboxylic acid. This one uses a ketone, and it is similar to the Baeyer-Villiger oxidation.

> In fact, the mechanism is slightly different—we will get to that in a moment.

We are going to use hydrazoic acid, HN$_3$, as the reagent. Since this is an acid, we can safely protonate the ketone carbonyl group.

> This is **Basics 2** again!

We will then add hydrazoic acid as a nucleophile.

419

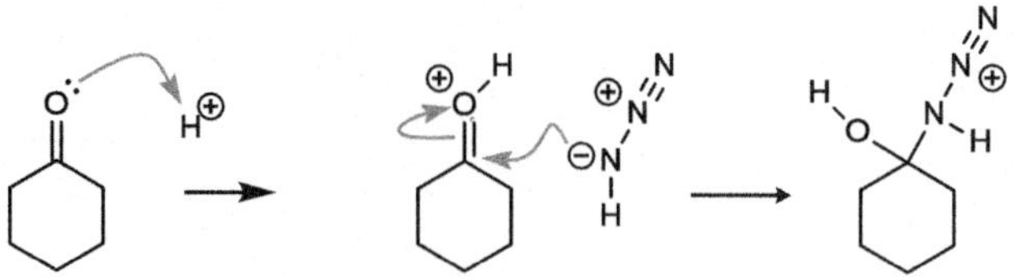

SECTION 5 REARRANGEMENT REACTIONS

> Have a look at how I have drawn hydrazoic acid, HN$_3$. You can't draw a structure where each nitrogen atom has three bonds and no charge.

From this point, there are two possible mechanistic pathways, and it appears that both pathways can operate, depending on the reaction conditions. At this level, all we are going to do is draw the mechanisms and try to consolidate our understanding of what is reasonable.

Here is the first mechanism. It is essentially a 'pinacol rearrangement to nitrogen' (or a Bayer-Villiger oxidation with nitrogen).

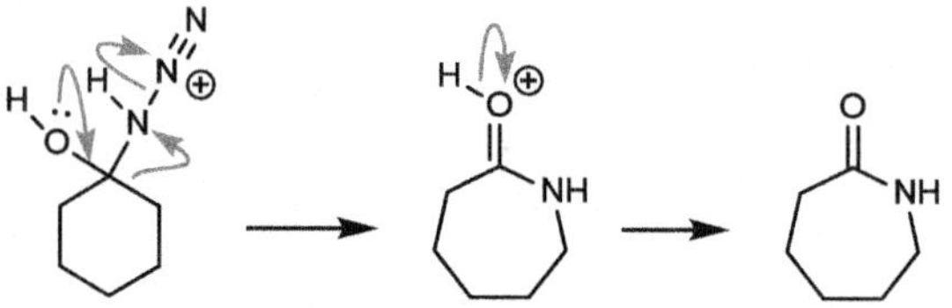

> Draw this mechanism out a couple of times, alongside the corresponding mechanisms for the pinacol and Baeyer-Villiger reactions.

You might think this is tedious, but it is a lot more effective than thinking of these as three entirely different reactions that you simply have to learn.

> We have an atom (C, N, O) with a good leaving group. We have a lone pair on an oxygen atom which is ready to 'help' the alkyl group migrate.

There's just one complication in this case. It turns out that while this mechanism operates under some reaction conditions, there is another mechanism that is actually more common. The preferred pathway involves an elimination reaction to form an imine (**Applications 2**). Sure, it has a very unusual substituent on nitrogen, but it is still an imine of sorts—there are more similarities than differences!

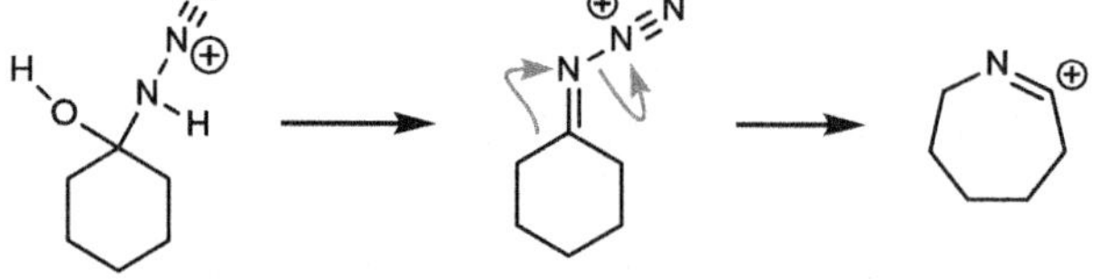

We then get migration from an sp^2 carbon to an sp^2 nitrogen.

> This is the rearrangement part, but it isn't quite the end. You knew that, right? You saw that the product was a carbocation, and you recognized that this isn't the thing you are going to put in a bottle.

Let's see how we finish this off. We have a carbocation. It is electrophilic, and there is water present.[94] Perhaps water could act as a nucleophile. Here is the overall mechanism. After attack of water, followed by loss of a proton, we get something that looks remarkably like an enol, apart from the fact that it has a nitrogen atom. It then undergoes a tautomerization reaction. The mechanism isn't quite the same as keto-enol tautomerization (**Fundamental Reaction Type 4**), but only because the nitrogen atom has a lone pair, so this is the site of protonation.

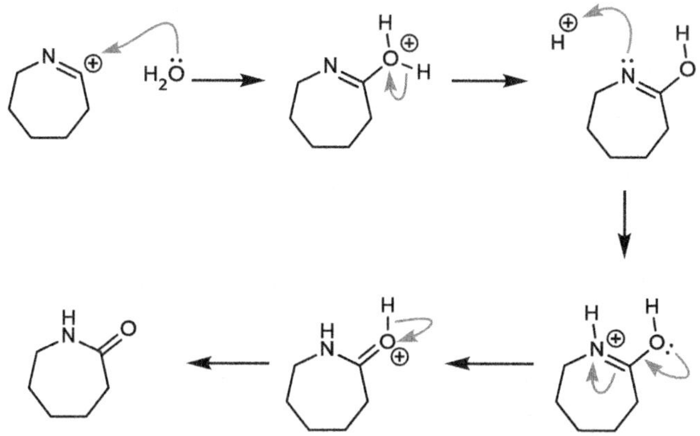

> The product of this reaction is caprolactam, used in the synthesis of nylon-6. It is a very useful compound, more than 4 million tonnes of which is produced annually. However, most of this is made using the Beckmann rearrangement, which is the next rearrangement reaction we will cover.

A BRIEF INTERLUDE

It is so easy to get bogged down in learning reaction mechanisms, and you can lose sight of the bigger picture. Caprolactam is a very important industrial chemical. We can make it from cyclohexanone using a rearrangement reaction that starts with a carbonyl addition reaction.

We can make cyclohexanone from phenol, using chemistry similar to that in **Reaction Detail 7**. However, most cyclohexanone is made by oxidation of cyclohexane (we won't cover that reaction in this book), which is made by reduction of benzene (**Reaction Detail 7** again).

Throughout this chapter, there are many links to previous chapters/reaction mechanisms. Although the rearrangement is new, how we get there is not.

THE BECKMANN REARRANGMENT

The Beckmann rearrangement is the rearrangement of an oxime under strongly acidic conditions. Formation of an oxime is shown below. This is a lot like an imine, so look back at **Applications 2**. An acid catalyst is used.

---

[94] There must be water present. It was eliminated to form the imine above!

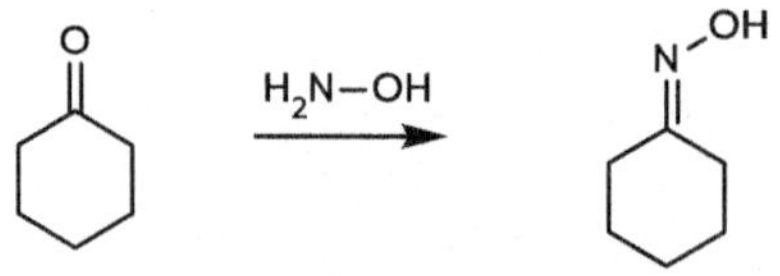

SECTION 5 REARRANGEMENT REACTIONS

cyclohexanone + H₂N–OH → cyclohexanone oxime (N–OH)

> Draw the mechanism of oxime formation. Pay careful attention to proton transfer steps.

> *You may feel that you know how to draw an imine formation, so you don't have much to gain by drawing the same mechanism again. You would be wrong!*

With more acid, the oxime rearranges as shown below. The oxime oxygen atom is protonated. Then, one of the alkyl groups migrates as the water leaves.

> The only different between the Schmidt rearrangement (second mechanism) and the Beckmann rearrangement is the leaving group used.

This gives us the same carbocation intermediate as we saw in the Schmidt rearrangement. The hydrolysis step has already been drawn above.

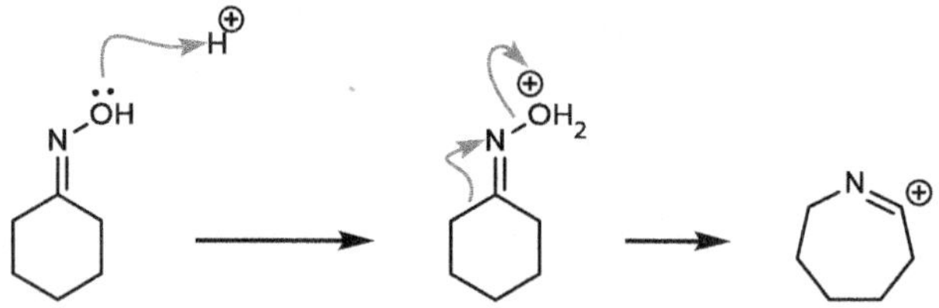

oxime + H⁺ → protonated oxime (N–OH₂⁺) → seven-membered ring cation

So, why is the Beckmann rearrangement used industrially instead of the Schmidt rearrangement? Why would industry prefer to use the (relatively) safe reagent 'hydroxylamine' rather than the highly explosive hydrazoic acid on a scale of many tonnes?

> When it's put like that, I'd like to think the answer is rather obvious.

We are going to look at one more aspect of these rearrangement reactions, and then we have a couple more reactions to consider.

REGIOSELECTIVITY IN REARRANGEMENT REACTIONS

In the Baeyer-Villiger reaction, the regioselectivity depends on the migratory aptitudes of the migrating groups, and is the same as the pinacol rearrangement (Reaction Detail 31). This can be seen in the following two examples.

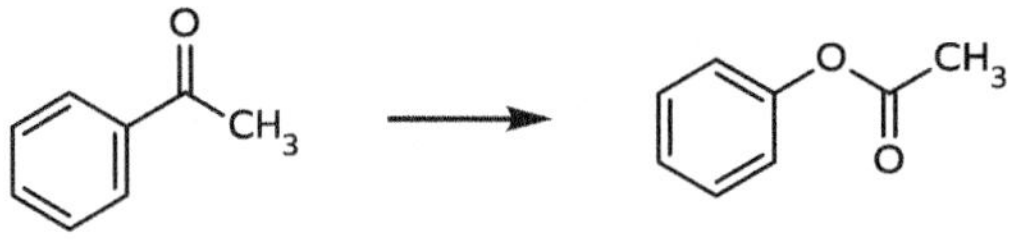

SECTION 5 REARRANGEMENT REACTIONS

aryl migrates

more substituted alkyl migrates

> Go back to Reaction Detail 31 and check that you understand why some groups migrate better than others. These two examples cover two distinct scenarios. If you just relate this to carbocation stability, you will then start thinking that a phenyl cation is particularly stable (which it is not!).

The outcome of the Beckmann rearrangement gives us direct insight into the mechanism. Oximes are double bonded functional groups. They can exist as (E) and (Z) isomers.

(E)-acetophenone oxime   (Z)-acetophenone oxime

> Did you expect this? Did you think that the two isomers would be interconverted? If so, what would the mechanism be? Would the mechanism be feasible, in terms of having stable intermediates/low activation barriers?

When these two isomers are subjected to the Beckmann rearrangement, each one gives a different product as shown below.

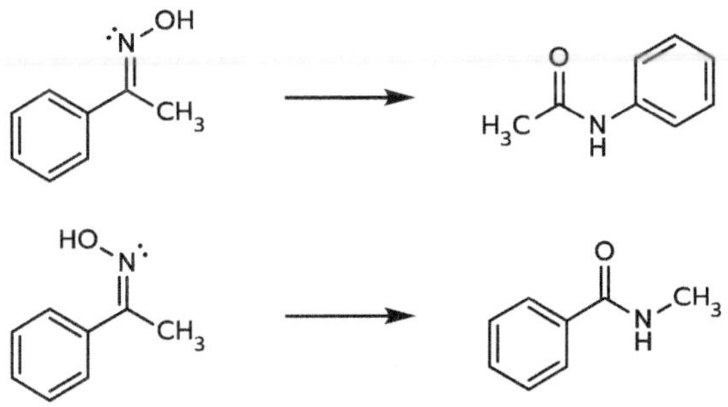

Clearly, this outcome cannot be attributed to migratory aptitude. If it was, then the same group would migrate in both cases. The group *anti* to the water leaving-group always migrates.

> This is how we know that the migration and loss of water both take place in the same step.

SECTION 5 REARRANGEMENT REACTIONS

If water was lost first, we would get the same intermediate, and hence the same outcome, in each case.

> I've given you enough information to draw the mechanisms in both cases. Take the opportunity to draw them out fully. Make sure you understand what *anti* means. The practice is where you will learn.

ANOTHER INTERLUDE

Before we look at the next group of rearrangement reactions, let's start with a transformation and a short philosophical aside. Here is a transformation in which a carboxylic amide is converted into an aliphatic amine with one fewer carbon atom.

$$R-C(=O)-NH_2 \longrightarrow R-NH_2$$

It turns out that there are far more organic compounds with an even number of carbon atoms than an odd number of carbon atoms. There are a number of reasons why this is the case, but the most relevant one, when considering simple carboxylic acid derivatives is that the biosynthesis of these compounds occurs by joining two-carbon (acetate) units together. Up to a point, the biosynthesis uses the Claisen condensation which we encountered in **Reaction Detail 23**. The fatty acids in your body all have an even number of carbon atoms.

> Therefore, a transformation that allows you to add or (as in this case) remove one carbon atom is very useful. You can use it to make things that are less available.

Of course, while some transformations are used much more than others, every synthetic transformation is useful, if it happens to be the reaction that will allow you to make the compound you need to make!

THE HOFMANN, CURTIUS, LOSSEN AND SCHMIDT REACTIONS

I'm not quite sure why, but I found these reactions difficult to learn. There are four related reactions, and it took me a long time to remember which was which.

> The most important thing is that up to a point, it doesn't matter!

When four reactions are variations on a theme, the most important thing is to understand what they all do, and how they all do it. Once you've been looking at them for a while, you'll find that you do remember them.

In each case, they key intermediate is formally an anionic nitrogen species. The nitrogen atom also has a potential leaving-group, which we will simply refer to as "X" for now.

The mechanism can be drawn in two different ways. We are going to look at the **nitrene** mechanism first. A nitrene is a reactive intermediate which has a neutral

SECTION 5 REARRANGEMENT REACTIONS

nitrogen atom with only one bond. The nitrene nitrogen atom is electron-deficient as it only has six outer electrons.

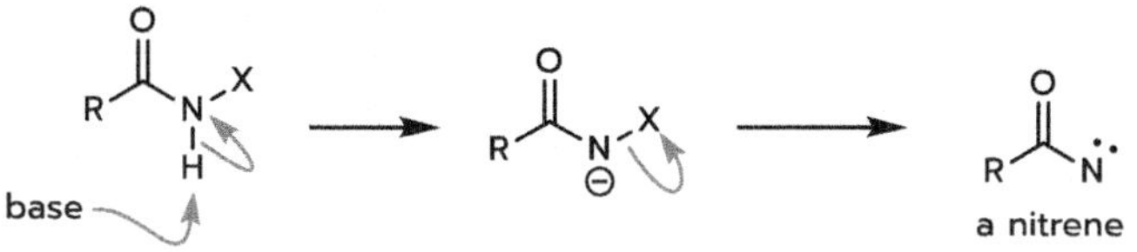

Now we can look at the rearrangement mechanism, and the product. If we migrate the "R" group to the nitrogen atom, this will leave the carbonyl carbon with a positive charge and the nitrogen atom with a negative charge. We can get around this by sharing the non-bonded pair of electrons on the nitrogen atom with the carbon atom.

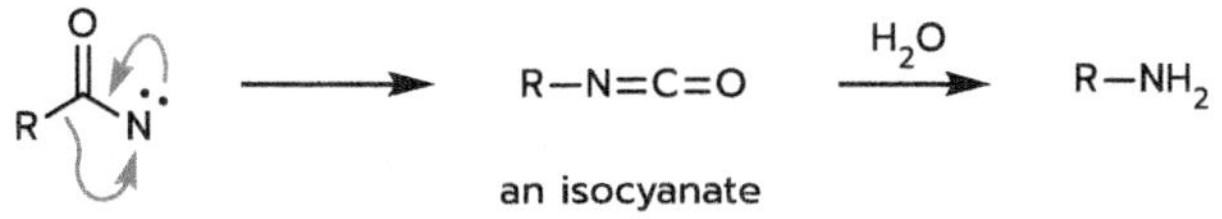

> Draw this mechanism for yourself, with only the curly arrow for migration. Check that you can see why the above statements are correct.

Have another look at the pinacol rearrangement reaction in **Reaction Detail 31**. We have a heteroatom lone pair 'kicking in' and pushing the migrating group 'away'. The difference here is that the migrating group is going to the atom with the lone pair.

The initial product of the reaction is an **isocyanate**. This functional group looks unusual at first. The carbon atom in the middle is sp hybridized. Have another look at **Recap 1** and make sure you understand why this is.

> *Isocyanates might look a little strange, but they are very important industrially. The old name for an isocyanate is a urethane. You've heard of polymers called polyurethanes. Guess what they are made from!*

This will probably be the first time you have encountered a nitrene, and they seem quite unusual. In fact, they are well-established intermediates. They undergo some rather useful reactions.

> *What this means it that if we add certain compounds to the above reaction, we might get interception of the nitrene.*

It turns out that we don't! No study has been able to prove that a nitrene is actually an intermediate in any of these reactions. Instead, we can draw the reaction with simultaneous loss of X and migration of R. The curly arrows look quite confusing at first.

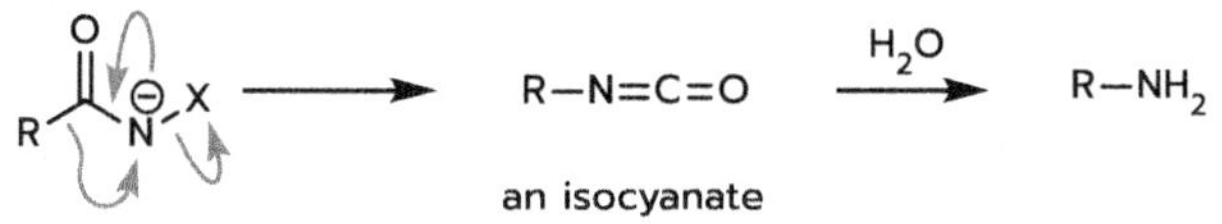

425

SECTION 5 REARRANGEMENT REACTIONS

> *That's why I drew the nitrene mechanism first, so you could get used to it. Now focus on the additional curly arrow, and check that it makes sense to you.*

In both schemes above, there is a second step—hydrolysis and loss of a carbon atom. I'm not going to worry about a "detailed correct" mechanism for this process. I'd rather focus on a "sensible plausible" mechanism. In many cases the detailed mechanisms of reactions have not actually been investigated, so this is all you will get from the chemical literature anyway!

Let's start with protonation of the nitrogen atom. There is a lone pair of electrons here, and the product has two hydrogen atoms attached to this nitrogen atom, so this is sensible. The reaction is done in (or in the presence of) water, so there will be protons!

> *Perhaps protonation takes place on oxygen, and then the nitrogen atom is protonated/oxygen deprotonated later. It doesn't really matter!*

Water can then attack the isocyanate carbon atom as shown. This is very much like a carbonyl group, so we are back in **Basics 2** territory. Now, we need a couple of proton transfer steps followed by decarboxylation. The mechanism isn't quite the same as in **Reaction Detail 24**, but the driving force (entropy) is the same.

The mechanisms above cover all four reactions in the heading for this section. So, what is the difference between these four reactions?

The only real difference is the nature of "X" and how it is prepared.

The Hofmann rearrangement uses an N-bromoamide (X = Br). The Lossen rearrangement uses an N-acetoxyamide. We aren't going to worry about how these are prepared.

The Curtius and Schmidt reactions both use acid azides as substrates. Have a look at the scheme below, which summarizes all these reactions.

> *This is the second Schmidt rearrangement, which is done at the carboxylic acid oxidation state. Apart from that it's the same as the one we just looked at.*

SECTION 5 REARRANGEMENT REACTIONS

[Scheme showing Hofmann rearrangement (from R-C(=O)-NH-Br) and Lossen rearrangement (from R-C(=O)-NH-O-C(=O)-CH₃) both leading to R-NH₂, with Curtius and Schmidt rearrangements from acyl azide resonance structures.]

When we draw a resonance form for the azide, we find that we do not need to deprotonate to give the negative charge. Additionally, we have a stunning leaving group attached to the left-hand nitrogen atom—nitrogen gas![95]

In the case of the Curtius rearrangement, the acid azide is generally prepared by reaction of an acid chloride with sodium azide. The mechanism is similar to amide formation from an acid chloride (**Reaction Detail 16**). The Schmidt reaction makes the same intermediate from a carboxylic acid and hydrazoic acid (HN₃). Again, we have seen similar mechanisms for ester formation from a carboxylic acid and an alcohol (**Reaction Detail 16** again).

> Have a go at drawing these mechanisms![96]

There's just one more point to make. With the Curtius rearrangement, we tend to isolate the acid azide. We don't need a base to initiate the rearrangement. All we need to do is heat the acid azide. Therefore, it is easier to isolate the isocyanate if that is what we want.

CARBENES—YET ANOTHER REARRANGEMENT

We can always add more reactions. I'm trying to be careful here, and to not add reactions without good reason.

In **Section 7**, we will look at examples from natural product synthesis.

> You will see what an amazing range of compounds we can make using the reactions in this book.

We also don't want to leave too many 'mechanistic gaps'. If we add one more reaction that slots in nicely, the natural coherence of the subject is emphasized.

---

[95] Entropy again!!!

[96] As always, there is a point to not giving you too much detail. It makes you work for it!

SECTION 5 REARRANGEMENT REACTIONS

The final reaction in this chapter falls into both categories. It is definitely useful, but the mechanism fits nicely with the previous group of reactions. The reaction we are going to talk about is the Wolff rearrangement. It is used most commonly in the Arndt-Eistert homologation of carboxylic acids.

> That's three more names to remember. But as before, your lecturers are **very** unlikely to ask you to draw these reactions by name. They are much more likely to give you starting materials, reagents and conditions.

In the last section, we saw the Curtius rearrangement, and we initially drew this with a nitrene intermediate. It turns out that there is a corresponding intermediate called a carbene, in which a carbon atom only has two bonds and no charge. Here is a typical reaction scheme, starting from the carbene. We will worry about how we form this in a moment.

The product is called a **ketene**. It's like an isocyanate, but with a carbon atom instead of nitrogen. It might look a bit strange, but that's only because you're not used to seeing it. As before, the atom in the middle (now carbon) is sp hybridized (**Recap 1**).

Ketenes undergo a range of useful reactions, but all we are going to look at is addition of water.

> You saw the reaction of an isocyanate with water in the previous section. Draw the corresponding mechanism for reaction of a ketene. Here, you will need to protonate on oxygen. Make sure you can see why this one doesn't undergo decarboxylation.

The Wolff rearrangement and the Wolff-Kishner reduction (**Applications 4**) are the same Wolff.

Now let's see how we get to the carbene. We can start with a carboxylic acid, and convert it into the acid chloride using thionyl chloride (**Reaction Detail 17**). We then react the acid chloride with diazomethane.

Remember, our focus here is on building your skills. We can draw resonance forms of diazomethane as follows.

SECTION 5 REARRANGEMENT REACTIONS

$$\left[H_2C=\overset{\oplus}{N}=\overset{\ominus}{N} \quad \longleftrightarrow \quad H_2\overset{\ominus}{C}-\overset{\oplus}{N}\equiv N \right]$$

> Using the right-hand resonance form, draw a curly arrow mechanism for the reaction of diazomethane with the acid chloride above.

Next up, we can draw the same resonance forms for the acyldiazo compound.

$$\left[\begin{array}{c} R-C(=O)-CH-\overset{\oplus}{N}=\overset{\ominus}{N} \end{array} \quad \longleftrightarrow \quad \begin{array}{c} R-C(=O)-\overset{\ominus}{C}H-\overset{\oplus}{N}\equiv N \end{array} \right]$$

> Add the curly arrows to this scheme.

Now we just need one curly arrow from the right-hand resonance form to show loss of nitrogen gas to form the carbene.

> Draw it. Look back at the previous section if you need help.

Now let's add a little detail. With the previous rearrangements, we saw that the nitrene intermediate wasn't actually involved, and instead we had a concerted mechanism where the migration of the "R" group and loss of leaving group happened at the same time. It turns out that there is also a concerted mechanism for this rearrangement.

In fact, the stepwise mechanism involving a carbene operates with some compounds and under some reaction conditions. In other cases, the concerted mechanism operates.

You've seen the carbene rearrangement mechanism above.

> Now draw the concerted mechanism. Look back at the Hofmann/Curtius/Lossen/Schmidt rearrangement mechanisms if you need a hint.

Now let's look at the overall transformation. We start with a carboxylic acid. We end up with a carboxylic acid, but with one more carbon atom. We have already seen that compounds with an even number of carbon atoms are more readily available than those with an odd number of carbon atoms.

$$R-C(=O)-OH \quad \longrightarrow \quad R-CH_2-C(=O)-OH$$

A one-carbon homologation allows us to make something that we might not be able to simply buy.

It's a useful transformation.

SECTION 5 REARRANGEMENT REACTIONS

REACTION DETAIL 33

Borate Rearrangements

Here is a type of rearrangement reaction that isn't generally covered in undergraduate chemistry courses.

> I don't understand why this is!

FINISHING OFF HYDROBORATION

In Reaction Detail 5, we saw the following last step in the hydroboration of an alkene.

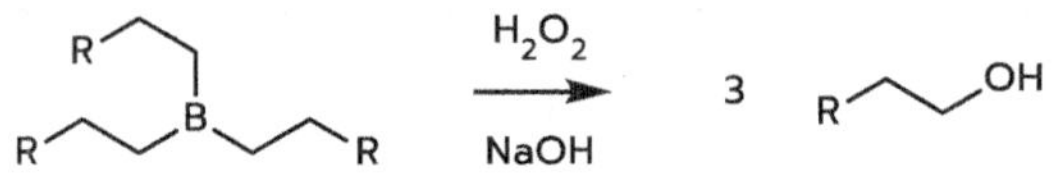

It would be easy to just say this is an oxidation of the borane, and hydrogen peroxide is an oxidizing agent.

> Both statements are, of course, true.

But every reaction needs to have a mechanism. It's time to look at this one.

We have the trialkylborane, hydrogen peroxide and hydroxide.

> I didn't say 'sodium hydroxide'. We expect the sodium ion to coordinate to an oxygen atom at some point, but it isn't usually critical when we draw the mechanism.

Recall that the defining characteristic of any borane is that it is electron-deficient. It can accept two electrons, as it did from the alkene in the original hydroboration reaction. We could draw the following reaction in which the borane accepts a pair of electrons from hydroxide.

We get a trialkylborate, but ultimately this doesn't get us anywhere.

I still think it's important to mention this, though. When you only see the correct mechanism drawn, it is perfectly reasonable to wonder why 'something else' doesn't happen instead. In this case, it almost certainly does, but this would be better drawn as an equilibrium, and there is a more productive equilibrium.

SECTION 5 REARRANGEMENT REACTIONS

The equilibrium above favours the left-hand side. Hydrogen peroxide has a pK_a of 11.6. The hydroperoxide anion is more reactive than the hydroxide anion.

> As is so often the case, I have made three different, but equivalent, statements.

In **Applications 4**, we saw that hydrazine is more nucleophilic than ammonia. We have the same effect here. Repulsion between the negative charge on oxygen (electrons) and the lone pairs on the adjacent oxygen atom (more electrons!) means that the negative charge is less stable/more reactive.

> So, we will have more hydroxide than hydroperoxide, but hydroperoxide is a better nucleophile, and even if it wasn't, the hydroxide doesn't really do anything.

Now let's see what happens when hydroperoxide is the nucleophile.

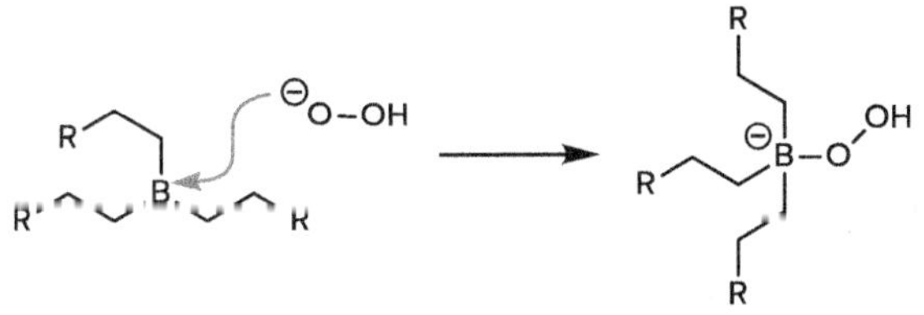

Now we have reached the actual rearrangement step.

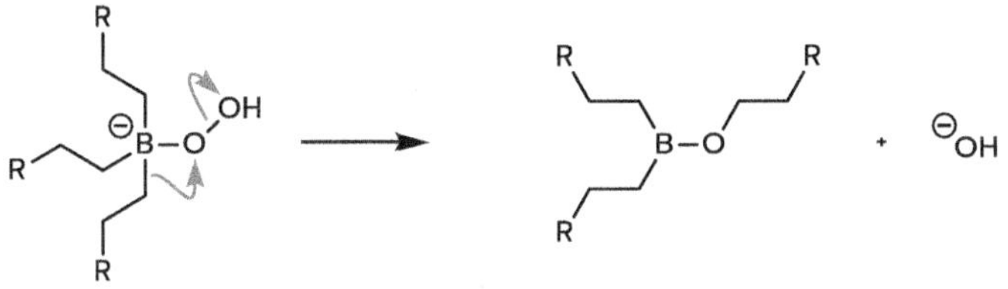

Compare this with the following combined scheme from **Fundamental Reaction Type 5**.

Here, the alkyl group is migrating to carbon rather than to oxygen. And we have drawn this as breaking the C–Cl bond first, rather than both processes happening at the same time.

> We did say, all the way back in **Reaction Detail 9**, that this might actually happen at the same time.

And of course, we did see migration to oxygen in **Reaction Detail 32**, so what we are seeing here is not new.

SECTION 5 REARRANGEMENT REACTIONS

MAKING CONNECTIONS

We have seen alkyl groups migrate from carbon to carbon, from carbon to nitrogen, from carbon to oxygen, and now from boron to oxygen. The precise order of steps can vary.

You won't form an unstable carbocation, or indeed any unstable intermediate, so there might be a faster (lower activation energy) process in which the migration proceeds in a single step.

> Look at S_N1/S_N2 substitution mechanisms again (HTSIOC Reaction Detail 1).

Here, we have a different reaction type, but with all the same considerations. In fact, you could even consider these rearrangements to be nucleophilic substitution reactions in which the nucleophile is delivered directly from the adjacent atom.

So what's missing? We haven't seen rearrangement from boron to carbon. Of course we can, in effect, predict the existence of these reactions. It is really easy to form a C–B bond. If we can make two different C–B bonds, and then migrate one of the carbons onto the other, then we have a way of making C–C bonds.

> Of course, it has been done.

MIGRATION FROM BORON TO CARBON

I'm just going to give one example, to show the power of this chemistry. Reaction of compound **1** with ethylmagnesium bromide (**Basics 5**) gives compound **2**.

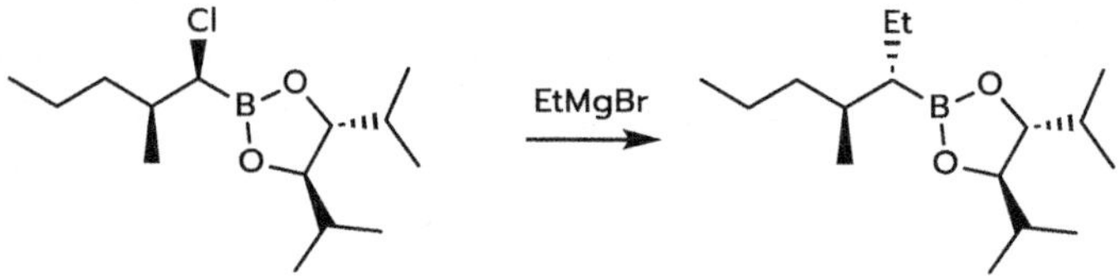

This looks, for all the world, like an S_N2 substitution, in which the nucleophilic ethyl group replaces the chloride with inversion of configuration.

> The problem is, ethylmagnesium bromide *is* a nucleophile, and the chloride is replaced with inversion of configuration. But it's not as simple as S_N2.

The ethyl group initially becomes bonded to boron, exactly the same as hydroperoxide did in the previous example.

SECTION 5 REARRANGEMENT REACTIONS

In **Reaction Detail 31** we saw an orbital diagram for a carbocation rearrangement. Here is it again.

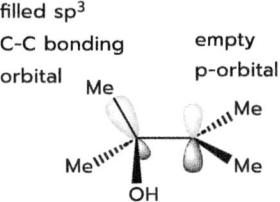

Now we can draw the same type of diagram for the borate rearrangement above. In this case, the migration terminus is not an empty p-orbital. It is a σ* orbital of a C–Cl bond.

In these two rearrangement reactions, we have a direct parallel between S_N1 and S_N2 substitution. In the former, we get a carbocation, because it is a relatively stable carbocation. In the latter case, we would never get a carbocation, but we have a "better nucleophile".

> *It's a better nucleophile because it is part of a negatively charged borate anion. In the former case, the methyl group that migrated was perfectly happy on carbon, right until we formed the carbocation.*

THERE'S MORE

There's always more. In this case, the boron compound **1** has a chiral auxiliary (**Stereochemistry 13**) on boron. It turns out that there are stereoselective organoborate rearrangements.

> *The one we have looked at is stereospecific rather than stereoselective (**Stereochemistry 2**).*

Donald Matteson, at Washington State University, has spent much of his career developing stereoselective organoborate rearrangements, and (in my opinion) doesn't get the recognition he deserves. This is really powerful chemistry.

But we are going to leave it with this one example.

IN CLOSING

Don't lose sight of the bigger picture. You can't just memorize every reaction, and your lecturers don't expect you to. What I'm trying to do here is show you examples

SECTION 5 REARRANGEMENT REACTIONS

of each type of reactivity, so that if you are asked 'what will **X** do with **Y**?', you can consider all options.

> *You'll only get the right answer first time with practice and experience.*

But with a logical and methodical approach, you'll get a solid pass.

> *Much more so than simply trying to memorize a list of reactions without making the connections.*

SECTION 6

PERICYCLIC REACTIONS

SECTION 6 PERICYCLIC REACTIONS

INTRODUCTION

There are a handful of reactions where drawing of curly arrows and consideration of stabilization of charge is not quite enough. We need to use more rigorous molecular orbital theory.

In this section, we will look a little deeper into the molecular orbital interactions that govern structure and reactivity. In some cases, we will use this to provide alternative explanations of aspects we have already covered. This doesn't mean that we will dispense with the previous explanations entirely. You don't stop using a bicycle as soon as you buy a car!

In other cases, we will see that we cannot use curly arrows to fully explain the outcome of reactions. In these cases, we will have to rely on the molecular orbitals. It is important that you know which cases these are, so that you apply the correct approach.

SECTION 6 PERICYCLIC REACTIONS

FUNDAMENTAL REACTION TYPE 6
Pericyclic Processes

So far, the reactions we have seen involve a nucleophile reacting with an electrophile. These make up the vast majority of organic reactions. In **Reaction Detail 6**, we encountered the ozonolysis reaction for oxidation of alkenes. We found that we were unable to consider one reacting partner as the nucleophile and the other as the electrophile. Instead, this reaction involves a cycloaddition.

> A cycloaddition is a **pericyclic** reaction. In a pericyclic reaction, the bonds are all formed in a single (concerted) step, and all bonds lie in a ring.

It turns out that cycloaddition is not the only type of pericyclic reaction. What we will do in this chapter is define the pericyclic reactions. Each will then be considered in turn in subsequent chapters.

CYCLOADDITIONS

In a cycloaddition, two components react together with the formation of two new σ-bonds, and with a reduction of the length of the conjugated system of orbitals in each component. Overall, we lose two π-bonds and gain two σ-bonds.

Here is the classic example, known as the Diels-Alder reaction after its inventors, Otto Diels and Kurt Alder. This particular Diels-Alder reaction doesn't work very well at all. We will see why this is in due course.

The Diels-Alder reaction is a 4π + 2π cycloaddition. We have a diene with 4 π electrons reacting with an alkene with 2 π electrons. We need some definitions. We have established that one of the components in a Diels-Alder reaction is a 1,3-diene. The other is an alkene, but because it is reacting with a diene, we call it a **dienophile**.

> I'll be honest. This is pointless terminology. If you understand the reaction, it doesn't matter whether you call the alkene a dienophile or not. What you should absolutely **not** do is call every alkene a dienophile. It's **only** behaving as a dienophile when it reacts with a diene. Don't worry too much about the terminology. If you immerse yourself in organic chemistry, you'll find you are talking about dienes and dienophiles as if it is the most natural thing in the world.

The Diels-Alder reaction is the reaction of a diene and a dienophile, two separate π-systems, to give a six-membered ring containing a double bond.

437

SECTION 6 PERICYCLIC REACTIONS

> The reaction forms two σ-bonds at the expense of two π-bonds.

The Diels-Alder reaction is only one example of a cycloaddition. Here is another.

We call this one a **1,3-dipolar** cycloaddition. The starting material on the right is forming bonds from atoms 1 and 3, and there is a positive and a negative charge. It's a bit of a nebulous definition, but it works. In this case, the starting material on the left is described as a **dipolarophile**.[97]

Let's count the bonds. The dipolarophile has two π-bonds (a triple bond, an alkyne). The dipole (in this case a nitrile oxide) also has a triple bond, so that is a total of four π-bonds. The product has only two π-bonds, so we have lost two π-bonds. We have also formed two σ-bonds.

Here is the reaction we encountered as a step in the complex ozonolysis mechanism in **Reaction Detail 6**.

molozonide

Once again, we are losing two π-bonds and gaining two σ-bonds.

At a glance, a reaction which does not have a nucleophile reacting with an electrophile appears to 'break the rules' we have been building.

> Actually, it doesn't break the rules. We just need to think a little more carefully about what is happening when a nucleophile reacts with an electrophile.

A BRIEF DIGRESSION

When a nucleophile reacts with an electrophile, we draw a curly arrow going from the nucleophile to the electrophile. Remember what this is showing us—an occupied orbital of the nucleophile is overlapping with an unoccupied orbital of the electrophile.

This is the rule—orbital overlap. I'm not going to suggest that nucleophiles and electrophiles aren't important—they are! But the ***really*** important thing is to understand why they behave as they do.

---

[97] What next—Grignard reagentophile? Wait, that would be silly! In this case, the dipolarophile is simply an alkene that is reacting in a particular way.

SECTION 6 PERICYCLIC REACTIONS

Fundamentally, a nucleophile is a species that has an occupied molecular orbital with the 'right energy' to overlap with an unoccupied molecular orbital (also with the 'right energy') of an electrophile.

What do we mean by the 'right energy'? Molecular orbitals that have similar energies will provide more stabilization when they overlap. In **Recap 1** we defined the terms HOMO (highest occupied molecular orbital) and LUMO (lowest unoccupied molecular orbital).

When two species react together, both will have a HOMO and a LUMO. The only exception is a proton.

You can't have a HOMO if you don't have any electrons!

Here is a hypothetical molecular orbital energy diagram.

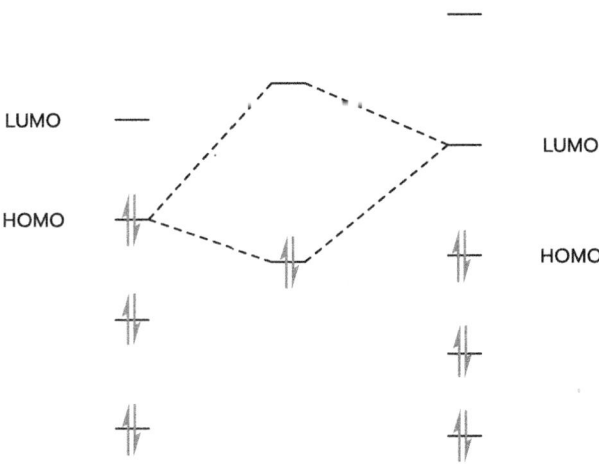

What we have is the HOMO of the species on the left overlapping with the LUMO of the species on the right. This arrangement gives a new bonding orbital and a new antibonding orbital. There is net stabilization.

Now consider the following alternative arrangement. The LUMO of the species on the left overlaps with the HOMO of the species on the right. We still get a new bonding orbital and a new antibonding orbital.

Here is the important point. We get more stabilization from the overlap of orbitals that are closer in energy. The first scenario is 'better'.

So, the generalization is that, in principle, we can obtain an energetically favourable interaction by overlap of the HOMO of either component with the LUMO of the other component, but very often one of these interactions will dominate.

This bias defines nucleophiles and electrophiles in molecular orbital terms.

SECTION 6 PERICYCLIC REACTIONS

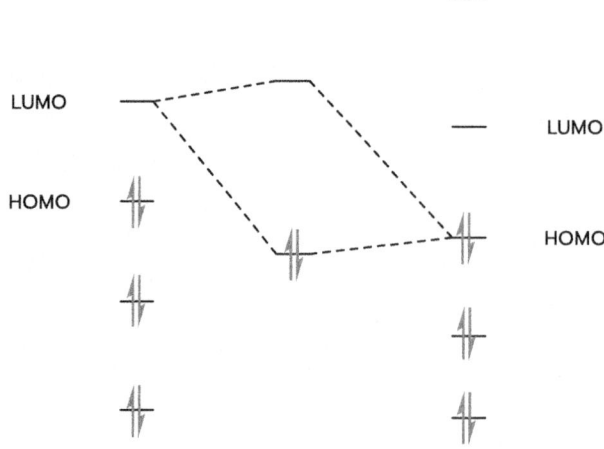

In most reactions, this orbital overlap only forms one bond. In a cycloaddition, it forms two bonds, and we will see the orbital overlap in Reaction Detail 34. We will find that the symmetry of the orbitals is a major factor. For now, let's look at the other types of pericyclic reaction.

ELECTROCYCLIC REACTIONS

An electrocyclic reaction results in the formation of a ring from an open chain conjugated system. A σ-bond is formed across the ends of the conjugated system, with a loss of a π-bond at each end. Overall, one π-bond is lost and one σ-bond is formed.

This simple case has no stereochemical consequences. However, when we use a reactant with substituents, we can observe very specific stereochemical outcomes. We will look at these in Reaction Detail 35.

SIGMATROPIC REACTIONS

These reactions are rearrangements that involve the movement of a σ-bond from one position to another. Effectively, one σ-bond is broken and another one is formed. We describe them according to the number of atoms between the two σ-bonds. Here is one example, in which the bonds broken/formed are both C–H bonds. We would describe this reaction as a 1,5-hydride shift.

In Reaction Detail 19 we saw reactions in which hydride appeared to be acting as a leaving group. There, we found that this isn't quite the case. Here, we have a similar situation. Although we appear to be drawing a curly arrow that shows H⊖ being transferred, we could just as easily draw the curly arrows going in the opposite direction to show this as a proton shift.

> *The convention is to draw it as above and to call it a hydride shift. We will see, in Reaction Detail 36, that there are some stereochemical consequences which become apparent when we draw the molecular orbitals.*

There is another type of sigmatropic reaction we should consider, in which C–C bonds are broken and formed. Here is an example. Again, we classify the reactions according to the number of atoms between the σ-bonds formed/broken. This is a [3,3]-sigmatropic reaction.

Note that in both of the sigmatropic reactions in this section, the starting material and the product are the same.

> *The reactions have a transition state and no intermediate. Draw a reaction energy profile.*

Once we start adding substituents, we find that there are significant chemical and stereochemical consequences which we will explore in **Reaction Detail 37**.

WHERE IS THIS GOING NEXT?

Pericyclic reactions are fantastic. They have very clearly defined transition states, so that when a stereogenic centres are formed, it can be formed with high levels of stereoselectivity. To explain these reactions, we need to look at the symmetry of the molecular orbitals involved. Each of the reaction types requires a slightly different approach. We will look at the methodology in the following chapters.

We are only going to do this at a very basic level. Even at this level, though, we will be able to explain why some reactions work and other reactions do not.

> *We would not be able to do this if we only considered the curly arrow mechanisms.*

We will also be able to explain the preferred (or in some cases the only) stereochemical outcome of pericyclic reactions. The orbital symmetry informs the shape of transition state we will draw.

SECTION 6 PERICYCLIC REACTIONS

REACTION DETAIL 34
Allowed and Forbidden Cycloadditions

In the previous chapter, we saw the following cycloaddition reaction.

It turns out that this reaction does not work well. We will come to the reasons very soon. However, this reaction does work! Some cycloadditions, for which we can draw entirely sensible curly arrows, can never happen. We describe cycloadditions as either **allowed** or **forbidden**.

A forbidden reaction cannot happen! The orbital symmetry is wrong. It isn't a matter of this being a slower reaction, or a less favourable outcome. Similarly, just because a reaction is allowed (according to orbital symmetry) doesn't mean it will happen rapidly. There are many allowed cycloadditions that are so slow that you can consider that they do not happen.

DIMERIZATION OF CYCLOPENTADIENE

When cyclopenta-1,3-diene (**1**) dimerizes, it gives product **2**. In fact, we cannot buy cyclopenta-1,3-diene, since product **2** is more stable and this reaction is very favourable. We buy compound **2** instead and heat it to form compound **1**. Cycloadditions are reversible.

Calculate the enthalpy change for the formation of compound **2** from compound **1**. I'm not going to tell you where to find the data you need.

But this isn't the only cycloaddition of compound **1** that we could draw curly arrows for. We could just as easily draw curly arrows for the formation of compound **3** or compound **4**. They are shown on the next page. But the simple fact is that these reactions do not work. At all!

I don't mean that you don't get a high yield, or that compound 2 is the preferred product. After all, we saw in Basics 6 that formation of a six-membered ring is better than formation of a four-membered ring or an eight-membered ring.

What we are talking about here is something much more fundamental. These two reactions are **symmetry forbidden**. Let's see why this is.

SECTION 6 PERICYCLIC REACTIONS

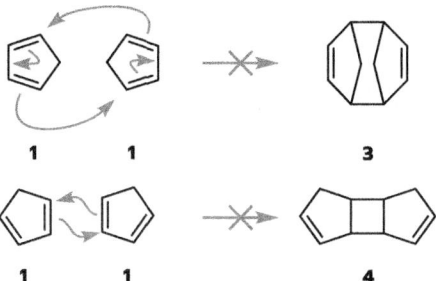

We need to consider the interaction of an occupied orbital of one component and an unoccupied orbital of the other component. There are two occupied π-orbitals, and two unoccupied π*-orbitals. In this case, both components are the same, so that the energy diagram looks like this.

> When thinking about the orbital interactions, there is one really important point. The closer in energy the interacting orbitals, the more favourable the interaction will be. We saw one implication of this in **Fundamental Reaction Type 6**. Here is another. While we *could* consider the interaction of ψ_1 of one cyclopentadiene with ψ_3 or ψ_4 of the other, the dominant interaction will be ψ_2 of one with ψ_3 of the other.

With one of these interactions added to the diagram, it looks like this:

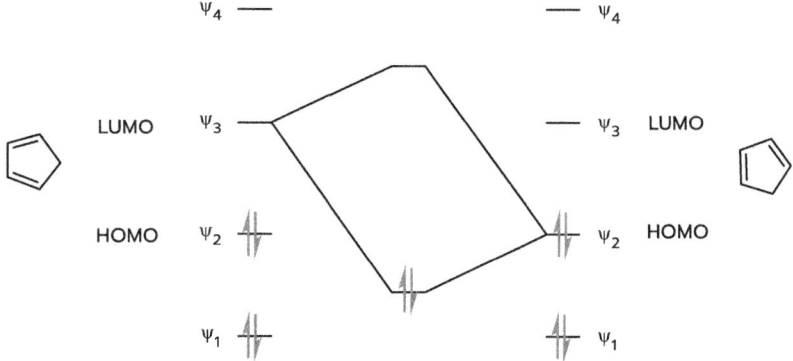

We can see that this leads to a net lowering of energy compared to ψ_2.

443

SECTION 6 PERICYCLIC REACTIONS

> However, this is only any good if the orbitals have the correct symmetry to interact.

We saw the molecular orbitals of buta-1,3-diene in **Recap 6**. The fact that cyclopenta-1,3-diene is cyclic and has an extra CH_2 group doesn't change anything. Here is the diagram again.

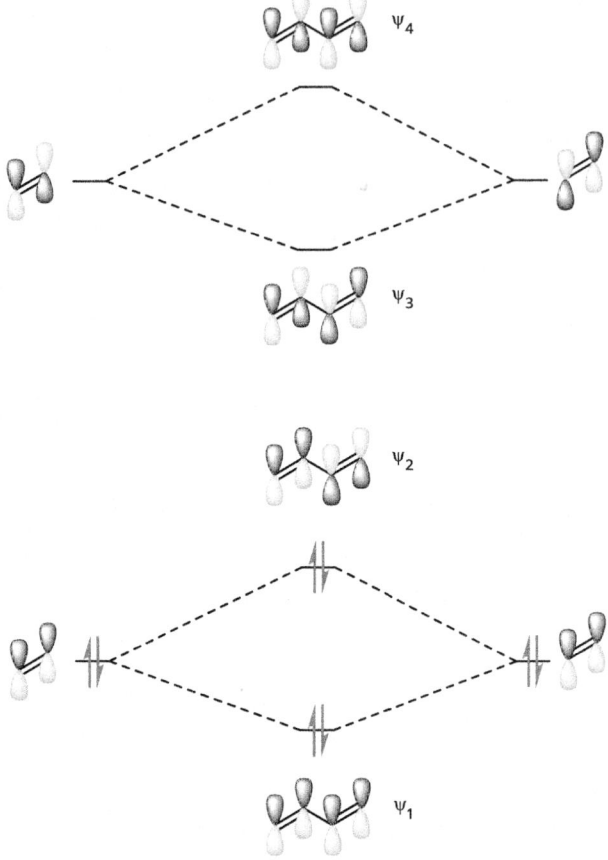

Based on this, here is the HOMO of cyclopentadiene.[98] We need this side-view to see the orientation in which the two molecules of cyclopentadiene react.

HOMO

Now let's add the second one. We need the LUMO of this one.

---

[98] I have dispensed with '1,3-'. We don't really need to specify.

444

SECTION 6 PERICYCLIC REACTIONS

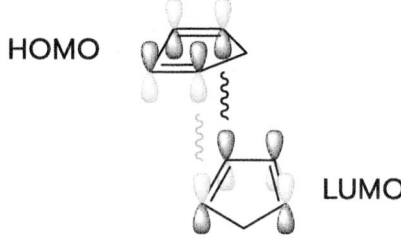

Look at the wiggly lines that indicate overlap of orbital lobes with the correct symmetry. The black line shows correct symmetry between the black lobes of the HOMO and the LUMO. The light gray line shows correct symmetry between the light gray lobes.

> It doesn't matter whether we take the HOMO of the diene and the LUMO of the dienophile (which in this case is also a diene!), or the LUMO of the diene and the HOMO of the dienophile.

> Check that you can see this! Draw as many variations as you need to. Number the atoms, if necessary to ensure that you can see which atoms are becoming bonded.

IF THEY ARE BOTH WRONG, THEY ARE BOTH RIGHT!

There is something else we need to consider. There is no absolute for the black and light gray lobes. Suppose I had drawn the dienophile LUMO as follows.

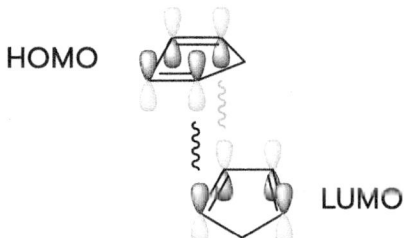

Both interactions look like they have the wrong symmetry. But that was a choice I had. I can simply swap the shades of all lobes within one reacting partner.

To see how this differs from a **forbidden** reaction, let's look at the [4 + 4] cycloaddition.

FORBIDDEN [4 + 4] CYCLOADDITION

Now if we consider the alternative possibilities, we can see the problem. For the [4 + 4] cycloaddition, if the symmetry at one end matches, then the symmetry at the other end will not. This is shown with the **X** below.

SECTION 6 PERICYCLIC REACTIONS

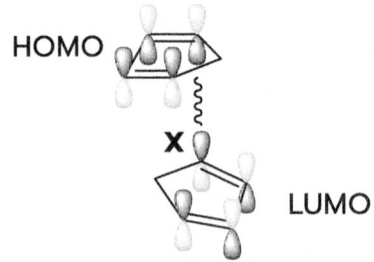

The same applies to a [2 + 2] cycloaddition process. Draw it!

Once again, it would not matter which of the two components we chose the HOMO of, and which the LUMO. We would always reach the same conclusion.

Just remember the following steps to work out if the reaction is allowed.

1. Draw the two reactants in a sensible orientation.
2. Draw the HOMO on the first component.
3. Start to draw the LUMO on the second component by ensuring that the first point of interaction has the correct symmetry. Carry on drawing and see what happens at the other end.

It is perfectly okay to start drawing it 'correctly'. After all, we are talking about 'relative' symmetry here, not absolute.

DIMERIZATION OF ETHENE

Let's take a closer look at a [2 + 2] cycloaddition, just considering ethene reacting with ethene. We can see that, as with the [4 + 4] cycloaddition, if we have overlap of the orbital at one end (black wiggly line) then we do not have the correct symmetry at the other end (**X**).

The HOMO–LUMO interaction is not symmetry-allowed. We can still draw an orbital energy diagram for this interaction. From the orbital energies, it does appear that this would be bonding in nature. However, the orbitals have to be able to overlap, which they cannot.

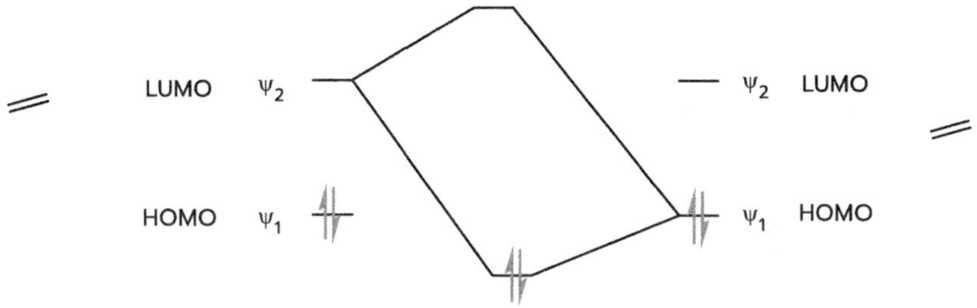

SECTION 6 PERICYCLIC REACTIONS

We could also consider the HOMO–HOMO interaction, as below. From a symmetry point of view, this looks fine.

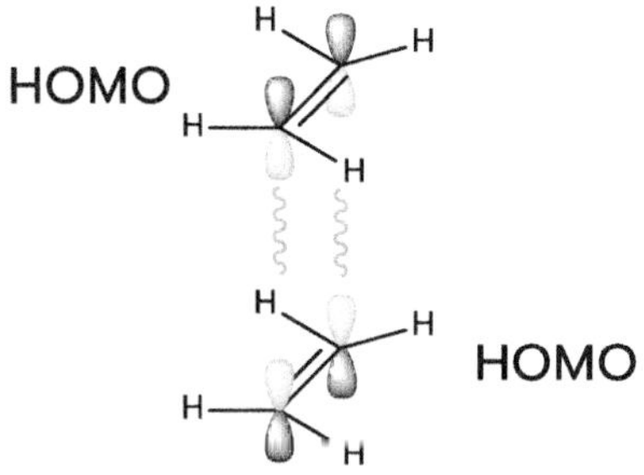

However, when we look at the energy level diagram, it is not so good.

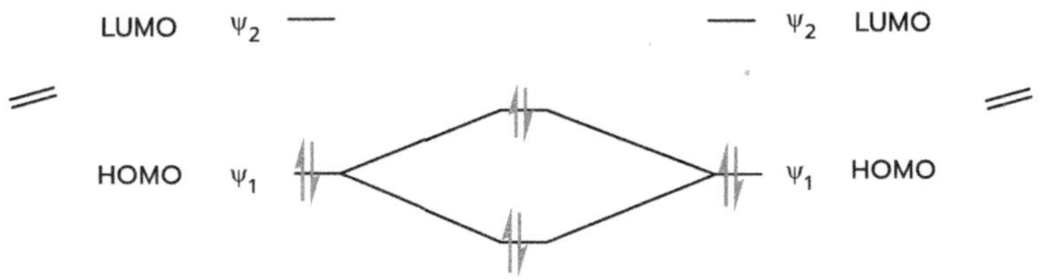

Overall, it looks as though we might not be gaining any energy, but equally well we are not losing any energy. Would this be so bad?

> It turns out that by considering only the HOMO and LUMO, we can make a reliable prediction, but in order to fully explain why a reaction is forbidden, we would need to consider the symmetry of all molecular orbitals. For this, we would use a correlation diagram. They are beyond the scope of this book.

So, the take-home message is that a reaction can proceed if the HOMO–LUMO interaction has the correct symmetry. It doesn't matter which component you choose to use the HOMO for and which the LUMO (assuming they are different).

There is one cautionary note. This discussion only applies to thermal reactions. By this, we mean reactions that we heat, or perhaps do not even need to heat—they just happen.

PHOTOCHEMICAL CYCLOADDITIONS

We have just seen that ethene does not dimerize thermally to give cyclobutane.

SECTION 6 PERICYCLIC REACTIONS

The difference in energy between the HOMO and the LUMO of ethene corresponds directly to a frequency of electromagnetic radiation. This is generally in the UV region of the spectrum, but for highly conjugated compounds, it can extend into the visible region of the spectrum.

> *These compounds would be coloured!*

If we shine a light of the correct frequency onto a reaction, electrons are promoted to higher energy levels, so that different orbitals are occupied. These reactions would be described as 'photochemical' and we will look at one of them now, just to establish the process.

When we carry out the above reaction photochemically, we will promote one electron (highlighted) from one of the ethene molecules. This gives the following situation.

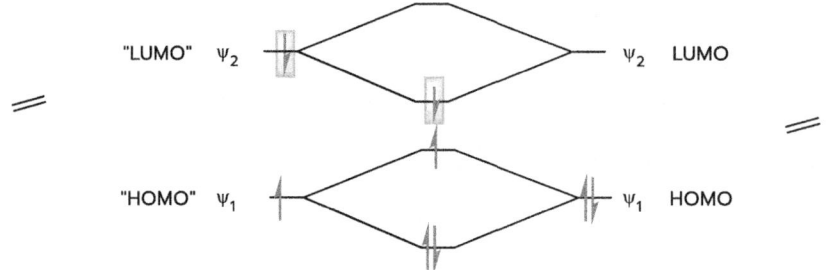

Now, from the left-hand side, I am putting HOMO and LUMO in inverted commas since they are now both singly occupied orbitals.

The "HOMO"-HOMO interaction now involves three electrons. As a result of the interaction (which we established above has the correct symmetry) two electrons are lowered in energy and one is raised. Therefore, there is a net stabilization. The "LUMO"-LUMO interaction only involves one electron (highlighted), and as a result is it lowered in energy, which is also a net stabilization.

> *Therefore, the net effect of these two interactions from the excited state produced by photochemical excitation is that there is stabilization from a symmetry-allowed process.*

Therefore a [2 + 2] cycloaddition is photochemically allowed, although it was thermally forbidden.

Once again, this is a simplification. Again, the correlation diagram that we would need for a fuller explanation is beyond the scope of this book.

GENERALIZATIONS

As a generalization, a thermal cycloaddition reaction is symmetry allowed if there are 4n + 2 π electrons involved, where n is an integer. A photochemical cycloaddition is allowed if there are 4n π electrons involved.

SECTION 6 PERICYCLIC REACTIONS

> *Any cycloaddition that is forbidden thermally will be allowed photochemically (and vice-versa).*

Remember that a [4 + 2] cycloaddition is allowed, and you can work out all the rest very quickly.

ONE MORE POINT

Orbital symmetry determines whether a reaction can take place at all. This doesn't mean that we can dispense with all the other aspects of our theory.

We saw that the reaction of buta-1,3-diene with ethene is symmetry allowed, but doesn't work well in practice.

Here is buta-1,3-diene again. The problem, in this case, is that the central bond can rotate (**Stereochemistry 5**). We call the two conformations *s*-cis and *s*-trans.

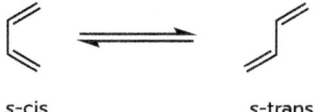

s-cis   s-trans

For a Diels-Alder reaction to take place, we need the *s*-cis conformer. Unsurprisingly, the *s*-trans conformer is more stable.

> *Now we can see why cyclopentadiene is so good in Diels-Alder reactions—it is locked into the s-cis conformer.*

SECTION 6 PERICYCLIC REACTIONS

STEREOCHEMISTRY 16
Diels-Alder Reactions

BEFORE WE START
Have a look at the following Diels-Alder reaction. We have two things to consider.

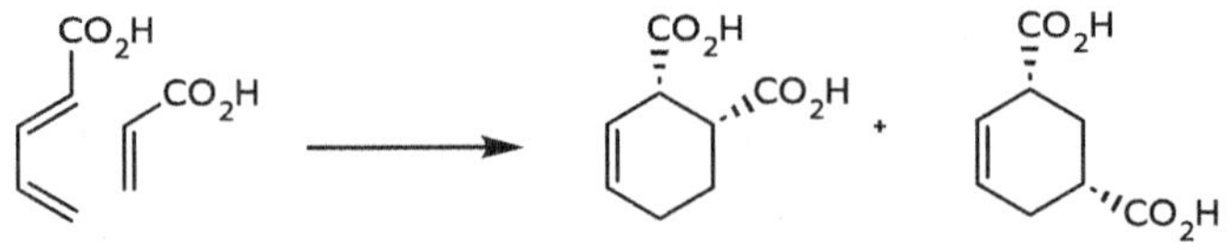

98:2 ratio

> In the major product, why do the carboxylic acid groups end up on adjacent carbon atoms?

This is regioselectivity (**Recap 4**). We are actually not going to explain the reasons for the observed regioselectivity. We need a bit more theory than we would usually encounter at this level.

> In both products, why do the two carboxylic acid groups end up on the same side of the ring?

This is stereoselectivity, and we are going to deal with it here. From here on in, we will only draw the outcome from the major regioisomer.

STEREOSELECTIVITY IN DIELS-ALDER REACTIONS—THE *ENDO* RULE
The stereoisomer we form has the two substituents on the same side of the ring. It is the more crowded, and it is also formed *via* the most crowded transition state.

> There must be something that is favouring it.

As always, when we are faced with a problem like this, we need to identify a process that we can follow. In **Reaction Detail 34**, we established the orientation in which a diene and dienophile must react in a Diels-Alder reaction. We need overlap of the HOMO and the LUMO.

Before we address the stereoselectivity in the example shown above, we will look at a simpler example. We will consider the reaction of butadiene (**1**) with dienophile **2**. There are two possible orientations for the dienophile **2**. It could have the carboxylic acid group close to the diene, or away from the diene. The dienophile **2** could approach the top face of the diene or the bottom face of the diene. These four possibilities are summarized as follows.

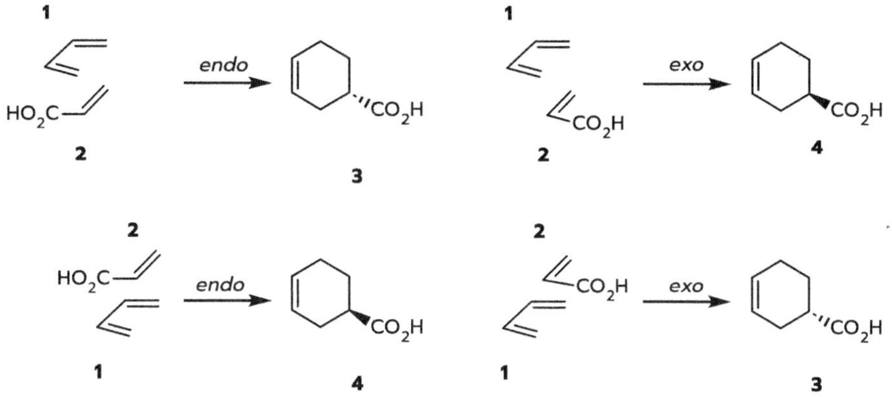

SECTION 6 PERICYCLIC REACTIONS

In orientations where the dienophile carboxylic acid group is directly underneath or above the diene, we refer to the orientation as *endo*. The alternative is *exo*. For the reaction above, there are no observable consequences. The two 'different' *endo* orientations are mirror images (enantiomers) with equal energy.

> We would expect to get a 1:1 mixture of enantiomers of 3 and 4.

The same is true for *exo*. Similarly, we won't know how much *endo* transition state and how much *exo* transition state we have in the reaction, even though they are different (diastereomeric).

> This reaction will produce a racemic mixture of enantiomers 3 and 4. This is not because *exo* and *endo* are equally likely. It is because attack from the top face or the bottom face of the diene is equally likely.

Now let's add a substituent to the diene as well. Now, *exo* approach and *endo* approach produce diastereomeric products. Here are the first two possibilities.

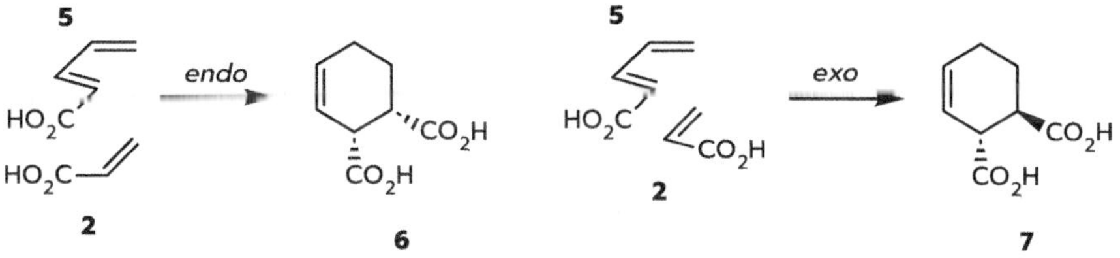

Structures **6** and **7** are diastereoisomers. We would not (**Stereochemistry 2**) expect them to be formed in equal amounts, although we have not (yet) explained why isomer **6** is formed almost exclusively.

Now we can look at the second two possibilities.

451

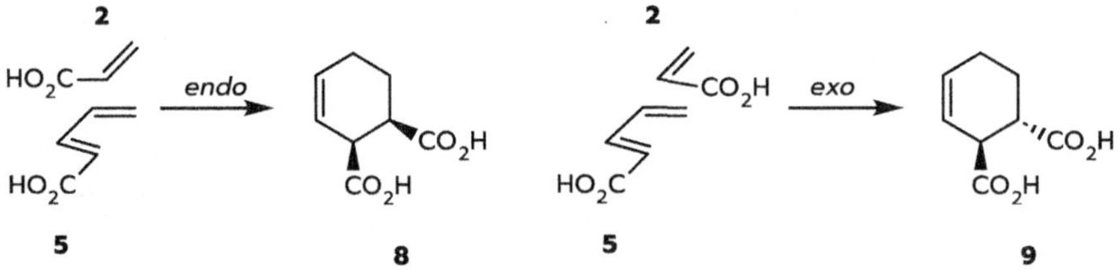

SECTION 6 PERICYCLIC REACTIONS

The same applies. Structures **8** and **9** are also diastereoisomers, so we should not expect them to be formed in equal amounts either.

On the other hand, structures **6** and **8** are enantiomers, and they **will** be formed as a 1:1 (racemic) mixture. The same applies to structures **7** and **9**.

ENDO IS FAVOURED

In a moment, we will see the molecular orbital interaction that favours the *endo* transition state. First of all, though, we need to be able to **see** that the above product *is* the *endo* product.

> You will probably find that you look at the structures above, and cannot see that *endo* approach gives isomers **6** or **8**. It isn't easy! With more complicated structures, it becomes even more difficult. You need a strategy!

APPROACH 1—TRY TO DRAW IT!

If you try to draw the structure over and over, you will get better at it. There are some refinements you can make that will increase your chance of getting this right.

Possibly the simplest way is to transpose the diene and dienophile structures onto a cube. This is shown below, although I did have to stretch it. I also added the hydrogen atoms on the key carbon atoms. If I didn't explicitly draw them, I wouldn't have been able to see exactly where they were.

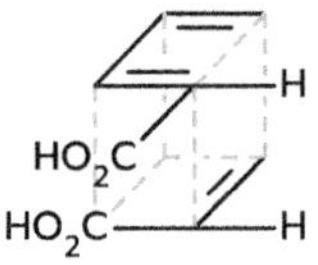

Hopefully you can see that in this orientation, the two hydrogen atoms will end up on the same side of the ring.

> Personally, I don't think it is a given that you will be able to see this. On to Approach 2....

APPROACH 2—MAKE A MODEL!

If you make a model of the diene and dienophile in the above orientation, but replace the sp² carbons with sp³ carbons so you can form the bonds, then you can simply 'unfold' the structure and you will see which substituents are on the same side of the ring. I can't draw a diagram to show you this process. You have to just do it!

> When you've done this, the diagram above will be easier to interpret.

APPROACH 3—LEARN ONE EXAMPLE

For me, this is the fall-back. I remember the following outcome, with generic substituents,

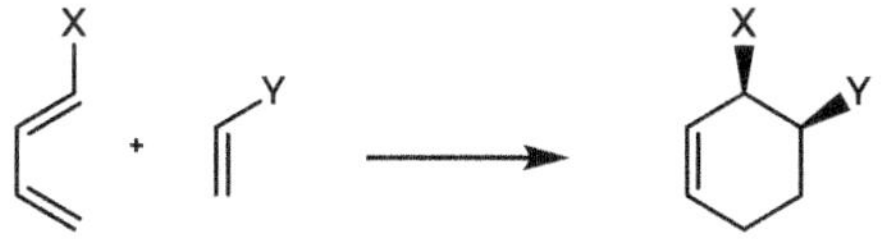

Whenever I see a new example, I consider the substituents in regard to those above. This is easiest to see with an example. We are going to look at a rather tricky one.

> We have said before that it is common at first to find that examples that include rings are particularly challenging.

Here is the example we are going to use.

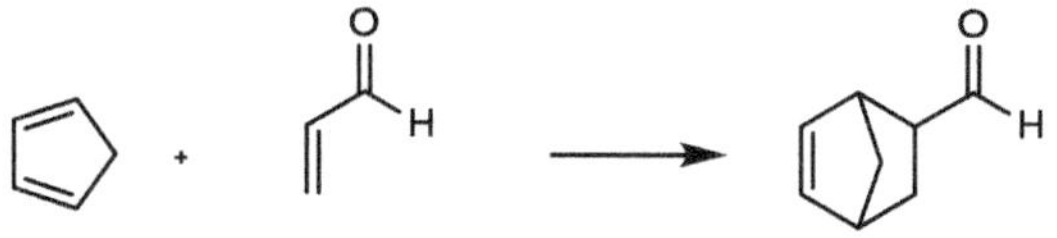

The diene is cyclic, and the only 'substituent'—the CH_2 group in the ring—is cis. This is different to the example we looked at above. The dienophile is easy. The aldehyde group is 'Y'. For the diene, 'X' must be a hydrogen atom. Here it is, with the groups explicitly shown.

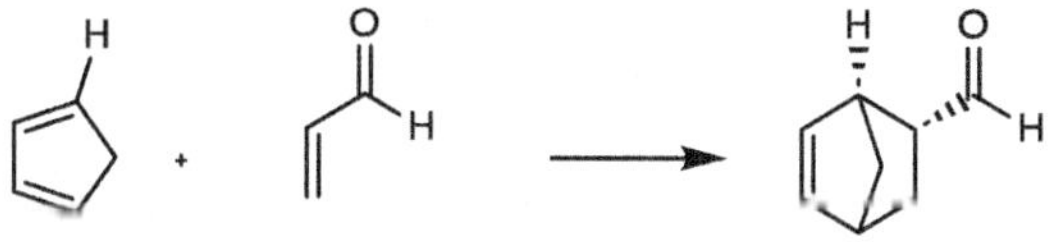

Perhaps it isn't so easy to see the shape of the molecule. I prefer to draw this as shown below.

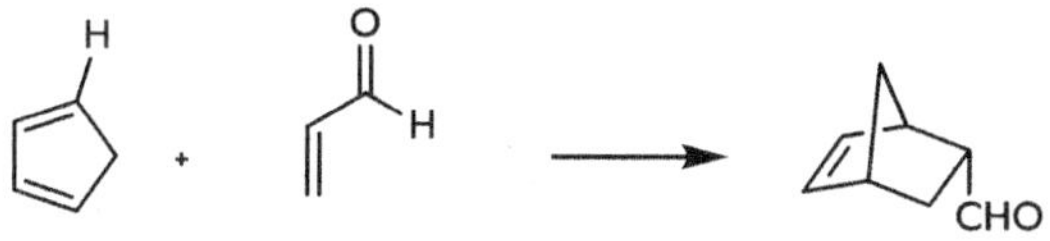

Of course, this begs the questions 'why do I prefer this?' and 'how do I know to draw it like this?' You already know the answer! I prefer this because it is clearer. I know to draw it like this because I've seen it a thousand times before.

SECTION 6 PERICYCLIC REACTIONS

> *When you learn a language, if you have reached a reasonable level of competence, and you encounter a new word, it is very likely that you will remember it. You have a framework in which to put it. If you try to learn 20 new words at the same time, you will probably forget all of them.*

BUT *WHY* IS *ENDO* FAVOURED?

We have established that the *endo* product is normally favoured. We established that the transition state leading to the *endo* product is actually more crowded than the *exo* transition state. If this was a steric effect, we might expect more *exo* product.

> If it ain't steric, it must be electronic.

The conventional explanation involves a "secondary orbital effect".

To show this, I have chosen a simplified reaction with just an aldehyde substituent. This is the example we have just seen.

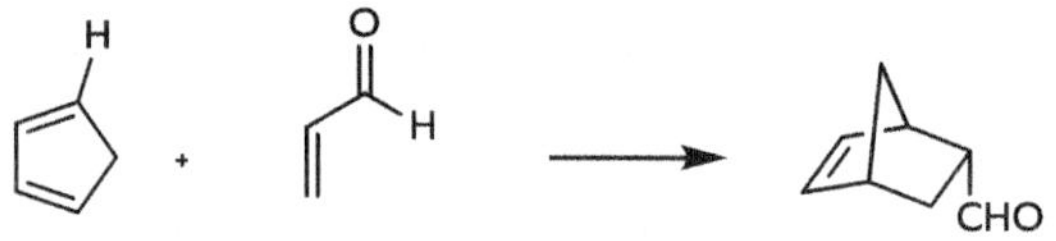

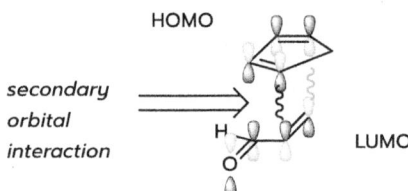

The diene HOMO and dienophile LUMO are shown below. The key point here is that the dienophile is not a simple "ethene", but it is (with one conjugated double bond) more like a butadiene itself. It has orbital coefficients on the carbonyl group.

You can see that there is a lobe on the diene close to a lobe on the dienophile. These have the correct symmetry to overlap, and yet are not on atoms that are becoming bonded. Therefore, it is claimed that they provide stabilization of the transition state while not being directly bonding. For this reason, it is referred to as a **secondary orbital interaction**.

> *There isn't really anything special about a secondary orbital interaction. It's just two orbitals with the correct symmetry interacting in a way that lowers the energy of the transition state during the reaction.*

Of course, now we know that this is an orbital effect, we know which substituents on the dienophile are likely to give *endo* products.

> *A methyl group would not do this!*

SECTION 6 PERICYCLIC REACTIONS

NOW A PROBLEM

This is a tricky one.

> Predict the stereochemical outcome of the intramolecular Diels-Alder reaction of the following compound.

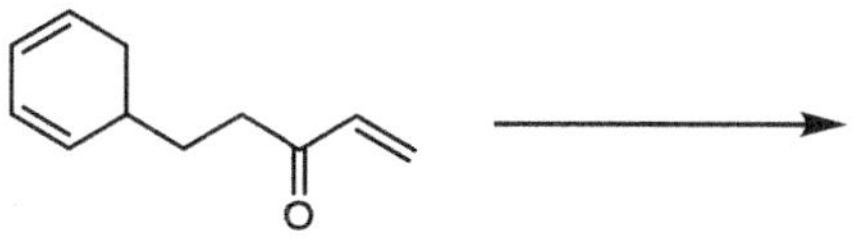

Hopefully you can identify the diene and the dienophile quite quickly. You should then work out which atoms become bonded.

> *You need a strategy. I recommend numbering the atoms in the starting material and in the product.*

The next question is 'can it be *exo* or *endo*?' If you can exclude one option, it must be the other.

Let's cut to the chase. This is the product.

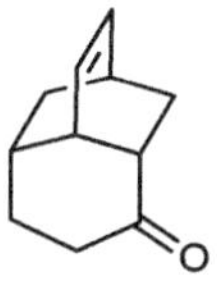

> *The carbonyl group is definitely exo if you compare it with the previous examples. However, it cannot possibly be endo. Make a model, and try to force it. Just don't try too hard.*

I didn't include this example because I thought you should be able to do it. I included it because it is different.

WHAT DO WE DO WITH THIS?

Once you know about secondary orbital effects in the Diels-Alder reaction, it's a one-trick pony! If you are asked to explain the stereochemical outcome, if it is *endo*, draw the orbitals.

Of course, we could easily imagine that adding bulky substituents to the diene or the dienophile would make the *endo* transition state more crowded and less favoured. We can over-ride the electronic effect with a steric effect.

If you have to check the stereochemical outcome of a Diels-Alder reaction, look at **everything**! Don't just assume it will automatically be *endo*.

SECTION 6 PERICYCLIC REACTIONS

REACTION DETAIL 35
Electrocyclic Processes

INTRODUCTION
This is the sort of thing we mean by an electrocyclic reaction. It is a bit like a cycloaddition but all in one component. We defined electrocyclic reactions in **Fundamental Reaction Type 6**.

> The π system cyclizes to lose one π-bond and gain one σ-bond.

STEREOCHEMISTRY IN ELECTROCYCLIC REACTIONS
If we add substituents to this system, we can see that there are stereochemical possibilities. In the following example, the methyl groups in the product could be on the same side, on opposite sides, or a mixture could be formed.

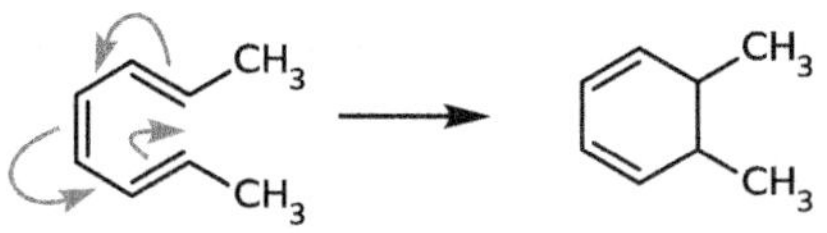

It turns out that these reactions are stereospecific (only a single outcome is observed) and in order to rationalize the outcome, we need to know how to construct the orbitals.

CONSTRUCTION OF ORBITALS FOR ELECTROCYCLIC REACTIONS
Electrocyclic reactions are slightly different to cycloadditions, and indeed different to sigmatropic reactions. In electrocyclic reactions, only a single π-system is involved, so we only need to consider one orbital. Since the reaction 'uses' electrons, this needs to be an occupied orbital.

> For thermal electrocyclic reactions, we need to think about the symmetry of the **HOMO**.

I will draw the HOMO of this system, showing the compound in a different view so that it is easier to see what happens.

SECTION 6 PERICYCLIC REACTIONS

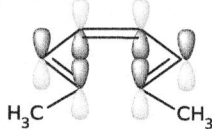

> Make a model of this compound. Note that there are also two hydrogen atoms on the carbon atoms with the methyl groups.

Yes, the representation above is not very realistic. Bear with it—we will deal with this problem once we introduce the stereochemical terminology associated with electrocyclic reactions.

First of all, look at the ends of the π-system. You can see that the light gray lobe is on the lower face and the black lobe is on the upper face at both ends.

We describe the stereochemical outcome of the reaction in terms of rotation of the two ends. In order to have an overlap with the correct symmetry, we need to rotate the ends of the π system. It doesn't matter whether we choose to overlap the light gray or black lobes, but let's make a choice and select the black lobes.

If we do this, we need to rotate the left-hand carbon atom clockwise and the right-hand carbon atom anticlockwise. This is shown in the reaction below, with the stereochemical outcome also shown.

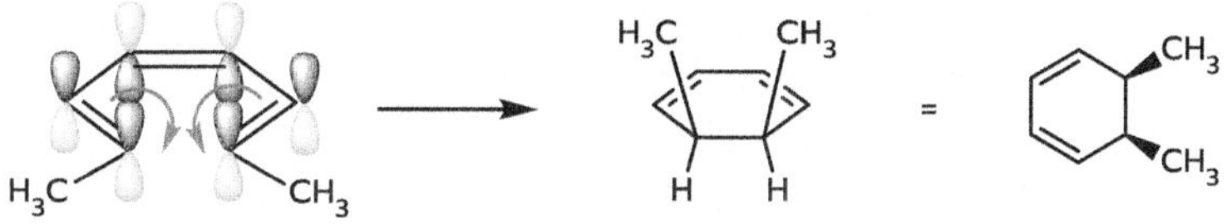

When the two ends of the π-system rotate in opposite directions, this is known as **disrotatory**.

Since there are three double bonds (6 electrons) involved in this reaction, it is known as a 6π electrocyclization.

We can look at other electrocyclization reactions. If we think about a 4π electrocyclization, involving two alkene bonds. We see, from the symmetry of the HOMO, the following situation:

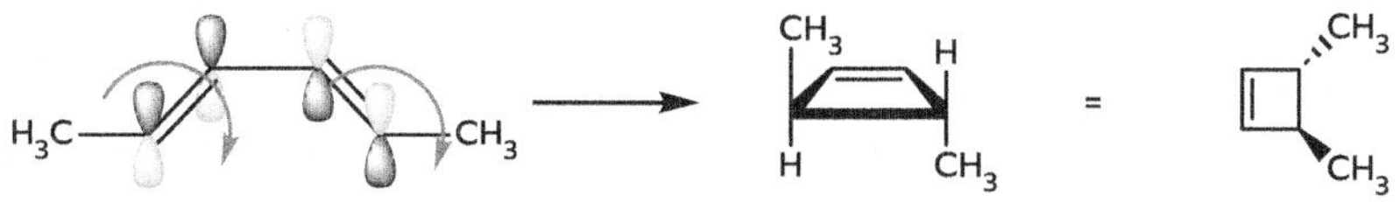

Now, the ends of the π-system have opposite symmetry, so that in order to overlap the black (or light gray) lobes, we need to rotate both ends in the same direction. This is known as **conrotatory** and gives the stereochemical outcome shown.

So, to summarize to this stage, a 4π thermal electrocyclization is **conrotatory** and a 6π thermal electrocyclization is **disrotatory**.

457

SECTION 6 PERICYCLIC REACTIONS

What about an 8π electrocyclization? Well, if adding two electrons changes the direction of rotation from 4π to 6π then we would expect another change in going from 6π to 8π. Therefore, an 8π thermal electrocyclization should be conrotatory. It is! Let's see the relevant orbital (HOMO) to convince ourselves of this.

All we need to know in order to determine whether a thermal electrocyclic reaction is **conrotatory** or **disrotatory** is the symmetry of the HOMO.

PHOTOCHEMICAL ELECTROCYCLIC REACTIONS

When we looked at cycloadditions, we found that a reaction that was thermally forbidden was photochemically allowed. Promotion of an electron from the HOMO to the LUMO changed the outcome of the process.

When we photolyze a polyene, we promote one electron from the HOMO to the LUMO.

This results in a potentially complicated situation. We now have a single electron in each of (what was) the HOMO and the LUMO. Which orbital do we then use to determine the outcome of a photochemical electrocyclization?

> *The rigorous way to determine the outcome is to draw a correlation diagram that considers the symmetry of all molecular orbitals involved in the reaction. As we noted in Reaction Detail 34, these are beyond the scope of this book.*

For now, we will just apply the simple rule of thumb that for a photochemical electrocyclization, we use the LUMO.

This makes things easier to learn. The LUMO will always have one more node than the HOMO. Therefore, the relative symmetry at the ends of the HOMO and LUMO will be different. I will only include a single example, the 6π photochemical electrocyclization, here.

Now we have the LUMO, so in order to overlap the light gray lobes, we need to rotate both ends in the same direction (clockwise) so we get the outcome shown. This reaction is therefore **conrotatory**.

> *Whatever stereochemical outcome we observe in a thermal electrocyclic reaction, a photochemical reaction will give the opposite outcome.*

SECTION 6 PERICYCLIC REACTIONS

ELECTROCYCLIZATION REACTIONS ARE REVERSIBLE

Remember that all these reactions are reversible, at least in principle.

We looked at the stability of various size rings in **Basics 6**. We should not be too surprised that an electrocyclic ring-opening of a cyclobutene to give a buta-1,3-diene is actually a pretty good process.

> We can use buta-1,3-dienes in Diels-Alder reactions. This is one way to make them.

If you know what product is produced from a particular polyene undergoing electrocyclization, then you can work out the stereochemistry of the reverse reaction. Here is an example.

It is generally more convenient to work out the stereochemistry of the electrocyclic ring-closure, and then work backwards from the stereochemistry of the cyclic compound to the geometry of the polyene.

> The 4π electrocyclic ring-closure will be conrotatory (we are assuming a thermal process unless we are told it is photochemical) so the cyclization of this double bond isomer of product will give this stereochemistry of cyclobutene.

It *is* possible to work out the orbitals for the electrocyclic ring-opening of the cyclobutene directly, but in this case we need to take a different approach. We need to define a HOMO and a LUMO.

In an electrocyclic ring-opening, we effectively have a σ-bond interacting with a π-bond. For the relevant HOMO-LUMO interaction, I will choose the HOMO of the σ-bond and the LUMO of the π-bond. These are as follows:

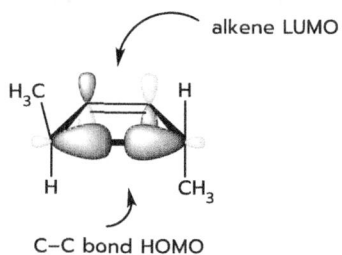

C–C bond HOMO

We must rotate the left-hand side anticlockwise so that the black lobe of the C–C bond HOMO can overlap with the black lobe of the π-bond LUMO. Similarly, we have to rotate the right-hand side anticlockwise so that the black lobe of the C–C bond HOMO can overlap with the black lobe of the π-bond LUMO. That is, the reaction is **conrotatory**, exactly as it was for the electrocyclic ring-closure.

SECTION 6 PERICYCLIC REACTIONS

So, which direction does a given electrocyclic reaction go in? Well, σ-bonds are generally stronger than π-bonds, so the cyclization is generally favoured. However, 4-membered rings are quite strained (Basics 6), which will disfavour a 4π electrocyclization. Substituents may also make a difference. Ultimately it is down to the thermodynamic stability of the starting material and product.

> *All that frontier orbital theory tells us is what the stereochemical outcome will be if the reaction takes place.*

SUMMARY OF STEREOCHEMICAL OUTCOMES

Thermal electrocyclization reactions can be rationalized from the HOMO symmetry. Photochemical electrocyclization reactions can be rationalized using the LUMO.

For a given number of π electrons, the thermal and photochemical reactions will have different outcomes (conrotatory or disrotatory).

BUT WAIT, WHAT DO CONROTATORY AND DISROTATORY *REALLY* MEAN?

Here is the diagram we saw before. We recognize that we cannot get all three π-bonds in a plane, because there are hydrogen atoms in the way.

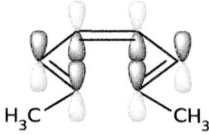

The idea of rotation of the ends of the π-system is a useful one for determining the stereochemical outcome of a particular reaction, but it is slightly misleading. It would be incorrect to say that the two ends of the π-system line up as shown, and then rotate to allow orbital overlap.

Instead, the ends of the π-system align so that overlap is achieved. Here is the shape of the transition state for a 6π disrotatory ring-closure. It's a boat! This brings the top faces of the two double bonds into close proximity, giving the observed stereochemical outcome.

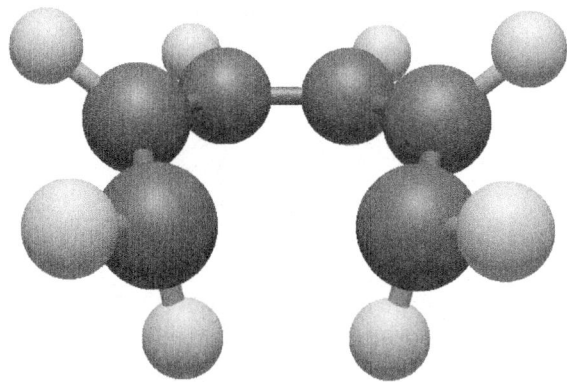

SECTION 6 PERICYCLIC REACTIONS

Don't over-think this. The terms 'conrotatory' and 'disrotatory' are a useful mnemonic for determining/remembering the stereochemical outcome of electrocyclic reactions. Be aware that the terms do have limitations.

> *But then crack on and use them anyway!*

WHAT DO YOU NEED TO DO WITH THIS?

When you know what the orbitals look like, you can predict the stereochemical course of any electrocyclic ring closing/opening reaction.

The more important thing is what the stereochemical course of the reaction ***means*** for the outcome. Usually in a conrotatory reaction, both ends 'rotating' clockwise is energetically the same as both ends 'rotating' anticlockwise.

> *It will be energetically the same if the products are enantiomers. It will not be energetically the same if the products are diastereoisomers.*

Just because we are now looking at reactions controlled by orbital symmetry, the basic rules (**Stereochemistry 2**) still apply.

From these general ideas of stereoselectivity, you should be able to determine and rationalize the outcome of any electrocyclic process, whatever the double bond geometries involved.

SECTION 6 PERICYCLIC REACTIONS

REACTION DETAIL 36
Sigmatropic Hydride Shifts

INTRODUCTION
Here is an example of a sigmatropic reaction. We defined these in **Fundamental Reaction Type 6**.

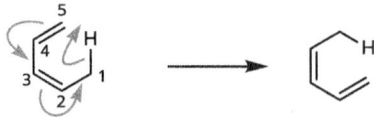

> In this case there is no overall reaction, since the starting material and product are identical. However, this reaction allows us to understand the nature of these processes, and to classify them.

This one is a 1,5-sigmatropic reaction.

> Look at the C–H bond that is being broken and the C–H bond that is being formed. Count clockwise from the bond that is being broken.

There are five atoms (all carbon) before the bond that is being formed. This is where the 5 comes from.

> Now count anticlockwise to the bond that is being formed.

Here there is just one atom (hydrogen). Therefore, this is a 1,5-sigmatropic reaction or, since it is H that is moving, a 1,5-hydride shift.

CONSTRUCTING THE FRONTIER ORBITALS
For the purposes of the frontier orbitals, we 'split' the molecule into two different parts, the π system and the C–H bond. We need to identify a suitable HOMO and LUMO. Since we are thinking of this as a **hydride** shift, it makes sense to use the electrons in the C–H bond. Therefore, we choose the HOMO of the C–H bond. The other component is a butadiene, and we choose the LUMO of this.

> Note the word 'choose'. We could use the butadiene HOMO and the C–H bond LUMO, and we would reach exactly the same conclusions.

There are two possible ways to set this up, and only one of them is correct. Let's look at both of them.

The difference between **A** and **B** is what happens when the butadiene LUMO and the C–H bond HOMO "meet". Look at the reaction scheme at the top of the page. We need to form a π bond between C1 and C2. Therefore, we need the correct

462

symmetry for these to overlap, as in **A**. In **B**, we have a node between C1 and C2, so this would be incorrect.

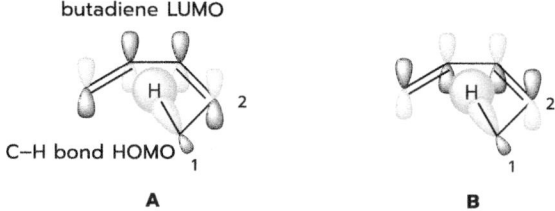

Now we can think about the stereochemistry of the reaction. That might seem like a rather odd way to think about is, because neither the starting material nor the product are chiral. Bear with me. Let's focus on the symmetry of the orbitals at C1 and C5.

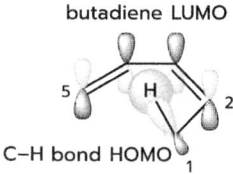

We can see that there is a light gray lobe on the upper face of the π system at both C1 and C5. Therefore the "hydride" can just "hop across". It is being transferred from the top face to the top face. There is a term for this—**suprafacial**.

> We could just as easily (and correctly) draw it being transferred from the bottom face to the bottom face.

Let's compare this with the next possible π system, a hexatriene. This will be a 1,7-sigmatropic reaction, or a 1,7-hydride shift.

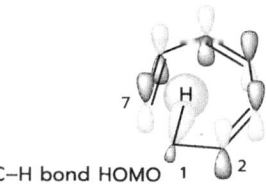

We have a hexatriene LUMO and a C–H bond HOMO, and the symmetry is arranged to give overlap between C1 and C2 to form a π bond (see the reaction below). Now, the light gray lobe of C7 is below the plane of the π system, so that for the hydride to transfer, it needs to go from the top of the π system on C1 to the bottom of the π system on C7. This is described as **antarafacial**.

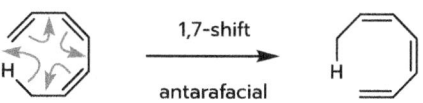

SECTION 6 PERICYCLIC REACTIONS

> *How do we know that this works? Indeed, how do we know that any reaction has taken place?*

For the specific examples shown, we would not know, since the starting material and the product are the same. However, for cases where substituents are added so that the starting material and product are different, this outcome can be confirmed.

It still isn't straightforward. One useful way is to use a cyclic system.

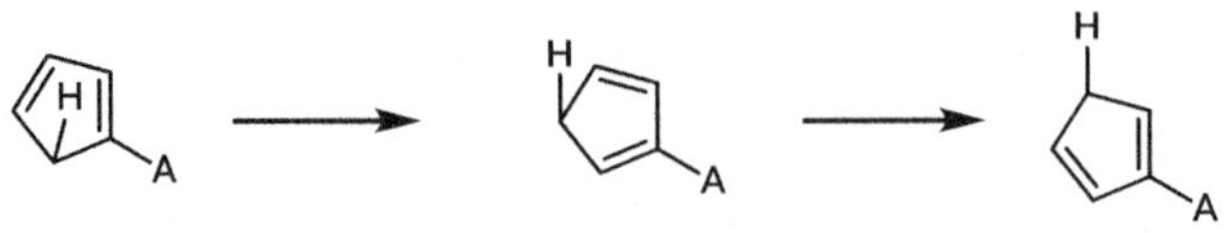

If a substituent "A" is used as a "marker", a 1,5-shift can take place as shown above. Actually, the second and third compounds are identical, and a further hydride shift will get back to the starting material.

In fact, there is a simpler experiment that we can do. Cyclopenta-1,3-diene has three chemically distinct hydrogen environments.

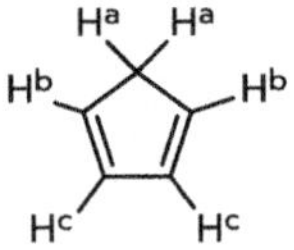

A ¹H NMR spectrum of cyclopentadiene should show three peaks, one for each hydrogen environment.

> *Don't worry that we haven't talked in detail about NMR spectroscopy. Different H = different peak. That's all we need for now.*

If we heat the cyclopentadiene in the NMR spectrometer (heat = faster reaction!!), we only get one peak. The hydrogen atoms all become equivalent.

> Starting with the structure above, draw a series of sequential 1,5-hydride shift mechanisms. Keep track of which 'type of H' they started as. Convince yourself that any of the hydrogen atoms can end up on any of the different carbon atoms.

The wording above is intended to give you 'just enough' information to do the job, but not so much that you don't have to work a little to understand the point.

Now let's compare this with the corresponding 1,7-hydride shift in a cycloheptatriene. This process needs to be antarafacial, which is not possible since there is a C–C bond between the carbon atoms at the migration origin and migration terminus.

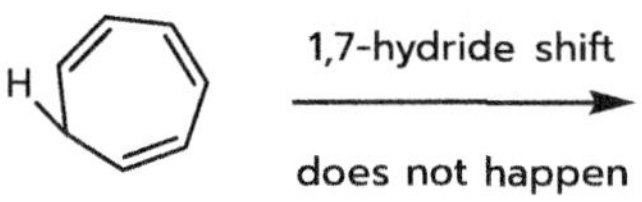

SECTION 6 PERICYCLIC REACTIONS

However, at least in principle a suprafacial 1,5-hydride shift is possible, although the hydrogen atom cannot get as close to the relevant carbon atom, so it isn't quite as easy. Two sequential 1,5-hydride shifts could give the same product as a single "suprafacial" 1,7-hydride shift.

> Draw this process, along with the curly arrows.

What about a 1,3-hydride shift? Well, if a 1,7-shift is antarafacial, and a 1,5-shift is suprafacial, you won't be surprised that a 1,3-shift would need to be antarafacial.

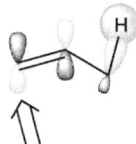

would need to transfer to here,
which is geometrically impossible

There is enough flexibility within the system to allow an antarafacial 1,7-hydride shift. However, there is no way that this can happen in 1,3-shift. Therefore, 1,3-hydride shifts **do not happen**.

PHOTOCHEMICAL HYDRIDE SHIFTS

We have already been developing a common theme—if you change reaction conditions from thermal to photochemical, there is a change in the outcome of pericyclic processes (cycloaddition, electrocyclic).

> *This is for a fundamental reason. Irradiation of a compound with UV light of the appropriate wavelength will promote an electron, normally from the HOMO to the LUMO. We have seen the molecular orbitals of butadiene in* **Recap 6**. *In general, we find that there will always be an additional node in the LUMO compared to the HOMO.*

Applying this to hydride shift reactions, any hydride shift that would be suprafacial thermally will be antarafacial photochemically, and *vice versa*.

In terms of drawing the frontier orbitals, instead of using the HOMO of one component and the LUMO of the other component, use the LUMO of both. Remember that photolysis will promote an electron from the HOMO (of the unsaturated component) to the LUMO.

ARE HYDRIDE SHIFTS THE ONLY SIGMATROPIC 1,n-SHIFTS?

No, but we are not going to cover the other types. I will mention briefly that there are 1,2- and 1,3-alkyl shifts. Because carbon has p-orbitals, these reactions look slightly different in terms of the symmetry of the orbitals. If you fully understand the orbitals for 1,n-hydride shifts, you won't have any trouble understanding alkyl shift reactions when you do encounter them.

REACTION DETAIL 37
[3,3]-Sigmatropic (Claisen and Cope) Rearrangements

[3,3]-SIGMATROPIC REACTIONS

There are two particularly important [3,3]-sigmatropic rearrangements. These are the Cope and Claisen[99] rearrangements as shown in their simplest possible form below.

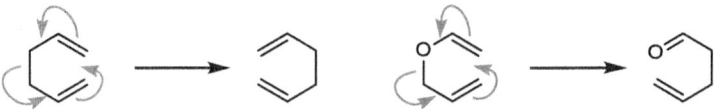

Cope rearrangement   **Claisen rearrangement**

Both reactions are widely used in synthesis, although the substrates and products are generally much more complex. The oxygen atom in the Claisen rearrangement doesn't affect the construction of the orbitals or the shape of the transition state, so we will focus our discussion on the Cope rearrangement. Don't confuse the Claisen rearrangement with the Claisen condensation (Reaction Detail 23)—same chemist, different reaction!

WHAT WILL WE NEED TO DO WITH THESE REACTIONS?

At this level, not much! All I want to do is introduce the reactions and their molecular orbitals. Like most pericyclic reactions, [3,3]-sigmatropic rearrangements often proceed with high levels of stereochemical control. The transition state of a Claisen or Cope rearrangement looks like a cyclohexane chair. Just as with aldol reaction transition states (Stereochemistry 14) everything you know about the drawing and conformational preferences of cyclohexanes can be applied here.

First of all, we will see one method of constructing the orbitals.

CONSTRUCTING ORBITALS

There are several different ways to draw orbitals that rationalize the outcome of [3,3]-sigmatropic processes. We are just going to look at one method. You will see different (but equivalent) representations of the orbitals in various organic chemistry textbooks.

For the method we will use, we will construct a HOMO-LUMO combination. We can't use the HOMO of just the C–C bond, because that would leave two different

---

[99] The report of the rearrangement of allyl phenyl ether in 1925 was Claisen's last ever publication.

components for the alkene bonds. Therefore, we have one component (conventionally the HOMO) of a C–C and C=C bond. This is effectively a butadiene-type system, and it is shown on its own below.

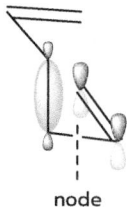

node

Effectively, this is a C–C bond HOMO and a C=C bond HOMO with a node between them to make it like a butadiene. We are making one key assumption here—that the transition state is a chair. We will justify this in a minute.

Now we need to add the other component, and bearing in mind that we are going to form a C=C bond, we need overlap between the HOMO and LUMO where this will happen. Remember, we did the same thing with the orbitals of a hydride shift in **Reaction Detail 36**.

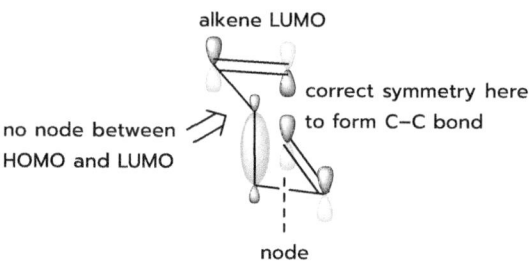

> We don't normally draw a cyclohexane chair like this. For some reason, if I draw the 'normal' orientation of a cyclohexane, I cannot see the orientation of the orbitals clearly. But if I draw this representation, I cannot see the stereochemical outcome clearly. Before we see how I handle this, there's one more thing to look at.

In the above diagram, we have shown that the symmetry is correct, which is why the reaction works. Let's look at the same reaction but with a boat transition state.

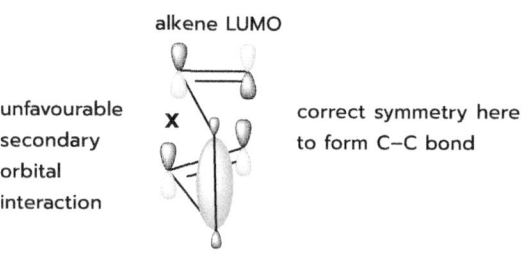

The only difference here is that there is an unfavourable secondary orbital interaction as shown. We encountered favourable secondary orbital interactions in **Stereochemistry 16**.

CHAIR VERSUS BOAT—DEEPER THINKING!

I've got to be honest, this is quite a weak argument. The chair was always going to be more favourable than the boat on steric grounds, so do we really need to justify why it is a chair transition state? Maybe not, but at least we can see the orbitals as well.

We should take a step back and remind ourselves why the chair form is normally favoured over the boat form. Here are calculated chair and boat transition states for the Cope rearrangement of 1,5-hexadiene. In the boat transition state, we do not have the unfavourable flagpole interactions that we identified in **Stereochemistry 1**. The boat isn't all that bad in this case.

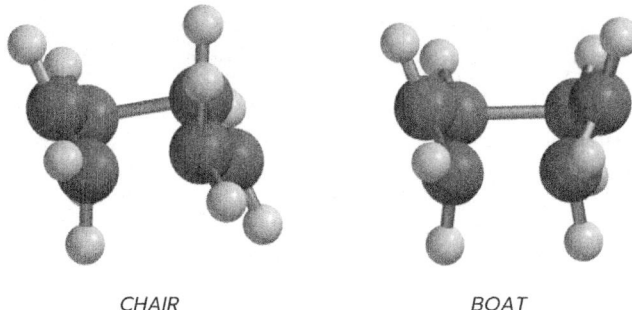

CHAIR   BOAT

However, when we look down the axis of the bonds we are forming/breaking, we can see that there is considerable eclipsing of bonds (torsional strain) in the boat transition state.

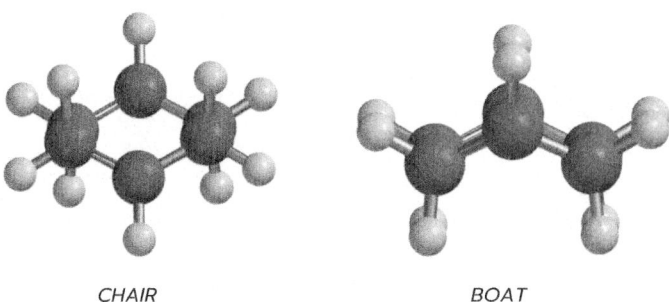

CHAIR   BOAT

> How much of the destabilization of the boat transition state is due to the secondary orbital effect, and how much is due to torsional strain?

It isn't easy to tell. The point is, you should question everything, and consider that the explanation may be very fundamental. Don't over-complicate things.

A SIMPLE STEREOCHEMICAL EXAMPLE

We need to understand how to use this theory. Once we've established that a chair transition state tends to be favoured, we don't need to draw the orbitals.

SECTION 6 PERICYCLIC REACTIONS

Let's look at this in the context of a simple problem. If the compound below on the left undergoes a Cope rearrangement, what will the stereochemistry of the product be?

> In the structure on the right, the wiggly bonds mean that the stereochemistry is not specified.

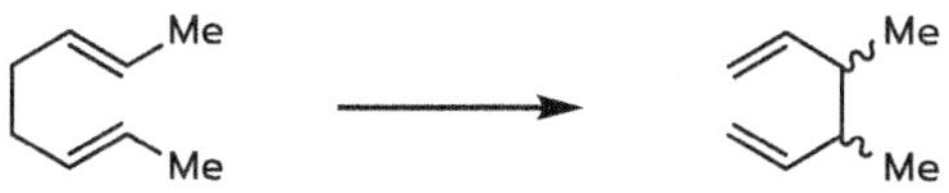

Effectively, all we have to do is work out where the substituents will be in a chair transition state. Bulky substituents would ideally be equatorial, but this is constrained by double bond geometry so it might not always be possible.

> We saw the same situation in transition states for aldol reactions in **Stereochemistry 14**.

In this case, the *trans*-alkenes do lead to equatorial substituents, as shown below. I have added a dashed bond to indicate the C–C bond that will be formed.

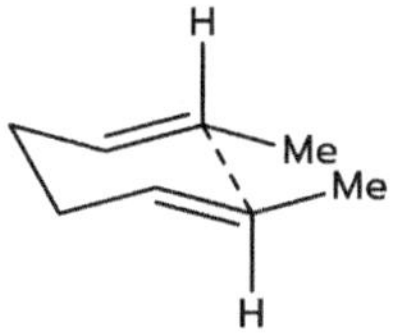

Therefore, the outcome is as follows.

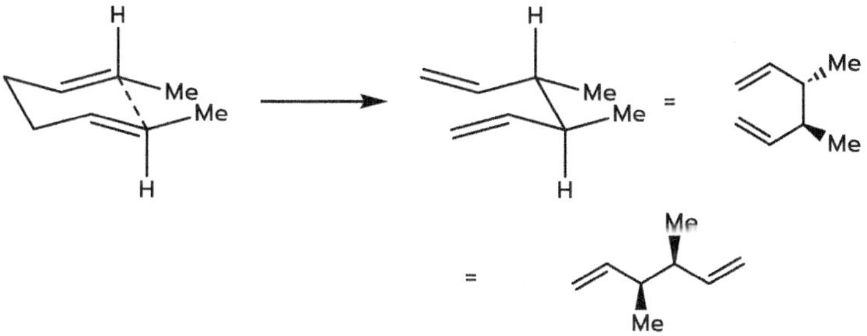

> I have shown the product in three different representations. By now, you should already be familiar with the interconversion of different stereochemical representations.

469

SECTION 7

TOTAL SYNTHESIS

SECTION 7 TOTAL SYNTHESIS

INTRODUCTION

Organic chemists like to make molecules. There are many reasons for doing this. At a fundamental level, organic molecules have very useful properties, including potent biological activity.

> *It is no exaggeration to say that organic molecules save lives!*

As organic chemistry was developing as a science, chemists looked to 'interesting' biological molecules for inspiration. It's the organic chemistry equivalent of climbing Mount Everest. In many cases, a chemist tackling a challenging target will find that they need to 'invent a new reaction' in order to meet the target.

> *Of course, 'invent' is a strong word—any reaction that an organic chemist 'invents' must follow the general principles that we have outlined. They do, but they still permit a great deal of creativity!*

Sometimes, the structure of a compound is not known. By making a proposed structure using reliable reactions, it is possible to determine the structure of the compound.

> *Spectroscopic methods are generally used these days, and they are very powerful.*

What we will do in this section is look at a few syntheses of interesting molecules, and show how the reactions you have already encountered have been applied in a more complex setting. This section will not be easy going. At first, you will look at the larger structures, and you won't be able to 'see the reaction' within it.

> *That's normal! You only get good at this stuff with practice and experience. After a while, you'll be spotting aldol reactions everywhere!*

You will get to this point by drawing out the mechanisms as directly applied to more complex structures. Don't be tempted to draw a simplified version of the molecule, even if it would be quicker.

TOTAL SYNTHESIS 1
Synthesis of Strychnine

INTRODUCTION

There are many names associated with the development of total synthesis as a science and as an art. Robert Burns Woodward (1917–1979) is arguably the 'biggest' of these names. Woodward was awarded the Nobel Prize in Chemistry in 1965 for his contributions to total synthesis, and was one of the two pioneers of the molecular orbital theory described in Section 6. His collaborator, Roald Hoffmann, shared the Nobel Prize in 1981 for this work, and Woodward would have undoubtedly also been a co-recipient had it been possible to award a Nobel Prize posthumously.

The synthesis of strychnine, reported in 1954, was a landmark. Bigger molecules have been made since, but this is still a complex target. The synthesis is quite long, and we will not go through all of it.

> The point of including it here is that you have already encountered the mechanisms of most of the reactions.

Possibly a bigger point is 'building skills'. The key skill is to be able to look at a complex molecule and to identify the bits of it that are undergoing reaction and the bits of it that don't change. You then need to be able to draw the mechanism when it is directly applied to the more complex compound.

The way we will do this is by looking at key steps in the synthesis, and by signposting you to the reactions that are involved. You should take this opportunity to revise the basic mechanism, and then draw it out for the specific example in this chapter.

MANAGING EXPECTATIONS

You should not try to memorize the structure of strychnine, and you should definitely not expect to remember the sequence of steps. That would be pointless at this stage.

We are focused on developing your skills in the drawing of reaction mechanisms. You should draw every step in each mechanism carefully. You should check that you have drawn the rest of the structure clearly.

> Don't abbreviate the structures, and especially don't leave off the stereochemistry.

The 'next level' would be to address aspects of selectivity. In a complex synthesis, a reaction takes place at one functional group and leaves another unchanged. We will only address this at a very basic level here.

SECTION 7 TOTAL SYNTHESIS

In subsequent chapters, the syntheses will be shorter (or we will only look at selected steps), which will allow us the opportunity to look at selectivity in more detail.

MAKING A START

Here is the structure of strychnine. I have labelled each of the rings so we will be able to keep track of them through the synthesis. It isn't easy to draw this structure. In particular, we have some very strange looking bonds in the 'G' ring.

> Make a model of the compound. It's not as bad at this representation would seem to indicate.
>
> Determine the molecular formula of the compound, and the number of double bond equivalents (**HTSIOC Habit 2**). Make sure you can identify the double bond equivalents on the structure.
>
> Identify the stereogenic centres in strychnine, and assign their stereochemistry (**HTSIOC Habit 6**).
>
> Name as many of the functional groups in strychnine as you can.

None of this will help you understand the synthesis. The purpose of these exercises is to get you looking **very carefully** at the structure.

> Oh, and I'm not going to give you the answers. You **need** to be able to get the answers right, and it doesn't matter if it takes a few attempts.

I will just give you one hint—the number of nitrogen atoms is even, so the number of hydrogen atoms will also be even.

STARTING THE SYNTHESIS

The first step is the synthesis of indole **3** from phenylhydrazine **1** and ketone **2**.

Phenylhydrazine has two nitrogen atoms with lone pairs.

> Which nitrogen atom will be the better nucleophile?

SECTION 7 TOTAL SYNTHESIS

Now it's time to develop your process. Identify the key difference. There is only one—the phenyl group. Does this stabilize or destabilize the lone pair on the left-hand nitrogen atom?

> Can you draw resonance forms to show any stabilization?

We don't actually have an example in **Recap 7** of a carbanion with a negative charge conjugated with a phenyl group, but we do have examples that rely on the same principles.

> Hang on! Why is he talking about carbanions when this is about amines?

Electrons are electrons! If something stabilizes a pair of electrons on carbon (a negative charge), it will probably stabilize a pair of electrons on nitrogen (which could be a lone pair, or it could be a negative charge).

> Isn't it wonderful? One idea that you apply consistently, everywhere!

The nitrogen atom on the right is a better nucleophile. Amines react with carbonyls (**Applications 2**). Have a look at that chapter. We form an imine.

> Okay, compound 1 is called a hydrazine rather than an amine, and the product you are trying to form is called a hydrazone rather than an imine, but it's the same reaction.

> Draw the mechanism of formation of the 'imine'.

Take your time. Make sure you have all the curly arrows going in the right direction. Make sure you have all the proton-transfer steps shown correctly. Make sure you don't have a negative charge on nitrogen, and you don't have hydroxide as a leaving group.

> Does this seem a bit familiar? We looked at the same problem in **Reaction Detail 29**.

I'm not going to give you the imine structure. In fact, the next step in the mechanism requires the enamine rather than the imine. These are in equilibrium, much like keto and enol tautomers (**Fundamental Reaction Type 4**).

> Here is the structure of the enamine. Draw the mechanism of the tautomerization.

4

475

SECTION 7 TOTAL SYNTHESIS

If this is really difficult, you might have got the imine formation wrong. Don't beat yourself up about it. Just use structure **4** to help you correct any mistakes you might have made.

> *If you are really struggling, number the atoms in compounds 1, 2 and 4, and check that you haven't missed anything. It's all about identifying a strategy that works for you.*

The synthesis of compound **3** is a 'classic' heterocycle synthesis called the Fischer indole synthesis. It is a really useful reaction, but it is truly horrible to learn.

> *If you look at compound 3 and try to draw a mechanism for its formation from compounds 1 and 2, you are doomed to fail!*

If you look at compounds **1** and **2** and ask yourself 'how might they react together?', you have a fair chance of getting to compound **4**.

The next step in the mechanism is the 'problem'. It is a [3,3]-sigmatropic rearrangement (**Reaction Detail 37**).

> Here is the product. Draw the curly arrows. We don't need the orbitals.

5

You might not believe it, but we are almost there. One of the imine groups in compound **5** tautomerizes to give an enamine.

> Which one? Think about the thermodynamics (Hint—**Recap 6**).

I don't want to draw any more structures from this point. Have another look at **Applications 9**. You should have a structure with an enamine and an imine. You have acid present.

> Cyclize the enamine onto the imine and lose ammonia to form product **3**.

As before, make sure the curly arrows are going in the right direction, and make sure you protonate/deprotonate appropriately to give you sensible structures without negative charges or unstable leaving groups.

> *Remember, if you need more help, this is a Fischer indole synthesis. You can look it up!*

The next step in the synthesis is as follows.

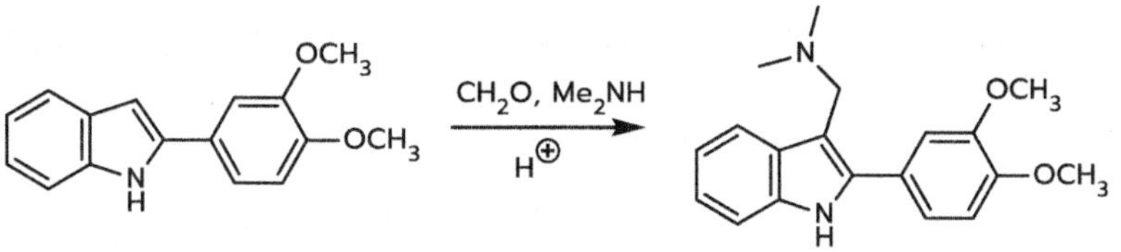

SECTION 7 TOTAL SYNTHESIS

This reaction involves formaldehyde (CH$_2$O) and dimethylamine, and an acid catalyst. There is a bond-formation to an aromatic ring, in this case the indole ring.

We have a *secondary* amine reacting with a carbonyl compound. We would normally expect to form an enamine (**Applications 2**) from a *secondary* amine, but you cannot do this with formaldehyde.

> Start drawing!!!

You should have drawn the formation of iminium ion **7**.

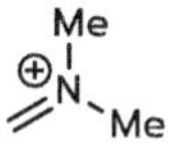

Indole **3** is aromatic. We could have aromatic electrophilic substitution (**Reaction Detail 8**) at any of the positions indicated on the structure below. Substitution at the 3-position is favoured.

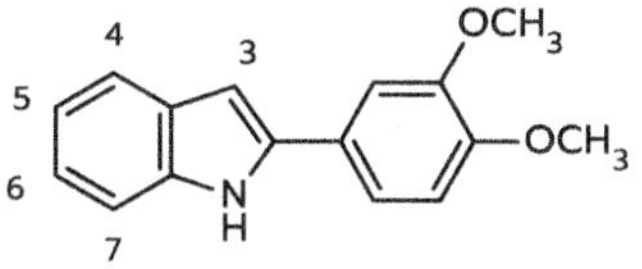

> Draw a mechanism for aromatic electrophilic substitution at each position. Draw the resonance forms for the intermediate in each case, and work out why substitution at the 3-position is preferred.

This is a lot of work, but it is important work! This is how you get good at organic chemistry. Remember, it isn't just a case of drawing the mechanism. Understanding which outcome is favoured is much more important.

Of course, in this case, substitution is not taking place on a benzene ring.

> That doesn't matter!

What matters is that you can apply fundamental principles (in this case carbocation stabilization—**Recap 2**) to solve a problem.

I'm going to skip a couple of steps. The next step I want to look at is the conversion of **8** into **9**. We use 4-toluenesulfonyl chloride. We haven't seen this reagent before, but it looks (and reacts) rather like an acid chloride.

477

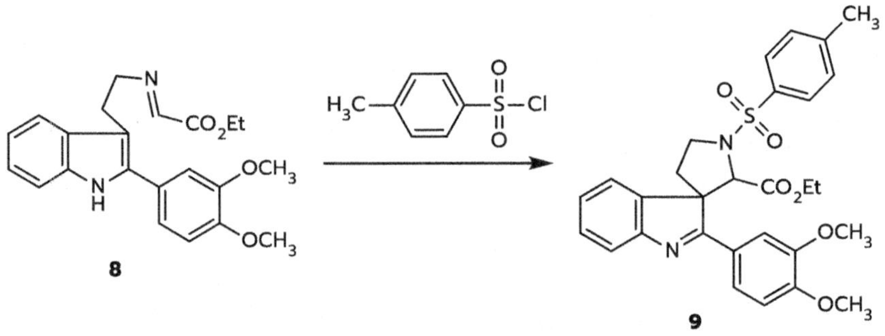

SECTION 7 TOTAL SYNTHESIS

> Draw a mechanism for the imine nitrogen in compound **8** reacting with 4-toluenesulfonyl chloride. Look at the end of **Reaction Detail 16** for inspiration. As a starting point, draw the lone pair on the nitrogen atom!

This won't get you all the way to compound **9**. You should get to an iminium ion (**Applications 2**) with sulfur attached to nitrogen.

> Now use the enamine to form the ring. This is very much like the aromatic electrophilic substitution reaction you drew to form compound **6**.

Now have another look at it. Imagine those nitrogen atoms were oxygen atoms instead. Then it would be an aldol reaction (**Reaction Detail 23**).

> *This is your real challenge—getting to the point where you see the fundamental types of reactivity, and you see past the superficial differences.*

I don't want to give you unrealistic expectations—it will take time and effort. It will take lots of 'drawing and looking'. Just reading about it won't make it 'stick'. But if you look at the above cyclization reaction, and compare it with the aldol reaction, it will help you build the patterns.

Compound **9** has two stereogenic centres. In the original synthesis, the stereochemistry was not determined, and it didn't really matter. Here is compound **9** as it is (as of 2023) drawn on a well-known web site.

> Assign the stereochemistry (**HTSIOC Habit 6**) of stereogenic centres 'a' and 'b'.

Take your time and have a good go at this.

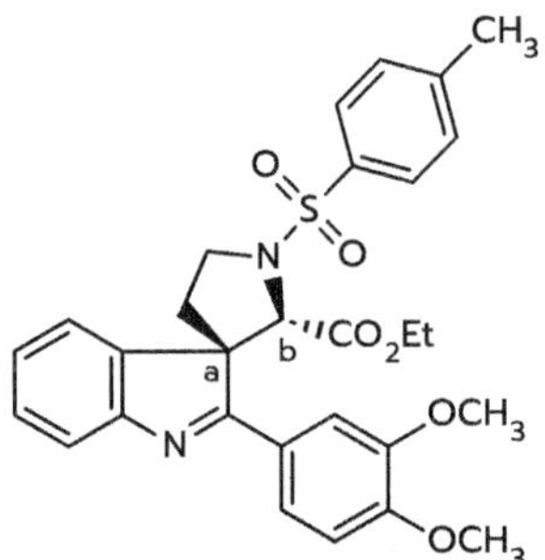

478

SECTION 7 TOTAL SYNTHESIS

Now read the comment at the bottom of the page.[100]

We now have two steps that are relatively straightforward. Reduction of compound **9** using sodium borohydride (**Reaction Detail 14**) was followed by reaction with acetic anhydride to give the amide (**Reaction Detail 16**) **10**.

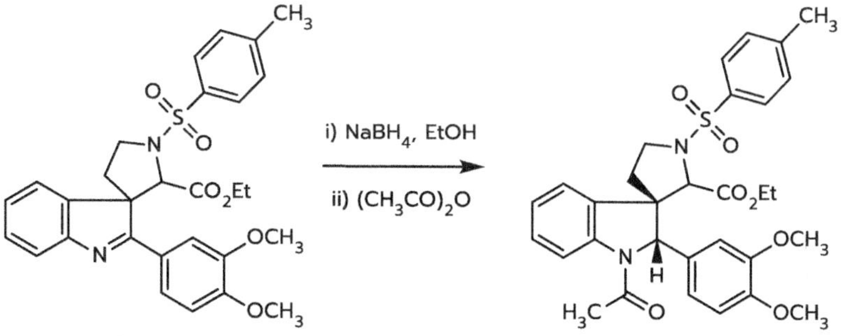

> Draw plausible mechanisms for both steps.

We now have a tricky step. Compound **10** was reacted with ozone to give diester **11**

> Draw a mechanism for this process (**Reaction Detail 6**), as far as the ozonide.

The last time we saw ozonolysis, it was 'just' an alkene. Here, we have a very electron-rich benzene ring, and ozonolysis works here as well.

> It's just a question of reactivity. Compared to an alkene, the same reaction of benzene will be slower, and may ultimately give a different outcome for reasons we saw in **Fundamental Reaction Type 3**. In this case, we are never going to get the aromatic ring back after ozonolysis.

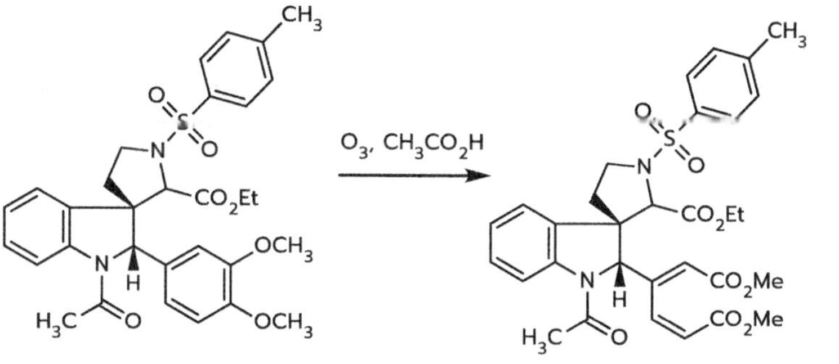

---

[100] Although stereogenic centre 'a' only has one wedged bond, the stereochemistry is unambiguous. This one is 'R' as drawn, although it isn't easy to assign. If you can work out where in space to add the 'H' on stereogenic centre 'b', you are doing better than me. This stereogenic centre is drawn ambiguously and cannot (in my opinion) be assigned.

479

SECTION 7 TOTAL SYNTHESIS

Oh, and did you notice that the second time, I drew a different resonance form for the benzene ring in compound **10**? That's fine, because we know what resonance forms mean!

Next up we have one of those steps when quite a lot happens, but each part of it is *relatively* simple.

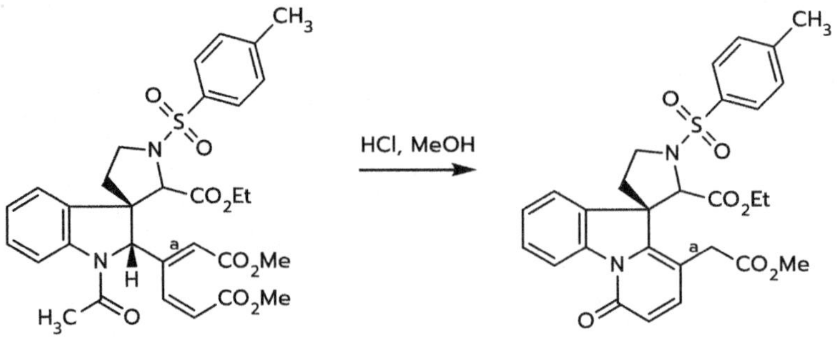

11 → HCl, MeOH → **12**

The first thing you need to do is work out which atoms in the starting material end up in the product. I would try numbering the atoms in both compounds.

> You are losing the acetyl (CH_3CO) group from nitrogen.

The next thing to do is work out what happens first.

> It would be difficult to form the new ring until you have lost the acetyl group.

> Draw a mechanism for the cleavage of amide **11** using methanol/HCl. You have a source of $H^\oplus$, but no water (so it is not a 'hydrolysis'). Use methanol as the nucleophile. Look to **Reaction Detail 16** for inspiration.

Make sure you didn't cleave the C–N bond to give a negatively-charged nitrogen atom. If you did, have another go and protonate the nitrogen atom first.

> You now need to form the **D** ring. To do this, you need to rotate some bonds (**Stereochemistry 5**) in structure **11** to bring the ester carbonyl group close to the nitrogen atom.

There is one more thing to note. The double bond 'a' in compound **11** has moved into the new **D** ring.

> Draw a mechanism for this process. Suggest why this might be favoured.

> **HINT:** Draw a resonance form for the ring we are focusing on. Is it 'special'?[101]

---

[101] This is a rubbish hint! But after a while you will see a hint like this and you will know exactly what the point is and which resonance form to draw.

SECTION 7 TOTAL SYNTHESIS

Take a break! Really! This is a long synthesis. We are only going to do a couple more steps. So far, there has been a lot of 'amine + carbonyl' chemistry. Sometimes it's an imine. Sometimes it's an enamine. Sometimes it's an amide. But if you draw correct curly arrows and mechanisms, the answers tend to 'drop out'. Before you carry on, I strongly recommend going back to the start of the chapter and going through the mechanisms up to this point one more time.

We are skipping a couple of steps now, and we get to compound **13**.[102] This underwent Dieckmann cyclization (**Reaction Detail 23**) to give compound **14**.

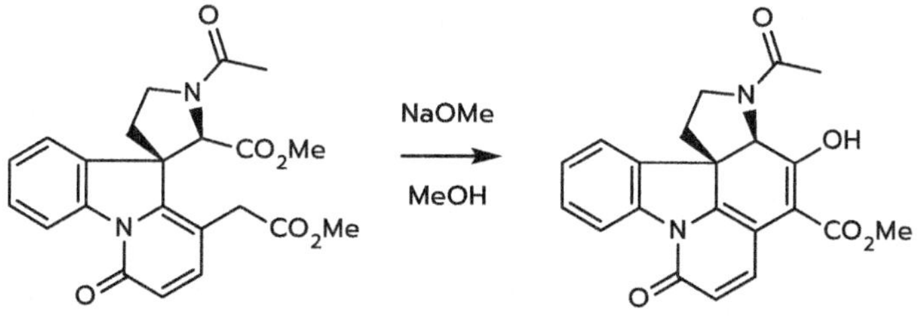

> Draw a mechanism for this reaction.

Remember that NaOMe is 'really' MeO$^\ominus$, and make sure you know what CO$_2$Me is—a methyl ester.

Compound **14** is an enol. This isn't quite what we saw in **Reaction Detail 23**.

> Suggest a reason why the enol is favoured in this case.

We are going to skip to the end of the synthesis now. I'm going to simplify the story a little, but basically, treatment of compound **15** with HBr and sulfuric acid leads to an isomerization to give compound **16**.

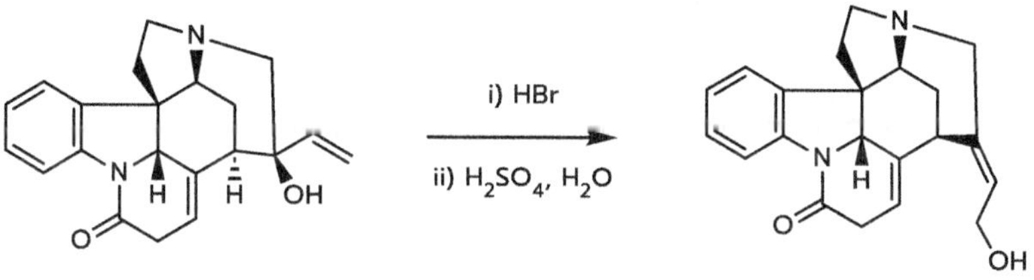

> Draw a mechanism for this process and explain why product **16** is favoured. Have another look at **Reaction Detail 4**. That chapter is about dienes. This example is not. But there are still some important common themes.

---

[102] There were a few challenges getting to this point. Woodward described compound **13** as being "free of the seed of potential decay with which its predecessor was afflicted". They don't write synthesis papers like that any more!

SECTION 7 TOTAL SYNTHESIS

Now the final step in the synthesis. Compound **16** was treated with sodium hydroxide (just draw $^{\ominus}$OH in the mechanism!) to give strychnine.

16 → KOH, EtOH → strychnine

These are not conditions for addition of an electrophile to an alkene bond. Something else must be happening!

> Draw a mechanism that fully explains this transformation.

> **HINT:** The ring is formed in a conjugate addition reaction (**Reaction Detail 28**).

LESSONS FROM THIS SYNTHESIS

This really is one of the classic syntheses from the leading organic chemist of the time.

> *A classic synthesis uses 'older' reactions, and most of the reactions covered in this book are older reactions.*

This is appropriate, as the fundamental types of reactivity were defined relatively early on.

Strychnine is a complex molecule. There are inherent challenges associated with the synthesis of complex molecules. These challenges often concern chemoselectivity, regioselectivity and stereoselectivity (**Recap 4, Stereochemistry 2**).

Fundamentally, though, simple molecules and complex molecules undergo the same reactions with the same reagents.

> *It takes longer to draw the more complex molecules, and you might make more mistakes at first.*

Just persevere. After a while you will look at the bigger structure and the reaction conditions, and you will home in on the part of the molecule that is going to react.

This is a good synthesis for us to start with. It's a complex molecule, but not too complex. There is lots of 'amine + carbonyl' chemistry, and a little 'aldol type' chemistry.

> *This is the chemistry that needs a lot of practice, so that you can fully internalize the common threads.*

SECTION 7 TOTAL SYNTHESIS

Of course, organic chemists have made strychnine many times since. The more modern approaches are often much shorter and more efficient.

> They use reactions that had not been invented when Woodward did his work.

Perhaps they are more elegant, but they are only possible as a result of the earlier pioneering work.

I will leave you with one more thought. All the work carried out in this chapter was done without recording a single NMR spectrum!

Woodward wrote very long and elegant papers describing the synthetic efforts of his group. The strychnine paper is 42 pages, and really does go into the thought process used in designing such a synthesis. The citation is R. B. Woodward, M. P. Cava, W. D. Ollis, A. Hunger, H. U. Daeniker and K. Schenker, *Tetrahedron*, **1963**, *19*, 247-288. It's well worth a look!

SECTION 7 TOTAL SYNTHESIS

TOTAL SYNTHESIS 2
Synthesis of Prostaglandin F2α

INTRODUCTION

Prostaglandins are a group of really important biological molecules. There are many of them, and they are found in almost every tissue in your body. They have a diverse range of biological activity, including roles in inflammation and blood platelet aggregation.

> *They are worthy targets for chemical synthesis.*

The first prostaglandin to be synthesized was prostaglandin F2α, in 1969, by the group of Elias J. Corey. Corey was awarded the Nobel Prize in Chemistry in 1990 for his contributions to chemical synthesis.

Here's the structure of prostaglandin F2α.

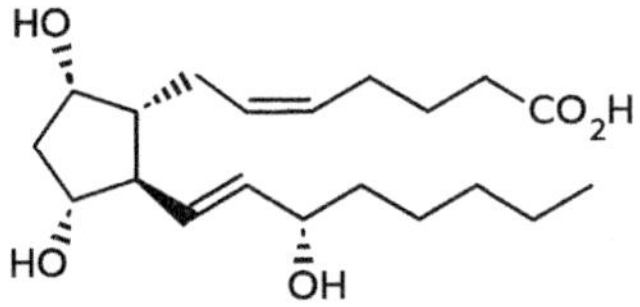

Let's establish some facts. This compound has twenty carbon atoms, one ring[103] and five stereogenic centres. There are also two alkene bonds, one *cis* and one *trans*, that will need to be produced with defined geometry.

Of course, there is no 'single best' approach to the total synthesis of any complex target. The Corey synthesis produces the target in racemic form, and requires 17 steps. A more recent synthesis, which we will look at briefly at the end of this chapter, only needs 6 steps, and produces a single enantiomer of the compound.

> *This newer approach uses a compound similar to a key intermediate in the Corey synthesis.*

THE COREY SYNTHESIS

The first step is deprotonation of cyclopentadiene (**1**) with sodium hydride.

> Have a look at **Recap 7** and work out why the anion we get is particularly stable. You might also need to look at **Recap 6**.

---

[103] Don't tell Sauron!

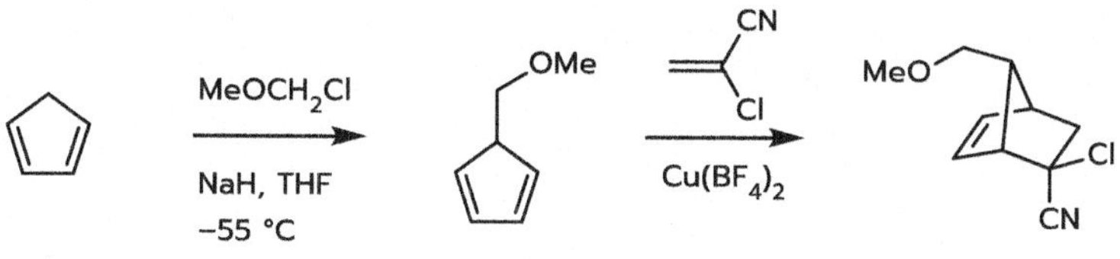

SECTION 7 TOTAL SYNTHESIS

The anion formed (draw it!) reacts with methoxymethyl chloride to give compound **2** in an S$_N$2 substitution (**HTSIOC Reaction Detail 1**). So far, so good.

The next step is a Diels-Alder reaction (**Reaction Detail 34**). There are two aspects to the stereochemistry of this reaction. Have a look at the product. I've drawn the *endo* isomer.

> Have a look at **Stereochemistry 16** and make sure you can see why this is *endo*. This will involve drawing it!

In fact, a mixture of *endo* and *exo* stereoisomers was formed, and Corey did not report the ratio. As we will see in a moment, it didn't matter.

What did matter, though, is the stereochemistry of the methoxymethyl group. Here is compound **2** again. Now, I've drawn it with a wedged bond. The dienophile approaches the lower face of the cyclopentadiene ring because it is less hindered.

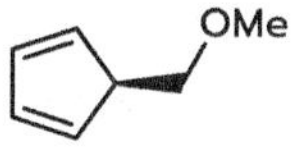

> A copper salt is used as a Lewis acid catalyst in this transformation. We aren't going to worry about that here.

Now we get to a step that we are going to spend a little more time on. Here is the transformation. It's not an 'obvious' reaction.

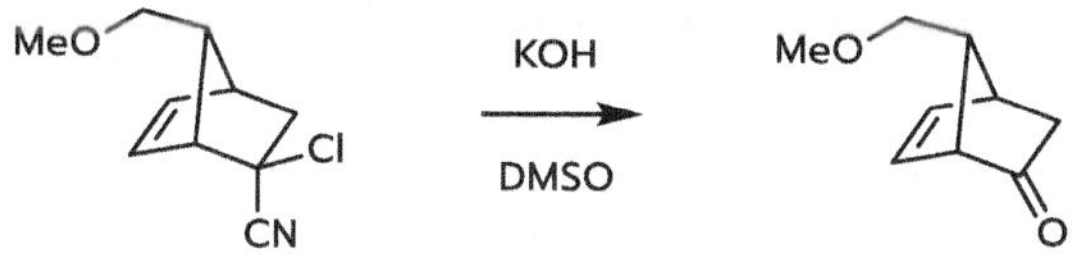

DMSO is dimethylsulfoxide. In this case it is acting as a solvent. It is polar, and it doesn't have any really acidic hydrogen atoms. Potassium hydroxide is a base.

> All bases are nucleophiles, all nucleophiles are bases.

We saw (**HTSIOC Reaction Detail 1**) that a leaving group next to a carbonyl is great in S$_N$2 substitution reactions.

> A leaving group next to a nitrile (cyano group, sp hybridized) is also great.

485

SECTION 7 TOTAL SYNTHESIS

We have a nucleophile and we have a leaving group. Fortunately, the cyano group is quite small, so it's not too hindered.

> Draw the S$_N$2 reaction mechanism.

What does this give us? Have a look at **Reaction Detail 2**.

How might you get from this compound to ketone **4**? Remember that all reactions are reversible.

> I'm not going to give you the answer here. It's not that I think it's too easy. But if you've got this far, I am confident that you can take this next step.

The next step is a Baeyer-Villiger oxidation (**Reaction Detail 32**).

$$\text{4} \xrightarrow[\text{CH}_2\text{Cl}_2]{m\text{-CPBA}} \text{5}$$

The reagent is *m*-CPBA. We saw this reagent in **Reaction Detail 32**. We looked at migratory aptitude in **Reaction Detail 31**.

> You could get two different Baeyer-Villiger oxidation products from compound **4**. Draw the mechanisms for formation of both of them. Can you see why compound **5** is the 'best' one?

This is regioselectivity, but there's also chemoselectivity (both in **Recap 4**). The reagent, *m*-CPBA, reacts with the ketone rather than with the alkene in **5**. At this level, there is no way you could have predicted that.

The next step is an ester hydrolysis (**Reaction Detail 16**). In this case, the ester is in a ring, so we call it a lactone, but that doesn't change anything. The other thing I have done is change how I draw the product **6**.

$$\text{5} \xrightarrow{\text{NaOH, H}_2\text{O}} \text{6}$$

This sort of problem is at the heart of organic chemistry. For a given structure, one representation 'works' better than another.

> Draw structure **6** in the same orientation as structure **5**. It doesn't look too bad, but as you get further through the synthesis, the same representation will look worse. Draw structure **5** in the same orientation as structure **6**. It will look very messy. Draw the mechanism for the ester hydrolysis reaction.

SECTION 7 TOTAL SYNTHESIS

The next step uses a slightly unusual reagent, KI_3. You are probably best to think of the tri-iodide anion as iodide plus molecular iodine, I_2. Have another look at the very end of **Reaction Detail 1**. In that case, we added bromine to an alkene to form a brominium ion, and then we added water as a nucleophile.

Here we have a poor nucleophile, but it's close to the double bond, so it reacts quickly. And it's on the lower face of the double bond, so this reaction is **stereospecific** (**Stereochemistry 2**). Then the next step is an ester formation from an anhydride. Pyridine is often used as a base in these reactions.

> Draw mechanisms for both steps. You'll notice I haven't told you what 'Ac' is. You could look it up, or you could simply draw the mechanism with the reagents given.

Compound **8** has eight of the carbon atoms in prostaglandin F2α. I am going to gloss over a couple of steps now, and pick up the synthesis with the aldehyde **9**. The next step is a Horner-Wadsworth-Emmons modification of the Wittig reaction (**Reaction Detail 26**) using reagent **10** and sodium hydride as base. This gives *trans* alkene **11**.

> As always, draw the mechanism out!

Have another look at the structure of prostaglandin F2α. The remaining five carbon atoms are attached *via* a *cis* double bond. It makes sense to form this bond last as *cis* alkenes can undergo isomerization to *trans* double bonds under some reaction conditions.

> *You wouldn't have known that, but it's worth mentioning the strategic point. The art of total synthesis is all about getting the steps in the best possible order.*

I've skipped a couple more steps, as I want to focus on an important principle. You will notice that in compound **8**, we put an acetate group onto the oxygen. We didn't do this because we wanted the acetate. We did it because we didn't want the alcohol to do anything else.

487

SECTION 7 TOTAL SYNTHESIS

> It is something we call a protecting group. We mentioned them in **Reaction Detail 15**, but there's no harm in reminding you.

Here is another one, a tetrahydropyranyl ether. Actually, it's an acetal (**Reaction Detail 15**), not an ether.

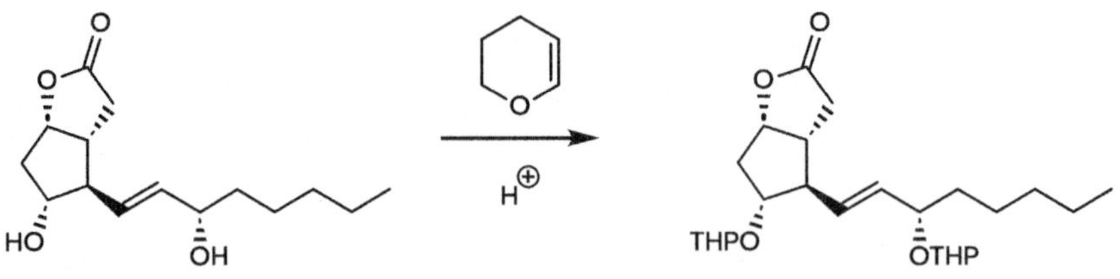

This is what the important part of the molecule looks like—remember, it has two of these acetals. The wiggly lines show where I have abbreviated the structure.

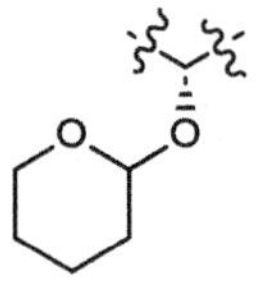

> Again, draw a curly arrow mechanism. **Hint**—the initial protonation is **not** on oxygen.

You will probably spend quite a bit of time in **Reaction Detail 15** working this out. That's okay, it will be time well spent. These are important mechanisms.

Next, we have a reaction that you have seen before, in **Reaction Detail 14** and in **Applications 3**. This one is slightly different. We reduce the ester, but only as far as the hemiacetal (**Reaction Detail 15**).

> For now, just draw the curly arrows. We are using a particular aluminium hydride that doesn't always reduce the ester right down to the alcohol. The main point is that the curly arrows for **all** hydride reductions will look the same.

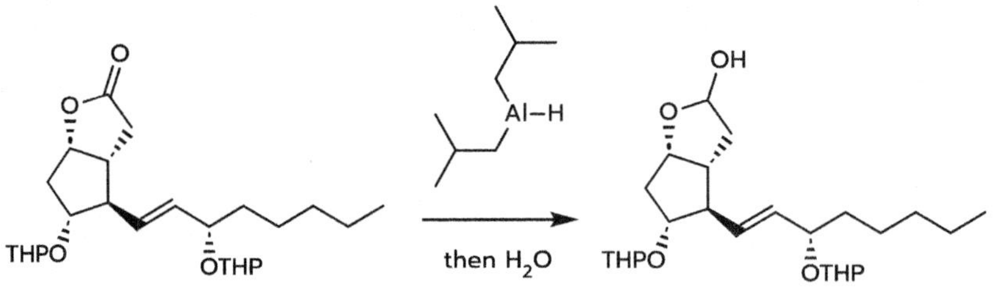

The next step is actually very elegant. It's really just a Wittig reaction (**Reaction Detail 26**), but I can't see an aldehyde in compound **15**.

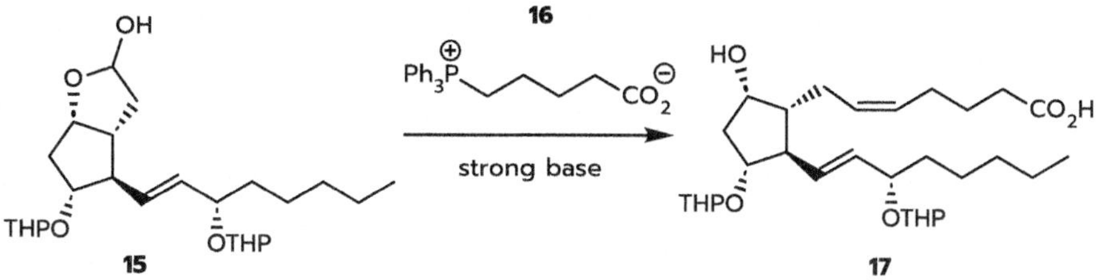

SECTION 7 TOTAL SYNTHESIS

Have another look at **Reaction Detail 15**, particularly the section 'The Role of Entropy'. The hemiacetal in compound **15** is in equilibrium with the corresponding hydroxyaldehyde.

> Draw it!

Compound **16** is a phosphonium salt. There is probably a halide counter-ion somewhere, but the original report is vague in this regard. Either way, adding a base will deprotonate the carboxylic acid group first.

> Think of this as loss of HBr.

There's just one final step. The tetrahydropyranyl ethers (acetals!) have done their job. We can remove them. They have 'protected' the alcohol functional groups through two steps.

> We call them protecting groups. In this case, I'm not convinced they are needed.

> Draw a mechanism for the hydrolysis of one of the THP (we generally use the abbreviation) ethers. Look at the previous page for the full structure. The stereochemistry in the starting material and product should tell you where the water does *not* attack.

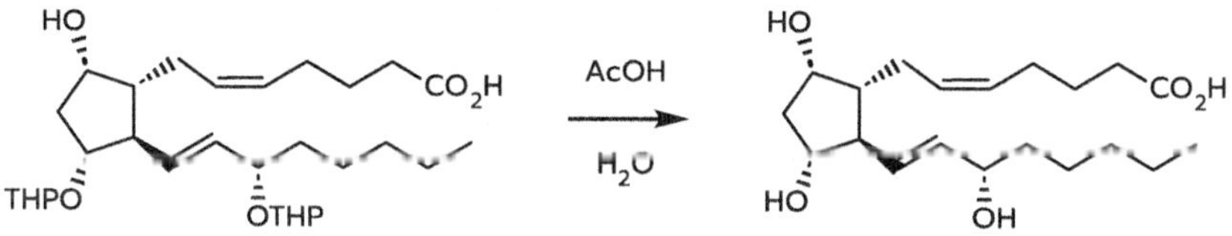

And that's the synthesis. We've looked at most of the steps.

THE CHEN/ZHANG SYNTHESIS

There's another nice synthesis of the same compound, reported in 2021. It is instructive to compare and contrast aspects of this approach with that of Corey.

SECTION 7 TOTAL SYNTHESIS

Chen and Zhang had more (new) reactions to use, so they were able to complete the synthesis in 6 steps, rather than 17.

The first step is an asymmetric hydrogenation (**Stereochemistry 9**). It uses an iridium complex and a chiral ligand, giving the product **19** with 94% enantiomeric excess (**Stereochemistry 7**).

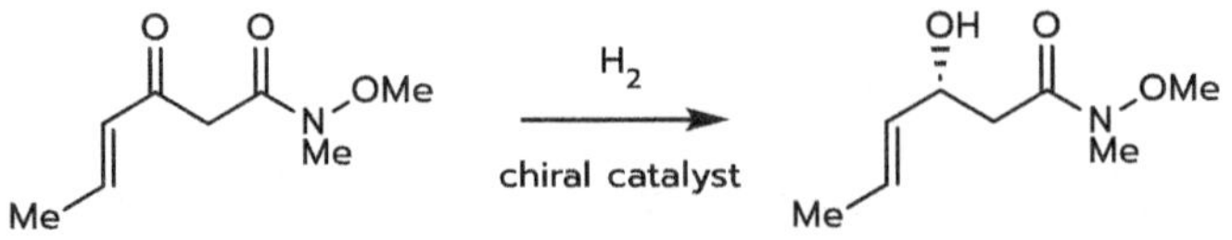

This is an immediate advantage over the Corey synthesis. Compound **3** (and hence, everything formed from it) was racemic (**Stereochemistry 7**). With compound **19**, we are producing essentially a single enantiomer.

> What ratio of enantiomers does 94% e.e. correspond to?

In the next step, the order of addition of reagents is important. We must add the butyllithium (**Basics 5**) to compound **20**.

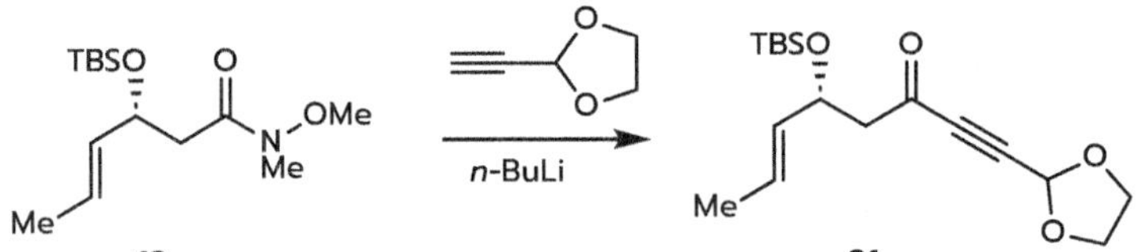

We couldn't mix compounds **19** and **20** and then add butyllithium.

> Why not?

There are a couple of things we need to discuss about this reaction. First up, we make the acetylide (**Reaction Detail 13**) from compound **20**. This is then added to the amide **19**.

Compound **19** is a bit 'special'. It turns out that a nucleophile only adds once to this type of amide. I wasn't planning to explain this, but I changed my mind. We saw, in **Reaction Detail 12**, that if you add a carbon nucleophile to an ester, you get a tetrahedral intermediate. We found that it reacts further. In this case, we find that the intermediate is stabilized by coordination from the OMe group to the Li atom.

> Whenever we draw a negative charge, it will be associated with 'something'. In this case, it is chelation forming a five-membered ring (**Basics 6**).

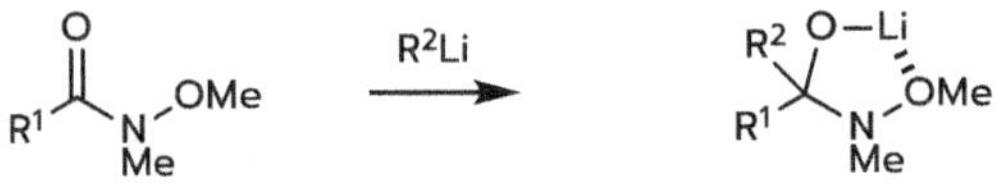

SECTION 7 TOTAL SYNTHESIS

You couldn't have predicted this (well, someone did!), but hopefully you can see that if we can stabilize an intermediate in a reaction, we can direct the reaction down a different pathway.

Now you've got a bit of work to do. You add acid and water at the end of the reaction which forms compound **21**.

> Draw a mechanism for this reaction. Make sure you don't have a negatively charged nitrogen leaving group.

The next step is catalysed by a transition metal complex. It forms the five-membered ring and another stereogenic centre. You can do some amazing transformations with transition metals, but they are mostly beyond the scope of this book. We aren't going to explain this one.

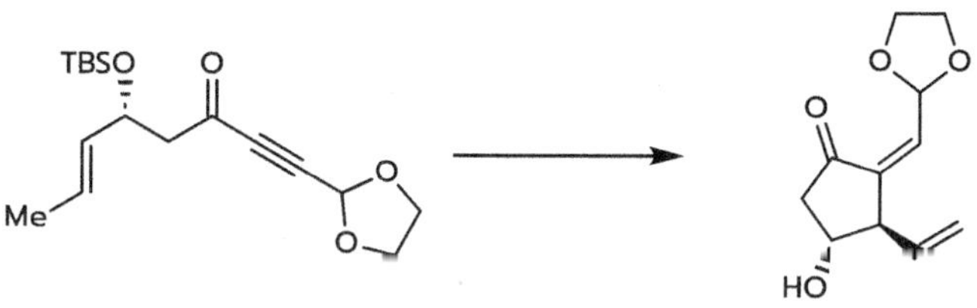

We now have a step that we can explain. There are actually three steps in one here—it was all done in one reaction. First of all, we add the reagents after 'i)'. You haven't seen these reagents before.

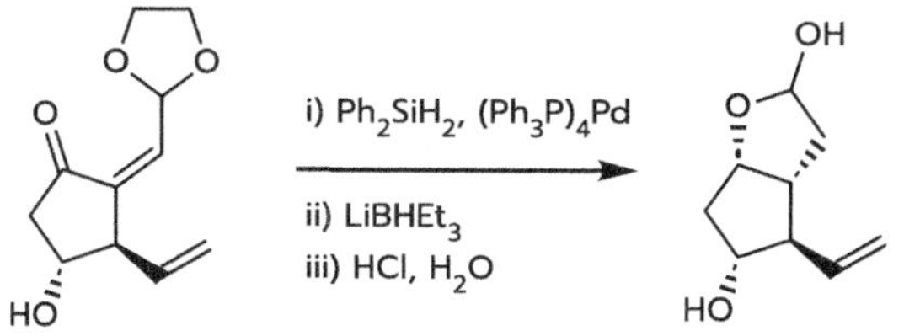

This is a fancy way of getting a hydride nucleophile. It's a conjugate addition reaction (**Reaction Detail 28**).

> Draw a mechanism, using H$^\ominus$ as the nucleophile. As this is a conjugate addition reaction, make sure the nucleophile attacks at the correct carbon atom.

Don't worry about *why* this particular reagent combination gives this outcome.

Then we add lithium triethylborohydride. It's got a B–H bond. This is **Reaction Detail 14**.

> Draw a mechanism. It's really just two curly arrows. Don't worry about trying to explain the stereochemistry.

The authors didn't comment on why they used this reagent instead of sodium borohydride.

491

SECTION 7 TOTAL SYNTHESIS

> *It probably just worked better.*

By this point you should have a product with only one double bond and two hydroxyl groups (although one of them is better drawn as O–Li). We now add acid and water. We are back in **Reaction Detail 15**.

> Draw a curly arrow mechanism that gets you all the way to compound **25**.

Almost there now. The penultimate step is an alkene cross-metathesis (**Reaction Detail 27**). The catalyst, **27**, looks complicated. You can draw a mechanism, but it's probably best to start with a simplified catalyst as shown below, where 'L' represents a ligand on ruthenium.

$$L_3Ru=$$

There are lots of opportunities within this mechanism to give products that are not **28**. But catalyst **27** works really well. A large excess of compound **26** was used.

> *Most molecules of compound 25 will give the cross-metathesis product with compound 26, even if most molecules of compound 26 also give the cross-metathesis product with compound 26!*

> Have a go at drawing a mechanism for this reaction. Try to work out the structures of some of the other products that might be formed.

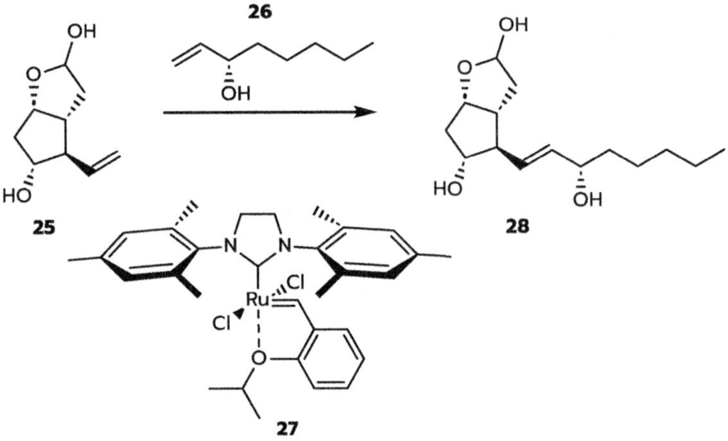

But don't worry too much about this one. Remember, this is *not* a book about transition metal catalysed reactions.

The final step is essentially the same as in the Corey synthesis, but without the protecting groups.

> *This is why I think the protecting groups were not really needed, although Corey did get a higher yield in this step, so there is a distinction between 'needed' and 'beneficial'.*

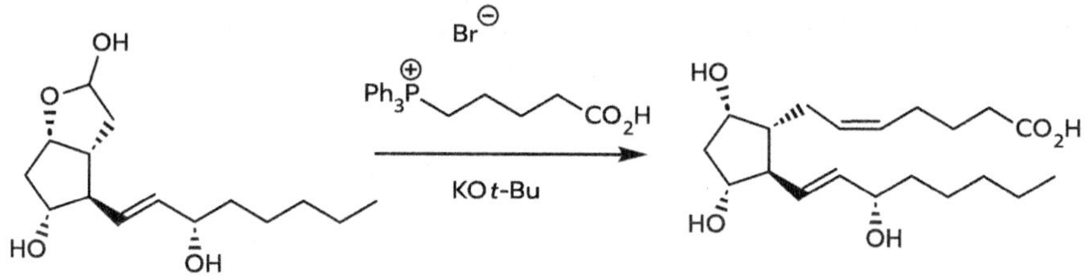

Prostaglandin F2α has a molecular weight of 354. The by-product in this reaction, triphenylphosphine oxide, has a molecular weight of 278. We can also consider that we are producing HBr as well. By the time we consider this, as well as the potassium t-butoxide base, less than half of the weight of reagents actually ends up in the product.

> This reaction is not very 'atom economical'. Atom economy is an increasingly important consideration in synthesis design.

But the bigger advantage is that with fewer steps, you use less reagents, solvents, energy, and time.

IN CLOSING

These are two really neat total syntheses of an important compound. The Chen/Zhang synthesis produces a single enantiomer, and is **much** shorter. Chen and Zhang had access to reactions that were not available at the time of Corey's work.

The key point, for us right now, is that most of the reactions used in these syntheses are 'standard' organic transformations. Even when they use reagents that you haven't seen, you can still draw a sensible curly arrow mechanism.

The citations for these two syntheses are as follows:

E. J. Corey, N. M. Weinshenker, T. K. Schaaf and W. Huber, *J. Am. Chem. Soc.*, **1969**, *91*, 5675.

F. Zhang, J. Zeng, M. Gao, L. Wang, G.-Q. Chen, Y. Lu and X. Zhang, *Nature Chemistry*, **2021**, *13*, 692.

I have followed the convention of referring to these syntheses by the names of the corresponding authors on the papers, although the co-authors would have done the actual reactions and contributed to the development of the chemistry described.

SECTION 7 TOTAL SYNTHESIS

TOTAL SYNTHESIS 3
Synthesis of Laurenene

INTRODUCTION

Laurenene is compound **1**. Here, we are going to talk about Leo Paquette's synthesis of laurenene. In fact, we are going to talk about a 'formal' total synthesis.

1   **2**

> *During a previous synthesis, Itō and co-workers made compound **2** and converted it into laurenene. Paquette devised a new approach to compound **2**. Since it had already been converted into laurenene, Paquette didn't need to 'finish the job'.*

Laurenene is pure hydrocarbon. It has four rings, and they all meet at a single carbon atom. This is a really challenging target. I don't think it has any really useful biological activity, but organic chemists like to push the limits of what they can do.

> Make a molecular model of this compound and spend a bit of time familiarizing yourself with it. You're not trying to learn the structure. You just need to see how densely functionalized it is.

GETTING STARTED

The starting point is going to feel like a bit of a cop-out. Compound **3** was reasonably available as it was used in a different synthesis. We are going to spend a little time talking about how this compound itself was prepared.

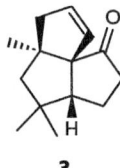

3

This does raise an interesting and relevant question.

> *When planning a synthesis, how do you decide where to start?*

This isn't really a book about synthetic strategy. There are plenty of those! The short (but probably unsatisfactory) answer is that as you get more used to organic reactions, you will begin to see possibilities. You will then search chemical suppliers

494

SECTION 7 TOTAL SYNTHESIS

for possible compounds. With yet more experience, you get better at 'guessing' which compounds are commercially available.

> *In all honesty, looking back, I'm not sure how I learned all this stuff. It kind-of just happened.*

What I didn't do is spend hours flicking through chemicals catalogues and trying to memorize what they included!

PREPARATION OF COMPOUND 3

Enone **4** reacts with Grignard reagent **5** to give compound **6** as intermediate. You probably expected the Grignard reagent to react at the carbonyl carbon atom in compound **4** (**Reaction Detail 12**) but it turns out that addition of copper bromide forms an organocopper reagent. We saw, in **Reaction Detail 28**, that these are soft nucleophiles and undergo conjugate addition reactions.

Compound **6** wasn't isolated. Instead, it was directly treated with acid and water to give compound **7**. This reaction is hydrolysis of an acetal (**Reaction Detail 15**) followed by acid-catalysed aldol reaction (**Applications 7**).

> Draw the mechanisms for these transformations.

> *I've drawn compounds 6 and 7 as if they were single enantiomers, but in reality they are racemic (Stereochemistry 2).*

Let's make a couple more connections. Ketone **6** could form two different enols, so we have a potential regioselectivity (**Recap 4**) issue. In reality, only one of the enols can reach to react with the aldehyde formed by hydrolysis of compound **6**.

> *This isn't quite a Curtin-Hammett situation (Basics 8) as the barrier to the aldol reaction would not necessarily be high compared to that for interconversion of the two different enols.*

> Draw the mechanism for the interconversion of the two enols. Make sure you draw **all** α- hydrogen atoms, and your mechanism isn't just one curly arrow!

We saw that acid-catalysed aldol reactions tend to result in elimination of water. This one doesn't, so another couple of steps were required to form compound **8**. We aren't going to look at these steps.

495

SECTION 7 TOTAL SYNTHESIS

Compound **8** reacts with methyllithium to give compound **9** (**Reaction Detail 12**). Now the organometallic reagent **does** react with the carbonyl carbon atom.

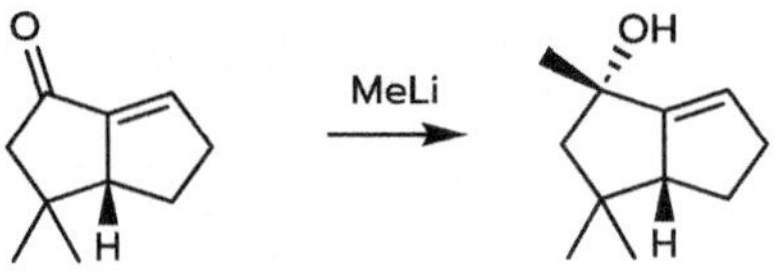

The 'new' stereochemistry in compound **9** wasn't given in the paper, so I'm guessing.

> Make a model of compound **8** and work out which face of the carbonyl is more exposed. This is the first part of **Stereochemistry 10**.

Now we have a step that you haven't actually seen. PCC is pyridinium chlorochromate. We met this reagent in **Reaction Detail 20**. Here's the transformation.

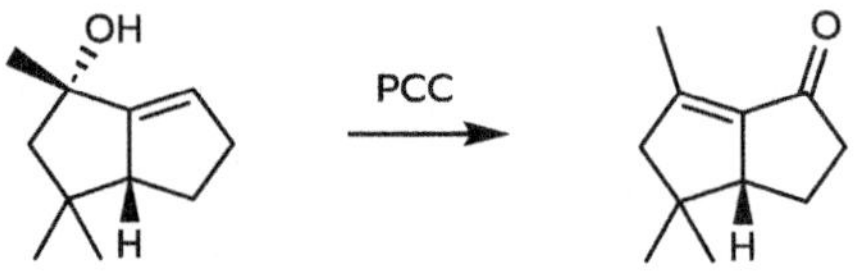

The oxygen atom appears to have 'moved'. The carbon-carbon double bond has moved. Let's try to work out how this might have happened.

Before we can do that, we should work out what hasn't happened. The alcohol in compound **9** is *tertiary*.

> It cannot be oxidized.

But there's no reason why the first step in the alcohol oxidation mechanism (**Reaction Detail 20**) cannot take place. We can draw the reaction of alcohol **9** with chlorochromate to give compound **11**.

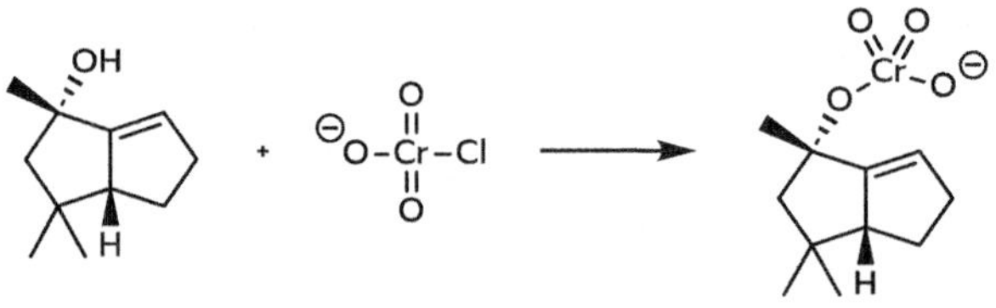

> Draw the mechanism for this.

What we have here is a really important principle. Don't draw garbage! When you see a reagent apparently doing something different, make sure it reacts according to its fundamental nature.

SECTION 7 TOTAL SYNTHESIS

> Back to you. From compound **11**, draw three sensible curly arrows to 'move the oxygen atom'[104] Then, carry on with the oxidation mechanism (**Reaction Detail 20**) to give compound **10**.

What I hope I am showing you here is that with a good level of understanding, you can draw reasonable mechanisms for reactions you haven't seen before.

I don't recall seeing this exact transformation before, so I really did have to follow this process.

The next few steps are largely a repetition of what we have seen. We need another cyclopentane ring, so why not do it the same way?

The stereochemistry in compound **12** supports our assignment of the stereochemistry of compound **9**.

> Draw the mechanisms out again, for the formation of this third ring.

The final step is 'just' an elimination reaction. We looked at these in **HTSIOC Reaction Detail 4**.

> Just use H⊕ as the reagent.

COMPLETING THE SYNTHESIS

From this point, it's actually quite a short synthesis. We saw cyanohydrin formation in **Reaction Detail 2**. We now see a variation where, instead of putting a hydrogen atom onto the carbonyl oxygen atom, we put a trimethylsilyl group on.

---

[104] You aren't really moving the same oxygen atom from one carbon atom to a different carbon atom!

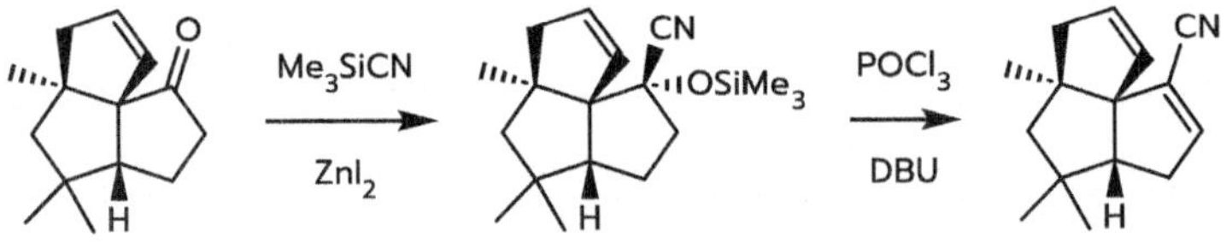

SECTION 7 TOTAL SYNTHESIS

> Have a go at drawing a mechanism for this reaction.

The next step isn't a reaction we have covered here, but I would draw substitution of OSiMe₃ for chloride, and then elimination. You can bet the phosphorus bonds to oxygen to help the substitution along.

> Have a look at **Reaction Detail 18** for inspiration, and then have a go at drawing something plausible.

Here's the structure of DBU. It's a reasonably strong base—not quite as strong as hydroxide, but not so far off.

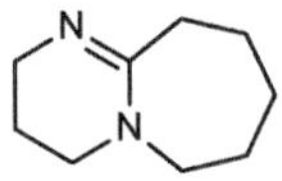

Now we have a problem. The next step is the reduction of a double bond in compound **15**, to give compound **16**. The problem is, if we use hydrogenation (**Reaction Detail 7**) we would reduce the wrong (more electron-rich) alkene bond. In that same chapter I mentioned, in a footnote, that chemists have developed clever ways to partially reduce benzene rings. While this isn't a benzene ring, the same principles used in these reactions apply. Magnesium in methanol gives the required chemoselectivity (**Recap 4**).

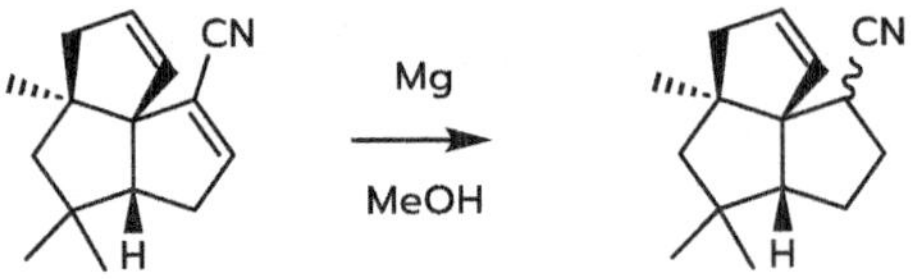

> Don't worry about the details, but in effect the reducing agent here is an electron.

We need to form a carbon-carbon bond, and this is done using an enolate alkylation reaction (**Reaction Detail 21**). Strictly speaking, it's not an enolate as we don't have an anion α- to carbonyl group. But the cyano group does the same thing.

> Draw the anion and the resonance forms! Then draw the alkylation reaction to give compound **17**.

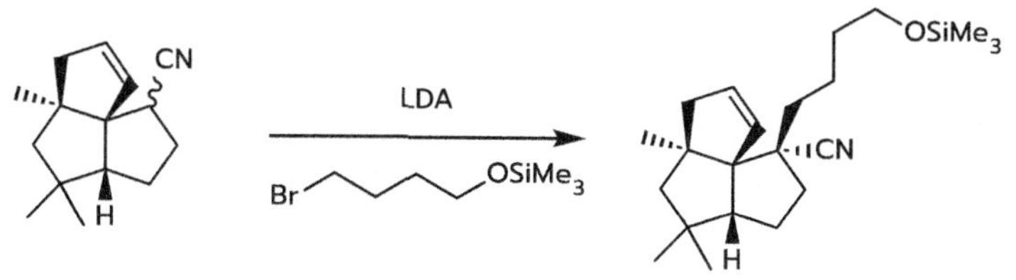

SECTION 7 TOTAL SYNTHESIS

16 → LDA, Br~~~OSiMe₃ → **17**

Of course, this alkylation reaction is diastereoselective because there are two possible products, and they are diastereoisomers. It's not quite the same as in **Stereochemistry 13** but the same underlying principles apply.

> Make a model of compound **16** and check that you can see why the top face of the enolate is more exposed.

The next step is a hydride reduction. We are reducing a nitrile, which is at the carboxylic acid oxidation level (**Applications 3**). If we reduced the cyano group with LiAlH₄, we would form an amine. But there's a different aluminium hydride, DIBAL-H, which is more selective, and only reduces it part of the way.

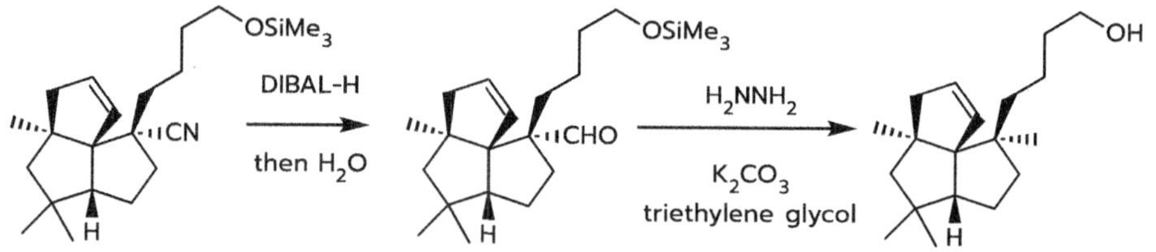

17 → DIBAL-H, then H₂O → **18** → H₂NNH₂, K₂CO₃, triethylene glycol → **19**

Here's the structure of DIBAL-H again. In fact, we saw this reagent doing something similar in **Total Synthesis 2**.

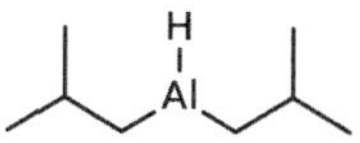

> Don't over-think this. Draw a mechanism for the reduction of compound **17** with DIBAL-H, to give the imine, and then the hydrolysis of the imine (**Applications 2** in reverse).

We aren't explaining **why** we get this selectivity.

Finally, we have a Wolff-Kishner reduction (**Applications 4**) to form compound **19**.

> Draw this mechanism. The trimethylsilyl group on the other oxygen atom was removed during this transformation. Don't worry about that bit.

We are almost done now. Just three more steps. The next step is hydroboration of the alkene bond in compound **19**. The dimethylsulfide complex of borane is commonly used—just treat it as BH₃. In **Reaction Detail 5** we considered the regiochemical outcome of the hydroboration reaction, but this alkene is

SECTION 7 TOTAL SYNTHESIS

1,2-disubstituted so the arguments we used about developing positive charge don't apply here.

> Draw the mechanism giving both possible regiochemical outcomes. Make sure you can see why they are the same from an electronic perspective.

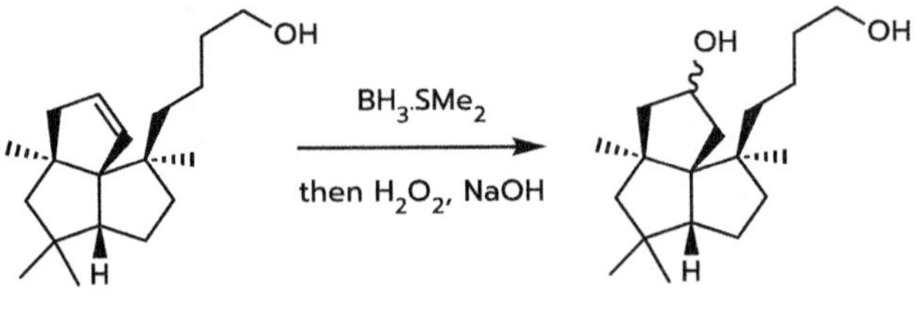

19 → (BH$_3$.SMe$_2$, then H$_2$O$_2$, NaOH) → **20**

In this case, the reason for the regioselectivity (**Recap 4**) must be steric. Some of the other product was obtained, but not enough to hamper the synthesis.

> Don't forget to draw the borate rearrangement part of the mechanism which we encountered in **Reaction Detail 33**.

The penultimate step is a Swern oxidation (**Reaction Detail 20**). Actually, it is two Swern oxidations. We oxidize the *primary* and *secondary* alcohol groups in compound **20**, giving keto-aldehyde **21**. I've abbreviated oxalyl chloride as (COCl)$_2$.

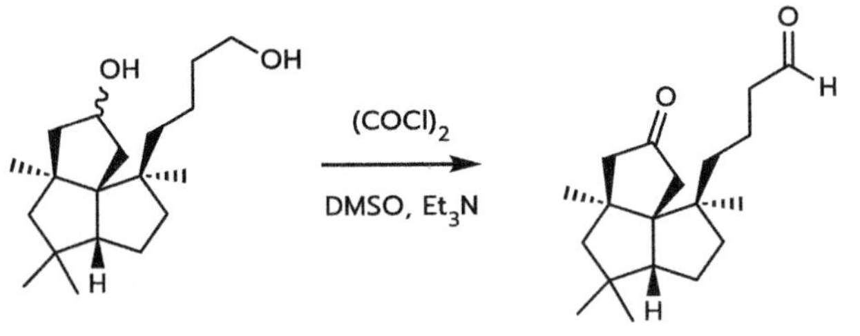

20 → ((COCl)$_2$, DMSO, Et$_3$N) → **21**

> Have a look at the full structure in **Reaction Detail 20**. Is there another structure you could draw based on this abbreviated formula?[105] Draw the mechanism of the Swern oxidations to give compound **21**.

Finally, we have another aldol condensation. This is also with acid (**Applications 7**) but this one does eliminate under the aldol reaction conditions.

---

[105] There wouldn't be much point using an abbreviated formula if it wasn't clear.

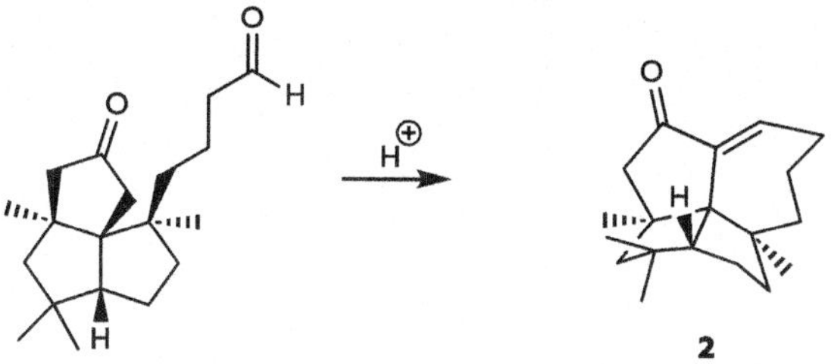

SECTION 7 TOTAL SYNTHESIS

21 → (H⁺) → **2**

IN CLOSING

This is a neat synthesis. In some respects, it is repetitive. Organic chemists do have their own 'style', so it's not too surprising to see the same reactions being used to make different rings in the target molecule.

When we apply simple reactions to complex structures, particularly those with rings, we do sometimes see subtle differences in reactivity. We have seen three acid-catalysed aldol reactions. Two of them didn't spontaneously eliminate, but the third one did.

> *The reasons for that are beyond the scope of this book.*

For now, we draw curly arrow mechanisms for reactions we have seen, but with more complex molecules. Reflect on what you have drawn. Discuss mechanisms with your colleagues. Make sure you don't have any five-valent carbon atoms along the way.

And gradually, imperceptibly, you will get better at organic chemistry.

The reference for this synthesis is L. A. Paquette, M. E. Okazaki and J.-C. Caille, *J. Org. Chem.* **1988**, *53*, 477.

I'm not expecting you to go away and read it, but I think it is worth including the reference so that you can access it if you want to.

SECTION 7 TOTAL SYNTHESIS

TOTAL SYNTHESIS 4
Synthesis of Ionomycin

INTRODUCTION

Here is the structure of ionomycin.

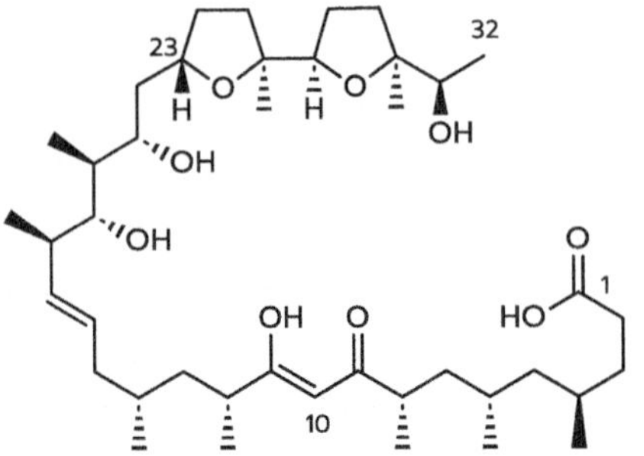

It's an absolute beast of a molecule, with 41 carbon atoms and 14 stereogenic centres. And yet, compared to some molecules that have succumbed to total synthesis, this is small. It's all relative.

We are looking at the Evans total synthesis, which is a showcase for his diastereoselective enolate and aldol chemistry. We won't look at the complete synthesis—it is too long.

Instead, we will look at a few fragments/steps that exemplify some of the chemistry we have discussed.

GETTING STARTED

We will begin with the synthesis of compound **1**, which represents C1-C10 of ionomycin (we are counting the carbon chain, not including the appended methyl groups).

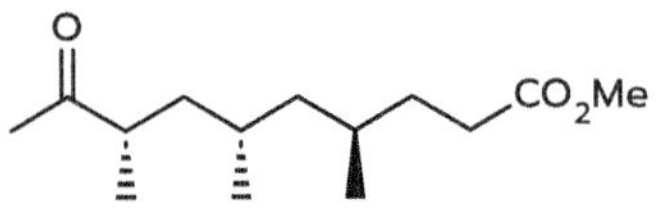

Even this fragment is a challenge. There are three stereogenic centres to be controlled. We are going to start with the aldol reaction (**Reaction Detail 23**) of oxazolidinone **2** (**Stereochemistry 14**, **Applications 8**—I did say this was a synthesis from the Evans group!) with ethanal.

The stereochemistry is controlled by the methyl group on the oxazolidinone, but you can bet the phenyl group is 'helping' (**Stereochemistry 15**).

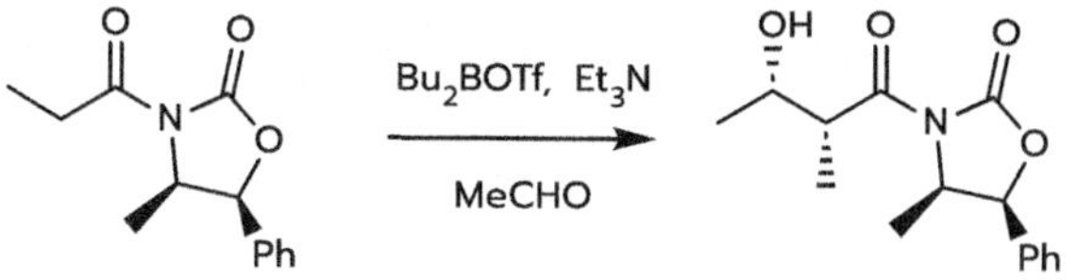

SECTION 7 TOTAL SYNTHESIS

2 → **3** Bu$_2$BOTf, Et$_3$N, MeCHO

> Draw a curly arrow mechanism for the reaction. Draw a 'suitable' transition state for the reaction and hence explain the stereochemical outcome.

> Words like 'suitable' are horrible at first. With practice, there's only one thing you would draw.

The next two steps are basically 'housekeeping'. We put a silicon protecting group onto the alcohol (don't worry about this step—we haven't covered it) and we remove the oxazolidinone chiral auxiliary. In **Stereochemistry 13**, we saw hydroperoxide used for this step. In this case, the benzyloxy anion (PhCH$_2$O$^{\ominus}$) worked fine.

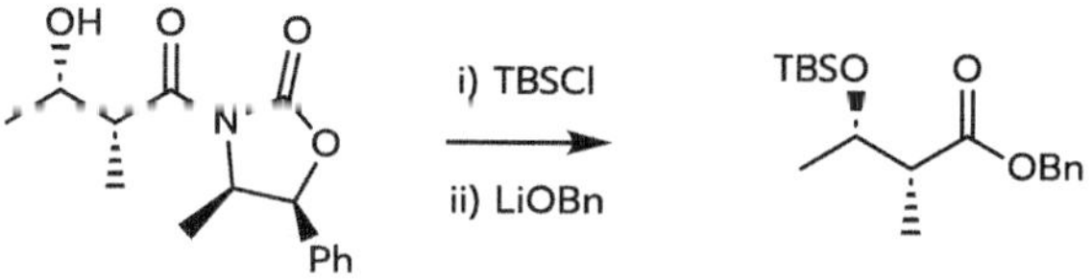

3 → **4** i) TBSCl ii) LiOBn

> Draw a mechanism for this second step. You might want to have a look at **Reaction Detail 16** if you need help.

Now, we reduce the ester to the alcohol (**Applications 3**). We saw in **Total Synthesis 3** that this reagent can be used to reduce a nitrile to an aldehyde. It can also be used to reduce an ester to an aldehyde, or all the way to the alcohol. In this case it's the latter. It's all about controlling the reaction conditions.

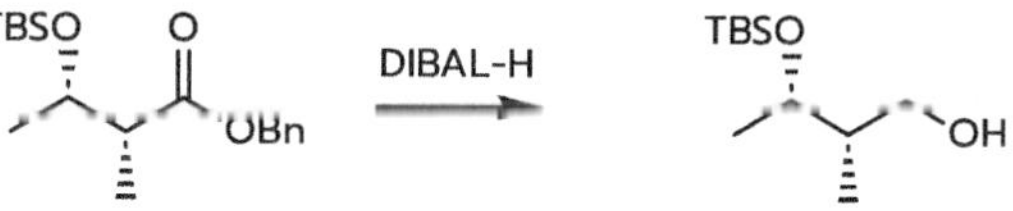

4 → **5** DIBAL-H

We've seen the next two steps before in **Section 7**.

> Find the first step, then draw out the mechanism.

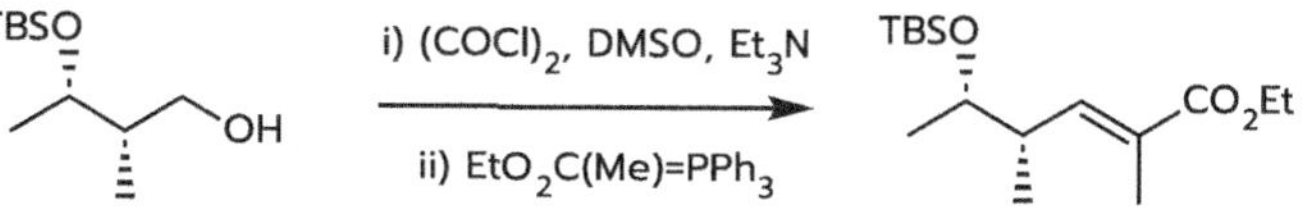

5 → **6** i) (COCl)$_2$, DMSO, Et$_3$N ii) EtO$_2$C(Me)=PPh$_3$

The second step is a Wittig reaction (**Reaction Detail 26**). I've been very 'economical' in my drawing of the reagent below the arrow.

503

SECTION 7 TOTAL SYNTHESIS

> Look at this representation, then look at the product, and make sure you can see what the full structure of this phosphorane/phosphonium ylide is. Then draw the mechanism and explain the stereochemical outcome (= double bond geometry!).

Compound **6** is then converted into compound **7**. We aren't going to discuss these steps. This is 'just' an alkyl iodide, and it reacts with oxazolidinone **2** to give compound **8**. The base is similar to LDA, but with a couple of minor differences.

> Draw out the structure of the base. Compare it with LDA (**Fundamental Reaction Type 4**). The stereochemistry at the α-carbon in compound **8** is different to that in compound **3**, despite using the same chiral auxiliary. Have a look at **Stereochemistry 13** and **Applications 8**, and make sure you can see why this happens.

Let's be clear. I'm not getting you to do this because the Evans aldol and alkylation chemistry is *that* important.

> *Sure, it is a great way to control stereochemistry in these reactions.*

I'm getting you to do it because you will get better at drawing structures, and you will then get better at drawing **any** stereoselective reaction.

I'm skipping another couple of steps again. The preparation of compound **9** from compound **8** uses the same reactions we have already seen.

> See if you can work out how this might have been done.

Then, hydrogenation (**Reaction Detail 7**) of compound **9** gave compound **10**. A rhodium(I) catalyst was used here, but that's a minor detail.

In the case of the left-hand alkene, the stereochemistry of hydrogenation was controlled by $A^{1,3}$ strain (**Stereochemistry 6**).

> Now let's flip the question a bit. Does the OH group in compound **9** direct this hydrogenation?

SECTION 7 TOTAL SYNTHESIS

Don't over-think this. Identify the conformer in which A^{1,3} strain is minimized. If you then get hydrogenation on the same side as the OH group, it's probably directing rather than blocking. You don't **need** to know anything about the catalytic species.

> Stereochemistry 6 was all about epoxidation. But we are able to explain other types of stereoselective alkene reaction using the same ideas.

From compound **10**, an oxidation reaction was required to give ketone **1**. On this occasion, a Swern oxidation was not used.

ANOTHER BIT!

Don't worry, we aren't going to do the whole natural product. I just want to show you enough of the steps to exemplify some of the reactions we have looked at.

The steps shown below didn't end up being used in the total synthesis, so they are not fully described in the paper. Lactone[106] **11** was converted into carboxylic acid **12**. This is an ester hydrolysis (**Reaction Detail 16**) followed by an ester formation.

> See if you can propose a reasonable way of doing this, and draw the mechanism.

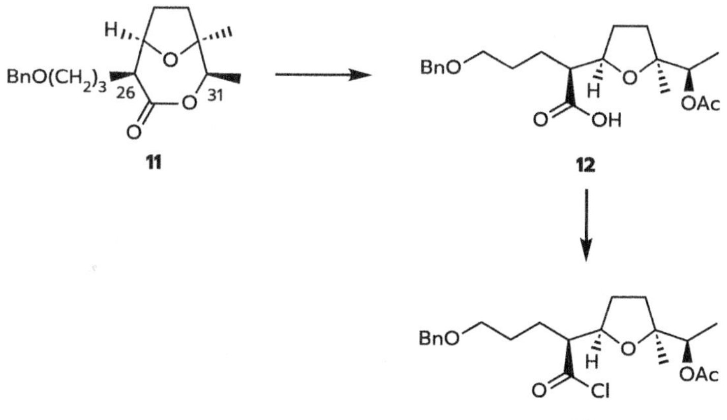

The we have formation of an acid chloride (**Reaction Detail 17**).

> Assume that thionyl chloride was used, and draw the mechanism.

Okay, we've seen how they **didn't** make this part of the natural product. Now let's have a look at a few of the steps that **did** get them there.

Starting from the same lactone **11**, compound **14** was formed. We aren't going to explain this step. It looks like the sort of thing that could be done using a Wittig reaction, but this doesn't work on esters.

> You wouldn't know that!

---

[106] A lactone is a cyclic ester. You could just call it an ester!

SECTION 7 TOTAL SYNTHESIS

11 → **14**

The next step used an acid catalyst to isomerize the alkene bond in compound **14** to give compound **15**. Remember from **Recap 3** that a more substituted double bond is more stable. There *must* be a reason for this to happen.

14 —H⊕→ **15**

> Draw a curly arrow mechanism for the double bond isomerization. You need to use the acid catalyst, and you need to draw more than one curly arrow!

Here's the final step we are going to discuss in this particular synthesis. It's 'just' an ozonolysis (**Reaction Detail 6**).

15 —O$_3$ then Me$_2$S→ **16**

But it's a very elegant ozonolysis. We form the ester on the right-hand side. Normally we would be in **Reaction Detail 16** territory here, but in this case the ester is formed as a direct consequence of the alkene undergoing ozonolysis.

> Have a go at drawing the mechanism for the ozonolysis. It's always a bit more difficult when the double bond is in a ring. Take your time with this one, and don't worry if it takes you a few attempts. Oh, and have a look at the stereochemistry on the right-hand side. We've gone from having a methyl group on the wedged bond to having the acetate ester on the wedged bond, but it's the same stereochemistry. Make sure you can see why this is correct.

THAT'S QUITE ENOUGH!

There's plenty more of the synthesis if you want to have a look. The reference is D. A. Evans, R. L. Dow, T. L. Shih, J. M. Takacs and R. Zahler, *J. Am. Chem. Soc.*, **1990**, *112*, 5290. It's a big synthesis, so it's a big paper.

For many of the reactions, you would be able to draw a plausible mechanism. In some cases, they use different reagents from the ones we have seen, and they just happen to give higher yields.

SECTION 7 TOTAL SYNTHESIS

> *The important thing, as always, is to understand the fundamental types of reactivity.*

If you can do this, learning new reagents for the same old transformations isn't such a challenge.

SECTION 7 TOTAL SYNTHESIS

TOTAL SYNTHESIS 5
Synthesis of (−)-7-Deacetoxyalcyonin Acetate

INTRODUCTION
This is the last one! It is quite a densely functionalized molecule, isolated from a soft coral.

(−)-7-Deacetoxyalcyonin Acetate

This synthesis uses a type of pinacol rearrangement developed by the group of Larry Overman at Irvine, California. I mentioned, in **Total Synthesis 2**, that while we often refer to papers by the name of the 'senior' author, the co-workers carry out the actual reactions and definitely make an equal contribution. In this case, the co-worker on this synthesis was David MacMillan, a future Nobel Prize winner who we mentioned in **Reaction Detail 25**.

This is a really challenging molecule, with three rings, seven stereogenic centres, two double bonds and a *tertiary* alcohol.

> Bad things would happen to this compound if you treated it with a strong acid.

THE SYNTHESIS
We won't be discussing many steps in this synthesis, as one of the steps will take a good while to work through fully. I'm going to start with iodide **1**, which is reacted with *t*-BuLi to give the organolithium reagent **2**.

In **Reaction Detail 12**, we saw the formation of a Grignard reagent by reaction of an alkyl halide with magnesium. Metal-halogen exchange with an existing organometallic reagent (usually Li) is also a common strategy. In **Basics 5** we saw that *t*-BuLi is very unstable (= reactive!).

Compound **2** would never be isolated. It was directly added to aldehyde **3**. This is an aldehyde with an α- stereogenic centre (**Stereochemistry 10**).

> Is this reaction under chelation control, or do we get the Felkin-Anh stereochemical outcome?

That isn't trivial to work out. You need to 'draw stuff' to work this out.

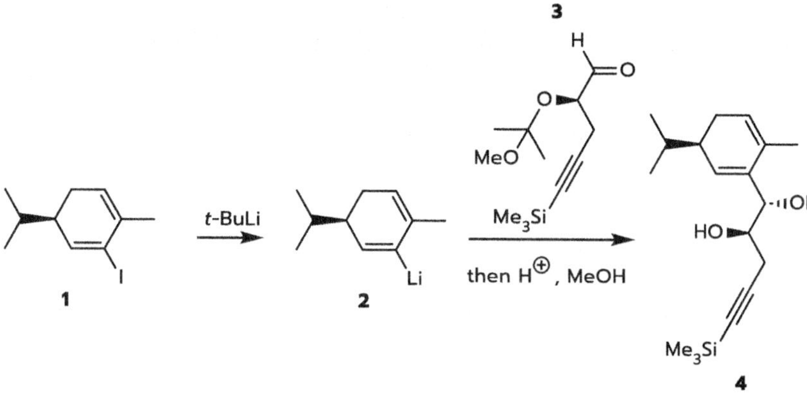

SECTION 7 TOTAL SYNTHESIS

Compound **3** has an acetal (**Reaction Detail 15**) group, which is removed using acid and methanol.

> Draw a plausible mechanism for this reaction. Don't use water!

Now we are straight into the 'key step'. It isn't all that easy to see exactly what is happening here. There's a cationic rearrangement (**Reaction Detail 31**) in there, but we have a bit of work to do before we get to that point.

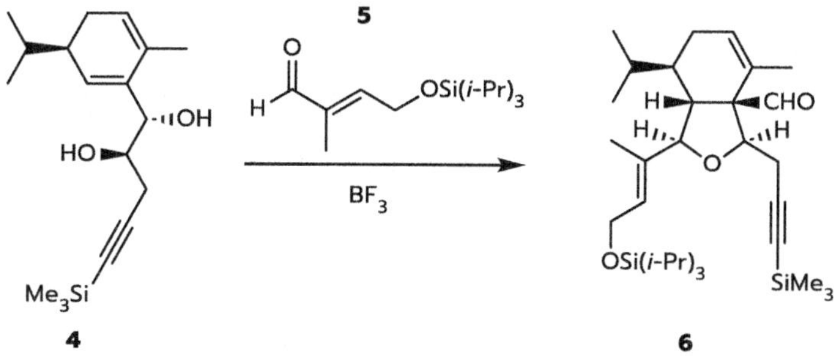

> Number the atoms in the starting materials **4** and **5**, and in the product **6**. I think you will get most of them right, but there are probably two that you won't be certain about.

We need to work out what is going on, and you know how we do that?

> We draw some curly arrows!

I've simplified the reagent, BF$_3$. It is normally supplied as a solution in diethyl ether, which complexes to it. After all, a boron atom with three bonds is electron-deficient.

> It's a Lewis acid. We are going to simplify things even further by drawing the mechanism with H$^{\oplus}$.

As always, we follow 'the rules'. We have an aldehyde, and a diol. We could possibly form an acetal, **7** (**Reaction Detail 15**).

509

SECTION 7 TOTAL SYNTHESIS

[Scheme: Compound **4** + **5** (aldehyde with OSi(i-Pr)₃), BF₃ → Compound **7**]

> Draw a mechanism. Keep it simple by using H⊕. I didn't include the stereochemistry, but as I've rotated a bond, it doesn't look quite like the starting material. Work it out and draw it correctly.

Along the way, you would have had one of the following two intermediates.

[Structures of intermediates **8** and **9**]

I'm not convinced this reaction makes it as far as the acetal. It's going to take a bit of work to explain why, so please bear with me. In order to do this, we will need to see how we form compound **6**. It can *only* be formed from oxocarbenium ion **9**, and *not* from **8**. First of all, we will see the curly arrows for this process.

[Mechanism showing **9** → **10**]

This is a 'new' reaction to us. But of course, it's not really new. We have an electrophile reacting with a C–C double bond (**Fundamental Reaction Type 1**). The regiochemical outcome is such that we get the more stable carbocation. It's also a six-membered ring transition state.

> I bet you can guess what shape this will be!

510

On the other hand, we have a nucleophile adding to a 'protonated carbonyl group'. It's not really a protonated carbonyl group, but it looks like one and it reacts like one.

> This is actually called a Prins cyclization, but you don't need to know that to draw the curly arrows, or even to predict that this might have happened.

> Draw a corresponding Prins cyclization of compound **8**. Which double bond will be attacked if we want a six-membered ring transition state *and* we want a *tertiary* carbocation to be formed?

Now another exercise for you.

> Look at structure **7**. Which oxygen atom will be more easily protonated? Is there any electronic difference between them? Perhaps it's just steric!

Now we can explain why I don't think we get to the acetal **7**. If we did, I think acetal opening would favour formation of oxocarbenium ion **8**. And this wouldn't give us the product. Since we know we form the product, and it is formed in high yield, this tells us what **must** happen.

On the other hand, we could think about which hydroxyl group of compound **4** is least hindered, and will therefore react with aldehyde **5**.

> I think formation of oxocarbenium ion **9** is reasonable. But it must cyclize to give intermediate **10** more rapidly than it cyclizes to give compound **7**.

This is quite complicated, and yet all we are doing is considering possible intermediates along the reaction pathway. In the interest of transparency, Overman does propose an acetal intermediate in at least some of these reactions.[107]

> Perhaps I'm wrong!

What we *can* say with absolute confidence is that *if* acetal **7** is an intermediate in the reaction, then its acid-mediated ring-opening either gives **9** exclusively, or if it does also give **8**, then the cyclization of **8** is slow and equilibration of **8** and **9** is rapid.

> Wait, that sounds like a Curtin-Hammett situation (Basics 8).

I wasn't actually looking for that, but it happened anyway.

MOVING FORWARD

Now let's look at the rearrangement step. In order to do this, I need to redraw structures **9** and **10**, and to simplify them with 'R groups'.

---

[107] This isn't the only target made with this chemistry.

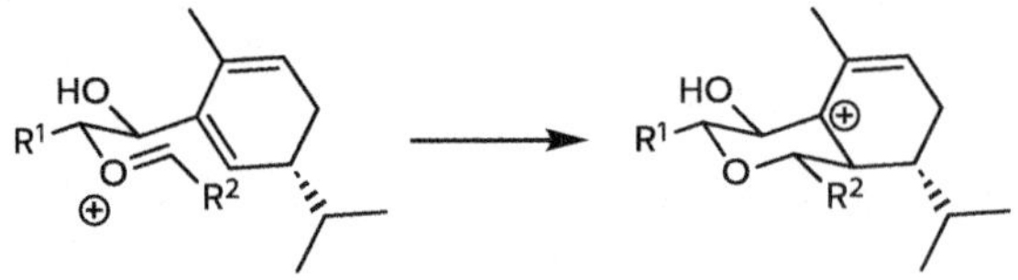

SECTION 7 TOTAL SYNTHESIS

One of the first things I said, at the start of this section, was not to simplify the molecules.

> Sometimes it's necessary!

We have a clear stereochemical outcome that comes from the chair transition state. One face of the alkene bond in **9** is blocked by the isopropyl group, so that it reacts on the other face.

Now we have our pinacol rearrangement (**Reaction Detail 31**) which will be followed by loss of a proton to get us back to a stable neutral molecule.

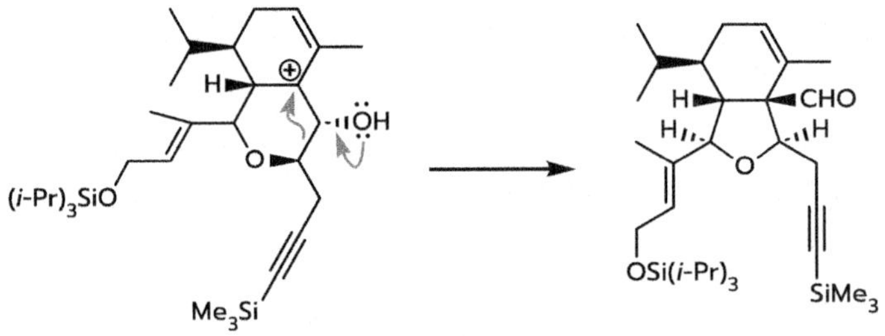

> Identify the migrating bond in the chair structure for **10** shown above. Convince yourself[108] that it is suitably oriented to overlap with the p-orbital of the carbocation. Convince yourself that it is the **only** bond that could migrate to give a more stable carbocation (**Recap 2**).

AND THAT'S IT

There are a few more steps required to get us from compound **6** to the natural product target. It would be instructive to go through a few more steps, but on balance, I think we've done enough here.

> Remember, we are not trying to learn reactions. We are building skills!

IN CONCLUSION

As before, I'm going to tell you where to look to read more about this synthesis. The citation for this work is D. W. C. MacMillan and L. E. Overman, *J. Am. Chem. Soc.*, **1995**, *117*, 10391.

---

[108] Molecular models?

SECTION 7 TOTAL SYNTHESIS

As a broader conclusion, we have come to the end of our short trip through total synthesis. We haven't included examples of every reaction type found in this book. And yet some reactions are covered several times. This wasn't by any design.

> *It just happens that some reactions are used more often than others.*

Along the way, we have seen a few reaction types that we have not covered, but we have seen the most important reactivity types of the most important functional groups.

Managing cognitive load is important. Once you've assimilated the material in this book, it will be a lot easier to add the handful of additional reactions, one at a time. At least you will be in a good position now to tell whether these **really** are 'different' reactions, or just different reagents for the same reaction types.

AFTERWORD

I have tried desperately not to make this a list of reactions to memorize, but even so there are quite a lot of reactions in here. I hope you have understood the curly arrows, the orbital interactions, and the other factors that allow these reactions to work as they do.

But there's a lot of organic chemistry that isn't in here! For a start, we haven't covered any free-radical chemistry. Free radicals are species with a single electron. Their reactions are useful and important, but right now we are building your skills with curly arrows and principles. Once you've got that, adding free-radical chemistry to your repertoire will be easy.

We haven't included any spectroscopy either. There isn't much point making compounds if you can't work out what you've made. But there are lots of good spectroscopy textbooks, and if you can look at a molecular structure and see 'everything', then interpreting the spectra will be a lot easier.

In fact, for each of the reactions we have seen, there are lots more ways of carrying them out. Amides are important. We looked at one way to form them, but there are lots of ways that require milder reaction conditions and give high yields.

The good thing is, all these ways conform to the principles we have seen.

So hopefully, this book will have helped you prepare for the next stage in your journey. You'll see a 'new' reaction and you will know exactly how it fits within the framework of your existing knowledge.

FURTHER READING

There are many resources out there to help you improve your knowledge and understanding of organic chemistry. There has definitely been a shift in recent years away from textbooks and towards internet resources, particularly YouTube.

I'm going to tell it like it is. There are lots of great organic chemistry web pages and YouTube videos. And there is a lot of rubbish. I could point to web pages and videos that have **major** errors in the science—incorrect reaction mechanisms and curly arrows, among other things. You **can** learn from a good YouTube video, but only when you pause the video and start drawing.

> *A book works even better! And the person who took the time to write the book probably has decades of experience in organic chemistry research and teaching.*

I'm going to keep the 'further reading' list brief and focused.

How to Succeed in Organic Chemistry, M. C. Elliott, Oxford University Press, Oxford, 2020.
Yep. More 'me'! If you like the style of this book, you might enjoy the first one.

Organic Chemistry, 2nd Edition, J. Clayden, N. Greeves and S. Warren, Oxford University Press, Oxford, 2012.
The Organic Chemistry Bible! It's a big book, with lots more examples and details relating to the reactions we have covered. And lots of other reactions. A lot of people find the larger textbooks intimidating, but I would strongly recommend owning one. And for me, this is the best of the bunch!

Chemistry of the Carbonyl Group, T. K. Dickens and S. Warren, Wiley, Chichester, 2018.
This is effectively a second edition of a book I used as a student. It is active learning at its best!

Molecular Orbitals and Organic Chemical Reactions: Student Edition, I. Fleming, Wiley, Chichester, 2009.
Again, I used a previous iteration of this book when I was a student. It's great for developing a deeper understanding, particularly of the reactions in Section 6.

PREVIOUS READING

Throughout the text, there have been lots of references to chapters in HTSIOC. In case you don't have that book, here are chapter numbers to the 2nd Edition of Organic Chemistry by Clayden, Greeves and Warren.

There are a few topics that are covered in HTSIOC that do not appear to be (directly) covered in 'Clayden'. However, it is a big book, and I may be doing the authors a disservice. If so, the error is mine.

Chapter in HTSIOC	Chapter in 'Clayden'
Applications 2	Chapter 16
Applications 3	Chapter 16
Applications 5	Chapter 17
Basics 5	Chapter 4
Basics 6	Chapter 4
Basics 9	Chapter 5
Basics 10	Chapter 7
Basics 12	Chapters 10 and 37 in part
Basics 13	Not covered
Basics 15	Chapter 12
Basics 16	Chapter 8
Basics 18	Chapter 8
Basics 29	Chapter 14
Basics 30	Chapter 16
Basics 32	Chapter 16
Basics 34	Chapter 17
Basics 35	Not covered

Previous Reading

Fundamental Reaction Type 1	**Chapter 15**
Fundamental Reaction Type 2	**Chapter 17**
Habit 2	**Chapter 3**
Habit 6	**Chapter 14**
Perspective 1	**Not covered**
Perspective 2	**Not covered**
Perspective 3	**Not covered**
Perspective 4	**Not covered**
Reaction Detail 1	**Chapter 15**
Reaction Detail 2	**Chapter 15**
Reaction Detail 4	**Chapter 17**
Worked Problem 5	**Chapter 15**

If you want chapter numbers for a different 'big organic chemistry book', ask on the Facebook group (**https://www.facebook.com/groups/159669516998134**) and I will see what I can do!

INDEX

This isn't a book that needs an index. But you're expecting one, so here it is. There are some terms that permeate the book, and giving page numbers would be pointless. You'll find these quickly enough. In other cases, I recommend finding the chapter where an idea is covered and reading that rather than trying to find individual pages. You'll probably find the index more useful for terms that only crop up once or twice. But it's your book now!

$A^{1,2}$ strain, 177, 179, 180, 184
$A^{1,3}$ strain, 177–179, 181, 183, 186, 396, 398, 504
acetal, 59, 237, 239, 240, 242, 245–247, 253, 269, 488, 489, 495, 509–511
acetylide, 42, 81, 82, 85, 209, 210, 217, 490
acid chloride, 49, 50, 259, 262–265, 268, 273, 282, 283, 285, 297, 357, 427–429, 477, 505
activation energy, 17, 23, 26, 45–47, 71, 73, 78, 79, 94, 124, 138, 140, 144, 147, 163, 187, 238, 247, 256, 258, 264, 271, 272, 276, 307, 308, 363, 379, 381, 392, 393, 423, 432
aldol reaction, 97, 291, 319–321, 323, 324, 333–344, 349, 367, 466, 469, 472, 478, 482, 495, 500–502, 504
alkene, everywhere!
alkyne, 7, 8, 82, 100, 138–141, 170–174, 209, 210, 438
allowed (cycloaddition), 442, 446, 448, 449, 458
allyl/allylic, 12, 15, 69, 72, 74, 171, 177–180, 188, 194, 291, 396
amide, 49, 50, 88, 154, 155, 250, 251, 256–260, 267, 295, 303, 357, 390, 424, 426, 427, 479, 480, 490
antarafacial, 463–465
anti, 175, 176, 337, 338, 340, 343, 423, 424
antibonding orbital, 5, 6, 24, 39, 64, 83, 371, 439

aprotic, 213
aromatic electrophilic substitution, 74, 77, 95, 100, 142, 146, 156, 159–161, 167, 168, 266, 268, 269, 371, 377, 383, 417, 477, 478
aromaticity, 64, 66, 67, 75, 76, 78, 140, 147, 156, 264, 279, 327, 377, 380, 381, 383, 385
asynchronous, 104
axial, 119, 120, 122, 136, 157, 184, 215–218, 337, 395

Baeyer-Villiger, 418–420, 422, 486
Baldwin's rules, 331, 332, 385
Beckmann rearrangement, 419, 421–423
benzene, everywhere!
boat conformer, 121, 460, 467, 468
bond dissociation energy, 7, 19, 38, 44, 59, 82, 170, 233, 234, 236, 241, 242, 251, 273, 300, 301, 316, 317
bonding orbital, 3, 5–7, 415, 439
borane, 104, 114, 174, 213, 261, 430
borohydride, 102–104, 108, 116, 141, 212, 213, 217, 223, 227, 261, 277, 279, 281, 479, 491
bromination, 30, 32, 33, 35, 71, 75, 76, 93, 94, 127, 130, 137, 156, 157, 312–314
bromonium ion, 31–33, 71–73, 102, 113, 127, 128, 130–132, 172, 311, 410
Bürgi-Dunitz angle, 214, 222, 385

519

Index

Cahn-Ingold-Prelog rules, 129, 131, 197, 227, 228
Cannizzaro reaction, 277, 278
caprolactam, 421
carbanion, 51–53, 55, 80, 82, 85, 87, 203–205, 208, 273, 284, 304, 305, 325, 336, 360, 361, 364, 371, 387, 406, 475
carbocation, everywhere!
carbonyl, everywhere!
carboxylic acid, 110, 248, 251–256, 258, 259, 262, 263, 293, 294, 297, 352–354, 419, 424, 426–429, 450, 451, 489, 499, 505
chair conformer, 119–121, 129, 184, 218, 338, 340, 341, 460, 466, 467, 468, 469, 512
chelation, 223, 343, 490, 508
chemoselectivity, 22, 34, 35, 167, 194, 300, 323, 347, 402, 482, 486, 498
chiral auxiliary, 294, 342, 400–402, 433, 503, 504
Claisen condensation, 324–326, 334, 348, 349, 424, 466
Claisen rearrangement, 466
concerted, 104, 112, 114, 429, 437
conformer, 119–121, 129, 130, 133–137, 175–188, 218–223, 295, 329–331, 340, 394–398, 401, 449, 504
conjugate addition, 263, 371, 372, 375, 387, 388, 396, 482, 491, 495
conjugation, 64–68, 79, 322, 355, 374, 437, 440, 448, 454, 475
conrotatory, 457–461
Cope rearrangement, 466, 468
Cram rule, 220–223
cross-coupling, 167, 168, 377
Curtin-Hammett, 392–397, 417, 495, 511
Curtius rearrangement, 424, 426–429
cyanide, 41–43, 45, 46, 61, 81, 82, 85, 205, 209, 397
cyanohydrin, 41, 43–45, 47, 81, 497
cycloaddition, 114, 367, 437, 438, 440, 442, 445–449, 456, 458, 465
cyclobutane, 447

cyclohexane, 119–121, 128, 129, 133, 137, 157, 184, 218, 242, 327, 329, 337, 338, 400, 421, 466, 467
cyclohexanone, 35, 215, 217, 289, 291, 304, 379, 403, 421
cyclohexene, 121–124, 127, 128, 130, 133–135, 137, 338
cyclooctane, 328
cyclopenta-1,3-diene, 442–444, 449, 464, 484, 485
cyclopentane, 327, 497
cyclopropane, 328, 329, 405, 406

decarboxylation, 345, 347, 348, 352, 353, 426, 428
delocalization, 66, 67, 89, 148, 305, 374, 383, 390
Density Functional Theory, 44
dianion, 278, 345–348
diastereoisomer, 36, 123, 125, 134, 135, 185, 192, 196, 214, 222, 225, 227, 230, 296, 335, 340, 363, 385, 451, 452, 461, 499
diastereomeric, 191, 197, 214, 229, 451
diastereomeric excess, 193
diastereoselective, 133, 183, 193, 214, 219, 220, 293, 299, 335, 336, 339, 399, 499, 502
diastereotopic, 225–227
Diels-Alder reaction, 437, 438, 449, 450, 455, 459, 485
diene, 64–66, 68, 69, 71, 374, 437, 442, 444, 445, 449–455, 459, 462, 465, 467, 481
dienophile, 437, 445, 450–455, 485
dihedral angle, 176, 179
dihydroxylation, 116, 117, 124, 196, 197
dimethylformamide, 168, 267, 303
dimethylsulfoxide, 112, 115, 283, 485
disproportionation, 278
disrotatory, 457, 458, 460, 461
double bond equivalents, 474

eclipsed, 121, 175, 176, 178, 179, 184, 186, 330
electrocyclic, 440, 456–461, 465

520

Index

electronegativity, 14–16, 36, 38, 40, 49, 50, 63, 85, 89, 106, 107, 109, 148, 149, 151, 203, 222, 259, 260, 268, 283, 313, 314, 329, 379, 380

electrophile, everywhere!

elimination, 281–285, 291, 338, 351–355, 357–359, 385, 389, 420, 495, 497, 498

enamine, 233, 235, 236, 244, 247–249, 383, 384, 386, 387–391, 475–478, 480

enantiomeric excess, 193, 229–231, 490

enantiomer, 36, 117, 123–126, 128–130, 132, 135, 139, 189–198, 226, 227, 229–232, 280, 293, 294, 298, 335, 339, 340, 342, 451, 452, 461, 484, 490, 493, 495

enantiotopic, 225–227

endergonic, 61, 242

endo, 332, 450–452, 454, 455, 485

endothermic, 18, 60, 61, 76, 147, 158, 234, 235, 242, 258

enol, 87, 90, 91, 96, 97, 172–174, 233, 234, 272, 288, 289, 291, 301, 303, 310–316, 318, 323, 333, 334, 347, 364, 368, 383, 388, 390, 391, 421, 475, 481, 495

enolate, 82, 83, 85, 87–91, 96, 97, 223, 228, 288–291, 293, 295–297, 299, 300, 302–306, 308, 310–312, 314, 315, 318–326, 333, 335–344, 350, 351, 356, 364, 372, 373, 399, 400, 405, 498, 499, 502

epoxidation, 110, 115, 117, 124, 127, 133–137, 177, 181–189, 192, 193, 195, 211, 219, 220, 343, 396, 418, 505

epoxide, 110, 111, 117, 118, 133–135, 185, 186, 189, 191, 192, 195, 211, 329

equatorial, 119, 120, 122, 136, 184, 215–218, 337, 340, 394, 395, 469

ester, 49, 50, 84, 168, 197, 206, 207, 210, 250–252, 254–258, 260, 263, 282, 283, 315, 324, 325, 330, 336, 346–350, 352, 354–356, 372, 403, 427, 479–481, 486–488, 490, 503, 505, 506

exergonic, 207, 216, 262, 284

exo, 332, 385, 451, 454, 455, 485

exothermic, 18, 28, 44–46, 61, 76, 138, 147, 207, 216, 224, 235, 285, 307

Felkin-Anh, 221, 222, 339, 340, 396, 398, 508

Fischer indole synthesis, 476

flagpole, 121, 468

forbidden (cycloaddition), 442, 445, 447–449, 458

Friedel-Crafts acylation, 264–267, 269, 270

Friedel-Crafts alkylation, 161–166, 264, 408, 409, 419

Grignard reagent, 38, 204–208, 210, 220, 222, 223, 226, 260, 263, 375, 495, 508

Halogenation, 28, 68, 156, 310

Hammond Postulate, 17, 18, 22, 45, 79, 94, 147, 158, 162, 258, 377, 379, 411

Hantzsch pyridine synthesis, 349, 386, 392

Hantzsch thiazole synthesis, 389

hemiacetal, 240, 241, 488, 489

heterocyclic, 67, 140, 144, 145, 168, 249, 279, 382–384, 386, 389–391, 476

HOMO, 8, 9, 29, 64, 374, 377, 378, 439, 444–448, 450, 454, 456–460, 462, 463, 465–467

homologation, 428, 429

homolytic, 82, 209

homotopic, 226, 227

Horner-Wadsworth-Emmons reaction, 487

hybridization, 3–8, 10, 11, 19, 24, 30, 37, 38, 42, 62, 81–83, 168, 170, 209, 215, 236, 328, 331, 397, 405, 425, 428, 485

hydrate, 59–63, 233, 237, 238

521

Index

hydration, 20, 21, 34, 62, 101, 103, 110, 171, 174
hydride shift, 411, 412, 440, 441, 462–465, 467
hydroboration, 103, 171, 173, 359, 430, 499
hydrogen bonding, 111, 136, 186, 187, 334, 382, 387, 392
hydrogenation, 138–141, 198, 490, 498, 504
hydroperoxide, 194, 298, 431, 432, 503
hyperconjugation, 11, 15, 24, 47, 69, 410, 415

imine, 233, 235, 244–246, 248, 249, 379, 383, 384, 391, 420–422, 475, 476, 478, 480, 499
iminium ion, 248, 267, 352, 355, 387, 477, 478
inductive, 14, 15, 49, 50, 63, 82, 84, 85, 148, 149, 151, 152, 373, 374, 379
intermolecular, 241, 331
intramolecular, 241, 242, 247, 330, 356, 405, 455
ionomycin, 502

Jacobsen-Katsuki epoxidation, 195

ketal, 59, 60, 233, 234, 236, 237, 239, 240, 245, 292
Knoevenagel condensation, 349, 352, 354, 387

laurenene, 494
L-DOPA, 198
Lewis acid, 156, 161, 165, 265, 485, 509
Lindlar catalyst, 140
lithium diisopropylamide (LDA), 88, 89, 289, 290, 295, 303, 304, 323, 336, 504
Lossen rearrangement, 424, 426, 429
LUMO, 8, 9, 29, 30, 64, 302, 374, 377–379, 439, 444–448, 450, 454, 458–460, 462, 463, 465–467

Meerwein-Pondorf-Verley reduction, 275, 277, 280
Meerwein-Wagner rearrangement, 415
mercurinium ion, 102, 172
mesomeric, 14, 15, 85, 148, 149, 374
meta-chloroperbenzoic acid (*m*-CPBA), 110, 123, 135, 136, 177, 183, 187–189, 194, 418, 486
metathesis, 367–370, 492
molozonide, 113, 114

naphthalene, 67, 157, 158, 159
nitration, 34, 146, 147, 149, 151–157
nitrene, 424–426, 428, 429
nitronium ion, 146, 148
node, 4–6, 64, 65, 458, 462, 465, 467
nucleophile, everywhere!

Oppenhauer oxidation, 277, 280
orbital, everywhere!
organometallic, 51, 80, 168, 203–205, 224, 229, 263, 375, 496, 508
oxaphosphetane, 362, 363, 365
oxazolidinone, 293, 294, 342, 343, 502–504
oxidation, 106–110, 115–117, 174, 193, 212, 248, 270, 275, 277, 280–283, 293, 324, 359, 384, 403, 418–421, 426, 430, 437, 486, 496, 497, 499, 500, 505
oxymercuration, 101, 171–173
ozone, 112, 114, 116, 479
ozonide, 113–115, 479
ozonolysis, 112, 116, 118, 437, 438, 479, 506

Paal-Knorr furan synthesis, 384
phosphonium ylide, 85, 361, 362, 504
phosphorane, 361, 367, 504
photochemical, 447–449, 458–460, 465
pinacol rearrangement, 414, 419, 420, 422, 425, 508, 512
pK_a, 36, 37, 41, 53, 80–84, 86–88, 203, 209, 210, 244, 250, 251, 253, 257, 261, 268, 303, 305, 306, 311, 316–318,

321, 323–325, 345, 346, 348, 350, 351, 355, 357, 360, 362, 431
Prins cyclization, 511
prostaglandin, 484, 487, 493
protecting group, 240, 488, 489, 492, 503
protic, 213, 406
pyridine, 168, 249, 279, 349, 352, 353, 386, 387, 392, 487
pyrrole, 249, 383

racemic, 117, 125, 129, 139, 189, 193, 227, 231, 337, 339, 451, 452, 484, 490, 495
rearrangement, 164–166, 334, 403, 404, 409–412, 414, 415, 417–433, 440, 466, 468, 476, 500, 508, 509, 511, 512
regioisomer, 31, 101, 103, 150
regioselectivity, 20, 28, 34, 35, 74, 77, 100, 101, 104, 142, 167, 168, 173, 291, 303, 369, 422, 450, 482, 486, 495, 500
resonance, everywhere!

Schmidt rearrangement, 419, 422, 424, 426, 427, 429
Schrödinger equation, 3
Sharpless asymmetric epoxidation, 193–195
Sharpless asymmetric dihydroxylation, 196, 197
sigmatropic, 440, 441, 456, 462, 463, 465, 466, 476
S$_N$1 (substitution), 32, 88, 161, 162, 239, 273, 380, 411, 412, 432, 433
S$_N$2 (substitution), 6, 9, 23, 32, 33, 88, 91, 118, 127, 128, 161, 165, 238, 239, 246, 252, 254, 273, 290, 291, 329, 380, 390, 404, 405, 412, 432, 433, 485, 486
staggered, 176, 242, 327, 329, 330

stereoisomers, 117, 123–125, 130, 133, 139, 175, 185, 186, 188, 191, 220, 222, 231, 293, 294, 296, 297, 322, 335, 341, 396, 400, 401, 450, 485
stereoselectivity, 35, 122, 125, 126, 133, 135–137, 167, 183, 185, 186, 188, 192–196, 211, 214, 215, 223, 229, 280, 291, 293, 297, 298, 319, 324, 335, 342, 352, 399, 400, 402, 433, 441, 450, 461, 482, 504, 505
steric, 14, 15, 25, 47, 48, 51, 70, 72, 148, 149, 150, 152, 157, 158, 176, 183, 184, 215, 216, 218, 236, 305, 312, 328, 329, 331, 363, 364, 400, 401, 454, 455, 468, 500, 511
Stille coupling, 168, 169
Stobbe condensation, 354, 355, 358
strychnine, 473, 474, 481–483
suprafacial, 403–405
Suzuki coupling, 168
symmetry, 4–6, 9, 13, 24, 47, 64, 65, 367, 374, 440–442, 444–449, 454, 456–458, 460–463, 465, 467
syn, 336–338, 343, 344

tautomerization, 90, 172–174, 272, 311, 316, 318, 333, 347, 390, 421, 475, 476
tetrahydrofuran, 289, 290, 297, 303
thionyl chloride, 262, 267, 428, 505
torsional strain, 157, 158, 176, 216, 327, 329, 468
transition state, everywhere!

Vilsmeier-Haack reaction, 267–269

Wittig reaction, 359, 363, 364, 366–368, 487, 488, 503, 505
Wolff-Kishner reduction, 270, 271, 428, 499

ylide, 361, 362, 364, 365, 367, 504

zwitterionic, 57

Printed in Great Britain
by Amazon